D0812169

BIOLOGY OF INTERTIDAL ANIMALS

Biology of Intertidal Animals

R. C. NEWELL
Department of Zoology, Queen Mary College,
University of London

AMERICAN ELSEVIER PUBLISHING COMPANY, INC.
New York

Standard Book Number 444–19712–5

Library of Congress Catalog Number 73–96861

Printed in England

CONTENTS

PREFACE

This book is an attempt to account for the distribution of intertidal animals in physiological terms. This is by no means an easy task, for although the diversity and distribution of such organisms is now comparatively well-known and has formed the basis of an excellent book by Lewis (1964), the ecological significance of much of what is known of the physiology of intertidal animals is more difficult to assess. For this reason I have tried first of all to give a brief account of the distribution of the organisms of rocky, sandy and muddy shores and of the environmental factors which might be expected to play a part in their distribution. Subsequent chapters then deal in some detail with the ways in which such organisms are adapted to the intertidal environment, special emphasis being placed on the mechanisms by which the characteristic zonation patterns are established by the site-selection of larvae and by the behaviour of the adults. Mechanisms of food collection and of respiration under varying environmental conditions are also discussed in detail, as well as the resistance to temperature extremes and to desiccation shown by intertidal animals.

It should be emphasised, however, that the distribution of such animals can only rarely be interpreted in terms of a single environmental parameter. More often a complex interplay of factors, which may vary with season and according to the age of the organism, as well as with position on the shore, play a part in controlling the occurrence of intertidal animals. Despite these difficulties, it is to be hoped that as our knowledge of the intertidal environment, and of the physiology of the organisms increases it will be possible to understand more clearly the ways in which environmental factors control the occurrence of animals and plants on the shore. Each chapter is intended as a complete review of a particular aspect of the biology of intertidal animals. Where necessary the appropriate background information is provided at the beginning of the chapter and reference is made to pages where more detailed discussions occur.

I should like to acknowledge my debt to Professor T. I. Shaw, Dr. G. B. Williams and Dr. V. I. Pye who have given me a great deal of encouragement during the writing of this book. They have also saved me from several errors and have helped me to clarify some of the ideas expressed in the text; for any errors or inconsistencies which remain, I must accept full responsibility. Miss Alison Anderson of Logos Press has helped a great deal in the correction and preparation of the manuscript and it is a pleasure to acknowledge both her help and that of the referee. Some of the work which is reported here has not yet been published and I am extremely grateful to several authors for generous permission to use their data, especially Professor H. Micallef, Dr. H. Barnes, Dr. M. Clark, Dr. G. B. Williams, Dr. P. F. D. Cornelius, Dr. M. Longbottom, Dr. C. P. Mangum, Dr. V. I. Pye, Mr. M. Daniel and Dr. M. Ahsanullah. All illustrations have been drawn specially for this book and the source from which they have been adapted is acknowledged in the appropriate legend. This book would not have been written without the encouragement of my father, the late Professor G. E. Newell, nor without the immense moral support and help given by my wife over the past eighteen months. This book is dedicated to both of them for their kindness and understanding.

R. C. Newell, Department of Zoology,
Queen Mary College, University of London

June, 1969

CHAPTER 1

Physical, chemical and biological features of the intertidal zone

A. THE PATTERN OF ZONATION

It is a matter of common observation that at the edge of any static body of water there is a gradation of organisms from fully terrestrial to fully aquatic forms. On the coast, however, the situation is complicated by the semi-diurnal rise and fall of the tide. This ebb and flow results in the periodic emersion and submersion of an intertidal zone, the extent of which depends upon a wide variety of factors including the slope of the shore (p. 23) and the extent of the rise and fall of the tides. The physical features of such a zone are of immense complexity, and vary not only from coast to coast but also within comparatively small distances on any one shore. Nevertheless, Stephenson and Stephenson (1949) (for reviews, see Lewis, 1955, 1961, 1964; Southward, 1958b) have recognised the occurrence of three major divisions of organisms, whose precise vertical limits vary according to locality but which appear to be of universal occurrence on rocky shores. Firstly they recognised a sublittoral (= infralittoral) fringe corresponding approximately with extreme low water of spring tides (i.e. the upper marginal belt of the sublittoral (=infralittoral) zone and corresponding with the lowest limit of the intertidal zone) and characterised by the presence of the alga *Laminaria digitata*; the latter may be replaced by a variety of other algae such as *Laminaria saccharina, Halidrys siliquosa, Himanthalia elongata* and *Saccorhiza polyschides.* Also present are red algae and a wide variety of animals (p. 6) which are unable to withstand prolonged exposure to air (see chapter 9). The second zone recognised by Stephenson and Stephenson (1949) is the midlittoral zone, characterised by barnacles and limpets. Finally, there is a supralittoral fringe characterised by blue-green algae, lichens and small molluscs such as *Melaraphe* (*Littorina*) *neritoides* and *Lasaea rubra* as well as the isopod *Ligia oceanica* (also p. 12).

Lewis (1961, 1964) has recognised the importance of this basic scheme of classification but has suggested two main modifications. Firstly, the subdivision of the midlittoral zone of Stephenson and Stephenson (1949) into an upper littoral fringe and a lower eulittoral zone. Secondly, he proposed that the supralittoral zone be renamed the maritime zone since its characteristic indicators were mainly terrestrial rather than marine organisms. The two schemes with some of the characteristic organisms found on British shores are summarised in tables I (a) and I (b).

The Stephensons' (1949) scheme defined the top of the littoral zone in terms of the upper limit of abundant barnacles. There is, in fact, a general correspondence between the tidal level and the occurrence of the universal zones described above, but the precise relationship between tidal level and the limits of the zones are influenced by such a wide variety of factors that Lewis (1955, 1961, 1964) has convincingly argued the case for the sole use of biological indicators of the intertidal zones rather than reference to tidal level. For example, one factor which greatly affects the level of the particular intertidal zones on the shore is the degree of exposure to wave action. Thus when the zones are defined in terms of their indicator organisms, and the levels of such zones are studied on exposed and sheltered shores, it becomes evident that all the zones are displaced up the shore in areas exposed to wave action compared with sheltered shores. (Fig. 1.1).

Many other factors influence the occurrence of the universal zones (Stephenson and Stephenson, 1949; Lewis, 1955, 1961, 1964), and it is clear that no single physical or chemical factor can be invoked to account for the precise limits of the indicator organisms. The zones are thus, as Lewis (1961, 1964) suggests, best defined in terms of their characteristic organisms, although we are now rapidly approaching a stage when an attempt must be made to account for the distribution of

Table I (a) Based on Stephenson and Stephenson (1949)

Tidal Level		*Zone*	*Indicator Organisms*
		SUPRALITTORAL ZONE	
Extreme high water			
of Spring tides	LITTORAL ZONE	SUPRALITTORAL FRINGE	*Upper limit of Littorinids* *Melaraphe* (=*Littorina*) *neritoides* *Otina otis* *Ligia oceanica* *Lasaea rubra* *Verrucaria spp.* *Xanthoria* *Petrobius maritimus* *Scolioplanes maritimus* *Lipura maritima* Bdellid mites
		MIDLITTORAL ZONE	*Upper Limit of barnacles* Barnacles Mussels Limpets Fucoids (plus many other organisms cited on p 9-11)
Extreme low water		SUBLITTORAL (=INFRALITTORAL) FRINGE	*Upper limit of Laminarians* Rhodophyceae Ascidians (plus many other organisms cited on p 6-9)
of Spring tides		SUBLITTORAL (=INFRALITTORAL) ZONE	

Table I (b) Based on Lewis (1961)

Tidal Level	*Zone*		*Indicator Organisms*
	MARITIME ZONE		Terrestrial vegetation, Orange and green lichens.
Extreme high water / of Spring tides	LITTORAL ZONE	LITTORAL FRINGE	*Upper limit of Littorinids* *Melaraphe* (=*Littorina*) *neritoides* *Ligia, Petrobius,* *Verrucaria etc.*
		EULITTORAL ZONE	*Upper limit of barnacles* Barnacles Mussels Limpets Fucoids. (plus many other organisms cited on p. 9-11)
Extreme low water / of Spring tides	SUBLITTORAL ZONE		*Upper limit of Laminarians* Rhodophyceae Ascidians (plus many other organisms cited on p. 6-9)

organisms of the intertidal zone in terms of their physiology and behaviour.

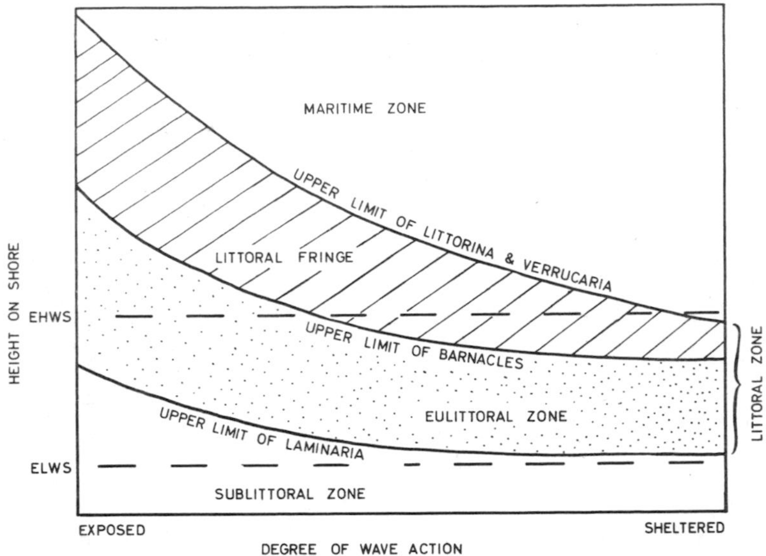

Fig. 1.1. Diagram showing the effect of exposure to wave action on the intertidal zones on rocky shores in the British Isles. (After Lewis, 1961.)

We may now consider some of these associations of animals and plants which typically occur in the sublittoral zone, the eulittoral zone and the littoral fringe on rocky shores in the British Isles. For the detail which this subject merits, reference should be made to the excellent account by Lewis (1964).

1. ORGANISMS OF THE SUBLITTORAL ZONE

The upper part of the sublittoral zone corresponds approximately with extreme low water of spring tides. It is therefore exposed for only a short time during the tidal cycle and is normally exposed only when the tidal oscillation is maximal during the spring tides occurring just after the time of the new and full moon. The extremes of physical conditions associated with exposure by the tide are thus small compared with higher levels on the shore and it is not surprising to find that both the animals and algae of the intertidal zone attain their maximal variety at extreme low water, although Colman (1933) has shown that there are fewer species in the middle of the intertidal zone compared with zones immediately above and below this (Fig. 1.2). This is perhaps associated with the rapid rate of change of the physico-chemical conditions in the middle shore compared with upper and lower levels where conditions are more constant. Nevertheless, it remains true that there is a marked decline in the number of species in the upper shore compared with the lower shore. Indeed, it is a general rule that as the physical or chemical conditions become more rigorous, the variety of organisms capable of withstanding such conditions is reduced; at the same time the population density of such species increases. We shall see that this generalisation is true not only for the higher intertidal zones but also for habitats such as muds where, for other reasons, the physico-chemical environment becomes more demanding (p. 41).

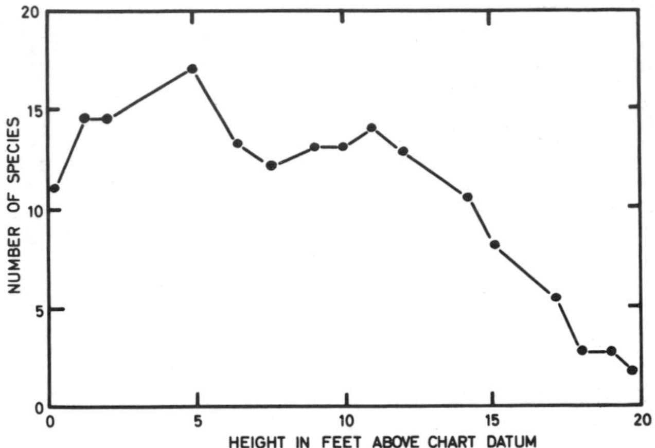

Fig. 1.2. Graph showing the total number of species as a function of height in the intertidal zone. (Based on Colman, 1933)

On shores of normal exposure, the upper sublittoral zone is dominated by laminarian algae, especially *Laminaria digitata, L. saccharina,* and *Saccorhiza polyschides.* Also present is a wide variety of red algae, some of which occur as epiphytes on the laminarians. Some of the more common red algae include *Rhodymenia palmata, Calliblepharis lanceolata, C.ciliata, Ptilota plumosa, Phycodrys rubens, Hypoglossum woodwardi, Plocamium coccineum, Heterosiphonia plumosa, Lomaria articulata* and *Griffithsia flosculosa,* many of which may become locally abundant on shores of moderate exposure. With increasing wave action, the nature and abundance of the algae of the lower shore varies, the laminarians being replaced by *Alaria esculenta* and *Himanthalia elongata* and by calcareous algae such as *Lithophyllum, Lithothamnion* and

Corallina officinalis. Under more sheltered conditions such as rock crevices, however, *Laminaria* tends to be replaced by *Chorda filum, Membranoptera alata, Rhodymenia palmata* and *Plumaria elegans.*

Such changes in the algal assemblage with variations in the degree of exposure to wave action are paralleled by comparable changes in the fauna. This is partly because the animals themselves vary in their tolerance of wave action, but also because many of the animals are dependent upon the presence of particular algae which provide a suitable substratum for growth and reproduction (also Chapter 3). The blue-rayed limpet *Patina pellucida,* for example, normally lives attached to *Laminaria* (also p. 133) whilst the sea hare *Aplysia punctata* is commonly also found amongst *Laminaria.* Again, the dorid opisthobranch *Archidoris pseudoargus,* which feeds on the breadcrumb sponge *Halichondria,* is commonly found under rock overhangs in association with its food. *Onchidoris muricata* is often found associated with the ectoproct polyzoan *Membranipora* upon which it feeds, whilst *Acanthodoris pilosa* often occurs with its foods *Alcyonidium* and *Flustrella.* Similar instances of close association between predator and prey could be cited for a variety of opisthobranchs, many of which have specific food preferences including both algae and other animals (also p. 115).

Despite such factors which influence the local distribution of the lower shore animals, we may review some of the more common animals which are likely to be encountered in the upper sublittoral zone before considering some of the changes in faunal assemblage which occur further up the shore. The fauna living at or about the extreme low water mark has two main components. Firstly there are typically sublittoral organisms which are able to survive intertidally in areas exposed to extensive wave action. Secondly, under more sheltered conditions where the corresponding universal zones are lower on the shore (Fig. 1.1), the animals living at or near the low water mark comprise organisms which typically inhabit the eulittoral zone but which may have extended their range down the shore under the sheltered conditions. Nevertheless, under intermediate degrees of exposure representatives of both components occur and may be used as a basis for comparison with upper intertidal organisms.

Animals occurring in the upper sublittoral zone include a wide variety of organisms with few obvious structural or behavioural adaptations to resist the stresses of the intertidal environment. The keyhole limpet *Diadora apertura* and the slit-limpet *Emarginula reticulata,* for example, both have an aperture in the shell through which extensive water loss would occur in the drier conditions of the upper intertidal zone. The ormer *Haliotis tuberculata,* which occurs in the Channel Isles, has a shell penetrated by a series of openings and is restricted to extreme low water, whilst the whelk *Buccinum undatum* does not withdraw into the shell when exposed by the tide even when under such conditions its rate of water loss must be great (Gowanloch, 1926). Again, the dorid and aeolid nudibranch opisthobranchs, as well as a variety of anemones such as the snakelocks anemone *Anemonia sulcata,* the dahlia anemone *Tealia felina,* and the plumose anemone *Metridium senile,* are without a shell into which they can withdraw and are confined to damp situations in the lower intertidal zone. Many of the hydroids such as *Syncoryne eximia, Clava* sp., *Tubularia* spp., *Plumularia* spp., *Sertularia* spp., *Campanularia* spp., and *Obelia* spp., as well as the stalked scyphozoan Stauromedusae such as *Haliclystus, Lucernaria* and *Craterolophus* are commonly attached to the algae of the lower shore, especially to *Laminaria, Rhodymenia* and *Fucus serratus* where the latter extends onto the lower shore in sheltered areas. The dense algal beds are also inhabited by large numbers of tanaids and isopods including *Idotea* spp., *Jaera* spp., amphipods such as gammarids, *Hyale* spp., and *Caprella linearis,* as well as the prawns *Leander serratus, Hippolyte* spp., and *Athanas nitescens.* The algae and the damp clefts in the rocks provide a suitably moist situation for the purse sponges such as *Grantia compressa* and *Sycon coronatum* as well as encrusting sponges such as *Halichondria panicea* and *Hymeniacidon.* Encrusting ascidians such as *Botryllus schlosseri, Botrylloides leachii, Dendrodoa* and *Polyclinum* are often abundant along with solitary forms such as *Ascidiella* spp., *Ciona intestinalis* and *Clavelina lepadiformis.* The encrusting polyzoans *Membranipora* spp. and serpulid polychaetes such as *Spirorbis borealis, S. spirillum, Pomatoceros triqueter* and *Hydroides norvegica* may locally form dense patches which cover the surfaces of the rocks and algae on which they occur.

Elsewhere, where the algal cover is perhaps sparse, the lower shore limpet *Patella aspera,* the saddle oyster *Anomia ephippium* and the barnacles *Verruca stroemia, Balanus perforatus* and *B. crenatus* may cover the surface of the rocks. In exposed situations the calcareous algae and *Alaria* may dominate the algal assemblage and form a suitable habitat for the brittle-star *Ophiothrix fragilis* and the intertidal echinoid *Psammechinus miliaris.* In such areas too, particularly sheltering under rocks or in crevices, one might expect to find a variety of crabs such as the velvet crab *Macropipus* (= *Portunus*) *puber, Xantho incisus,* and the edible crab *Cancer pagurus,* as well as the common shore crab *Carcinus maenas* and perhaps also the spiny spider crab *Maia squinado.* Other spider crabs such as *Macropodia rostrata* may be found, as well as the squat lobster *Galathea* spp., the porcelain crabs *Porcellana longicornis* and *P. platycheles,* and hermit crabs *Pagurus* spp.

Gastropods such as the painted top-shell *Calliostoma zizyphinum,* the tortoiseshell limpets *Acmaea* spp., *Gibbula magus,* the common wendeltrap *Clathrus clathrus,* and the cowries *Trivia* spp. are all characteristic lower shore animals. So too are the starfishes *Asterina gibbosa, Marthasterias glacialis, Henricia sanguinolenta* and *Solaster papposus* as well as the rocky shore fishes such as the rocklings *Onos* spp., the cornish sucker fish *Lepadogaster gouani,* the blennies and the gobies, all of which may also be found in the eulittoral zone when suitable rock pools occur.

This brief survey of the organisms of the upper sublittoral zone gives an impression of the diversity of the organisms which thrive on the lower shore, although many local variations in the types of animals occurring will be found according to the area studied and the season of the year. Nevertheless it emphasises the importance of the algae which provide food and shelter for a wide range of organisms which would not otherwise occur in the intertidal zone. Such factors, as we shall see, play an increasingly important part in enabling the colonisation of the higher shore levels by animals which otherwise would be unable to resist the rigours of the intertidal environment.

2. ORGANISMS OF THE EULITTORAL ZONE

The eulittoral zone is the lower and major division of the littoral zone, the majority of characteristic intertidal organisms being restricted to this area with only the hardier species extending their range up into the littoral fringe. Lewis (1961, 1964; also Table I), following Stephenson and Stephenson (1949), regards the upper limit of the barnacles as indicating the upper limit of the eulittoral zone, whilst the lower limit is set by the laminarians comprising the sublittoral fringe. Once again, as in the sublittoral fringe, the characteristic organisms forming the eulittoral community vary greatly with local conditions, particularly with the geographical situation and with the degree of exposure to wave action. Nevertheless, barnacles and limpets may be regarded as the characteristic indicators of the zone; these organisms are replaced by dense growths of fucoid algae under more sheltered conditions and by red algae and *Mytilus* under exposed conditions.

The floral succession typically consists of an upper band of the black lichen *Lichina pygmaea* which can extend down into the middle shore in some south-western areas. Also present in the barnacle zone, particularly on rocks where barnacles and fucoids are scarce, are lichens such as *Verrucaria microspora* and *V. mucosa.* This zone intermingles with, and is replaced by, the brown alga *Pelvetia canaliculata* followed downshore by *Fucus spiralis* or by *F. ceranoides* which occurs particularly in areas where the salinity may be reduced by drainage from the land. Here too the green algae *Ulva lactuca* and *Enteromorpha intestinalis* may be present in damp situations. The *Fucus spiralis* zone tends to be overlapped and replaced by *Ascophyllum nodosum* in sheltered situations, this alga commonly having the epiphytic *Polysiphonia lanosa* forming red tufts on its surface. *Ascophyllum* and *F. vesiculosus* often occupy similar levels but on different shores. The *Ascophyllum/F. spiralis* zone is normally followed downshore by *Fucus vesiculosus,* then by *F. serratus.* Throughout this zone, the tufted *Laurencia pinnatifida, Chondrus crispus,* and the calcareous alga *Corallina officinalis* become increasingly common together with *Lomentaria articulata, Gigartina stellata, Porphyra umbilicalis, Rhodymenia palmata* and *Himanthalia elongata* which border on the upper limits of the sublittoral zone.

Just as a distinct pattern of vertical zonation occurs amongst the algae of the eulittoral zone, so too the animals may be arranged in a characteristic vertical sequence, the level of which may be displaced up or down the shore according to local conditions. The littorinid gastropods, for example, occur in the following sequence: *Melaraphe* (= *Littorina*) *neritoides* which lives, along with the minute bivalve *Lasaea rubra* (p. 240), amongst the barnacles and in crevices at and above the limit of the eulittoral zone; *Littorina saxatilis* (= *rudis*) which rarely occurs below mid-tide level; and *Littorina littorea* which overlaps with *L. saxatilis* and extends right down to the low water mark. Where fucoids, especially *F. serratus,* are dominant the flat-topped winkle *Littorina littoralis* (= *obtusata*) may become abundant. Similarly, the top shells (Trochacea) are to be found in the sequence: *Monodonta* (= *Gibbula*) *lineata* living at and below the barnacle line at a comparable level with *Littorina saxatilis; Gibbula umbilicalis* living at and below mid-tide level; *G. cineraria* in damp situations under boulders; and finally the painted top shell *Calliostoma zizyphinum* in damp situations under rock overhangs. This species characteristically extends down into the upper sublittoral fringe where it is replaced by *Gibbula magus* in some areas. The fishes, too, whilst always confined to damp situations under boulders and in rock pools, may be arranged in a definite vertical zonation pattern which is, however, often profoundly modified by local conditions. The gobies, especially *Gobius minutus,* are common inhabitants of pools in the upper eulittoral zone, this species overlapping with *Gobius niger, G. pictus* and *G. paganellus* at somewhat lower levels. Also abundant at mid-tide level on many rocky shores are the blennies of which the most abundant species is the common blenny *Blennius pholis,* which occurs along with rocklings *Onos* spp., the butterfish *Centronotus gunellus,* and the eel *Anguilla anguilla* in damp situations under boulders and in rock pools. On western and south-western coasts the Cornish suckerfish *Lepadogaster gouani* may be common, this species being replaced by *Liparis* spp. in the north of the British Isles. In the lower parts of the eulittoral zone and extending down into the sublittoral zone, fishes such as the worm pipe-fish *Nerophis lumbriciformis,* the greater pipefish *Entelurus aequoreus,* and the wrasses may be found in the deeper pools.

Of the barnacles, *Chthamalus stellatus* is a southern species which extends round the south and west of the British Isles from the Isle of Wight to north east Scotland and occupies a higher zone on the shore than *Balanus balanoides. Chthamalus stellatus* commonly replaces *B. balanoides* on exposed headlands where the two genera overlap, and generally in the extreme south west of the British Isles. *Elminius modestus* also extends high in the intertidal zone and is confined to the southern half of the British Isles. This Australian barnacle was first recorded in Chichester harbour and in the Thames estuary at the end of the last war (Bishop 1947; Stubbings, 1950; Crisp, 1958) and has since spread northwards. *Balanus balanoides* extends low in the eulittoral zone in many areas, but may be replaced by *Balanus perforatus* on the lower shore in the south and west of the British Isles or by *Balanus improvisus* in the south where reduced salinities occur. *Balanus crenatus,* too, may be locally common especially under overhangs and under boulders on the lower shore. Limpets are characteristically most common on such barnacle-dominated shores rather than where the fucoids dominate or where there is extreme exposure to wave action and a *Mytilus* community is established. Here again, there is an overlapping series of species comprising *Patella vulgata* high in the barnacle zone on some shores, *P. depressa* (= *intermedia*) below mid-tide level on many shores (although the latter may extend as high as spring tide level in exposed areas) and in the lower part of the eulittoral zone and towards the upper sublittoral fringe, *P. aspera* (= *athletica*).

Other organisms are of more widespread occurrence throughout the lower part of the eulittoral zone, and their presence or absence may be correlated primarily with local variations in the substratum or in the availability of suitable food organisms. The anemones *Actinia equina* and *Sagartia elegans,* for example, tend to be of general occurrence wherever suitable damp situations occur, whilst the dog whelk *Thais* (= *Nucella*) *lapillus* is commonly to be found near the patches of barnacles upon which it preys. Other special habitats such as crevices, rock pools and the holdfasts of some of the larger algae present special physico-chemical conditions and support characteristic faunas. For this reason they are discussed separately in Chapter 2, although it should be remembered that their species too, help to comprise the faunal assemblage which characterises the eulittoral zone, and crevices from higher shore levels would be expected to support different organisms from those occurring lower on the shore.

3. ORGANISMS OF THE LITTORAL FRINGE

The littoral fringe is a zone which extends from the upper limit at which marine animals occur down to the upper limit of the barnacles. The organisms are characteristically few in number and all are well-adapted to resist the physical extremes associated with only intermittent immersion coupled with extremes of temperature and desiccation. The zone is dominated by the black lichens *Verrucaria maura* and *V. microspora,* the silvery grey *Ramalina* spp. and the orange *Xanthoria parietina.* The lower part of the littoral fringe is often dominated by *Lichina pygmaea* which occurs in the barnacle zone and thus contributes to the upper part of the eulittoral zone. Also, certain algae occur at a sufficiently high level on some shores to be included not only in the eulittoral zone but also in the littoral fringe. *Pelvetia canaliculata* and *Fucus spiralis,* especially the form *F. spiralis 'nanus'* which is very resistant to prolonged emersion, are both commonly found at the lower edge of the littoral fringe together with the red algae *Porphyra umbilicalis, P. linearis, Bangia fuscopurpurea* and *Hildenbrandia* sp., and the green alga *Enteromorpha.*

The animals represent two basic components. Firstly there are marine animals which have extended their range into the upper littoral zone. Such animals include the molluscs *Lasaea rubra, Littorina neritoides* and *L. saxatilis* as well as the isopod *Ligia oceanica,* all of which characteristically occur in rock crevices, in empty barnacle shells and amongst the lichens. The second component comprises terrestrial organisms such as the insects *Petrobius maritimus* and *Lipura* (= *Anurida*) *maritima,* the centipede *Scolicoplanes* (= *Geophilus*) *maritimus,* and small red bdellid mites. The fauna is a restricted one which reflects the complexity of the environmental stresses occurring at this level on the shore.

Throughout this review of the distribution of some of the more common intertidal animals, it has become clear that the upper intertidal zones are inhabited by relatively few species compared with the sublittoral fringe. Further, within any one zone the organisms are arranged not only in a vertical sequence down the shore, but some species replace one another laterally according to the conditions of exposure to wave action. It is therefore appropriate to consider in more detail some of the physical, chemical and biological factors which play a part in determining such variations in the distribution of intertidal animals.

B. THE CAUSES OF ZONATION

As has been mentioned on p. 1, and as discussed by Stephenson (1942), Stephenson and Stephenson (1949) and Southward (1958b), zonation of animals and plants would occur even on the edge of a body of water in which tides and waves were absent. The zonation pattern would then correspond with the maritime (= supralittoral) zone and with the sublittoral zone. The occurrence of tides results in the establishment of the eulittoral zone and accompanying littoral fringe whilst wave action may not only raise the effective level at which intertidal animals can survive on the shore but also allow the survival of sublittoral organisms in the lower intertidal zone (Lewis, 1964). It is thus clear that the prime cause of zonation within the littoral zone is the differing ability of intertidal organisms to withstand the physical and chemical stresses set up by exposure and immersion by the tide, although the relative importance of particular physico-chemical factors may vary from species to species and according to the stage of development of the organisms. Indeed, as is stressed by Stephenson (1942) and by Lewis (1961, 1964), the distribution of intertidal organisms is not necessarily correlated with any one environmental parameter but is influenced by a whole complex of factors, the relative importance of which may vary both spatially and seasonally.

1. DIRECT EFFECTS OF THE DURATION OF EMERSION AND SUBMERSION

The coasts of the British Isles are subjected to two tides per day; that is, the tides are semi-diurnal. The tides are, moreover, subjected to a progressive change in amplitude of rise and fall according to the phase of the moon. This change in tidal amplitude follows a cycle of approximately 14·5 days duration; it is at a maximum just after the new or full moon ('spring tides') and at a minimum at intermediate phases of the moon ('neap tides'). Further, the actual magnitude of rise and fall during the spring and neap tides varies according to more complex factors (Pilkington 1957; Russell and Macmillan 1952) which result not only in the well-known equinoctial spring tides of especially large amplitude but also in large diurnal inequalities between the two tides of the day in some parts of the world. After such factors have been taken into account, it is possible to relate the percentage emersion/submersion to tidal level on any one shore. When such a curve is plotted as in Fig. 1.3 it is apparent that the amount of exposure to air does not increase uniformly from low to high water. Instead, there are two main points at which there is a relative increase in the percentage exposure. These points coincide approximately with the mean low water and high water of neap tides. Such changes in the percentage exposure of the intertidal zone to air were found to coincide with faunistic boundaries (for example, Colman, 1933; David, 1943; Doty, 1946; Evans 1947a,b) and suggested that 'critical levels' in the zonation

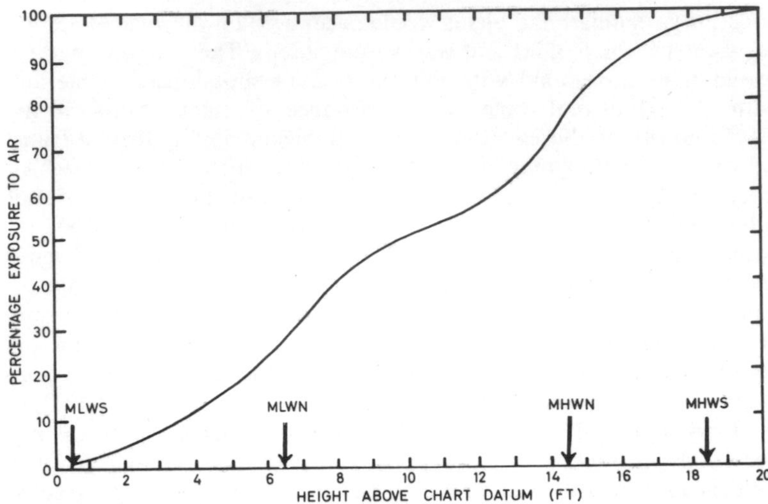

Fig. 1.3. Diagram showing the percentage emersion relative to height above chart datum on a shore with a semi-diurnal tidal rise and fall. (Based on Colman, 1933)

of animals and plants were related to sudden increases in the percentage exposure to air at certain tidal levels. Whilst it is true that many organisms of the eulittoral zone have their upper or lower limits at low water neaps or high water neaps, this is by no means true of the majority of the organisms of the eulittoral zone. Also, the magnitude of the changes in percentage exposure to air which occur at these two levels does not seem to be large enough to account for the observed faunistic breaks. Lewis (1964) has suggested that the period of *continuous* exposure to air (emersion) or immersion by the tide (submersion) may account for the coincidence of zonational critical levels with the mean high and low water of neap tides. He points out that organisms living at the mean high water of neap tides are normally covered twice daily by the tide but fail to be covered at those times of the year when the neap tides are small and do not reach their mean high water mark. Under these circumstances adaptations which allow life under conditions of exposure to air for not more than 12 hr may then fail to ensure survival when the organism is unwetted for several days. Such factors may account for the failure of some species to extend higher in the intertidal zone than mean high water of neap tides, since lethal conditions, even if they exist for only a short period during the year, will affect the zonation pattern for much longer.

The situation is further complicated by the fact that modifying factors are rarely, if ever, completely absent. Such factors include the aspect of the shore, the nature of the substratum, the degree of exposure to wave action, and biological factors such as the presence or absence of fucoids or barnacles whose own presence may in turn be dependent upon other variables. It seems safe to assume, however, that the general zonational level occupied by a particular intertidal species is a product of the physico-chemical factors set up by the alternate ebb and flow of the tide. Superimposed upon such a general zonational sequence, there are often sharp boundaries between successive species as well as lateral variations in the intertidal fauna. Such variations may be accounted for in terms of a variety of modifying factors some of which are considered in more detail below, together with those limiting factors which are imposed as a result of the rise and fall of the tides.

2. SECONDARY FACTORS SET UP BY THE EBB AND FLOW OF THE TIDE

The most obvious effect of the semi-diurnal emersion/submersion sequence is that it sets up a whole series of secondary effects which may themselves play an important part in limiting particular intertidal species to their characteristic zones on the shore. One such factor, which is discussed further in Chapter 6, is the duration of time available for feeding. It is clear that the time available for suspension-feeding organisms to obtain their food is dependent upon how long they are covered. Thus, in the absence of any compensatory changes in the rate of feeding (p. 239), the upper limit of distribution of a particular suspension-feeding organism may be set by the level at which sufficient food can be obtained for growth and reproduction. Again, some scavenging and browsing organisms (p. 242) feed mainly when uncovered and their lower limit of distribution in the intertidal zone may be controlled by the time available for feeding.

A second factor which may play a part in controlling the level on the shore at which a particular species is able to survive is the extremes of temperature to which particular zones are subjected. Obviously the temperature of the upper part of the sublittoral zone approximates to the sea temperature throughout the year, whereas the higher shore levels approach the air temperature or may even exceed this on bare rock surfaces exposed to the sun. Equally, during the winter such zones may be subjected to extremely low temperatures during periods of frost. Thus, both the maximum and minimum temperatures may set a limit to the upper level of distribution of organisms in the intertidal zone; the shelter afforded by algal growth may, however, greatly reduce such thermal stress. The influence of thermal stress on intertidal animals is discussed in more detail in Chapter 9. The humidity of the upper intertidal zone varies too in a general way with shore level. It is at a maximum at the sublittoral fringe and reaches minimum values during dry weather on bare rock surfaces high on the shore; but here again local factors such as the presence or absence of algae or of crevices largely determine the degree of desiccation to which a particular part of the shore is subjected. Although many motile animals tend to evade such environmental stresses, it remains true that desiccation may play an important part in limiting the distribution of both sessile and motile animals in the upper part of the littoral zone. The effects of such desiccation and the ways in which intertidal animals may resist it are also considered in Chapter 9 (p. 451).

Finally, the rise and fall of the tide may indirectly play a part in influencing the zonation of organisms in the littoral zone by limiting the time during which animals of the upper shore may pass oxygenated water over the surface of the body. High level animals continue to respire even when the tide is low, and under these conditions may either breathe air, as in a variety of intertidal gastropods (Micallef and Bannister, 1967; Sandison, 1967) or undergo anaerobic respiration as in the bivalves *Mya arenaria* (van Dam, 1935), *Venus mercenaria* (Dugal, 1939) and *Mytilus edulis* (Schlieper, 1957) (for review, von Brand, 1946). Some may be able to either breathe air or respire anaerobically as in the barnacles *Balanus balanoides* and *Chthamalus depressus* according to the conditions prevailing in the habitat (Monterosso, 1928a,b,c, 1930; Barnes and Barnes 1957; Barnes, Finlayson and Piatigorsky, 1963; Grainger and Newell, 1965; Colman and Stephenson, 1966). The differing abilities of intertidal animals to undertake either aerial or anaerobic respiration may play an important part in determining the level on the shore at which they can survive. Such factors are discussed in more detail in Chapter 7 (p. 265), but it is worth emphasising at this point that the tidal-dependent factors mentioned above rarely, if ever, operate independently of one another. High temperatures and extremes of desiccation, for example, normally operate together in the upper littoral zone, and an animal which is unable to withdraw into its shell and respire anaerobically may place itself in more danger of death from desiccation than one such as the mussel *Mytilus edulis* (Schlieper, 1957) or the barnacle *Balanus balanoides* (Barnes *et al.*, 1963) which can close the shell and incur an "oxygen debt" until more favourable conditions prevail. The differing abilities of intertidal animals to withstand the physico-chemical stresses set up by the ebb and flow of the tide have been the subject of a considerable amount of experimental analysis which is considered in more detail in later chapters.

3. MODIFYING FACTORS UNRELATED TO THE EBB AND FLOW OF THE TIDE

(a) Exposure to wave action

Perhaps the most important single factor which modifies the height of a particular zone on the shore and the nature of the organisms living there, is the degree of exposure to wave action. This factor has recently been reviewed by Ballantine (1961) and by Lewis (1964) to whom reference should be made for discussions on the use of exposure scales on rocky shores in the British Isles. As has been mentioned above, there is a variety of tidal-dependent factors such as limited feeding time for suspension-feeding organisms, desiccation, and extremes of temperature which tend to limit the upward distribution of animals and plants in the littoral zone. (for discussions on the causes of algal zonation, see Moore, 1958; Lewis, 1964; Boney, 1966). Exposure to wave action, however, operates in the reverse direction and tends to moisten the upper tidal levels either by wave splash or by spray, thus increasing the effective submersion period and allowing an upward extension of the distribution of intertidal organisms. It is therefore clear that under sheltered conditions the zonation patterns will be most easily related to tidal level and to tidal-dependent factors, whilst on shores exposed to extensive wave action the distribution of littoral organisms is related less obviously to tidal dependent factors.

The upward displacement of algal zones on exposed coasts was described by Brenner (1916) and by Du Rietz (1940) and has also been observed by Colman (1933) at Wembury as well as by a large number of other workers including Evans (1947a,b), Burrows, Conway, Lodge and Powell (1954) and Ballantine (1961) (for reviews, see Moore, 1958; Lewis, 1964). Burrows, *et al.* (1954) compared the distribution of a number of intertidal algae in relation to tidal level on Fair Isle, and found that not only were the algal species different in exposed and sheltered areas but that the corresponding algal zones were displaced upwards by as much as 12 ft (3·65 m) on a shore exposed to extensive wave action. These results are shown in Fig. 1.4. *Ectocarpus fasciculatus, Fucus inflatus f.distichus, Rhodymenia palmata, Polysiphonia urceolata, Corallina officinalis* and *Alaria esculenta* were all absent on the sheltered coast at North Haven; on the other hand, species such as *Ascophyllum nodosum, Polysiphonia lanosa, Fucus vesiculosus* and *Cladophora rupestris* were present on the sheltered coast at North Haven but absent on the exposed coast at North Gavel. Evans (1947b) studied the vertical distribution of a number of common intertidal organisms in relation to exposure in the Plymouth area; he found that both the upper and lower limits of distribution of the lichen *Lichina pygmaea* and the alga *Pelvetia canaliculata* extended higher on

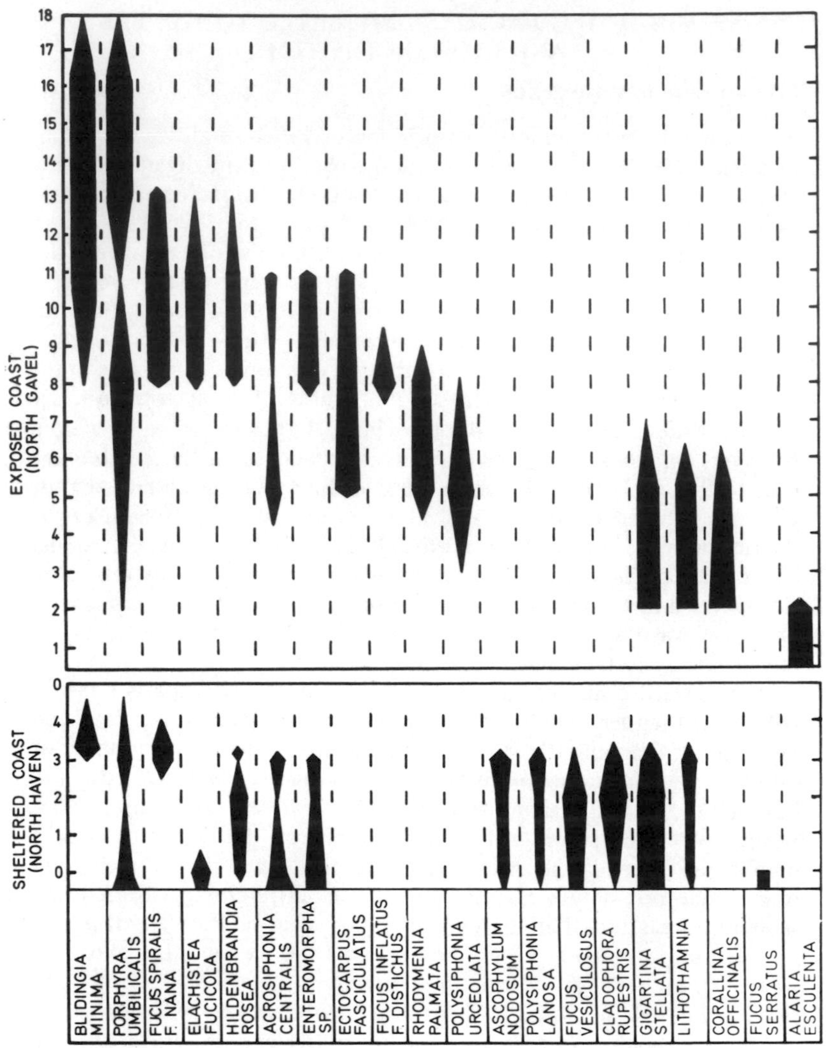

Fig. 1.4. Diagram showing the vertical distribution (feet above chart datum) of intertidal algae on an exposed coast (North Gavel) and sheltered coast (North Haven) on Fair Isle Scotland. (After Burrows, Conway, Lodge and Powell, 1954)

the shore in exposed than in sheltered areas (Fig. 1.5). Similar results have been obtained by Lewis (1964) on a series of intertidal organisms on a reef near Stoer Bay, Sutherland (Fig. 1.6).

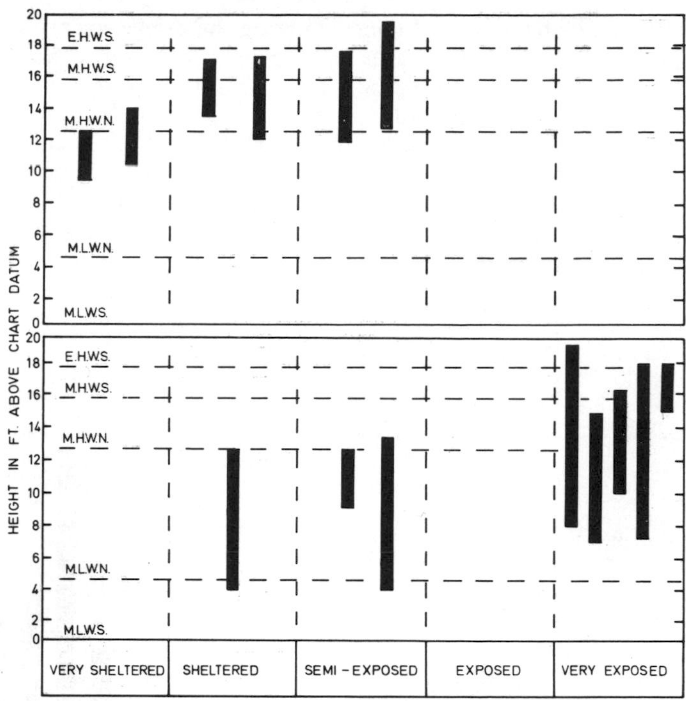

Fig. 1.5. Diagram showing the vertical distribution of the brown alga *Pelvetia canaliculata* (above) and the lichen *Lichina pygmaea* (below) in relation to exposure to wave action. (Based on Evans, 1947)

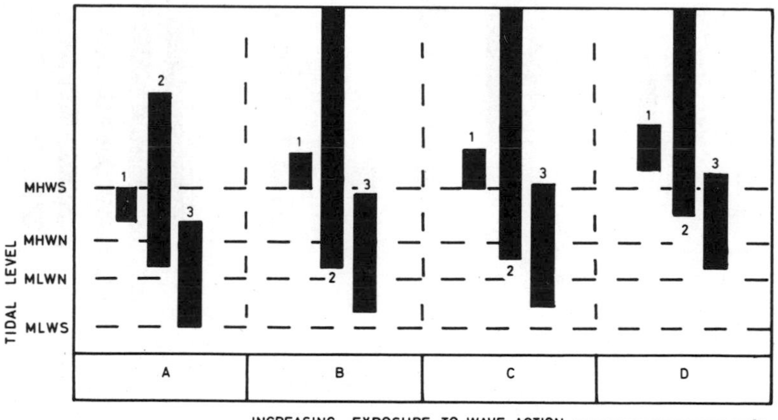

Fig. 1.6. Diagram showing the vertical distribution of *Verrucaria*, Myxophycae and *Gigartina* in relation to exposure to wave action. 1 = *Verrucaria*, 2 = Myxophycae, 3 = *Gigartina*. (Based on Lewis, 1964)

Similar results have been obtained showing the displacement of the zones of intertidal animals towards higher shore levels under conditions of increasing exposure to wave action. Moore (1935a), for example, showed that the upper limit of distribution of *Balanus balanoides* on the shore at Plymouth (U.K.) could be correlated with the intensity of wave action although here the lower limit of distribution was extended downshore under exposed conditions, perhaps because of the removal of competing algae under these conditions (Lewis, 1964). Evans (1947b), also working in the Plymouth area, has shown that a similar pattern of distribution in relation to exposure to wave action exists in *Melaraphe* (= *Littorina*) *neritoides* (Fig. 1.7), whose upper limits of distribution are displaced up the shore under conditions of exposure to wave action. Lewis (1964) has confirmed these observations and

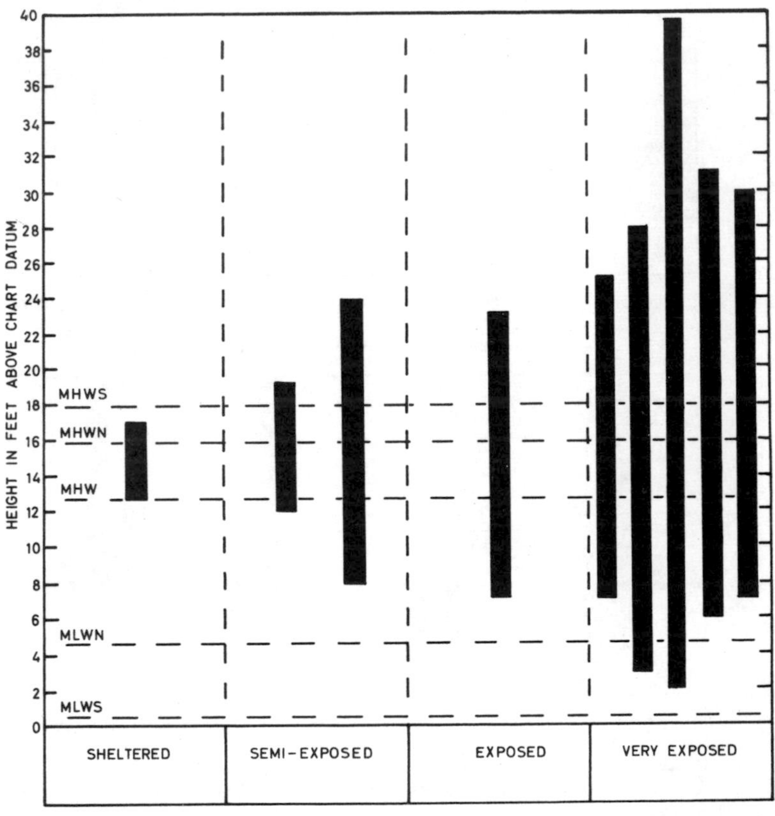

Fig. 1.7. Diagram showing the vertical distribution of *Melaraphe (= Littorina) neritoides* in relation to exposure to wave action. (Based on Evans, 1947)

surveyed the distribution of littorinids, barnacles, mussels and limpets in relation to exposure to wave action near Stoer Bay, Sutherland. These results are summarised in Fig. 1.8 (also Fig. 1.1).

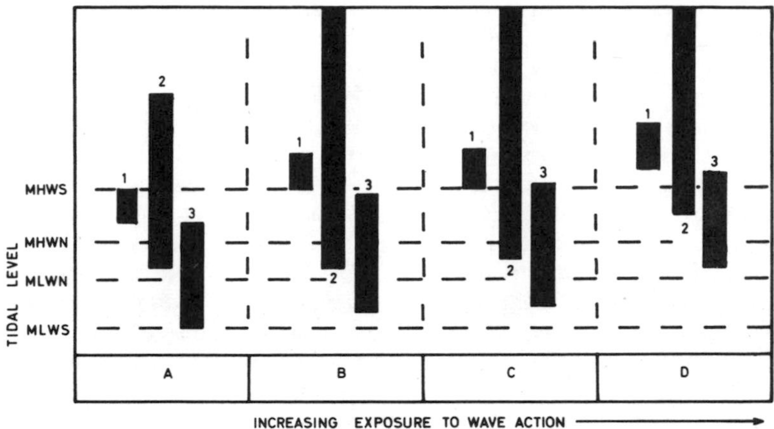

Fig. 1.8. Diagram showing the vertical distribution of *Chthamalus*, Littorinids and *Balanus* in relation to exposure to wave action. 1 = *Chthamalus*, 2 = Littorinids, 3 = *Balanus*. (Based on Lewis, 1964)

Exposure to wave action thus has profound effects on the levels at which plants and animals may be found on the shore, but as mentioned on p 17 and shown in Fig. 1.4, there are not only distributional changes but also qualitative changes in the plants and animals. Such differences may in part be due to the differing susceptibilities of organisms to mechanical damage by wave action and by the boulders and sand which may be hurled against the rocks by waves (for example, Johnson, 1919) but is also due to more complex factors involving larval recruitment, competition between species such as barnacles and fucoids, the presence of organic debris and silt in more sheltered areas, as well as the interdependence of the animals and plants, some of which may be absent not because of wave action itself but because of the absence of particular organisms upon which they depend. The winkle *Littorina littoralis* (= *obtusata*), for example, lives in association with fucoids and does not occur on exposed shores where fucoids are absent, irrespective of its own tolerance of wave action. Other species with a broader pattern of distribution under varying conditions of exposure, appear to be able to modify the form and thickness of the shell to accommodate the increased stresses associated with exposed shores. The limpet *Patella vulgata* living high on the shore has a taller shell than individuals living under damper conditions lower on the shore (Orton 1933) although

Davies (1969) has more recently accounted for this in terms of the resistance of tall-shelled limpets to desiccation. Limpets high in the intertidal zone under the damp conditions associated with exposure to wave action are correspondingly flatter in shape and better adapted to resist wave shock than their counterparts in sheltered conditions (Moore, 1934b). Again, specimens of the sea urchin *Echinus esculentus* have a thicker shell when collected from areas of strong wave action than specimens from sheltered areas (Moore, 1935b).

It might be supposed that physical criteria such as the strength or duration of wave action would be sufficient to indicate the degree of exposure of a particular shore. Moore (1935a) assessed exposure in terms of the percentage number of days in which any wind blew into the area over more than three miles of sea, but this method of estimation ignored the wind speed, the length of sea over which the wind had travelled (fetch), as well as the height of waves generated by winds outside the area (swell). In some areas, as on the Atlantic coasts of the British Isles, the swell may account for a considerable degree of wave action even in the absence of local winds, whereas in other areas where a swell is absent the strength and fetch of the wind plays the dominant part in controlling the degree of exposure. Again, although Southward (1953) assessed exposure in areas bordering the Irish Sea in terms of the swash above predicted tide levels under known wind conditions, his method has been criticised insofar as swell and fetch were not accounted for (Ballantine, 1961). In this particular area there is very little swell, but in other areas it is clearly extremely difficult to arrive at a physical expression of exposure which accounts for waves generated outside the area. Both Ballantine (1961) and Lewis (1964) have supported the use of biological indicators of exposure which, although giving little indication of the physico-chemical conditions responsible for the presence of the indicator organisms, do provide a convenient series of reference points for comparison of rocky shores around the British Isles. Ballantine (1961) suggested the division of rocky shores into the following eight categories:—

(1) Extremely exposed;
(2) Very exposed;
(3) Exposed;
(4) Semi-exposed;
(5) Fairly sheltered;
(6) Sheltered;
(7) Very sheltered;
(8) Extremely sheltered.

Each of these categories is defined in terms of the abundance of the barnacles *Balanus balanoides* and *Chthamalus stellatus*, the limpets *Patella vulgata*, *P. aspera* and *P. intermedia* (= *depressa*), the winkles *Littorina littorea*, *L. littoralis* (= *obtusata*) and *L. neritoides*, the top shells *Gibbula umbilicalis* and *Monodonta lineata* and the dog whelk *Thais* (= *Nucella*) *lapillus*. He also used the algae *Pelvetia canaliculata*,

Fucus spiralis, Ascophyllum nodosum, F. vesiculosus, F. serratus, Laminaria digitata, Alaria esculenta, Laminaria saccharina, Porphyra and *Lithothamnia* as well as the lichen *Lichina pygmaea* as indicators of exposure. Lewis (1964) on the other hand, recognised the occurrence of only five categories of exposure which again are characterised by the presence of particular indicator organisms. The categories include:–

(1) Very exposed shores;
(2) Exposed shores;
(3) Semi-exposed shores;
(4) Sheltered shores; and
(5) Very sheltered shores;

Reference should be made to both Ballantine (1961) and Lewis (1964) for details of the abundance and nature of the indicator organisms which characterise the particular categories of exposure. More recently, Jones and Demetropoulos (1968) have described a simple dynamometer which makes possible the investigation of wave force as an ecological parameter. They showed that it was possible to interpret the zonation pattern of a wide variety of organisms on rocky shores on Anglesey, North Wales, in terms of maximum readings on the dynamometer, and that such values were in good agreement with the theoretical values for the maximum dynamic pressure produced by the waves concerned. This method for the estimation of the intensity of wave action should supercede the use of biological exposure scales and allow this important environmental parameter to be assessed quantitatively.

(b) Topography and aspect of the shore

The second factor which is unrelated to the ebb and flow of the tide but which plays an important part in determining the nature and distribution of the intertidal organisms, is the topography and aspect of the intertidal zone. The effects of topography are very complicated owing to the enormous variety of shores which range from shallow boulder-strewn shores to wide rock platforms and to steep vertical cliffs. Often one shore may have a wide wave-cut platform in the lower zones and steep cliffs above. Such variations in the topography have two main effects according to whether or not the shore is subject to extensive wave action. Under conditions of shelter from wave action where the zonation pattern is primarily determined by emersion/submersion factors (p. 17), a shore of shallow gradient will have a wide littoral zone which, owing to the poor drainage and extensive tidal pools, allows an upwards extension of the sublittoral fringe organisms. Such conditions are to be found in many parts of the British Isles and reach their extreme development under estuarine conditions where the slope of the shore may be as little as 1:600 and an intertidal zone of a mile or more may occur. Conversely, where the shore is steep, as on cliffs or on man-made structures such as jetties and piers, the whole of the littoral zone may be condensed to a relatively narrow band corresponding with the height of the tidal rise and fall (Fig. 1.9).

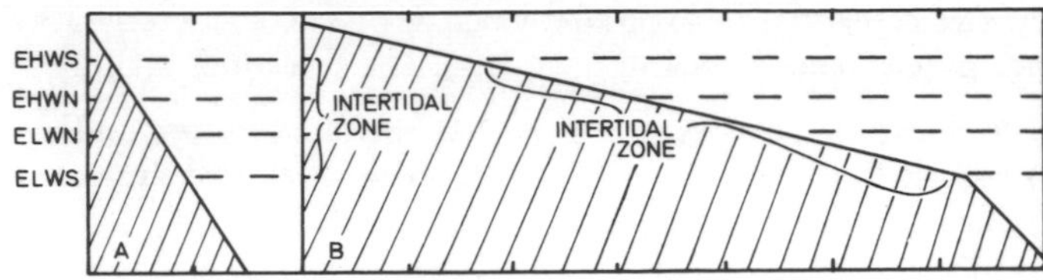

Fig. 1.9. Diagram showing the effect of the slope of shore on the width of the intertidal zone. (A) Very steep shore with correspondingly narrow intertidal zone, (B) Shallow sloping shore with broad intertidal zone. Approximate tidal levels are indicated by a broken line.

Under exposed conditions, the topography of the shore plays an important part in modifying the effects of wave action. Where the shore is shallow, even under very exposed conditions there may be little uplift of the higher intertidal zones. Only the sublittoral fringe may be affected, after which much of the remaining energy of the waves is expended before reaching the upper shore. Where the shore slopes steeply down to the low water mark, wave surge may result in an extension of the sublittoral and eulittoral organisms to higher levels on the shore than in sheltered areas, whilst the presence of cliffs and abundant spray facilitates the development of a wide *Littorina*/*Verrucaria* belt far above the extreme high water of spring tides. Thus one of the important effects of the topography of the shore is to modify the effects of wave action; the direct effects of topography on drainage, which allows an upward extension of the range of organisms of the lower shore, play an equally important part under both exposed and sheltered conditions.

The aspect of a shore, or of a particular region on a shore, is also an important factor in determining the upper limits at which intertidal animals can live. Typical organisms of the lower eulittoral zone or of the sublittoral fringe, for example, are commonly found higher in the intertidal zone on the shaded sides of boulders or gullies than on the side exposed to the sun. Shaded positions to some extent offset the rigours of desiccation and thermal stress, and when coupled with exposure to wave action allow a great upshore extension of the sublittoral organisms (for example, Lewis, 1954a,b). Aspect also affects the organisms of the upper shore, but since these are more resistant to extremes of desiccation and thermal stress the most pronounced effects of aspect are normally confined to the lower shore. Evans (1947b), however, demonstrated that the limpet *Patella vulgata* occurred at higher levels on the shore on shaded surfaces than on illuminated ones, and that the effect was much more pronounced when the shaded surface coincided with conditions of exposure to wave action. On very exposed shores aspect had little effect on the upper limit of *P. vulgata* presumably because situations of all aspects were suitably dampened by wave action (Fig. 1.10). Thus the effects of aspect, like those of

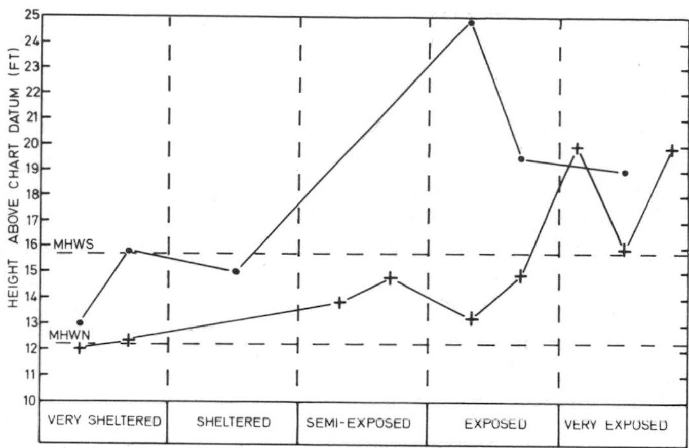

Fig. 1.10. Graph illustrating the effect of exposure to wave action on upper limit of distribution of *Patella vulgata* on shaded surfaces (.–.) and on illuminated surfaces (+–+) in the Plymouth area. (Based on Evans, 1947)

topography, can rarely be considered in isolation from those of wave action, and under sheltered conditions probably exert their effects by minimising the tidal dependent factors of desiccation and thermal stress (Chapter 9). Stephenson and Stephenson (1961) have made a detailed study of the intertidal fauna and flora of Brandon Island, in Departure Bay, British Columbia. The island is largely sheltered from wave action and the northern side consists of steep shaded cliffs whilst the southern coast is sunny and of gentle slope. The zonation on each of the two coasts was found to be markedly different despite the similarity of tidal range and wave action on each side of the island. As Stephenson and Stephenson (1961) pointed out, such differences in zonation pattern must therefore be largely due to the slope and aspect of the respective shores, although tidal level must also play a part in controlling the distribution of the organisms. The situation on Brandon Island is thus a good example of the modifying effect of aspect and slope on tidal-dependent zonation patterns.

(c) The nature of the substratum

The substratum of the intertidal zone falls into two main categories each of which supports a characteristic assemblage of organisms. Firstly, it may be solid as on rocky shores and provide a suitably stable situation for the attachment of algae. Secondly the substratum may be particulate, in which case it may be extensively disturbed by wave action and the algal cover is correspondingly sparse. Moore and Kitching (1939) have shown that minor variations in the abundance of algae and of the barnacles *Balanus balanoides* and *Chthamalus stellatus* on rocky shores are often associated with variations in the roughness of the rock surface, whilst the larvae of *Phoronis hippocrepia* select limestone rocks for settlement (Silèn, 1954). However, as pointed out by Moore (1958), the algae are merely attached to the rocks and do not derive nourishment from the substratum as in the case of land plants. Thus the marked variation in flora which occurs with changing substratum on land is not paralleled by comparable changes in the algae of the shore. Much the same applies to the barnacles whose presence depends not upon a particular type of rock, but upon any rough and stable substratum such as is provided by the majority of rock types occurring in the intertidal zone. Therefore, other factors such as the tidal level, exposure to wave action, and aspect of the shore play a major part in determining local variations in the abundance of animals and plants on rocky shores.

Particulate substrata (including gravels, sands and muds), represent an abrupt change in environment and the faunal assemblage changes correspondingly abruptly; algae, as we have seen above, also tend to be scarce owing to the absence of suitably stable objects for attachment. The absence of a stable substratum also involves a reduction of animals which cling to the surface of rocks to avoid being swept away by wave action. Such animals, constituting the epifauna, tend to be replaced by animals (the infauna) which burrow and thus avoid disturbance by wave action and other physical stresses of the environment. The stability of the substratum thus largely determines the overall nature of the organisms of the intertidal zone, but the size of particles constituting the deposit also play an important part in determining the type of animals living there. It is therefore of some importance to consider the factors responsible for controlling the grade of deposit on particulate shores as well as those physico-chemical features which influence the nature and abundance of the fauna.

(i) Physico-chemical features of intertidal deposits

The prime factor which determines the grade of deposit and corresponding slope of a particular shore is the exposure to wave action to which the intertidal zone is subjected. (for example Krumbein 1944/47.

Shepard, 1948; Bascom, 1951). Krumbein (1944/47) studied the physical features of a sandy shore at Halfmoon Bay, some 25 miles south of the Golden Gate on the Californian coast; he found that on shores with coarse deposits the slope of the beach was steep whilst on shores composed of fine sand the slope tended to be shallow, beach slope being a positive exponential function of particle size. (Fig. 1.11).

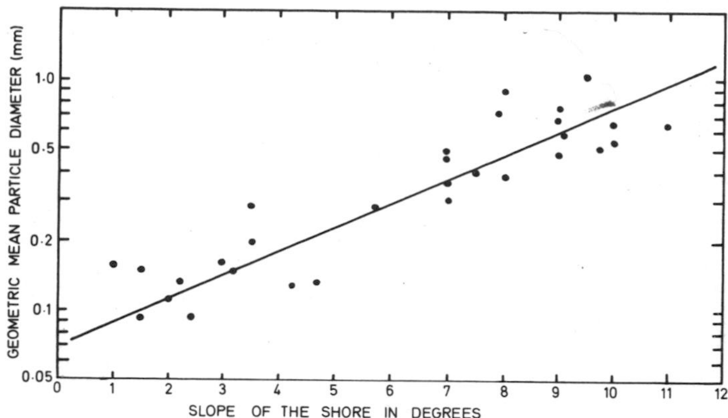

Fig. 1.11. Graph illustrating the relationship between the slope of the shore and particle size (geometric mean in mm) at Halfmoon Bay, California. (After Krumbein, 1944/47)

From this graph Krumbein showed that:–

$$r = R_o e^{bS} \tag{1}$$

Where r is the particle size associated with slope S degrees; R_o is the initial size for S = O, and b is a coefficient of size increase. When size is taken as the independent variable, then the function is logarithmic and

$$S = (\frac{1}{b}) \log_e (\frac{r}{R_o}) \tag{2*}$$

Having established this relationship between the shore slope and the particle size of the deposits, Krumbein (1944/47) then studied the relation of these factors to the degree of exposure to wave action to which the shore was subjected. He found that the foreshore slope was linearly related to the relative wave energy in any particular area of Halfmoon Bay where the two factors were measured at the same time.

* This is not the relationship expressed by Krumbein (1944/47) who appears to have in error omitted r from the final bracket.

This relationship is shown in Fig. 1.12, the equation of the line being:—

$$S = kE + K \tag{3}$$

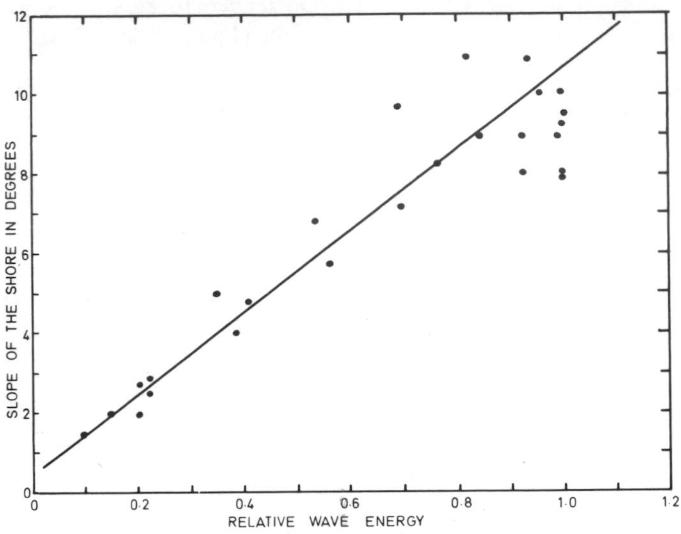

Fig. 1.12. Graph illustrating the relationship between the slope of the shore and relative wave energy. (After Krumbein, 1944/47)

Where S is the slope of the shore; E is the relative energy; constant k has the value 10·0 and K the value 0·6.

Since the relationship between particle size and slope of the shore is logarithmic (Equation (2) and Fig. 1.11) whilst that between slope and relative wave energy is linear (Equation (3) and Fig. 1.12), it follows that the particle size is a positive exponential function of relative wave energy. Thus by eliminating S from Equations (2) and (3) and solving for r we obtain:—

$$\left(\frac{1}{b}\right) \log_e \left(\frac{r}{R_o}\right) = kE + K$$

$$\text{and} \quad r = R_o e^{bK} e^{bkE} \tag{4}$$

$$\text{or} \quad r = R_o e^{b\,(kE + k)}$$

Krumbein (1944/47) found that $b = 0{\cdot}23$ and $R_o = 0{\cdot}07$ whilst from Fig. 1.12, $k = 10\ 0$ and $K = 0\ 6$.

$$\text{Thus} \qquad R_o e^{bK} = 0{\cdot}08$$

$$\text{and} \qquad bk = 2{\cdot}3$$

so that the predicted relationship between particle size r and the relative wave energy E is given by:—

$$r = 0{\cdot}08e^{2{\cdot}3E}$$

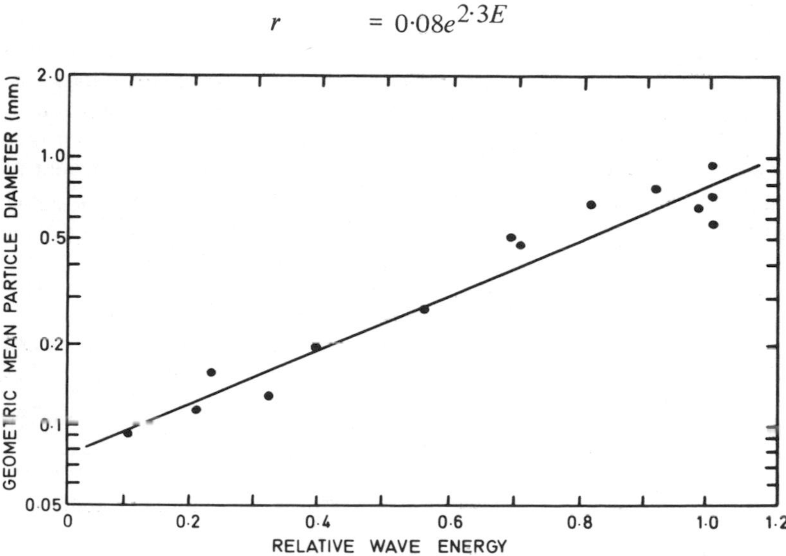

Fig. 1.13. Graph illustrating the predicted relation between particle size r and wave energy E according to the expression $r = 0{\cdot}08e^{2{\cdot}3E}$. Experimentally determined values are shown as dots. (After Krumbein, 1944/47)

Fig. 1.13 shows the experimentally determined values of particle size and wave energy together with the predicted line; these are in good agreement. Thus the slope of the shore S and the particle size r varies with the degree of exposure to wave action E, steep shores with coarse particles being associated with exposed situations and shallow shores with fine particles occurring under more sheltered conditions. This commonly-observed feature is thus placed on a quantitative basis.

Bascom (1951) has also found that particle size varies with the degree of turbulance, being coarsest at points of maximum turbulance and getting finer as wave action decreases. He showed in addition, that eroding beaches tend to be shallower in slope than accumulating beaches. Thus it is clear that under conditions of deposition the prime factor influencing particle size and beach slope is the degree of turbulance, and where there is a regular gradation in the degree of exposure to wave action there will be a corresponding alteration in the nature of the deposits and in their fauna.

Morgans (1956) (also Krumbein, 1939) has reviewed methods used in the analysis of marine sediments, and stressed the importance not only of determining the median particle diameter of the deposits but also of calculating the degree of sorting to the particles; that is, the proportion of particles which are of similar diameter to that of the median particle size. In general, it is found that coarse deposits are often poorly sorted with a high proportion of particles larger and smaller than the median particle size, whilst fine sediments deposited under more sheltered conditions tend to be of more uniform diameter. The degree of sorting profoundly modifies the physical features of the deposits (p. 34), and it is therefore of some importance to consider a simple method for assessing both the particle size and the proportion of particles greater and smaller than the median diameter. Morgans (1956) pointed out that the Wentworth scale of sieves divides the particles into unequal intervals; it is therefore convenient to convert the particle diameters in mm into phi (ϕ) units so that the unequal Wentworth units are then equal units. Fig. 1.14 illustrates a conversion graph based on Krumbein (1939) from which the corresponding ϕ units can be obtained. Now a cumulative curve relating the percentage of each grade

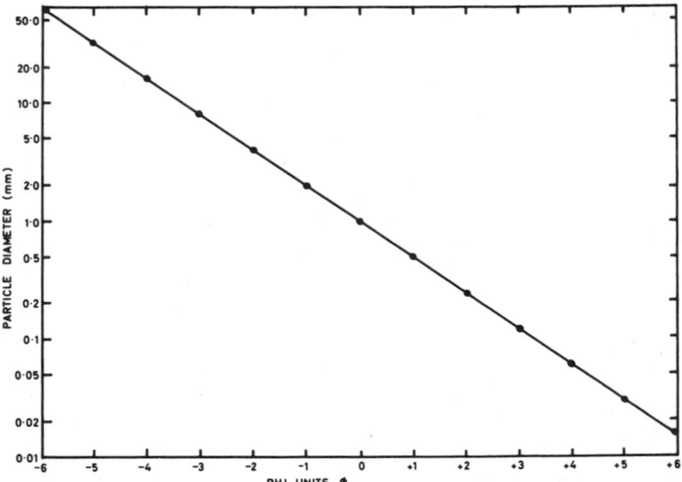

Fig. 1.14. Graph showing the relationship between particle diameter and phi (ϕ) units. (After Krumbein, 1939)

of deposit to particle diameter in ϕ units can be plotted as in Fig. 1.15, the median particle diameter of the deposit (Md_ϕ) being numerically equal to the value on the x axis corresponding to the 50% value on the

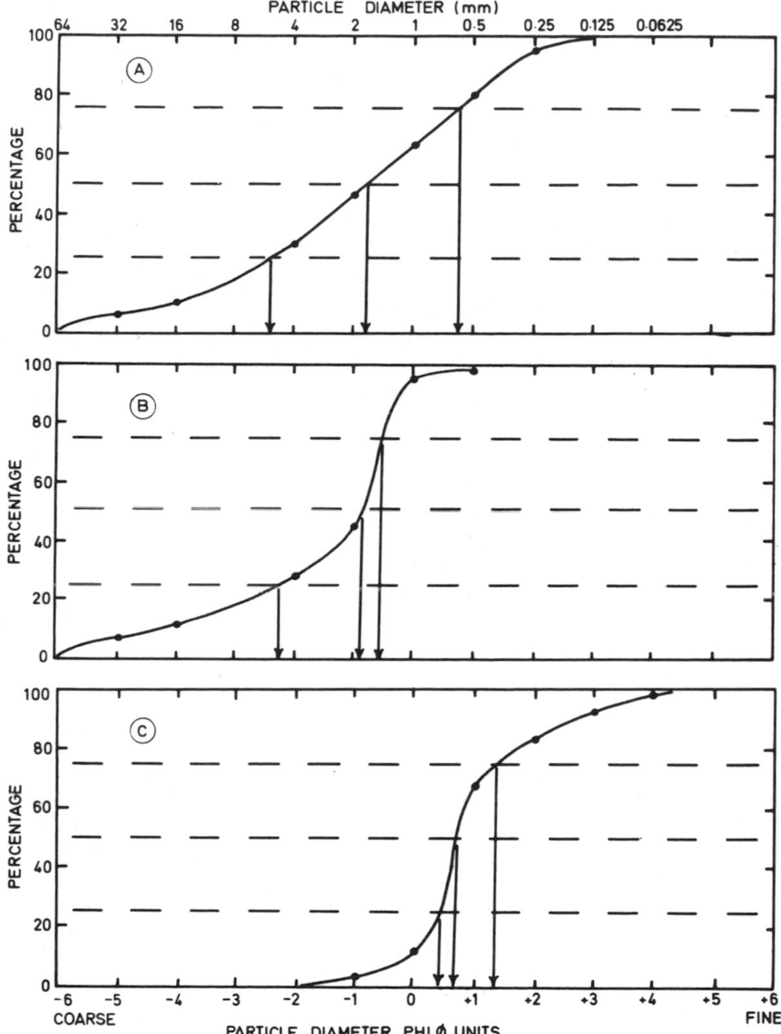

Fig. 1.15. Cumulative curves of three sediments plotted against particle size. Phi units are shown below and mm above. The 1st (25%), 2nd (50%) and 3rd (75%) quartiles are indicated by broken lines. The median particle diameter (Md_ϕ), phi quartile deviation (QD_ϕ), and phi quartile skewness (Skq_ϕ) of the three sediments are shown in Table II.

y axis. This value can then be expressed as mm by use of the graph shown in Fig. 1.14.

Now it is obvious that the deposit shown in Fig. 1.15(A) is composed of a large proportion of particles which are both larger and smaller than the median particle diameter. That is, the cumulative curve has a relatively shallow slope indicating that the deposit is ill-sorted. A measure of the slope is given by the phi quartile deviation (QD_ϕ) (Morgans, 1956).

$$QD_\phi = \frac{Q3_\phi - Q1_\phi}{2}$$

where $Q3_\phi$ and $Q1_\phi$ are the corresponding phi values for the 75% quartile and the 25% quartile respectively. Clearly, the ill-sorted deposit figured in Fig. 1.15 (A) has a higher value for the phi quartile deviation than curves (B) and (C). (See Table II).

The second factor which is of interest apart from the median particle diameter is whether the coarse and fine fractions are equally or unequally sorted. This is indicated by whether or not the cumulative curve is straight or curved between the 75% and 25% quartiles. In Fig. 1.15 (A) the cumulative curve is straight between the quartiles, particles larger and smaller than the median diameter being equally sorted. In Fig. 1.15 (B), however, the line is not straight, which indicates that the fine particles are better sorted than the coarser ones whilst in Figure 1.15(C) the coarse particles are better sorted than the fine ones. These factors may be conveniently expressed in terms of the phi quartile skewness (Skq_ϕ).

$$Skq_\phi = \frac{Q3_\phi + Q1_\phi - 2Md_\phi}{2}$$

A straight line, as in Fig. 1.15(A) gives a zero value for $Skq\phi$. A negative value is obtained for curve (B) since the mean values of the quartiles are to the left of the median particle size. This indicates that the smaller particles are better sorted than the coarser ones. A positive value is obtained for the curve illustrated in Fig. 1.15(C) since the mean of the quartiles is to the right of the median particle size. This indicates that the coarse particles are better sorted than the fine ones. The median particle diameter, deviation and skewness of the three deposits illustrated in Figure 1.15 are tabulated below. Details of this system of presentation, together with reviews of methods of mechanical analysis, are given by Krumbein (1939) and Morgans (1956).

It has been pointed out that on many beaches there is a gradation of particles from coarse ill-sorted deposits on exposed shores (where the wave energy is high) to fine well-sorted ones in sheltered areas. This difference in the degree of sorting of coarse and fine deposits is illus-

Table II Table showing the median particle diameter, phi quartile deviation and phi quartile skewness of the three deposits shown in Fig. 1.15.

DEPOSIT	*A*	*B*	*C*
Md_{ϕ}	-0·75	-0·75	0·5
QD_{ϕ}	$\frac{0{\cdot}75 - (-2{\cdot}25)}{2} = 1{\cdot}5$	$\frac{-0{\cdot}60 - (-2{\cdot}25)}{2} = 0{\cdot}825$	$\frac{1{\cdot}4 - (+0{\cdot}4)}{2} = 0{\cdot}5$
Skq_{ϕ}	$\frac{0{\cdot}75 + (-2{\cdot}25) - (-1{\cdot}5)}{2}$ = 0	$\frac{-0{\cdot}60 + (-2{\cdot}25) - (-1{\cdot}5)}{2}$ = -1·35	$\frac{1{\cdot}4 + (+0{\cdot}4) -1}{2}$ = 0·4

trated in Fig. 1.15. Such a regular gradation in particle size and degree of sorting results in important changes in the physico-chemical properties of the substratum, and these are reflected in the nature of the animals characterising the deposits. Amongst the more important of such parameters are the water content of the deposits, the mobility and depth of disturbance to which the deposits are subjected by wave action, the salinity and temperature of the interstitial water, the oxygen content and the quantity of organic matter present (p. 249 also Webb, 1969).

One of the important factors which affects the animals living in deposits is the water content of the sand. This water may be either retained in the interstices between the sand grains as the tide falls or be replenished from the water table below by capillarity. Bruce (1928a) has shown that in a system of uniform spheres the pore space accounts for 25·96% of the total volume irrespective of the absolute size of the spheres. It might be supposed, therefore, that the same water-retaining capacity would be found in natural sands. In fact Bruce (1928a) found that the water-retaining capacity of pure grades of sand between >0·54 mm diameter to < 0·09 mm diameter was always strikingly higher than the theoretical value of 25·96%, 35·8 vols of water being present in 100 vols of wet sand of particle size<0·54 mm, 44·7 vols of water/100 vols of 0·11 – 0·09 mm sand and 43·4 vols/100 vols of < 0·09 mm sand. This is due to the fact that sand grains are not spherical and therefore do not become packed in the closest possible array and may also form stable arcades with large interstitial spaces (also Webb and Hill, 1958). Natural ungraded sands, particularly those which are poorly sorted, have a much lower percentage pore space because the smaller sand grains pack into the interstices between the larger particles. Bruce (1928a) found that in samples collected from the shore at Port Erin, Isle of Man, U.K., only 20 vols of water were present per 100 vols of wet sand compared with the theoretical value of 25·96% for a system of uniform spheres. Thus on coarse sandy beaches in which the deposits are ill-sorted the porosity of the deposits is relatively low (approximately 20%) whereas under more sheltered conditions, where the deposits tend to be of more uniform diameter, the water retention may approach 44·7% under conditions of perfect sorting (i.e. $QD_{\phi} = O$; p. 32).

The rate of replacement of evaporative water loss by capillary rise from the water table below depends upon the diameter of the capillary channels between the sand grains. The width of such channels becomes less as the grains decrease in size irrespective of the pore space, so that it is in the fine deposits that capillary rise is greatest although, here again, the presence of fine particles in a poorly-sorted coarse deposit greatly alters its physical properties. Such an effect is shown in Fig. 1.16 from which it will be seen that the presence of a mixture of

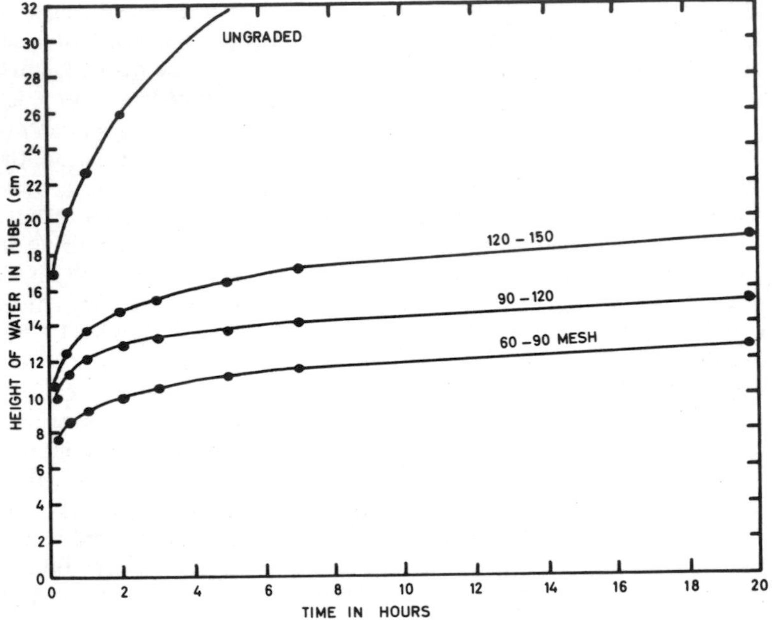

Fig. 1.16. Graphs illustrating the effect of grade of deposit on the rate of capillary rise of water through sands. The sands were contained in a tube of 13 mm internal diameter the lower end of which dipped into water taken from the black layer on the beach. The height of the saturated column of sand was then read off at intervals. (After Bruce, 1928a)

particles resulted in a vast increase in the rate of capillary rise of water through the ungraded sand (Bruce, 1928a). The net effect of these properties is that in fine deposits where the slope of the shore is shallow (p. 29) and the water retention (porosity) high, the substratum is permanently moist and may even be covered with a thin film of water between the ripple marks in some areas. In coarse deposits, however, where the porosity may be reduced by the presence of fine particles in the interstices between the larger grains and where the slope of the shore is steep (p. 29), the substratum contains less water. Such factors play an important part in controlling the stability of the substratum and the ability of intertidal animals to burrow in it.

Chapman and Newell (1947) and Chapman (1949), for example, have studied the effect of the water content of intertidal deposits on the ability of the lugworm *Arenicola marina* to burrow. They found that the deposits on the shore at Whitstable, Kent, show the properties of thixotropy or dilatancy according to the percentage of water which is present in the sand. When the water content was less than 22% by weight, an applied force was found to upset the close packing of the particles in such a way that the interstitial water was no longer able to fill the enlarged pores between the sand grains; the dilatant deposit then became harder and more resistant to shear (also Reynolds, 1885; Freundlich, 1935; Freundlich and Jones, 1936; Freundlich and Röder, 1938). When the water content was greater than 25%, dilatancy, as indicated by the whitening of the sand under the footfall, gave way to the property of thixotropy in which there was a reduction in the resistance with increased rate of shear. Chapman (1949) was later unable to obtain true dilatant properties with such sands under experimental conditions, but nevertheless demonstrated that the resistance to the passage of a sphere through water-saturated deposits was considerably lower than that of deposits containing only 21% water (Fig. 1.17). He also showed that the speed with which *Arenicola marina* can burrow into such deposits is dependent upon the hardness of the soil and hence upon the water content. Chapman and Newell (1947) showed that agitation of the sand by repeated applications of pressure caused its resistance to drop from an average of 412 g/cm^2 to 41 g/cm^2. Under these conditions, the maximum pressures available for worms to penetrate the sand were found to be sufficient to effect penetration. Clearly, if the sand remained in a semifluid state for long, the worm would be unable to grip the wall of the burrow and penetrate deeper into the soil, but it was then found that the sand regained its initial resistance in approximately 0·5 minutes and that this enabled further burrowing.

It is obvious, therefore, that the degree of water saturation of the deposits plays an important part in controlling the ability of the lugworm to burrow. Much the same system must be employed by other polychaetes such as *Nephtys hombergi* and perhaps also bivalves such as *Macoma balthica* and other common inhabitants of moist sands and muds. Nevertheless, other burrowing animals such as the razor-shells *Ensis* spp and *Solen* and the tusk-shell *Dentalium* are able to penetrate

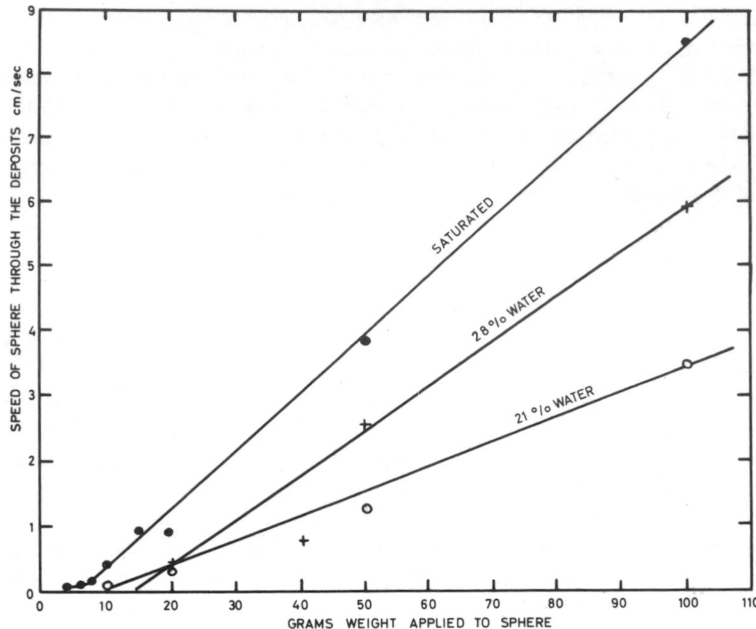

Fig. 1.17. Graphs showing the effect of water content of deposits on the resistance to the passage of a sphere. (After Chapman, 1949)

even dilatant deposits and must therefore employ a rather different method of burrowing. In *Solen* and *Ensis* (Fraenkel, 1927b) the foot is first extruded and then forms a terminal button. The longitudinal muscles then contract and the shell is pulled down into the substratum (Fig. 1.18.). A rather similar system involving the use of a terminal "anchor" against which longitudinal muscles can act has recently been described in the scaphopod *Dentalium* by Trueman (1968). In this animal the foot is poked downwards and the shell is prevented from being forced up out of the sand by its conical shape which forms a shell anchor and allows the penetration of the sand by the foot; the epipodial lobes are then expanded to form a pedal anchor against which the pedal contractor muscles can pull (Fig. 1.19). The use of a pedal anchor during burrowing is also described in *Glycymeris* by Ansell and Trueman (1967) and appears to be a common feature of animals which live in dilatant, as opposed to thixotropic, substrata.

Bruce (1928a) has shown that intertidal sands not only protect the infauna from desiccation but also from much of the thermal stress to which rocky shore animals are subjected (also Chapter 9). He recorded

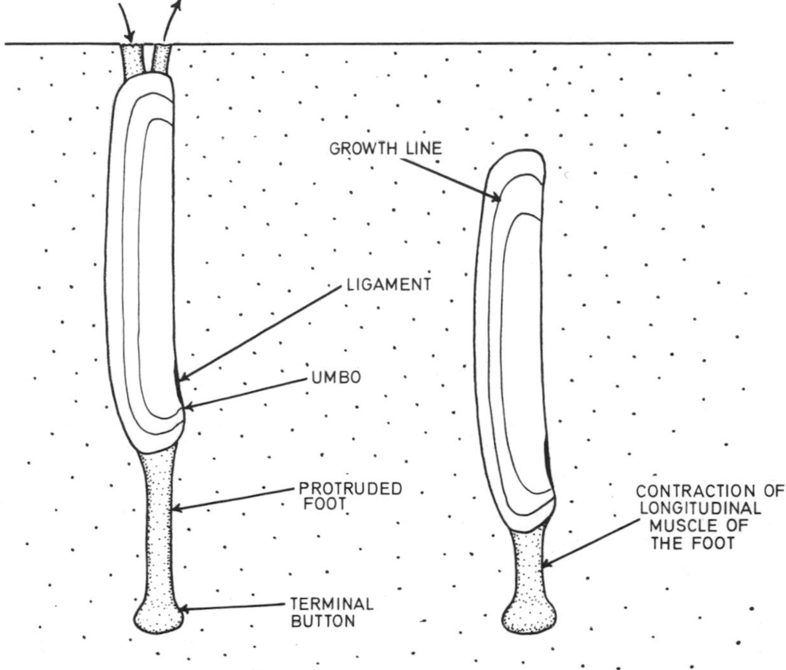

Fig. 1.18. Diagram showing the mode of burrowing of a razor shell *Ensis siliqua.*

average temperatures of 16·5°C in June, 18·8°C in August and 10·7°C in October in the upper 1 cm of sand at Port Erin. The highest was 21°C recorded in June. Such temperatures, however, rapidly decline with depth from the surface of the sand and approximate to the sea temperature at 20 cm depth. Fig. 1.20 shows the effect of depth on subsurface temperatures recorded by Bruce (1928a) on Port Erin beach in August.

In a similar way, the salinity of the interstitial water commonly approximates to that of the sea except for the surface layer which may be either more concentrated due to evaporation or diluted by rainfall. As has been demonstrated by Reid (1930, 1932) even when a stream of freshwater passes over the surface of intertidal sand, there is little effect on the salinity of the interstitial water at 10–12 inches depth at which the salinity approaches that of seawater. The salinity of the surface layers is anyway rapidly restored when the shore is covered by the tide. Bruce (1928a), however, has shown that under some circumstances freshwater streams may sink through the upper layers of a

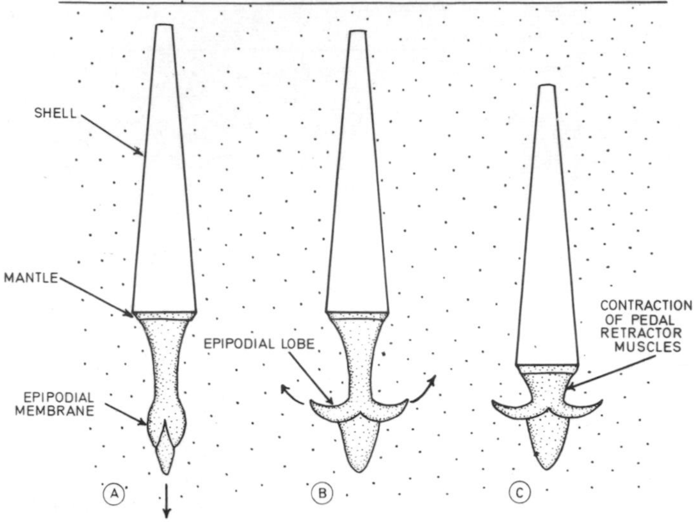

Fig. 1.19. Diagram showing the mode of burrowing of *Dentalium*. (A) Protrusion of the foot downwards, the shell being prevented from moving upwards by its conical shape. (B) Lateral movement of the epipodial lobes to form a pedal anchor. (C) Contraction of pedal retractor muscles after which the epipodial membrane loses its turgidity and returns the foot to the form shown in stage (A). (After Trueman, 1968)

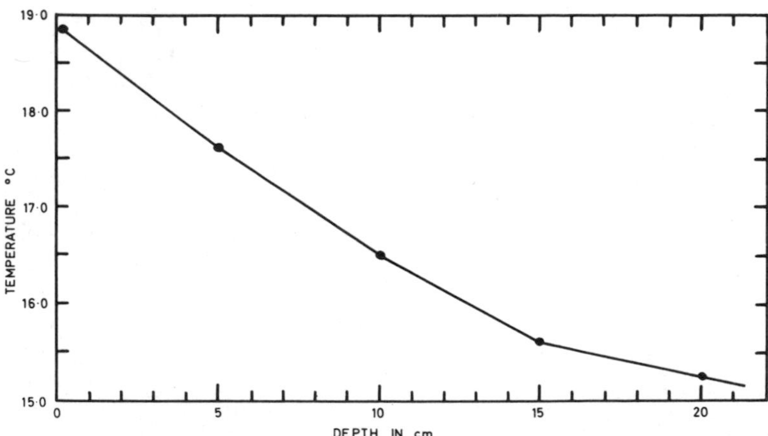

Fig. 1.20. Graph showing the effect of depth in the sand on subsurface temperature. (Data from Bruce, 1928a)

sandy shore and be drawn up to the surface from below by capillarity, thus giving rise to isolated patches of reduced salinity some distance from any visible source of freshwater. It seems likely that the effect of surface drainage depends among other factors on the slope of the shore and the relative position of the water table. On shallow sloping shores in which the water table is near the surface and in which a large quantity of interstitial water is retained by the deposits (p. 34), surface drainage may have little effect on the interstitial salinity (Reid 1930, 1932). Where the slope of the shore is steep and the water table low, surface drainage may percolate down until it reaches the water table and then be drawn up to the surface in other parts of the shore as described by Bruce (1928a). Such conditions, however, must be regarded as rather exceptional on the majority of sandy and muddy shores, and the animals living in intertidal deposits may be assumed to be exposed to relatively little change in salinity throughout the tidal cycle.

The factors controlling the concentration of dissolved oxygen in the interstitial water have been recently discussed by Brafield (1964) to whom reference should be made for a review of earlier work on this subject. He found that there was no clear correlation between the median particle diameter and quantity of dissolved oxygen in the interstitial water, and concluded that drainage was the chief factor controlling the oxygen concentration. Thus coarse deposits covered with surface water were found to have lower values of oxygen in the interstitial water than finer deposits where no surface pools persisted. Webb and Hill (1958) have found that the porosity and drainage time increases sharply when there is 20% or more of fine sand in the deposit (i.e. sand which will pass a 0·25 mm mesh sieve), and Brafield (1964) noted that the oxygen concentration in the interstitial water varied in a similar way with the percentage of fine sand (Fig. 1.21).

In poorly drained deposits such as are found at Whitstable, Kent, there is a marked vertical gradation of oxygen concentration from surface values which may exceed air saturation (due to the photosynthetic activity of naviculoid diatoms which migrate to the surface of the sand as the tide ebbs (Aleem, 1949)) to 1·4 ml $O_2/1$ at 2 cm and 0·3 ml/1 at 5 cm depth (Brafield, 1964). Similar data have been obtained by Pearse, Humm and Wharton (1942), the low values below 5 cm being due to the oxidation of black iron sulphides which are formed by the activity of the bacteria in the deposits (Bruce, 1928b). Nevertheless in some deposits, even where a pronounced black layer containing 24 vols per cent of hydrogen sulphide is present, the interstitial water has been found to contain as much as 1·81 ml $O_2/1$ which was equivalent to 25% of air saturation (Bruce, 1928b), and allowed the presence of a number of annelids, crustaceans and echinoderms in the deposits.

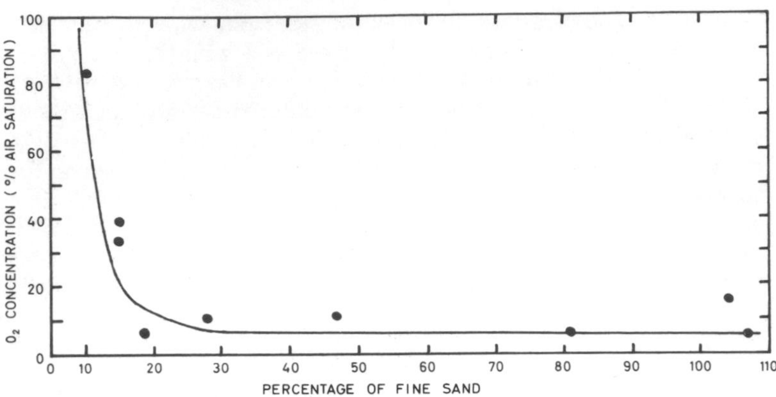

Fig. 1.21. Graph showing the relationship between the oxygen concentration of the interstitial water and the percentage fine sand in intertidal deposits. (After Brafield, 1964)

The rapid depletion of oxygen in poorly drained deposits in which there is more than 20% of fine sand has a profound influence on the mode of life of the animals living there. Such animals need to retain a connection with the oxygenated surface layers and maintain a flow of water to and from the body. Alternatively animals which do not maintain a connection with the surface are confined to the upper few centimetres above the black layer of fine deposits but can live rather deeper in coarse well-drained deposits where the levels of interstitial oxygen are higher. Animals retaining a connection with the oxygenated surface layers of the deposits include the lugworm *Arenicola marina* which lives in a U-shaped burrow (p. 235 and Fig. 5.48) through which a flow of water is maintained at high tide by means of peristaltic waves of contraction passing along the body (Wells 1945, 1949a,b, 1953, 1959). At low tide, when the water table falls below the surface of the sand, the animal is able to expose its gills to the moist air in the upper part of the tail shaft and so make use of atmospheric air. Other polychaetes such as *Chaetopterus* (Fig. 5.9), *Sabella* (Fig. 5.10) and *Clymenella torquata* (Mangum, 1964) live in U-shaped tubes through which an irrigation stream can be pumped, whilst bivalves living in anoxic sediments communicate with the surface by means of siphons as in *Macoma balthica, Mya arenaria* and *Scrobicularia plana* (Fig. 5.41). The burrowing crab *Corystes cassivelaunus,* although characteristically a sublittoral animal living in coarse deposits, also communicates with the surface from the safety of the deposits by means of a tube which in this case is formed from the elongated setose second antennae held together to form a tube down which oxygenated water is drawn.

There is one further factor which influences the mode of life of the animals living in intertidal deposits. It is well-known that coarse deposits are mobile and may even move shoreward at several metres per day under exposed conditions, whereas fine deposits are relatively static, only the surface centimetres being disturbed by wave action. For this reason the animals of sandy shores are active burrowers which do not retain connection with the surface by means of a permanent tube in contrast to the permanent burrows and tubes of animals living in fine deposits. The organic content of intertidal deposits also exerts an important influence on the nature and abundance of intertidal animals and is discussed in some detail in Chapter 6.

(ii) The fauna of intertidal deposits

Intertidal deposits are inhabited by a highly specialised interstitial fauna as well as by the more obvious macrofauna. We shall concern ourselves here with the nature and distribution of the macrofauna of sands and muds. We have seen (p. 1) that the zonation of animals and plants is an obvious feature of rocky shores. Zonation on intertidal sands and muds is, however, not immediately apparent for two main reasons. Firstly, as mentioned on p. 26, the majority of the organisms inhabiting such shores burrow in the deposits so that details of their distribution can be obtained only after careful collection. Secondly, particulate shores are commonly of a shallower gradient than rocky shores so that the corresponding intertidal zones are broader and less obvious. Nevertheless, zonation of the infauna does occur in sands and muds. Watkin (1939, 1941a, 1942) has shown that there is a well-marked zonation in the distribution of the amphipod *Bathyporeia* in the sand at Kames Bay, Scotland. The species inhabiting the highest shore level was *Bathyporeia pilosa,* this being replaced downshore by *B. pelagica.* This was overlapped and replaced downshore by *B. elegans* whose lower limit extended down to mean low water of spring tides. Finally, *B. guilliamsoniana* overlapped with *B. elegans* and extended into the sublittoral zone. Unlike the animals which inhabit muds, the small crustaceans of sands are active organisms which do not occupy permanent burrows in the substratum. The sand represents an environment where the physico-chemical stresses of the emersion period can be evaded and from which feeding excursions can be made when the intertidal zone is covered by the tide. Indeed, Watkin (1939; also Bossanyi, 1957) has shown that the distribution of *Bathyporeia* species in the inshore plankton of Kames Bay was similar to the distribution in the sand at low tide. Thus as the tide flowed *B. pelagica* and *B. elegans* became common on the edge of the incoming water but as the high water mark was approached, *B. pilosa* became common to the exclusion of *B. pelagica* and *B. elegans.*

Colman and Segrove (1955) have also studied the vertical distribution of *Bathyporeia* as well as a number of other amphipods in Stoupe Beck Sands, Yorkshire. Here again there was a distinct vertical sequence; *Haustorius arenarius* had an upper limit at approximately mean high water of neap tides, followed by *Bathyporeia sarsi, B. pelagica, B. elegans* and *B. guilliamsoniana.* (Fig. 1.22) *Pontocrates norvegicus* extended throughout much of the zone with a maximum at approximately mean low water of neap tides and was replaced downshore by *P. arenatrius*.

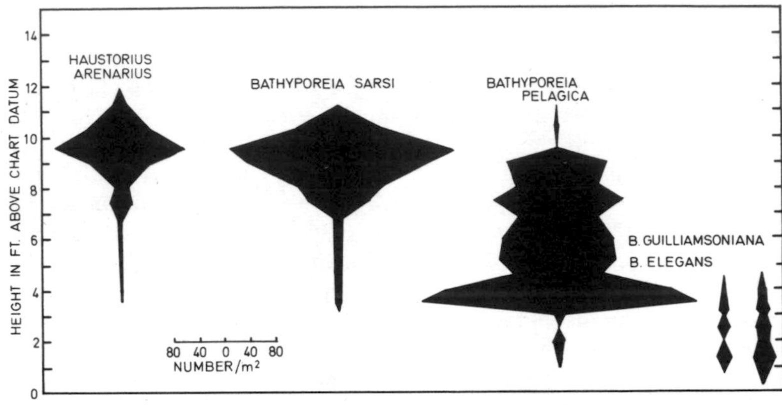

Fig. 1.22. Kite diagrams showing the numbers, and vertical distribution of Amphipoda in Stoupe Beck sands, Yorkshire. (After Colman and Segrove, 1955)

Different species of bivalves too have been shown to attain their maximum abundance at particular levels on the shore. Both *Tellina tenuis* and *Donax vittatus* are commonly the dominant bivalves on sandy shores and, although they occur at other levels on the shore, they attain their maximum abundance in the lower intertidal zone. *Tellina fabula,* however, occurs only in the lower intertidal zone and is more commonly found sublittorally (Stephen, 1929). However, other examples of apparent vertical zonation in burrowing animals are to be correlated with variations in the substratum rather than with tidal level. Thus Stephen (1930) showed that the upper shore of Loch Gilp, Scotland, was dominated by *Nereis diversicolor,* this being followed downshore by an extensive zone of *Macoma balthica* and *Cardium edule.* The midshore was dominated by *Scoloplos armiger* which was overlapped and replaced by *Nephtys caeca. Travisia forbesii* occurred in a narrow zone near the low water mark, but the dominant lower shore

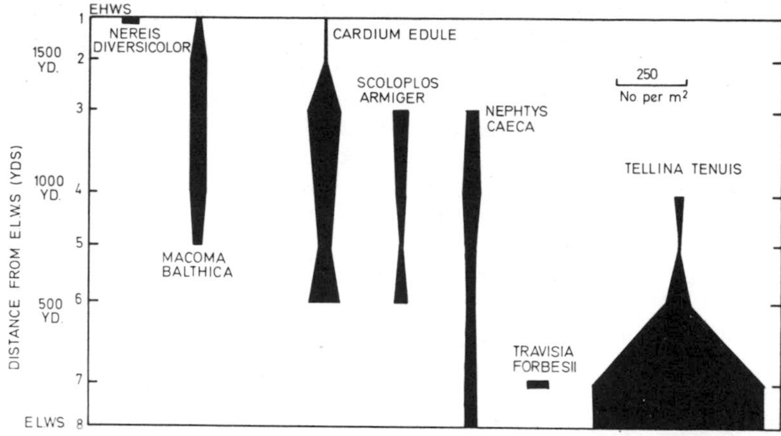

Fig. 1.23. Kite diagrams showing the abundance and vertical distribution of some of the macrofauna of Loch Gilp, Scotland. (After Stephen, 1930) The approximate distance in yards from the extreme low water mark is also shown.

animal was *Tellina tenuis.* (Fig. 1.23). There was, however, a corresponding variation in the grade of deposit from clean sand on the lower shore to fine muds at high tide level. Similar variations in the grade of deposit from low to high water mark are characteristic of many shores and account for many of the faunistic changes which occur at different tidal levels. Indeed, in a uniformly graded deposit it would be surprising to find marked changes in the fauna with tidal level, since the burrowing fauna to a large extent evades those stresses which are responsible for zonation on rocky shores. Differences in the fauna are therefore more associated with variations in particle size and associated factors some of which have been discussed on p. 33.

Sands which contain little silt are characterised by a variety of organisms including the bivalves *Mactra corallina, Donax vittatus* and *Tellina tenuis,* with occasional *T. fabula* at low tide levels. *Gari fervensis, Nucula nucleus* and the spiny cockle *Cardium echinatum* are often common and so too are the razor shells *Ensis ensis, E. arcuatus, E. siliqua* and *Solen marginatus.* Gastropods are scarce, but certain predatory forms such as *Natica catena* and *N. alderi* are responsible for the death of large numbers of the bivalves (p. 186). Small crustaceans such as *Haustorius arenarius, Bathyporeia* spp. *Pontocrates norvegicus, Eurydice pulchra* and *Talitrus saltator* (also p. 146) are often abundant as well as the ghost crab *Corystes cassivelaunus* and *Portumnus latipes.* Typical polychaetes include *Scolecolepis ciliata, Nerinides cantabra, Ophelia bicornis* and *Thoracophelia,* and the brittle-star *Ophiura albida* is also to be found near the low water mark.

As the deposits become finer the animals come to occupy permanent burrows in the substratum and increasingly utilise either the suspended plankton or the organic matter of the deposits as a source of food (p. 248). Deposit-feeders such as the lugworm *Arenicola marina,* the sand-mason *Lanice conchlega,* and *Macoma balthica* become abundant in sands containing appreciable quantities of fine material and organic debris. Here too, *Cardium edule, Scoloplos armiger* and *Phyllodoce maculata* are common whilst in very fine muds other animals such as the amphipod *Corophium volutator* (p. 233), the gastropod *Hydrobia ulvae* (p. 261), the terebellid worm *Amphitrite johnstoni* (p. 221), the nereid *Nereis virens,* and the minute spionid worm *Pygospio elegans*, are dominant. Some of the factors which account for these faunal differences are discussed on p. 33 and are also reviewed by Yonge (1950b, 1952), Stephen (1952), Moore (1958), and Southward (1965).

(d) Biological factors

Whilst it is true that the physico-chemical conditions in the environment control the overall nature and distribution of the organisms living in the intertidal zone, it is equally true that biological factors may profoundly influence conditions in the habitat. The intertidal algae, for example, considerably reduce the desiccation and thermal stress to which intertidal animals are subjected and would be expected to allow an upshore extension of lower shore species. Some animals such as *Littorina littoralis* (= *obtusata*) may indeed actively orientate towards such algal cover (van Dongen, 1956), whilst others such as the serpulid *Spirorbis borealis* orientate to *Fucus* in the larval stage (Knight-Jones, 1951b, 1953b,c,d; Ryland 1959; Wisely, 1960; for review, Knight-Jones and Moyse, 1961; also p. 107). On the other hand the presence of a dense cover of fucoids under sheltered conditions normally precludes the establishment of a barnacle/limpet community; these consequently tend to occur under more exposed conditions where the fucoid cover is less dense. Equally, the browsing activities of intertidal herbivores such as *Patella* spp. and *Littorina littorea* may greatly reduce the algal cover in some areas (for example Conway, 1946; Jones, 1948; Southward, 1956) and this has important repercussions on the distribution of organisms dependent upon the algae for protection from the physical extremes of the habitat. Predation by animals such as the dogwhelk *Thais* (= *Nucella*) *lapillus* on barnacles may account for an annual mortality of from 3–35% in *Balanus balanoides* on the Isle of Man (Moore, 1934a,) and from 3·3–46·5% in *Chthamalus stellatus* at Plymouth (Moore and Kitching, 1939); Fischer-Piette (1935) showed that *T. lapillus* may profoundly alter the relative proportions of a *Mytilus*/barnacle community by its predation. The effects of the algae in either sheltering or excluding intertidal animals, and of browsing and predation on the presence of other intertidal organisms, are thus extremely complex and may play an important part in modifying the effects of the physico-chemical features of the environment.

One of the most striking examples of the modification of the environment by intertidal organisms is that of mussel bed communities in estuaries. It will be recalled that organisms living in deposits are primarily burrowing forms and that characteristically the epifauna is scarce (p. 26). The infauna may therefore be regarded as a 'primary component' whose distribution and abundance is dictated directly by the physico-chemical conditions in the habitat. When such animals as, for example, *Cardium edule* or *Mya arenaria* die, their shells become exposed on the surface of the deposits and provide a stable substratum for the attachment of an epifauna. Since such areas for attachment are scarce, competition for space by larvae of the epifauna is intense and soon the shell becomes encrusted with barnacles which may be so crowded that they grow 2–3 cm tall (also Barnes and Powell, 1950b; Fig. 1.24A). Other organisms such as oyster spat, certain algae and *Mytilus edulis* may also settle on the shell. Such organisms would not occur unless the *Mya* had provided a suitably solid substratum, and they may be regarded as a "secondary component" insofar as their presence is to some extent independent of the physico-chemical nature of the environment. The third stage in the development of a mussel bed community involves an increase in the number of *Mytilus* which then attach to one another and deposit large quantities of consolidated silt (pseudofaeces). Such clumps of mussels may be torn up by storms and swept away, but in many cases the byssus threads attach the mussels to one another and to the dead shells below to form a compact mound covering a silt-like deposit of pseudofaeces and shells of dead mussels (Fig. 1.24 B and C).

Such a habitat is characteristically inhabited by the terebellid *Amphitrite johnstoni* and the ragworm *Nereis virens* whose presence is dependent upon the "secondary component" and whose abundance is largely independent of the original physico-chemical features of the environment which controlled the infauna. Finally, the burrow of *Amphitrite johnstoni* is often inhabited by the commensal scale worm *Gattyana cirrosa* (Fig. 5.35) whose presence is unrelated to either the nature of the deposits or the tidal level. We have, therefore, a hierarchy of faunal components, the first of which is directly dependent upon the nature of the environment but which suitably modifies the habitat to allow the presence of other organisms which are each dependent upon the presence of the other for survival. Occasionally the *Mytilus* population declines, perhaps due to predation by *Thais lapillus* (Fischer-Piette, 1935) or parasitisation by *Mytilicola,* and the absence of byssus threads allows the loss of the fine silt which characterises the

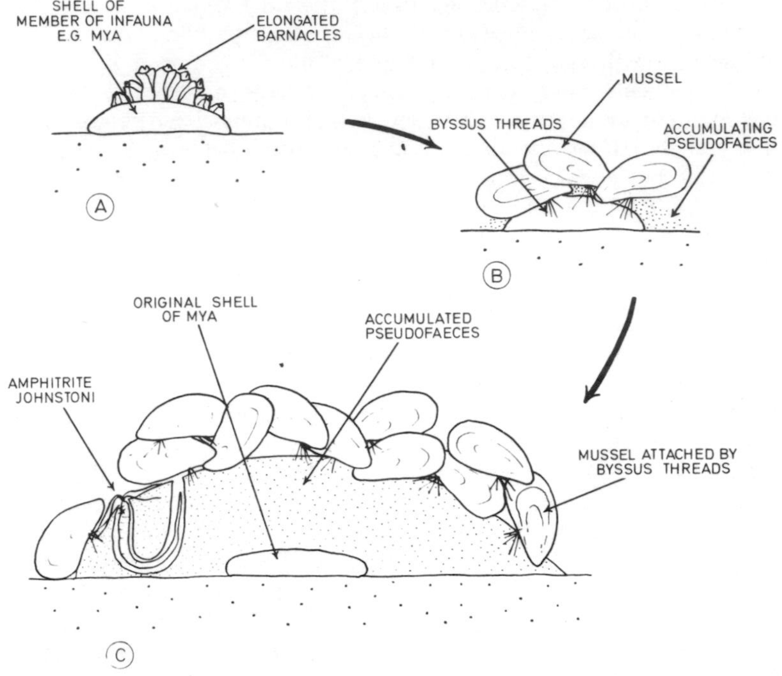

Fig. 1.24. Diagrams illustrating three stages in the development of a mussel bed. (A) Shell of a member of the infauna e.g. *Mya* encrusted with tall barnacles. (B) Attachment of mussels and accumulation of pseudofaeces. (C) A fully-developed mussel bed which has immersed the original *Mya* shell and which supports its own characteristic fauna e.g. *Amphitrite johnstoni.*

mussel bed. The *Nereis virens* and *Amphitrite johnstoni* then disappear together with their dependent organisms and the mussel shells become dispersed to be cast up upon the shore. However the mussel shells do form the basis for the attachment of other organisms, and many estuaries consequently support a substantial epifauna.

It is thus clear that biological factors often intervene in the control of the nature and abundance of intertidal organisms. Nevertheless it remains true that despite behavioural means to avoid the extremes of the environment, animals living high in the intertidal zone are in general better adapted to resist the extremes of the physico-chemical environment than their counterparts lower on the shore. Almost all intertidal animals are also found to be well-adapted for life in the particular zone in which they live; some of these adaptations are considered in Chapters 3–9.

CHAPTER 2

The faunas of special habitats

We have seen that although intertidal organisms occupy characteristic zones on the shore, the assemblage as a whole varies considerably according to the nature of the substratum, degree of exposure to wave action and aspect of the shore, as well as being influenced by a large number of tidal-dependent factors. It is not surprising to find, therefore, that in certain habitats on the shore physical and chemical conditions are sufficiently distinctive to account for the presence of special faunal assemblages which represent a relatively abrupt break from the surrounding fauna and flora. Such special assemblages include those which live in crevices, seaweeds, rock pools, and mussel bed communities (discussed on p. 44).

A. THE FAUNA OF INTERTIDAL CREVICES

The special nature of the fauna of crevices and the physico-chemical features of the habitat have been studied by several workers including Stelfox (1916), Badouin (1939, 1947), Glynne-Williams and Hobart (1952), Morton (1954) and Colman and Stephenson (1966). Glynne-Williams and Hobart (1952) studied the physical and chemical conditions within crevices on a rocky shore in Anglesey, North Wales. They showed that the prime factor which determines the presence of crevices on the shore is the structure of the rocks which outcrop in the intertidal zone. When such rocks are fissile, as in the mica-schist which Glynne-Williams and Hobart examined, the crevices of the lower intertidal zone were found to contain two distinct zones of deposit, although high on the shore such zones were not apparent. In crevices below mid-tide level the inner zone was found to contain brown muds with a high proportion of silt plus clay, whilst the outer zone was of coarser deposits which were black in colour and contained a high proportion of organic debris. The inner zone had less organic matter and a lower oxygen demand than the outer zone which took up a much greater quantity of oxygen from potassium permanganate than did the brown muds of the inner zone. Glynne-Williams and Hobart (1952) suggested that the differing colour of the zones was partly attributable to the high organic content of the outer zone which may filter organic debris from water draining through the crevice at low tide as well as retain material forced into the crevice by wave action, and partly to the presence of air being trapped and forced into the inner region of the crevice as the tide flowed. In a crevice at mean low water of spring tides, such air would be under an increased pressure of half an atmosphere at full tide; this would result in the oxygenation of the deposits of the inner zone.

Apart from such differences in physico-chemical conditions within the crevice compared with the surrounding environment, the habitat is in many respects sheltered from the rigours of the intertidal zone. Stresses such as desiccation and thermal fluctuation for example, are much reduced except in crevices high on the shore. Morton (1954) studied the temperature within crevices in Dartmouth Slate at Church Reef, Wembury, Devon during June and found that in crevices at, or below, a level corresponding with extreme high water of neap tides (approximately 11 ft above chart datum) (Fig. 2.1., crevice A), the temperature remained low and relatively uniform during the whole of the period of emersion. In crevices higher on the shore the temperature was found to rise more rapidly. A crevice approximately 15 ft above chart datum (Fig. 2.1., crevice B) had an average temperature which was 2°C higher than that of crevice A. The exposed rock surface on the shaded side of the reef had risen from the sea temperature of approximately 12°C to 15°C within 0·5hr of emersion by the ebbing tide, and after 4hr had risen to 20°C. A crevice on the southerly aspect of the reef, however, experienced much higher temperatures and reached 23°C even though this particular crevice was only 11 ft above chart datum (Fig. 2.1., crevice C). The position of the crevices and the results described above are shown in Fig. 2.1.

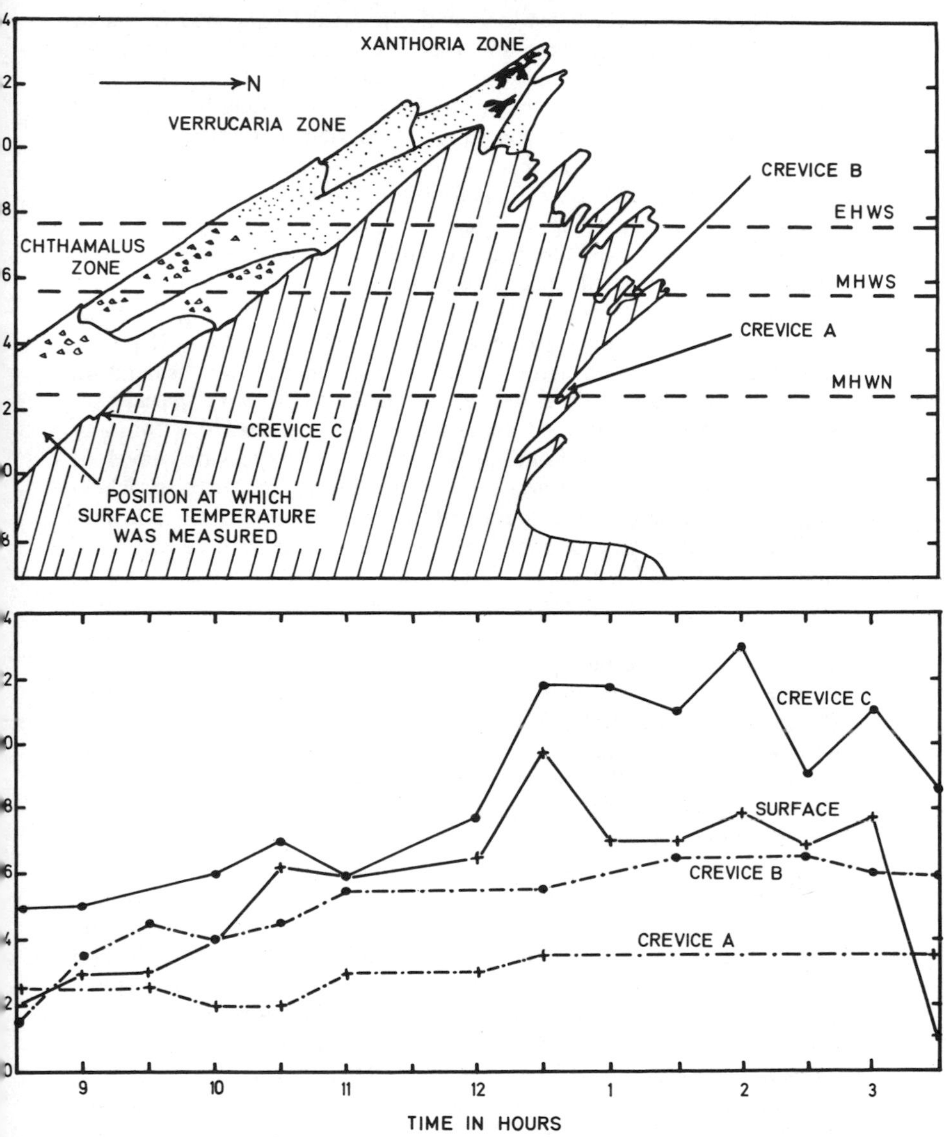

Fig. 2.1. Diagram showing the distribution of selected crevices on Wembury Reef (above), and the temperatures occurring in each over a complete intertidal period during June (after Morton, 1954). It will be noticed that the temperature of crevice C on the southern side of the reef exceeded that of the surface of the rocks on the southern side. This was due to the absence of cooling breezes within the crevice.

The humidity of crevices low in the intertidal zone was found to approach 100% in Anglesey, and even the surface of the rocks where there was a thick algal cover experienced little desiccation. Only in the region of mean high water of spring tides was the humidity low, and thus Glynne-Williams and Hobart (1952) inferred that desiccation may become a limiting factor at this level on the shore but probably was not so at lower levels. Morton (1954) found that at Wembury a crevice low in the intertidal zone (Fig. 2.1., crevice A) had a relative humidity of 95% (± 5%) at midday but reached 100% during the rest of the day, whereas on the exposed rock surface on the southern aspect of the reef the relative humidity was only 65%; crevices B and C (Fig. 2.1) had humidities not exceeding 90% and falling to 80% at midday. It is thus clear that there are two main features of importance in a consideration of the distribution of crevice-dwelling organisms. Firstly, spatial variations in the distribution of organisms would be expected within any particular crevice, corresponding with the inner oxygenated layer of brown muds and the outer coarser black layer; the latter is rich in organic matter and tends to be saturated with water but is low in oxygen when the tide has ebbed. Secondly, since the limiting factors of thermal stress and desiccation are to some extent diminished within crevices, an upward extension of crevice organisms might be expected.

Stelfox (1916) and Glynne-Williams and Hobart (1952) commented on the fact that the fauna of crevices comprises an aquatic component represented by marine animals extending up into the higher levels of the intertidal zone, and a terrestrial component represented by animals extending their range down into the intertidal zone. The aquatic component tends to be confined to the outer water-saturated regions of the crevice, whilst the terrestrial component is more common within the inner part of the crevice except when making feeding excursions into the outer zone and even out onto the surface of the rocks during the emersion period. Those terrestrial species which are normally found above mean high water of neap tides are commonly more uniformly distributed throughout the crevice. This may be associated with the fact that the crevices at this level on the shore show no division into zones, the whole crevice corresponding more with the inner part of lower shore crevices since periods of submersion are brief and infrequent. Typical crevice-dwelling organisms are listed in Table III. It has been pointed out by Glynne-Williams and Hobart (1952) that the aquatic component of crevices extends higher in the intertidal zone than similar species on the outer rock surface, but nevertheless declines in the higher crevices on the shore. The aquatic component of crevices is thus more diverse and numerous low on the shore. In contrast, members of the terrestrial component have well-defined upper and lower limits in the intertidal zone, and moreover there tends to be a more or less constant number of species present at each particular tidal level. There is, however, a peak in the number of lower limits between mean low water and extreme low water of spring tides, which implies that despite their adaptations to intertidal life members of the terrestrial component are unable to withstand continuous submersion and require periods of exposure even if these are infrequent and of short duration.

Table III. Table listing some of the more common organisms which inhabit crevices. (Data from Glynne-Williams & Hobart, 1952.)

Flora	Aquatic Fauna	Terrestrial Fauna
Xanthoria parietina	*Gromia oviformis*	*Anurida bisetosa*
Verrucaria maura	*Eulalia viridis*	*Anuridella marina*
Pelvetia canaliculata	*Perinereis cultrifera*	*Aëpophilus bonnairei*
Fucus spiralis	*Cirratulus cirratus*	*Aëpopsis robinii*
Ascophyllum nodosum	*Phascolosoma minutum*	*Hydrogamasus salinus*
Fucus vesiculosus	*Naesa bidentata*	*Cyrthydrolaelaps hirtus*
Fucus serratus	*Jaera marina*	*Halotydeus hydrodromus*
Laminaria digitata	*Lasaea rubra*	*Bdella longicornis* var.
Laminaria saccharina	*Littorina saxatilis*(= *rudis*)	*littoralis*
	Cingula cingillus	*Scolicoplanes maritimus*
		Ligia oceanica
		Leucophytia bidentata

Morton (1954) has divided the fauna of a typical crevice between extreme high water and mean high water of neap tides at Wembury into 6 zones, all of which are inhabited with the exception of the innermost zone. Zones C and D approximate to the outer water-saturated deposits of crevices at Anglesey, whilst zone E corresponds with the inner layer of brown muds described by Glynne-Williams and Hobart (1952). Morton (1954) thus considered that several other zones may be regarded as comprising part of the crevice environment. The outermost zone (Zone A, Fig. 2.2) is characterised by barnacles such as *Chthamalus stellatus* and the lichen *Lichina pygmaea,* but animals such as *Patella vulgata, Littorina saxatilis* (= *rudis*) and *Gibbula umbilicalis* may also be found. The second zone (Zone B, Fig. 2.2) is characterised by the serpulid *Spirorbis borealis,* the light intensity being still high enough to support the growth of the green alga *Enteromorpha* and the brown alga *Ralfsia.* Also present are the minute pulmonates *Otina otis* and *Leucophytia bidentata; Otina otis* may make browsing excursions into zone A when the tide is out.

A deeper zone (Zone C, Fig. 2.2) occurs when the crevice has narrowed to approximately ½ in diameter. Pieces of shell, pebbles and coarser sand particles are able to lodge here together with a rich accumulation of organic debris. This zone corresponds to the outer zone described by Glynne-Williams and Hobart (1952) and is characterised by the

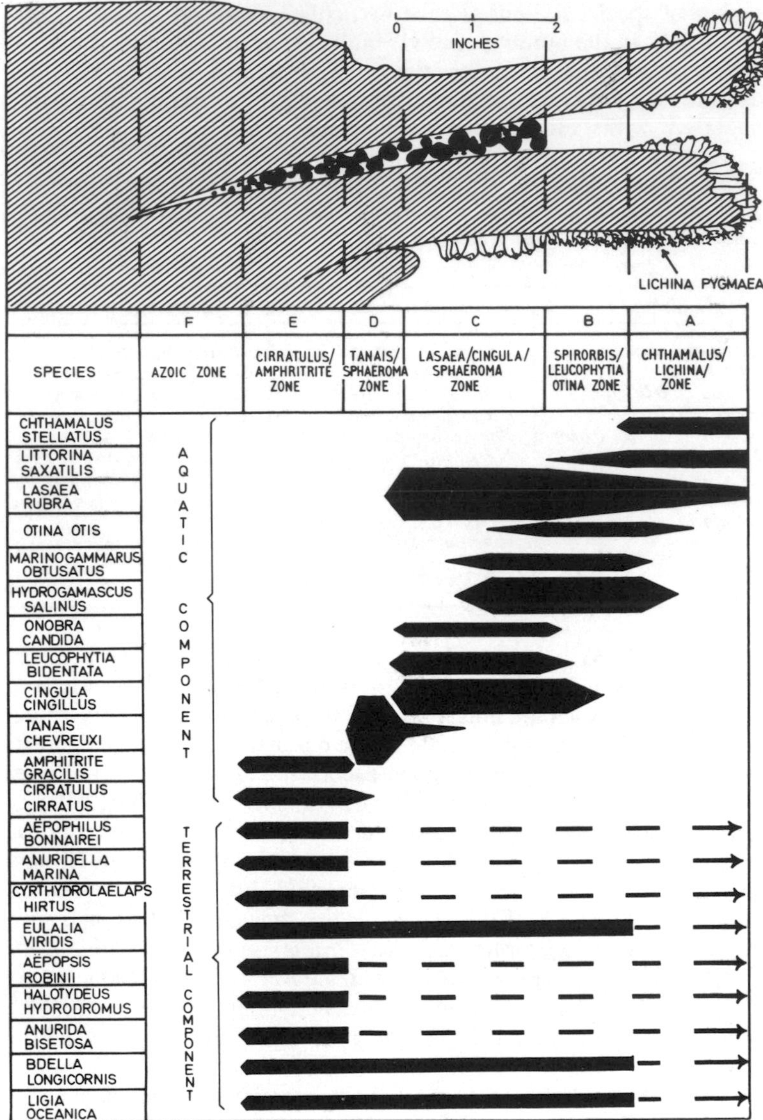

Fig. 2.2. Diagramatic section of a typical crevice (Location A, Fig. 2.1.) at Wembury, together with the approximate distribution of the organisms. Broken lines and arrows indicate feeding excursions made by the terrestrial component. (Based on Glynne-Williams and Hobart, 1952 and Morton, 1954.)

minute bivalve *Lasaea rubra* (also p. 240), the gastropod *Cingula cingillus*, *Leucophytia bidentata* and reduced numbers of both *Otina otis* and small specimens of *Littorina saxatilis*. The suspension-feeding *Spirorbis borealis* is entirely replaced by animals such as those above which all utilise the organic matter in the deposits of the crevice.

Zone C contains perhaps the largest variety of species; added to those listed above may be found the isopods *Sphaeroma serratum* and *Naesa bidentata*, the amphipods *Hyale nilssoni* and *Marinogammarus obtusatus*, rissoid gastropods such as *Onobra candida*, occasional *Mytilus edulis*, the nemertean *Lineus longissimus*, errant polychaetes such as *Eulalia viridis* as well as occasional specimens of the blenny *Blennius galeritus*. Also present are members of the terrestrial component (also p. 50) including the bluish-grey collembolan *Lupura* (= *Anurida*) *maritima*, the machilid *Petrobius maritimus*, the coleopteran *Aëpopsis robinii*, the pseudoscorpion *Neobisium maritimum*, the myriapod *Scolicoplanes maritimus*, and the red-brown mite *Hydrogamascus salinus*.

Zone D (Fig. 2.2) is a narrower region of the crevice than Zone C and the deposits trapped there are correspondingly finer. Typical species include the burrowing peracaridean *Tanais chevreuxi*, as well as *Sphaeroma serratum*, *Amphitrite gracilis* and *Cirratulus serratus*. Reduced numbers of *Lasaea rubra*, *Cingula cingillus* and *Leucophytia bidentata* also occur. Zone E (Fig. 2.2) has few organisms in it with the exception of the deposit-feeding polychaetes *Cirratulus cirratus* and *Amphitrite gracilis* which occur also in Zone D. Zone E contains the red-brown muds characteristic of the "inner zone" of the crevices described by Glynne-Williams and Hobart (1952). Zone F is represented only by the spaces between the laminae of the rocks and there is no space available for crevice-dwelling animals.

Apart from variations in the nature and abundance of organisms within any particular crevice, there is also a distinct pattern of vertical zonation amongst crevice-dwelling organisms. Morton (1954) recognised the presence of four main zones at Wembury in which the fauna of comparable crevices differed. (i) Above extreme high water of spring tides *Melaraphe* (= *Littorina*) *neritoides* is the dominant, and often the only, species represented. (ii) Between mean high water of spring tides and mean high water of neap tides *Melaraphe neritoides* and *Littorina saxatilis* are present, with *Lasaea rubra* becoming dominant. *Ligia oceanica*, *Otina otis* and the barnacle *Chthamalus stellatus* also occur. (iii) From mean high water of neap tides to the lower limit of the *Chthamalus* zone *Lasaea rubra* is dominant whilst the littorinids disappear. Species such as *Cingula cingillus* and *Leucophytia bidentata* occur as well as *Otina otis*, *Spirorbis borealis*, *Eulalia viridis*, *Hydrogamascus* and *Aëpopsis robinii*. (iv) Finally, below the upper limit of the

Fucus vesiculosus, the crevices are protected from desiccation and excessive wave action by the algal cover. Consequently there are less obvious faunal differences between crevices on the shaded and on the south facing side of the reef. In general, animals capable of withstanding the heavily sedimented parts of the crevice in the upper part of the intertidal zone are able to survive in crevices of the lower shore. Thus *Lasaea rubra* remains dominant and *Leucophytia bidentata* is abundant as well as *Aëpopsis* and *Hydrogamascus.* On the other hand, *Otina otis* and *Tanais* disappear whilst *Cingula cingillus* becomes scarce. New organisms such as *Porcellana, Gnathia, Hiatella,* and *Aëpophilus* occur together with low-level polychaetes such as *Amphitrite gracilis, Perinereis marioni* and *Cirratulus cirratus.*

Although the fauna dwelling within crevices is composed of marine and terrestrial components, it can be further subdivided into suspension-feeders and predators. But, as Glynne-Williams and Hobart (1952) point out, since light is absent within the narrower crevices, the crevice fauna does not represent an independent community. Food material must be trapped either by filter-feeding organisms such as *Lasaea rubra, Kellia suborbicularis* and *Mytilus edulis* operating at full tide, by wave-action carrying particles into the crevice, by particle-laden water filtering through the deposits, or by a combination of these factors. Alternatively some predators such as *Eulalia viridis, Bdella longicornis, Aëpopsis robinii,* and scavengers such as *Aëpophilus bonnairei* and collembolans, may feed when the tide has ebbed and return as the crevice becomes submersed. Once such organic matter has been brought into the crevice from outside, it becomes available to the scavengers and predators of the terrestrial component which migrate from the inner brown muds into the outer zones in search of food during the period of low tide, and retreat back into the inner regions when the crevice is submersed. This gives rise, as Glynne-Williams and Hobart (1952) point out, to two alternating food chains coinciding with the tidal cycle. When the crevice is submersed the organic content of the crevice is enriched and the filter-feeders are active, as well as those members of the aquatic component which are predators or scavengers. When the tide ebbs the filter-feeders stop feeding and become available as a food source for the terrestrial predators. There is also some evidence that superimposed on this rhythm there is a second cycle in which some crevice-dwelling organisms, such as the isopod *Ligia oceanica,* may emerge only at night during the ebb tide period.

B. THE FAUNA INHABITING INTERTIDAL SEAWEEDS

The fauna inhabiting intertidal seaweeds has been studied by Colman (1940), Dahl (1948) and by Wieser (1952) (also Kitching, Macan and

Gilson 1934). Colman (1940) surveyed the organisms inhabiting eight different species of intertidal seaweed at Church Reef, Wembury, and reference should be made to his paper for a detailed account of the species inhabiting such seaweeds. The species of seaweed which he studied were *Pelvetia canaliculata, Fucus spiralis, Lichina pygmaea, Fucus vesiculosus, Ascophyllum nodosum* with *Polysiphonia lanosa, Fucus serratus, Gigartina stellata* and holdfasts of *Laminaria digitata,* representing a series down the shore.

Pelvetia canaliculata is the highest of the algal series at Wembury and was found to support an average of only 44 animals per 100 g of damp weed, the dominant species being the isopod *Ligia oceanica,* the amphipod *Hyale nilssoni* and young littorinids. Polychaetes, ostracods, copepods and acarines were all absent. *Fucus spiralis* occurs lower on the shore than *Pelvetia canaliculata* and supported a correspondingly more varied fauna. The total number of animals living in the weed was also higher, averaging 99 individuals per 100 g of damp weed. The amphipod *Hyale nilssoni* was the dominant organism and comprised almost half of the total population, whilst copepods and acrines made their appearance at this level; *Ligia oceanica* was, however, absent.

The population of organisms living in *Lichina pygmaea* is unlike that of the fucoid algae insofar as the number of individuals is from 10 to 100 times greater than that of algae from comparable levels on the shore, and the number of species tends to be rather small. Colman (1940), for example, found an average of 13,716 animals per 100 g of damp *Lichina pygmaea,* of which more than 12,350 was made up of two species, the bivalve *Lasaea rubra* (p. 240) and the isopod *Campecopea hirsuta* which, with larvae of the dipteran *Geranomyia unicolor,* were found exclusively in the *Lichina.* Again, in contrast to the neighbouring fucoid algae, copepods, ostracods and nematodes were scarce in *Lichina pygmaea.*

The fauna inhabiting *Fucus vesiculosus* was found to be similar in composition to that of *Ascophyllum nodosum* (to which tufts of *Polysiphonia lanosa* were attached) except that the total number of animals living in *Fucus vesiculosus* was only 459 per 100 g of damp weed compared with 1,418 in *Ascophyllum nodosum* plus *Polysiphonia lanosa.* In both species of algae, ostracods, copepods, young littorinids and acarines were abundant as well as nematodes, polychaetes and small crustaceans, ostracods being much more abundant in *Ascophyllum nodosum* (353 individuals per 100 g damp weed) than in *Fucus vesiculosus* (16 individuals per 100 g damp weed) or in any other of the seaweeds studied. The only major groups which were rarely found in *Ascophyllum* or *Fucus vesiculosus* were the Porifera, Sipunculoidea, Cirripedia and Tunicata. *Fucus serratus* contained a wider variety of species including polychaetes, bivalves and polyzoans, but rather fewer individuals than *Fucus vesiculosus,* the bulk of the population inhabiting *Fucus serratus* being made up of copepods (178 animals per 100 g of damp weed), and acarines (76 animals per 100 g of damp weed).

In *Gigartina stellata* the average number of animals per 100 g of damp weed was as high as 3,145 of which 759 were acarines, whilst copepods in one sample reached 6,586 per 100 g of damp weed but were rare in others. Apart from these two groups, the fauna of *Gigartina stellata* is characterised by the absence of oligochaetes and insect larvae which are confined to higher levels on the shore, and by the increase in the numbers of sublittoral species as well as hydroids and polyzoa. Finally, the holdfasts of *Laminaria digitata* were found to contain a much larger variety of species than were found in the other seaweeds. Apart from the absence of insects and the scarcity of copepods and acarines, the fauna is characterised by sponges and tunicates which were absent from the other kinds of seaweed. Polychaetes were the dominant group comprising an average of 2,056 animals per 100 g of damp weed out of a total of 2,864 individuals present. Other groups such as tanaids, amphipods and nematodes were also much more abundant than in any of the other seaweeds studied. Thus at approximately mean sea level the most abundant groups are copepods, acarines, young littorinids and ostracods, the latter being especially characteristic of *Ascophyllum nodosum,* whilst at lower shore levels polychaetes largely replace these groups.

Wieser (1952) pointed out that the principal factor controlling the nature and abundance of organisms inhabiting intertidal seaweeds is the growth form of the particular species of weed. Since different species of seaweed are of different growth form they give differing degrees of protection from stresses such as wave-action, desiccation and temperature fluctuation. Thus animals living in a dense tuft of *Gelidium corneum* or *Lichina pygmaea* are much more sheltered from desiccation or wave-shock than those living on *Fucus serratus* or *Ascophyllum nodosum*. Assuming that both types of plants occur over a similar range, the upper limit of some intertidal animals might be much higher in *Gelidium* and *Lichina* than in *Fucus* or *Ascophyllum.* The vertical distribution of the fauna inhabiting seaweeds is therefore ideally to be studied within a single kind of seaweed which has a wide vertical range. Such a study cannot easily be made since the algae themselves are zoned and often have only a narrow vertical range, but Wieser (1952) found that a suitable alga to use was *Gelidium corneum* since at Plymouth this had a vertical range of as much as 1·65 m, and is also of a dense tuft-like shape which allows the presence of a very large fauna. It was found that the fauna inhabiting *Gelidium*, as well as the other tufted algae *Ceramium* sp., *Cladophora rupestris*, and *Lomentaria articulata*, could be grouped into three main components which, although they varied with the silt content of the locality, also varied in relation to tidal exposure. Firstly there were those species which declined in numbers downshore, amongst which may be included *Hyale nilssoni*, *Hyadesia* sp., *Jaera marina* and chironomid larvae, the latter being absent below the lower limit of *Gelidium corneum*. Secondly, there were those which declined in numbers towards higher levels on the shore; this group includes *Corophium* spp., *Stenothoë monoculoides* and all the errant polychaetes. Finally, there were those which were more or less evenly distributed throughout the intertidal zone, including *Mytilus edulis*, *Tanais cavolinii*, halacaridans, *Idotea neglecta* and the common nematodes such as *Anticoma limalis*, *Thoracostoma figuratum*, *Enoplus communis*, *Dolicholaimus marioni*, *Monoposthia costata* and *Chromadora nudicapitata*, as well as many others (Wieser, 1954).

It is obvious that although counts of the fauna inhabiting a single species of alga at different tidal levels are strictly comparable, the method has its limitations in that only a restricted range of the intertidal zone, corresponding with the vertical distribution of the alga, can be studied. Wieser (1952) argued that if it is the shape and consistency of the alga which determines the composition of the fauna at any particular tidal level, then any two species of alga of similar shape may be treated as a single algal substratum and allow a much wider range to be studied. He therefore took *Porphyra laciniata* and *Nitophyllum punctatum* as examples of two seaweeds with flattened uniform thalli, and compared the distribution of the fauna inhabiting these seaweeds with that inhabiting dense tufted forms. *Porphyra* extends down to approximately + 2·0 m whilst *Nitophyllum* is typically sublittoral extending down to – 3·0 m. This allowed a vertical range of comparable algal substratum of at least 5·0 m to be studied. It was found that the average number of animals in samples of the intertidal *Porphyra* was 111 per 50 g dried weight of weed, whereas in the sublittoral *Nitophyllum* the average number of specimens was 3,026 per 50 g dried weight of weed. The major increase in numbers occurred in the descent across the low water mark except where there was a dense protecting cover of tall algae such as *Laminaria*. Wieser (1952) suggested that the reason for the striking increase in the total number of specimens below the low water mark is that the flattened stipes of *Porphyra* and *Nitophyllum* offer little protection against thermal stress, desiccation and wave-action so that it is only where such factors are minimised, as below low water mark or beneath a protective canopy of taller algae, that a large fauna develops on algae with flattened, leaf-like stipes.

Apart from such variations in the number of species which can be related to the nature of the algal substratum (i.e. whether the algae are of leaf-like form or whether they form dense tufts) a general vertical zonation of the microfauna of seaweeds has been established by Wieser (1952). He expressed numbers of animals in terms of the average "dominance values" or percentage of one species of the total fauna inhabiting a wide variety of algae at that level on the shore, regardless of the species of algae from which they were collected. When such a

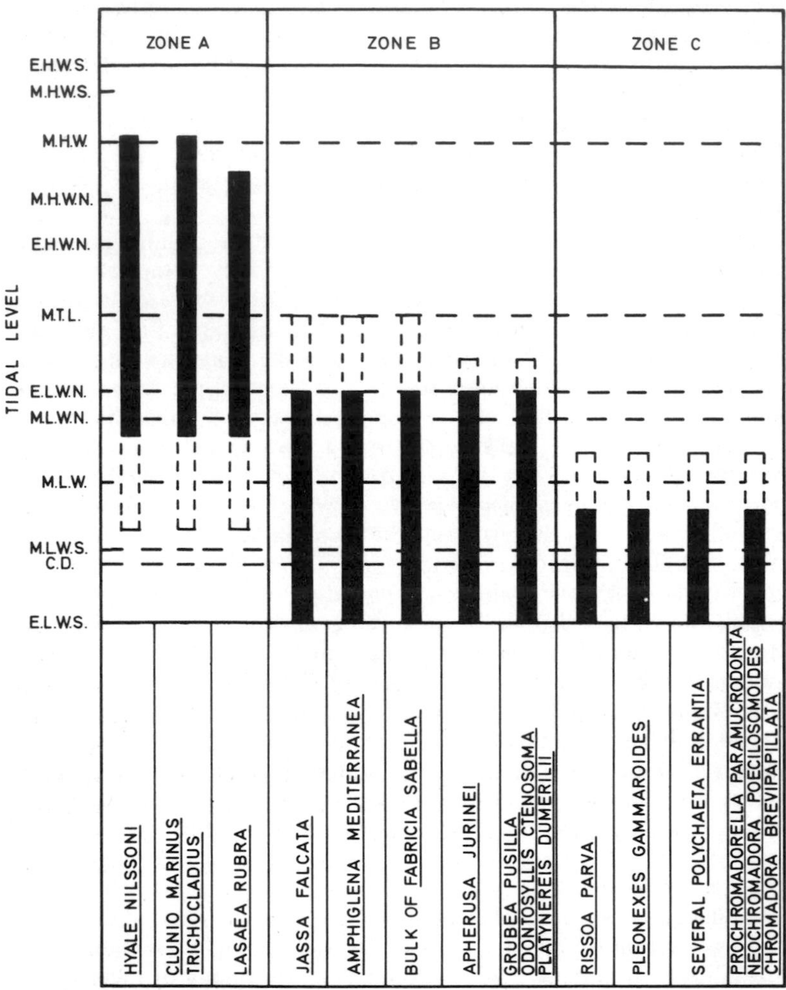

Fig. 2.3. Diagram illustrating the vertical distribution of certain members of the microfauna of seaweeds in relation to tidal level. (Based on Weiser, 1952.)

method was employed, it was found that there were three main zones of animals in the Plymouth algae. Firstly, there was an upper zone (Zone A, Fig. 2.3) between mean low water of spring tides and mean low water of neap tides, in which several intertidal species reached their lower limits on the shore. The characteristic species of this zone were *Hyale nilssoni, Clunio marinus, Trichocladius vitripennis,* and *Lasaea rubra* and are equivalent in tidal level to *Littorina littorea, L. littoralis* (= *obtusata*), *Patella vulgata, Monodonta lineata, Chthamalus stellatus, Ascophyllum nodosum* and *Fucus vesiculosus* (also Colman, 1933; Evans, 1947b; Yonge, 1949). The second zone (Zone B, Fig. 2.3) lay between extreme low water of neap tides and mid-tide level and was characterised by *Jassa falcata, Apherusa jurinei, Amphiglena mediterranea, Grubea pusilla, Odontosyllis ctenosoma* and *Platynereis dumerilli* and marked the upper limit of one set of sublittoral fringe organisms. The corresponding organisms inhabiting the shore at this level were *Gibbula cineraria, Rhodymenia palmata, Chondrus crispus* and *Gigartina stellata.*

Finally, there was a third zone (Zone C, Fig. 2.3) which, in the Plymouth laboratory area, lay between mean low water of spring tides and mean low water of neap tides where another set of sublittoral fringe organisms reached its upper limit. In this particular area, zones A and C coincided, but this is not necessarily so in other localities. Characteristic species included *Pleonexes gammaroides, Rissoa parva, Oridia armandi,* errant polychaetes, and nematodes such as *Prochromadorella paramucrodonta, Neochomadora poecilosomoides* and *Chromadora brevipapillata.* These animals are sublittoral (= infralittoral) fringe organisms (also p. 2) and correspond in level to *Laminaria digitata, Himanthalia lorea, Verruca stroemia* and *Calliostoma zizyphinum.* It should be noted, however, that the distribution of such organisms varied with the state of the tide in much the same way as has been described by Watkin (1941a) and Remane (1933) for other members of the intertidal fauna. Thus more copepods were found on the seaweeds during full tide than at low tide, indicating that the copepods left the seaweed cover with the ebbing tide and reoccupied it again on the flood. A similar situation was found in the amphipod *Stenothoë monoculoides* (Wieser, 1952), and the winkle *Littorina littoralis* (=*obtusata*) (also Haseman, 1911). This variation in the distribution of the fauna with the state of the tide is very similar to that described on p. 54.
for the fauna of crevices in which the distribution of the terrestrial component within the crevice varied with the ebb and flow of the tide (Glynne-Williams and Hobart, 1952), and is a general phenomenon shown by a large number of motile organisms of the intertidal zone (also Chapter 4).

C. INTERTIDAL ROCKPOOLS

Intertidal rockpools form an obvious habitat within the intertidal zone, and the specialised nature of their fauna and flora has long been recognised. Indeed, as long ago as 1897, Bohlin attempted a classification of rock pools based on the nature of the flora. He described three main types of pool including (1) Supralittoral pools containing water from peat moors, (2) Supralittoral pools containing rainwater, and finally (3) brackish pools which are invaded by the sea for variable periods according to their tidal level. Levander (1900) subsequently produced a scheme part of which is shown below and which has formed the basis of several other schemes (for example Kuenen, 1939; Lindberg, 1944; Forsman, 1951; Droop, 1953) which have recently been reviewed by Clark (1968). It will be noticed that Levander (1900) considerably amplified the scope of the classification of saline pools, but whilst he also subdivided the freshwater pools, we shall not consider the classification or the features of such strictly terrestrial habitats.

Levander (1900) recognised four different types of pools in which saline conditions occur and which together constitute a 'brackish water zone'.

(a) Sub-saline pools. These are rock pools which occur in the supralittoral zone and which are reached by wave splash only at extreme high water of spring tides. As would be expected, such pools are subjected to wide variations in salinity and may even become dry during the summer.

(b) Splash pools. Such pools are permanently filled with water but are never in continuity with the sea, water being replaced by wave splash.

(c) Intertidal or seawater pools. These pools are located in the upper parts of the eulittoral zone and are consequently inundated by spring tides but are isolated from the sea for most of the time.

(d) Algal pools. Such pools occur on sheltered shores and often contain decaying seaweeds deposited after storms.

The classification has its drawbacks since, as Clark (1968) points out, many intermediate types of pool exist. Further, no reference is made to the physico-chemical features which control the distribution of the fauna and flora. Klugh (1924), for example, stressed the importance of temperature fluctuations on the organisms living in intertidal rockpools, whilst Järnefelt (1940) used salinity as a basis for the classification of such pools, and also emphasised the importance of the nature of the substratum, as well as the composition of the plankton and of the water. Nevertheless, in essentials Levander's scheme provides a useful framework which forms the basis of many schemes of classification including that of Clark (1968) which is given on p. 82.

The fauna and flora of rockpools, together with some of the

physico-chemical factors which control them, have been discussed by Klugh (1924), Stephenson, Zoond and Eyre (1934), Fraser (1936), Pyefinch (1943), Naylor and Slinn (1958) and Ganning (1966, 1967) and by a large number of other workers (for review, see Clark, 1968). In summarising their work, it is convenient first of all to consider the special types of organisms and the physico-chemical features which characterise the rockpool environment before going on to discuss the factors affecting the nature and distribution of the organisms living in such habitats.

1. THE BIOTA OF ROCKPOOLS

The distribution of plants and animals in rockpools at different levels on the shore in general conforms to the relationship established in Chapter 1 in which it was found that as higher shore levels were approached, there was a reduction in the diversity of the biota but a corresponding increase in the numbers of those species which were able to survive. As will be seen on p. 63, rockpools of the upper shore are subjected to very large diurnal and seasonal fluctuations in temperature and salinity as well as oxygen content, and the fauna and flora inhabiting such pools would be expected to be a highly specialised one. The composition and distribution of such plants and animals has been studied in several parts of the world; reference should be made to the work of Klugh (1924) for the animals and plants inhabiting pools near St. Andrews, New Brunswick, Canada; Stephenson, Zoond and Eyre (1934) for the organisms of the Cape Peninsula, South Africa, and Ambler and Chapman (1950) for those on the shores of New Zealand.

Pyefinch (1943) was mainly concerned with the organisms inhabiting intertidal pools on Bardsey Island, North Wales. By definition (p. 60) these pools are invaded with moderate frequency by the sea and contain an increasing proportion of truly eulittoral organisms as lower shore levels are approached. Naylor and Slinn (1958), however, devoted more attention to the high-level brackish pools at Scarlett Point, Isle of Man, U.K., the lowest pool being some 5 ft above mean high water of spring tides (MHWS), whilst Ganning (1966, 1967) and Clark (1968) have been concerned exclusively with high-level pools characteristically containing *Enteromorpha* and a variety of highly specialised small crustacea, especially amphipods and harpacticoid copepods, as well as Acari.

The observations of Pyefinch (1943) are of interest since to some extent they overlap the review given in Chapter 1 of the nature and distribution of rocky shore organisms. He showed that in a pool above mid-tide level (pool 6 of Pyefinch, 1943) the fauna was well represented including a wide diversity of sponges, coelenterates, polychaetes,

Table IV. Table showing the main components of the biota of three rockpools on Bardsey Island, North Wales (after Pyefinch, 1943). (A) A pool just above mid-tide level (pool 6). (B) A pool just above MHWS (pool 3). (C) A pool some distance above MHWS (pool 7).

Pool A		*Pool B*		*Pool C*	
Algae	*Animals*	*Algae*	*Animals*	*Algae*	*Animals*
CHLOROPHYCEAE	PORIFERA	CHLOROPHYCEAE	PORIFERA	CHLOROPHYCEAE	PLATYHELMINTHES
Enteromorpha ramulosa.	*Leucosolenia botryoides*	*Enteromorpha ramulosa*	*Hymeniacidon sanguineum*	*Enteromorpha intestinalis*	*Procerodes ulvae*
E. intestinalis	*Sycon ciliatum*	*E. intestinalis*	CRUSTACEA		CRUSTACEA
Ulva lactuca	*Adocia cinerea*	*Chaetomorpha tortuosa*	*Melita palmata*		*Jaera marina*
PHAEOPHYCEAE	*Hymeniacidon sanguineum*	*C. linum*	*Amphithoë rubricata*		*Gammarus sp.*
Scytosiphon lomentarius	COELENTERATA	*Cladophora rupestris*	*Carcinus maenas*		MOLLUSCA
Asperococcus fistulosus	*Obelia geniculata*	RHODOPHYCEAE	INSECTA		*Littorina saxatilis*
Myriotricha filiformis	*Actinia equina*	*Chondrus crispus*	*Cricotopus fucicola*		
Fucus spiralis	ANNELIDA	*Ceramium rubrum*	*(Animals continued)*		
Ascophyllum nodosum	*Harmothoë spinifera*		*Pagurus bernhardus*		
Pelvetia canaliculata.	*Perinereis cultrifera*		*Carcinus maenas*		
RHODOPHYCEAE	*Lanice conchilega*		MOLLUSCA		
Chondrus crispus	*Pomatoceros triqueter*		*Patella vulgata*		
Polysiphonia lanosa	*Spirorbis borealis*		*Gibbula cineraria*		
Corallina virgata	*Marionina semifusca*		*G. umbilicalis*		
Ceramium rubrum	CRUSTACEA		*Littorina saxatilis*		
Furcellaria fastigiata	*Leander squilla*		*L. littoralis*		
Catanella repens			*L. littorea.*		

crustaceans and molluscs. In a pool just above mean high water of spring tides (pool 3), however, there was a great reduction in the fauna, whilst in a pool somewhat higher on the shore (pool 7) the fauna was still further reduced. Much the same situation applies to the algae which were reduced from a total of 16 species in the low-level pool to one species in the high-level pool. A comparison of the faunas of the three pools is shown in Table IV. Of these pools, that occurring lowest on the shore may be regarded as an intertidal pool, whilst the other two probably represent the splash pools of Levander (1900).

Clark (1968) made a detailed study of the fauna of high-level pools at Greg Ness, Aberdeen, and recognised the presence of permanent as well as accidental or intermittent members of the fauna of either marine or terrestrial origin. A list of the permanent inhabitants is given in Table V, and many of the same types of organisms, especially *Hyadesia fusca* (Acari, Sarcoptiformes), *Gammarus duebeni, Nitocra* sp. (Harpacticoida) as well as the ostracod *Heterocypris salinus* and the fish *Gasterosteus aculeatus,* have been found to occur in similar situations on the Swedish east coast by Ganning (1966, 1967).

Intermittent inhabitants of such high-level pools may be grouped according to whether they are of marine or terrestrial origin. As would be expected, many of the marine organisms are found mainly in the intertidal and splash pools, but some animals of both marine and terrestrial origin are to be found in all three types of rockpool. Amongst the commonest terrestrial organisms to be found are the insect *Lipura* (=*Anurida*) *maritima* and the mite *Hydrogamasus salinus* both of which, as we have seen on p. 51, are commonly also found in crevices of the intertidal zone. Marine larvae, especially of polychaetes and barnacles, are often to be found as well as the amphipod *Hyale pontica,* the gastropods *Patella vulgata, Littorina littorea* and the mussel *Mytilus edulis.*

2. PHYSICO-CHEMICAL FEATURES OF THE ROCKPOOL ENVIRONMENT

The principal physico-chemical and biological factors which have been investigated in rockpools are the dissolved oxygen content and *pH,* temperature, salinity, available food and tidal level. Although the relationships of such parameters to one another and to the rockpool biota are complex, and may vary according to the geographical locality, each factor may be assessed independently at this stage before we go on to discuss their interaction in the rockpool environment.

(a) Variations in the oxygen concentration and *pH* of rockpools

The oxygen content of rockpools containing algae has been known

TABLE V. Table showing the permanent inhabitants and their distribution in high-level rockpools at Greg Ness, Aberdeen, Scotland (based on Clark, 1968). Crosses indicate the presence of a particular animal, hyphens indicate absence.

Animals	*Intertidal Pools*	*Splash Pools*	*Subsaline Pools*
TURBELLARIA			
Amphanostoma diversicolor	–	+	–
Procerodes(=Gunda)ulvae	+	+	+
NEMATODA	+	+	+
ROTIFERA			
Encentrum sp.	+	+	+
Proales sp.	+	+	+
COPEPODA			
Harpacticoidea			
Tigriopus brevicornis	+	+	+
Mesochra lilljeborgi	–	+	–
Nitocra typica	+	–	–
Nitocra spinipes	–	–	+
Calanoidea			
Eurytemora velox	–	–	+
AMPHIPODA			
Gammarus duebeni	+	+	+
ISOPODA			
Jaera nordmanni	+	+	–
OSTRACODA			
Loxoconcha balthica	–	+	–
INSECTA			
Chironomus plumosus	+	+	+
ACARINA			
Sarcoptiformes			
Hyadesia fusca	+	+	+
Prostigmata			
Rhombognathus notops	+	+	–
R. magnirostris	+	+	–

to vary diurnally since the investigations of Powers (1920), Humphrey and Macy (1930), Orr (1933) and Stephenson, Zoond and Eyre (1934). The *pH*, too, is known to rise to values as high as 8·9 when photosynthesis is in progress compared with 7·6 in pools with large numbers of animals but few algae (Johnson and Skutch, 1928). Again, Gail (1919) found the *pH* of pools containing algae to rise to 8·8 during the afternoon from an average value before sunrise of 7·43. However it was not until the work of Stephenson, Zoond and Eyre (1934), Pyefinch (1943) and Clark (1968) that detailed investigations of diurnal as well as long-term fluctuations in the oxygen content and *pH* of rockpools were made.

Stephenson, Zoond and Eyre (1934) chose three main pools, the first of which (pool A) contained a mixed population of animals and plants. The principal algae were the brown *Gigartina radula* (1963 gm fresh weight), the green *Ulva lactuca* (220 gm fresh weight), and several genera of coralline algae including *Jania, Amphiroa* and *Corallina,* whilst the animals included a wide variety of phyla but consisted predominantly of sponges, polychaetes, gastropods and asteroids. The second pool (pool B) contained vastly more algae than animals, the principal algae being the brown *Pycnophycus brassicaeformis* and the red *Gigartina radula* which together represented a total fresh weight of 41,753 gm. *Ulva lactuca* made up 3,755 gm fresh weight and the final species of alga was the brown encrusting *Hildenbrandia pachythallos.* The animals were again represented by a wide variety of groups including sponges, anemones, polychaetes, echinoids, asteroids, gastropods, crabs and fishes. Finally, a third pool (pool C) was chosen which contrasted sharply with pool B. In pool C animals predominated, the algae being represented by only 3·3 gm fresh weight of the red alga *Pleonosporium* sp. The animals, however, included very large numbers of sponges, polychaetes, asteroids, gastropods and barnacles.

Clearly, comparison of the weights of algae and animals in the three pools must not be made in terms of the absolute abundance, since the area and volumes of the three pools were different. Stephenson, Zoond and Eyre (1934) therefore took the volumes of the pools into account and expressed the biomass of animals and plants in terms of the weight (or number) per 100 litres of water. They then found that pool A contained a total of 423 animals per 100 litres (excluding sponges) and a total of 619 gm fresh weight of algae per 100 litres. Pool B contained 189 animals per 100 litres (excluding sponges) and 5,688 gm fresh weight of algae per 100 litres; whilst pool C contained 3,618 animals per 100 litres (excluding sponges) but only 3 gm fresh weight of algae per 100 litres.

Records of the variations in the oxygen content and *pH* of pool B which contained predominantly algae were then compared with similar recordings of the oxygen content and *pH* of pool C which contained predominantly animals. The results are summarised in Table VI.

It will be noticed that the values of dissolved oxygen in the algal pool (pool B) started off higher than those in the animal pool (pool C). This was because measurements were not made immediately the pools were exposed by the ebbing tide, and by the time the measurements were started a difference had emerged between the two pools. Stephenson, Zoond and Eyre (1934) then showed that at night the oxygen content of both the algal pool and the animal pool was reduced, the lowest night record for oxygen in the algal pool being 1·2 mg/l compared with the 26·2 mg/l recorded on a sunny day (Table VI). Clearly then, the concentration of dissolved oxygen and the *pH* depends not only on the balance between the algae and the animals, but also on the conditions of illumination prevailing in the habitat.

Table VI. Table showing the variation of oxygen and *pH* in two pools containing (B) predominantly algae (C) predominantly animals. The oxygen content of the open sea was 8·7 mg/l and the *pH* 8·13. (Data from Stephenson, Zoond and Eyre, 1934.)

Time in hrs.	*Pool B containing predominantly algae*		*Pool C containing predominantly animals*	
	Oxygen (mg/l)	*pH*	*Oxygen (mg/l)*	*pH*
08·00	13·3	8·3	5·1	8·0
09·30	22·3	8·85	3·6	8·0
11·00	25·1	9·0	4·1	8·0
12·00	26·2	9·0	3·9	8·0
12·30	26·0	–	3·8	–

More recently, Pyefinch (1943) has made a detailed survey of the biota and physical features of rockpools on Bardsey Island, North Wales. His results for fluctuations in oxygen content confirmed those of Stephenson *et al.* (1934) and he was also able to make more detailed recordings of the diurnal changes of *pH.* Fig. 2.4 shows the changes in oxygen content, expressed as percentage of air saturation, made over a period of 14 hr. As would be expected, the pool containing the densest algal growth (pool No. 14 of Pyefinch, 1943) was found to contain the highest values of oxygen which reached 271% saturation. Pool No. 6 which contained fewer algae, although at the same tidal level as pool 14, yielded the intermediate curve in the figure, whilst the lower curve was obtained from the pool with the most scanty algal growth (pool No. 7). A variety of intermediate values, corresponding with intermediate densities of algae was recorded. Fig. 2.5 shows the corresponding variations of the *pH* in the three pools, values as high as 9·3 being reached for a period of 2 hr during the day in the algal pool (pool 14). Pool 7, however, which contained few algae, was subject to little change in *pH,* presumably because of the scarcity of animals. An intermediate value occurred in pool 6 which had an intermediate density of algal growth.

Pyefinch (1943) then went on to investigate whether particular species of algae which characterised the various pools influenced the range of *pH* values attained in fixed volumes of seawater. He found that comparable weights of a variety of species of algae including *Corallina officinalis, Enteromorpha intestinalis* and *Fucus serratus,* all produced comparable increases in the *pH* when placed in sealed tubes filled with seawater and exposed to light. Now it is obvious that since the maximal oxygen concentrations and *pH* values occur at approximately midday, then inundation by the tide at this time will have a more pronounced effect on the rockpool than inundation at times when the oxygen

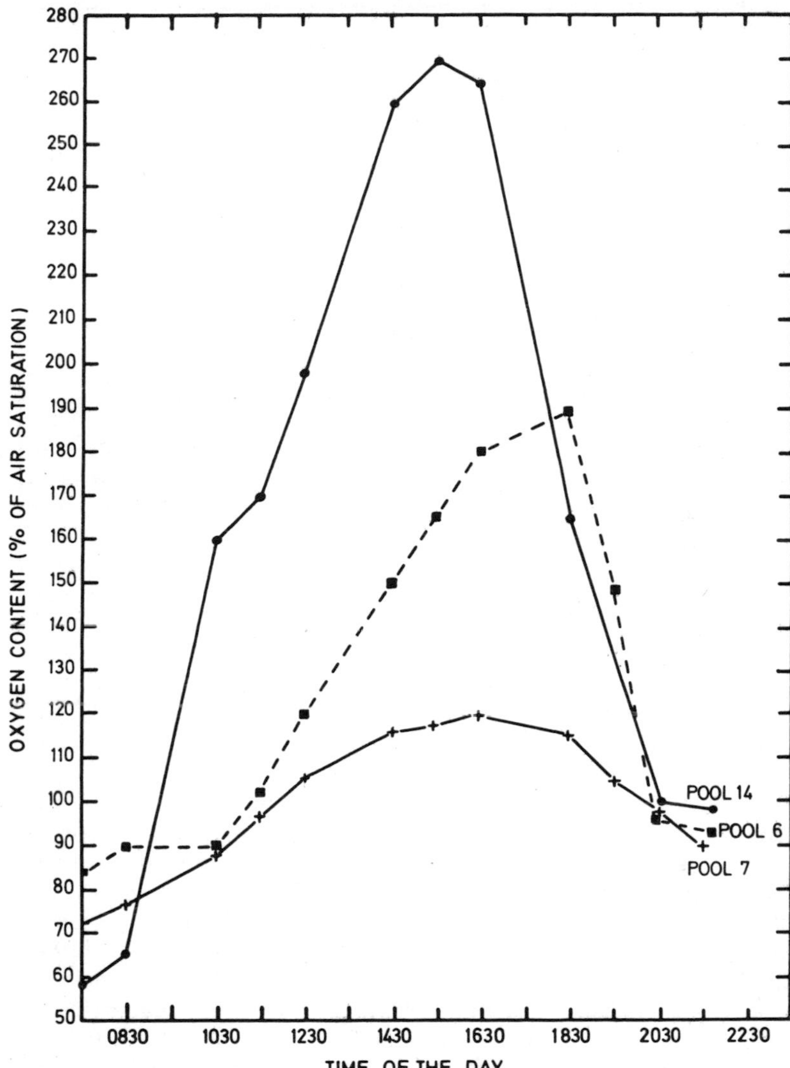

Fig. 2.4. Graphs showing the diurnal changes in the oxygen content of three pools containing varying quantities of algae. Pool 14, abundant algae, pool 6, an intermediate quantity and pool 7, with few algae. (Based on Pyefinch, 1943.)

content and *pH* approximate to that of the sea. The maximum variation in the oxygen concentration and *pH* depends primarily upon the balance between algae and animals. Where there is a preponderance of

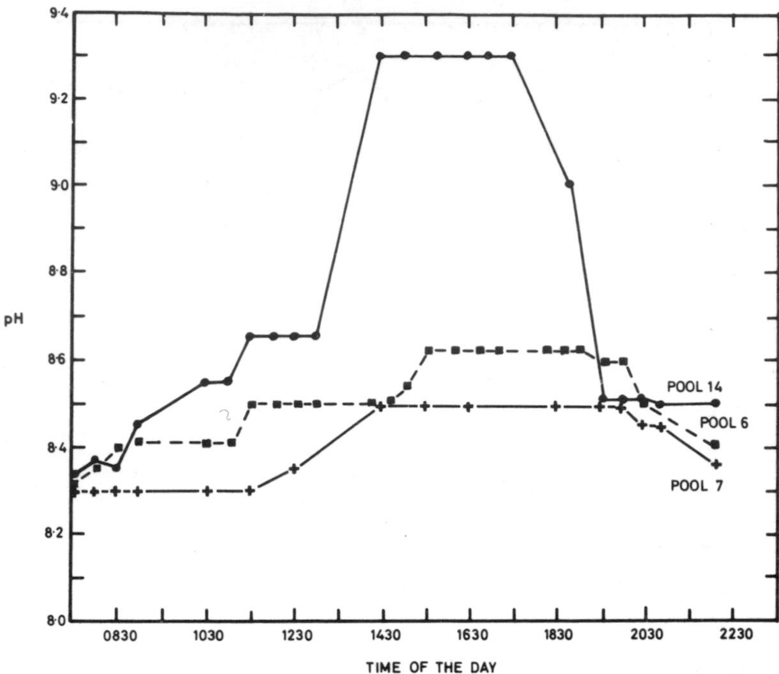

Fig. 2.5. Graphs showing the diurnal changes in the *pH* value of three pools containing varying quantities of algae. Pool 14, abundant algae, pool 6, an intermediate quantity and pool 7, with few algae. (Based on Pyefinch, 1943.)

algae, high *pH* and oxygen concentrations will occur on bright sunny days provided that the pool is sufficiently high in the intertidal zone to avoid long periods of inundation by the tide, particularly over the midday period. Rock pools high in the intertidal zone would therefore be expected to show more marked variations in oxygen concentration and *pH* than those lower on the shore although, as Pyefinch (1943) emphasised, the prime factor is the relative abundance of animals and plants. Fig. 2.6 shows a typical series of diurnal fluctuations in the oxygen concentration and *pH* of pool 6 which was just above mid-tide level on Bardsey Island, these results being comparable with those obtained in the other pools.

Much the same results have been obtained by Clark (1968) who also showed that the presence of cloud or sea fog during the daytime caused a marked reduction in the rate of increase of oxygen in high level rockpools. She also showed that there were seasonal variations in the levels of oxygen in the pools, high levels occurring during March and increasing to a maximum when the *Enteromorpha* cover was dense.

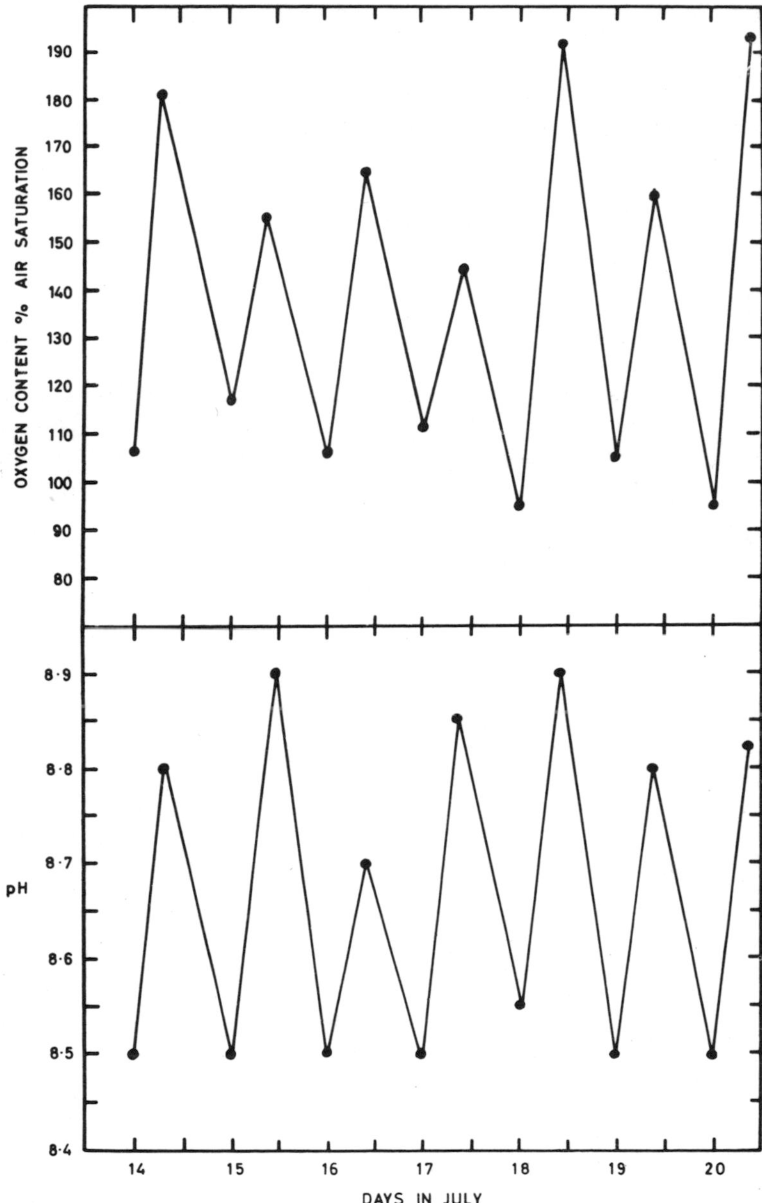

Fig. 2.6. Graphs showing daily variations in the oxygen content (above) and *pH* values (Below) of a pool (pool 6) during July. (Based on Pyefinch, 1943.)

Towards the end of the year, or after the pools had dried, the cover of *Enteromorpha* was reduced and there was a corresponding reduction of oxygen in the rockpools.

(b) Variations in the temperature of rockpools

The temperature of rockpools has been studied by a variety of workers including Klugh (1924), Pyefinch (1943), Naylor and Slinn (1958) and Clark (1968). In general, it has been found that throughout a wide range of levels on the shore, the temperature of the water of rockpools follows similar trends. It approximates to the air temperature until May and then until September the water exceeds the air temperature, maximum values being attained during June and July (Clark, 1968). Klugh (1924) has, however, emphasised the importance of temperature differences between high and low-level rockpools in controlling the biota at St. Andrews, New Brunswick, Canada. He chose a series of six rockpools ranging from near the low water mark to the high water mark, each pool being comparable in terms of the nature of the substratum, depth of water and exposure to wave action. It was found that in the three lower pools the temperature never rose to more

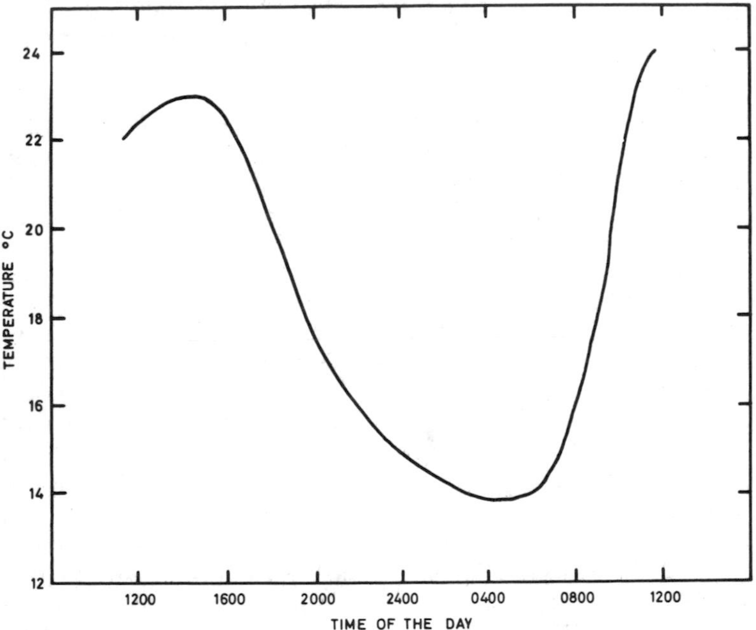

Fig. 2.7. Graph showing the diurnal changes in temperature in a rockpool on the Swedish east coast during May. (After Ganning, 1966.)

than 17·6° C, or about 3 °C above that of the seawater (14·7° C). The next two pools, however, reached 20·3 and 22·7° C respectively during an exposure period of 7·5 hr in bright sunshine, whilst the highest pool reached 24·3° C. Corresponding to such variations in the temperature of pools up and down the shore, it was found that the biota of the three lowest pools was that of the sublittoral zone, whilst no such species extended as high as the three upper pools. Klugh (1924), was unable to demonstrate a clear correlation between such a faunistic break and any other factors such as salinity and *pH,* and therefore concluded that temperature fluctuation was of overriding importance in controlling the biota of rockpools. Ganning (1966), too, noticed the extreme diurnal variations in temperature to which rockpools of the upper shore on the Swedish east coast are subjected. An example of such fluctuation is shown in Fig. 2.7.

Thus the rockpools of the upper intertidal zone are not only subjected to diurnal fluctuations in oxygen content and *pH,* but also to larger temperature changes than pools lower on the shore. Further, the degree of algal growth influences the temperature as well as the *pH* and oxygen content of rockpools. Clark (1968), for example, demonstrated that *Enteromorpha* conserved the heat of rockpools, so that the temperature rose 1–2°C above that of the open water or of pools at a comparable tidal level which contained only a sparse growth of *Enteromorpha.*

(c) Variations in the salinity of rockpools

Salinity variations in rockpools have been recorded by Klugh (1924), Pyefinch (1943), Naylor and Slinn (1958), Ganning (1966) and in some detail by Clark (1968). Clark pointed out that salinity variations of rockpools is primarily determined by three main factors:– (a) The amount of rainfall (b) the evaporation and (c) the frequency and extent of tidal inundation. Now it is obvious that the degree to which such factors influence the salinity depends upon the level of the rockpool on the shore, the depth and surface area of the pool, and the immediate topography of the shore around the rockpool; whether, for example, the pool is shaded from direct sunlight or from the evaporative action of the wind. Of all these factors, the major one controlling not only the extent but also the duration of salinity changes in the rockpools was found to be the access of seawater. The most pronounced effects of evaporation and rainfall was recorded in pools high on the shore where tidal inundation was rare. Fig. 2.8 shows how often seawater entered a series of pools between May and April, plotted as a function of their tidal level on the shore. As would be expected, the lower pools near MHWS were inundated more frequently than higher pools some distance from MHWS, some variation being experienced due to the local topography of the shore.

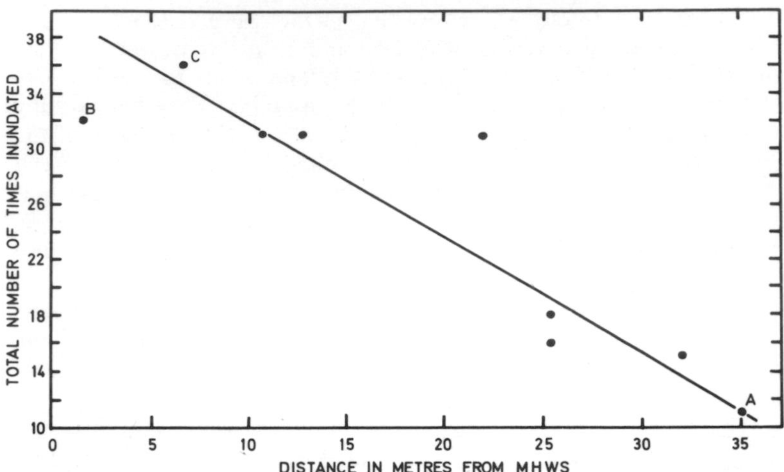

Fig. 2.8. Graph showing the relationship between the number of times pools at various tidal levels were inundated by the tide between May and the following April. Pools A, B, and C of Fig. 2.9. are indicated. (Data from Clark, 1968.)

Clark (1968) then plotted the percentage frequency of occurrence of salinities at 5‰ intervals for the series of pools and found that in the subsaline pools at the top of the shore rainwater profoundly influenced the salinity, values of 0·5 – 5·0‰ occurring with a frequency of 30 – 40%. In splash pools lower on the shore, however, the average salinities became higher, whilst those of the intertidal pools approximated to that of seawater for much of the time. Graphs showing the values for representative pools are shown in Fig. 2.9.

Thus on their salinity characteristics, the pools may be grouped into three types which correspond well with the distribution above the MHWS level and also agree with the scheme proposed by Levander (1900). Such a scheme is based primarily on studies of temperate rock-pools where precipitation often exceeds evaporation. However, in some regions and at certain times of the year salinities in excess of that of seawater will occur when evaporative loss from the pools exceeds that gained from rainfall. Again, pools high on the shore are subjected to greater salinity fluctuations and for longer periods than pools of similar volume lower on the shore, so that both low and high salinities reach their extremes and prevail for longer periods in high-level pools than in those frequently inundated by the tide.

(d) The sources of food in rockpools

A survey of the sources of food available to the inhabitants of rock-pools and the organisms utilising such sources has been made in detail

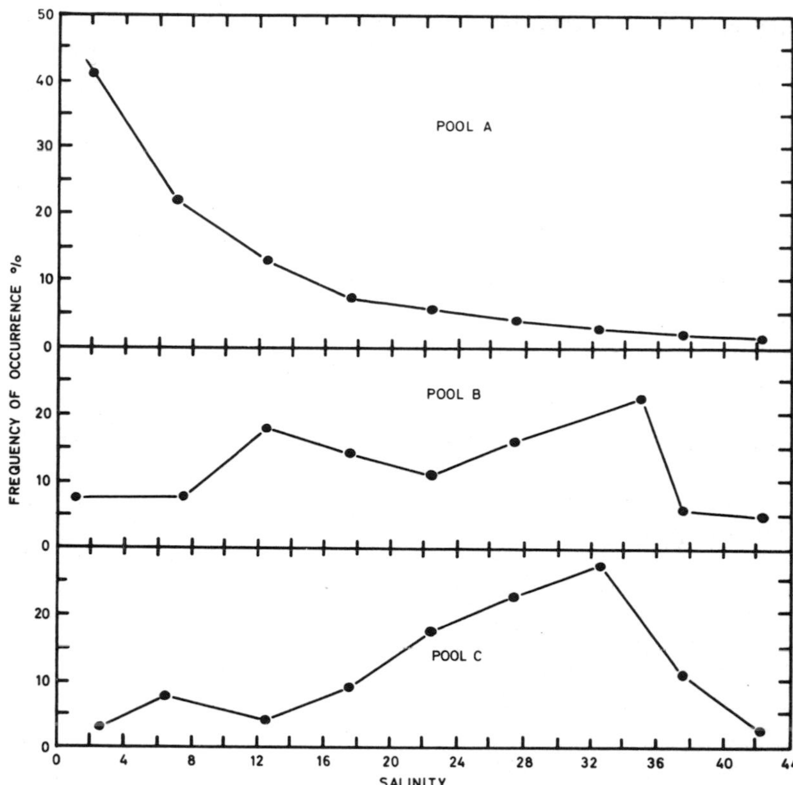

Fig. 2.9. Graphs showing the percentage frequency of occurrence of various salinities ‰ in three pools at Greg Ness, Aberdeen. A, subsaline pool; B, splash pool; C, seawater pool. (Based on Clark, 1968.)

by Clark (1968), although Järnefelt (1940) also stressed the importance of the nature of the deposits and the plankton in controlling the biota of rockpools. According to Clark (1968), there are three major sources of food available to the fauna of high level rock pools. Firstly there are the thalli of *Enteromorpha,* secondly there are the epiphytic micro-organisms attached to the surface of the *Enteromorpha,* and thirdly there are the micro-organisms associated with the substratum.

As discussed on p.259, micro-organisms play an important part in the nutrition of the estuarine deposit-feeders *Hydrobia ulvae* and *Macoma balthica* (Newell, 1965), as well as in the nutrition and distribution of the lugworm *Arenicola marina* (Longbottom, 1968). It is not surprising to find, therefore, that the appreciable quantities of protein which are available in the form of micro-organisms are utilised as a food source by some members of the fauna of rockpools. Clark (1968) found that there was a seasonal cycle in the abundance of such protein in rockpools, maximal values being attained during May, June and July owing to the enrichment of the bottom deposits by organic matter from a variety of sources, the most important of which was the green alga *Enteromorpha* (in which growth and reproduction are at a maximum during the summer), in addition to bird excreta and terrestrial vegetation (also Ganning and Wulff, 1969). From mid-September onwards, the pools near MHWS at Greg Ness, Aberdeen, were inundated by the sea which swept away the accumulated organic debris and deposits, although in general the higher the level of the pool on the shore, the less this effect was apparent.

Clark was able to demonstrate from laboratory experiments that the major source of food for the characteristic harpacticoid copepods of high-level rockpools, *Mesochra lilljeborgi, Nitocra spinipes* and *Nitocra typica* and possibly *Tigriopus brevicornis,* was the micro-organisms associated with the organic debris formed partly from the decay of *Enteromorpha* and partly from the faecal material of other inhabitants of the rockpool (cf *Hydrobia ulvae, Macoma balthica* and *Arenicola marina* p. 259). The amphipod *Gammarus duebeni* appeared to be the only member of the community able to ingest the thalli of *Enteromorpha* directly, but both *Tigriopus brevicornis* and the acarid *Hyadesia fusca* utilise the epiphytic organisms and possibly some portions of the thallus itself.

(e) The influence of salinity, temperature and oxygen concentration on the distribution of the fauna of rockpools

Both Ganning (1966, 1967) and Clark (1968) have recognised the importance of salinity and temperature in determining the nature and abundance of the fauna of rockpools. Since both of these factors have been shown to fluctuate more violently and for periods of longer duration in high-level pools than in low-level ones, it is not surprising to find that animals characteristic of high-level pools are more resistant to salinity and temperature change than those from low-level pools. Ganning (1966) studied the fluctuations in population density of the ostracod *Heterocypris salinus* living in rockpools in the northern Baltic in relation to salinity. He later correlated this with laboratory analyses of the resistance of this animal to a range of salinites below that of seawater (34·4‰) and to temperature fluctuation and differing concentrations of dissolved oxygen (Ganning, 1967). Clark (1968) studied the survival of the harpacticoid copepods *Tigriopus brevicornis, Nitocra spinipes, Mesochra lilljeborgi* and the ostracod *Loxoconcha balthica* in relation to temperature and to both hyper- and hyposaline conditions.

Ganning (1966), in his investigation of the resistance of the ostracod *Heterocypris salinus* to different salinities, found that the animals were well-adapted to life in an environment where wide fluctuations in salinity were likely to occur, since within the extensive range 1 ‰ to 20 ‰ the test individuals survived a similar length of time. The time taken for 50% mortality to occur in the different salinities is shown in Fig. 2.10 from which it will be noticed that overall survival time is comparable throughout a wide salinity range; the lowest mortality occurred in salinities of approximately 10 ‰ .

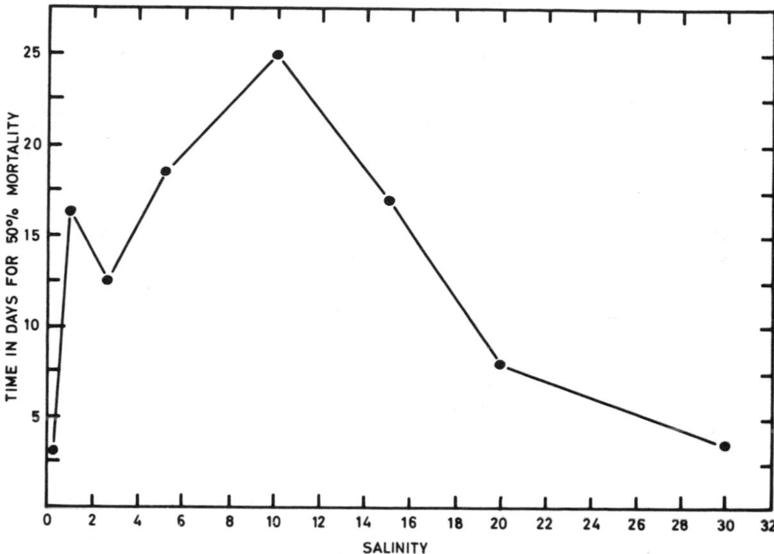

Fig. 2.10. Graph showing the time in days taken for 50% mortality of groups of *Heterocypris salinus* placed in various dilutions of seawater ‰ . The zero salinity was obtained with deionised water. (Data from Ganning, 1966.)

Ganning (1967) then went on to study both the survival time and the behavioural responses of *Heterocypris salinus* to salinity choices. He presented a group of 20 animals, which had been acclimated for five days in 5 ‰ salinity, with the choice of either 6 ‰ or 20 ‰ salinity and found that over a period of 1·75 hr a consistently higher proportion of the *Heterocypris* were to be found in the 6 ‰ compartment. The same result was obtained with animals stored for five days in 20 ‰ salinity and presented with the same choice (Fig. 2.11). When presented with a choice between 6 ‰ and 12 ‰ salinity, *Heterocypris* was also found predominantly in the compartment containing seawater of 6 ‰ , but appeared to be unable to distinguish

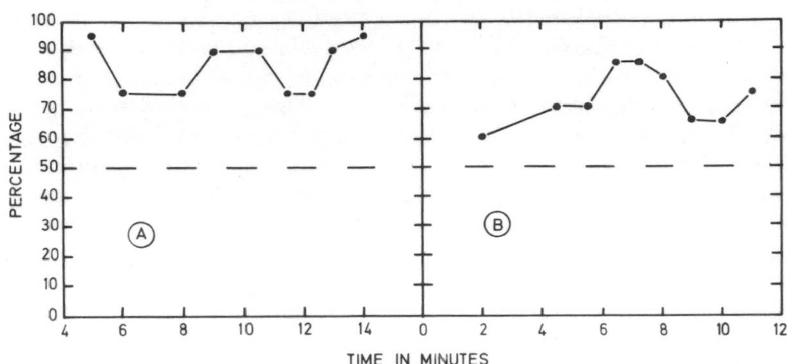

Fig. 2.11. Graphs showing the percentage of 20 specimens of *Heterocypris salinus* aggregating in a chamber containing water of 6‰ salinity when offered a choice between salinities of 6‰ and 20‰ A, animals reared for 5 days in 5‰ salinity; B, animals reared for 5 days in 20‰ salinity. (Data from Ganning, 1967.)

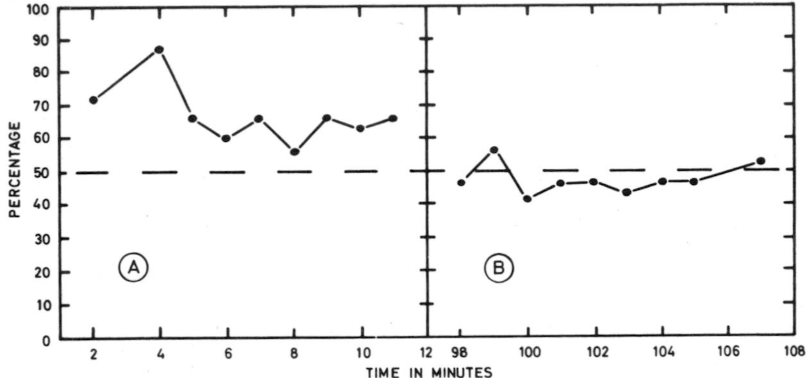

Fig. 2.12. Graphs showing the percentage of 30 specimens of *Heterocypris salinus* aggregating in a chamber containing water of 6‰ salinity when offered a choice between salinities of 6‰ and 1‰. A, initial response; B, response after immersion for more than 1 hr. (Data from Ganning, 1967.)

between the choice of 6 ‰ and 1 ‰ after more than 1 hr. (Fig. 2.12). Thus it may be concluded that *Heterocypris salinus* is tolerant of a wide range of salinities, but normally chooses the salinity which most commonly occurs in the rockpools in which it lives (approximately 6 ‰). Similar responses are shown by the amphipod *Gammarus duebeni* which also chooses salinities of approximately 6 ‰ when taken from pools whose salinity lies between 0·5 ‰ and 6·3 ‰ (quoted in Ganning, 1967).

The responses of *Heterocypris salinus* to temperature were less obvious, but nevertheless Ganning (1967) reported that the animal was able to distinguish between alternatives of 15°C and 22°C, more animals being found in the chamber at 15°C than at 22°C. Whether this aggregation at the lower temperature is a passive aggregation due to the lower rate of movement at the low temperature (i.e. an orthokinesis; Fraenkel and Gunn, 1940) is unknown, but it would tend to result in the animals avoiding the high temperatures found in the surface layers of rockpools during the daytime and would impose a diurnal rhythm of vertical migration on an animal. Apparently the harpacticoid copepod *Nitocra spinipes* aggregates in a chamber at 24°C when offered a choice between 24°C and 20°C even if the test animals are kept for as much as a week at other temperatures (Wulff, unpublished data quoted by Ganning, 1967). This animal differs from *Heterocypris* in its behaviour in the natural habitat, being found amongst *Enteromorpha* near the surface of rockpools even during the day, so that diurnal migrations are absent in *Nitocra spinipes.*

As has been shown on p. 63, the rockpool environment is subjected to large diurnal fluctuations in oxygen content, especially when there is a predominance of algae over animals. Ganning (1967) studied the survival time of *Heterocypris salinus* in seawater of various salinities which had been either partially deoxygenated with the appropriate gas mixtures or was fully aerated or else hyperoxygenated. He found as had Fox and Taylor (1955) with lake inhabitants, that survival was longer in hypo-oxygenated media than in aerated or hyper-oxygenated seawater. Nevertheless, there was no clear response to seawater of differing oxygen concentrations when the choice was between 34% and 410% air saturation, in contrast to another rockpool crustacean *Daphnia magna* (Ganning and Wulff, 1966) and to the bivalve *Macoma balthica* (Brafield, 1963).

Finally, Ganning (1967) showed that *Heterocypris* was negatively phototactic, a factor which undoubtedly plays an important part in confining the animal to the lower regions of the pool during the day whilst allowing a random distribution in the surface water during the night as in a large number of freshwater and marine planktonic organisms. Summarising this work, we may conclude that *Heterocypris salinus,* and by inference some of the other common inhabitants of rockpools, shows a vertical migration which is predominantly controlled by a negative phototaxis which has the effect of excluding the animal from the high temperatures and oxygen concentrations occurring in the surface waters during the day. Nevertheless, the animal is well adapted to resist salinity and temperature fluctuation and is also able to select those thermo-saline conditions which commonly occur in the biotope.

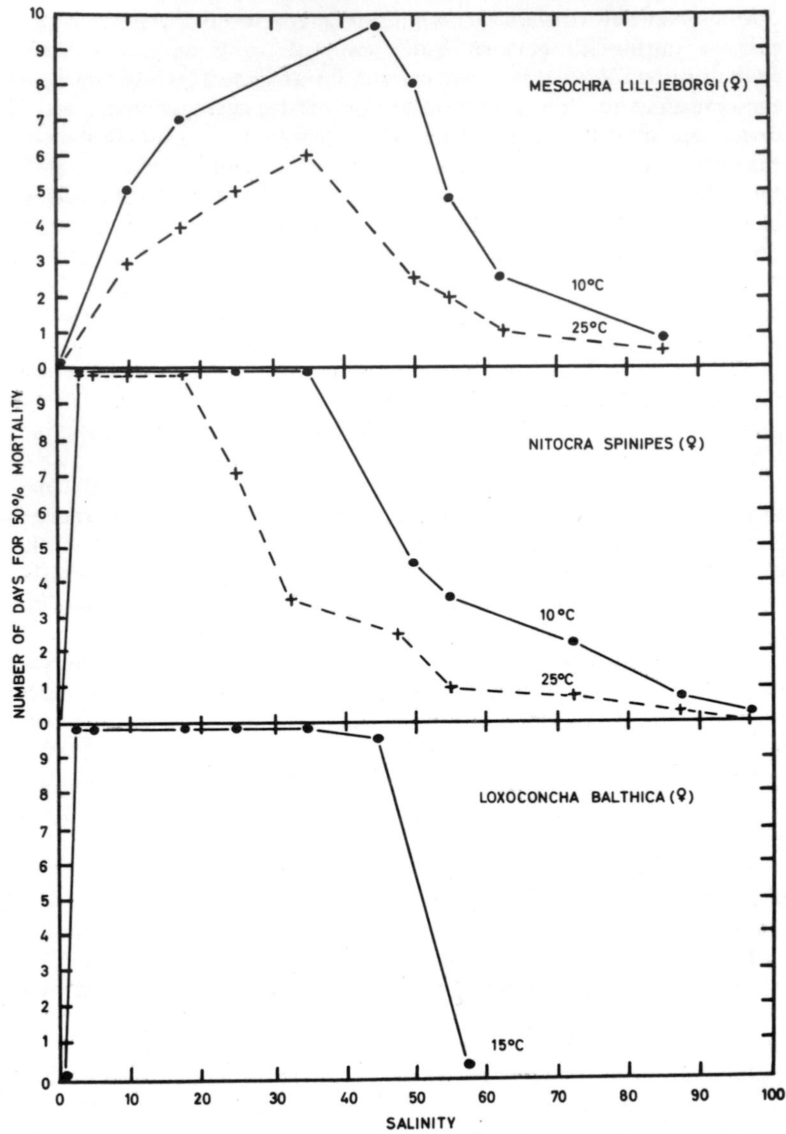

Fig. 2.13. Graphs showing the effect of temperature and salinity (‰) on the survival time of groups of female *Mesochra lilljeborgi, Nitocra spinipes* and *Loxoconcha balthica.* The experiment was ended after 10 days. (Based on Clark, 1968.)

Experimental work on other inhabitants of rockpools is confined to the observations of Clark (1968). She found that the copepod fauna of pools at different levels on the shore could be correlated with the salinity ranges of the pools, and carried out a series of measurements on thc temperature and salinity tolerances of most of the harpacticoid copepods, as well as the ostracod *Loxoconcha balthica,* found in the high-level pools at Greg Ness, Aberdeen. Survival in the different salinities was expressed in terms of the time taken for 50% of the animals to die at a particular salinity and temperature, 30 specimens commonly being used in each experiment. The results of such experiments on the harpacticoids *Mesochra lilljeborgi, Nitocra spinipes, Tigriopus brevicornis* and the ostracod *Loxoconcha balthica* are shown in Fig. 2.13 and 2.14.

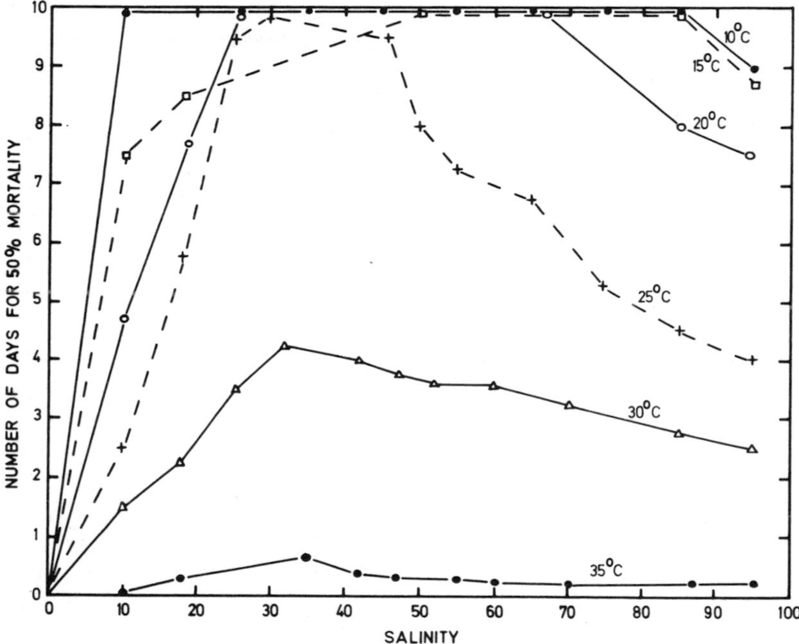

Fig. 2.14. Graphs showing the effect of temperature and salinity(‰) on the survival time of groups of female *Tigriopus brevicornis.* (Based on Clark, 1968.)

From such graphs, the effect of temperature on the mortality of the animals at various salinities may be shown as in Fig. 2.15 for the harpacticoid *Tigriopus brevicornis.* Clark (1968) carried out similar measurements on males and females and found that the females, especially of *Tigriopus* and *Mesochra,* were more tolerant to both low salinity and high temperature than the males; this result is similar to that reported for *Tigriopus japonicus* by Matutani (1961). The animals used in the above experiments were all stored at 10°C in 33·5‰ seawater and no measurements were made to determine the effect of storage at different dilutions and temperatures on the survival times in the test media. However, such adaptation is known to occur in other organisms such as the copepod *Acartia bifilosa* (Lance, 1963), in *Tigriopus japonicus* (Matutani, 1961), *Nereis diversicolor, Gammarus duebeni* and *Sphaeroma hookeri* (Kinne, 1958). It is possible, therefore, that the salinity and thermal tolerances of the rockpool organisms discussed above may be considerably extended by long periods of exposure to high temperatures and low salinities (also Chapters 8 and 9).

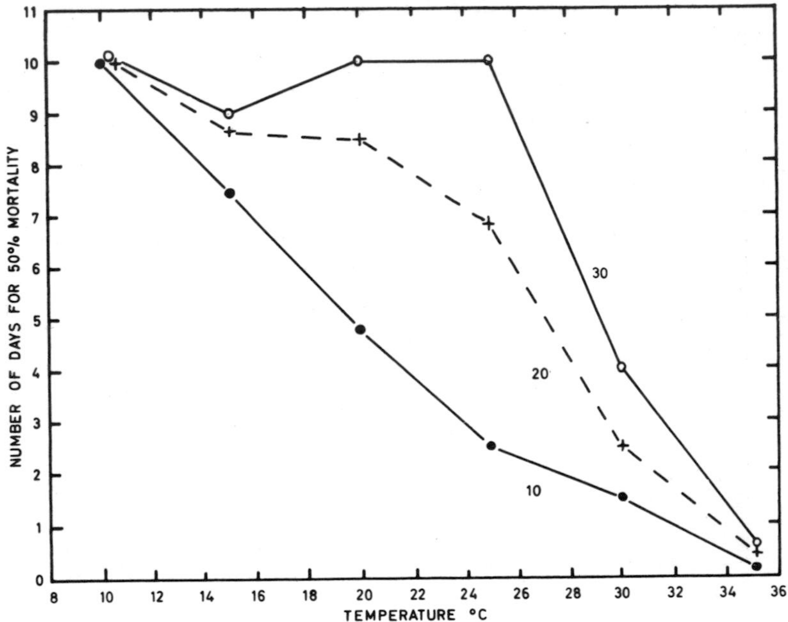

Fig. 2.15. Graphs showing the effect of temperature on the survival time of *Tigriopus brevicornis* in salinities of 10, 20 and 30 ‰ . (Compiled from Fig. 2.14.)

Despite the possible compliciting effects of acclimation to high temperatures and low salinities, it may be concluded from the work of Clark (1968) that except in the case of *Mesochra,* the maximum and minimum salinity ranges at which the animals could survive the duration of the experiment (10 days) at 10°C corresponded well with the range of salinities occurring in the pools in which such species occur. Widely-distributed species such as *Tigriopus brevicornis,* which occurs in seawater pools, splash pools and subsaline pools, show a wide tolerance of salinities from hypo- to hypersaline conditions at temperatures below 20°C. On the other hand species such as *Mesochra lilljeborgi,* which occur only in splash pools, have a much narrower salinity and thermal tolerance with an optimum at about 35‰. Finally, *Nitocra spinipes,* which is confined to the high-level subsaline pools, cannot tolerate hypersaline conditions for long periods.

Although there is a correlation between the resistance of the fauna to hypersaline and hyposaline conditions as well as to thermal stress and the distance at which they occur above MHWS, it is likely that such stresses make their effects felt mainly through their influence on the feeding and reproductive rates of the rockpool fauna. The population density of any particular species will be primarily determined by its ability to compete with other species at the particular salinity and temperature range of the pool. Clark (1968) has demonstrated a direct effect of salinity on the feeding rate of *Tigriopus salinus* and it seems likely that other rockpool organisms are adapted in their feeding and reproductive rates to the salinity range which normally occurs within the habitat. Any major change in the salinity or temperature may then result in an alteration in the population structure of the pool, not necessarily by its effects on the longevity of the dominant species, but by facilitating the growth and reproduction of another species whose salinity and temperature optima approach the changed conditions in the pool. Where no replacement species with an optimum approaching that of the pool exists, then the original species hierarchy persists but the overall population density of the pool is reduced. Food supply, in the form of organic debris plus micro-organisms and *Enteromorpha* and epiphytic organisms, is rarely limiting, so that salinity and, to a lesser extent, temperature fluctuation appears to play a dominant part in controlling the specialised fauna inhabiting rockpools.

Since salinity is of overriding importance, Clark (1968) has proposed a scheme for the definition and classification of high-level rockpools which is based partly on their salinity characteristics and partly upon the organisms which are commonly found in them. As she states, the framework of the scheme proposed by Levander (1900) is retained, but the faunal and salinity characteristics of the pools are greatly amplified and may well provide a scheme which is applicable to high-level rockpools around much of the British Isles. There are two main zones each of which is subdivided into two subzones. The zones are:–

(A) A *Tigriopus brevicornis* dominated zone in which the average salinity range is from 0·5‰ – 45‰ but in which higher salinities may occur.

(B) A *Nitocra spinipes* dominated zone in which salinities above 35‰ are rarely experienced.

This scheme is set out in more detail below:–

Zone A

Dominant species:– *Tigriopus brevicornis*

Salinity range:– Seawater and spray pools with an average salinity from 0·5 – 45‰ but with the occurrence of higher salinities occasionally.

(1) Pools in which salinities of up to 90‰, may occur and which contain euryhaline organisms which may be dominated by *Tigriopus brevicornis.* Other species often occur which have a wide tolerance of both hypo- and hypersaline conditions. e.g. *Hyadesia fusca* (Acarina, Sarcoptiformes) and *Chironomus plumulosus* (Insecta).

(2) Pools in which the salinity sometimes rises to 35 – 50‰. Dominated by *Tigriopus brevicornis* but occupied by two different groups of sub-dominant species.
   (a) Sub-dominant species which are euryhaline but have a limited hypersaline tolerance. e.g. *Gammarus duebeni* (Amphipoda) and *Procerodes* (= *Gunda*) *ulvae* (Turbellaria).
   (b) Sub-dominant species which are polyhaline with the narrow tolerance limits of 18 – 30‰. Such species include *Nitocra typica* (Harpacticoida) *Mesochra lilljeborgi* (Harpacticoida), *Jaera nordmanni* (Isopoda), *Loxoconcha balthica* (Ostracoda), *Rhombognathus notops* (Acarina, Prostigmata), *R. magnirostris* (Acarina, Prostigmata) and *Aphanostoma diversicolor* (Turbellaria).

Zone B

Dominant species:– *Nitocra spinipes*

Salinity range:– Subsaline rockpools in which salinities greater than 35‰ rarely occur. There are no polyhaline species.

(1) Pools in which *Nitocra spinipes* is dominant but in which the euryhaline species *Hyadesia fusca* (Acarina, Sarcoptiformes) and *Chironomus plumulosus* (Insecta) are sub-dominant. *Gammarus duebeni* (Amphipoda) and *Procerodes* (= *Gunda*) *ulvae* Turbellaria) also occur. *Tigriopus bevicornis* occurs but only when the salinity is maintained at approximately 35‰ for a prolonged period.

(2) Pools in which *Nitocra spinipes* is dominant and in which

Hyadesia fusca and *Chironomus plumulosus* as well as *Gammarus duebeni* and *Procerodes* (= *Gunda*) *ulvae* are to be found; but *Tigriopus brevicornis* never occurs.

CHAPTER 3

The establishment of zonation patterns

A. INTRODUCTION

It is obvious from the preceding chapters and from the references cited therein that intertidal organisms are not randomly distributed over the shore but occupy distinct zones and are often restricted to particular micro-habitats within those zones. Many of such animals, including serpulids, hydroids, barnacles, oysters and bryozoans, are sessile, whilst others such as the interstitial archiannelid *Protodrilus*, the deposit-feeding gastropod *Nassarius obsoletus* and the lugworm *Arenicola marina* are also restricted to distinct habitats within the intertidal zone and have at best only rather limited powers of movement from place to place. Some of these animals are viviparous (for review, see Thorson, 1946, 1950, 1957) whilst others such as *Arenicola marina* liberate larvae which do not become planktonic and thus remain in approximately the zone in which the adults occur (Newell, 1948, 1949). On the other hand a great variety of species liberate planktonic larvae, and it had been assumed that settlement was at random and was followed by a vast mortality of those individuals which had settled in unfavourable zones. Thorson (1946, 1950) and Wilson (1952) have, however, reviewed the evidence which suggests that many planktonic larvae have an active site-selection phase and that metamorphosis may be to some extent delayed until a suitable substratum for adult life has been found. The larval planktonic life may thus be regarded as consisting of (a) an initial and often brief phase following liberation during which dispersal occurs, (b) a phase during which testing of the substratum occurs and (c) an attachment and settlement phase when a suitable substratum has been encountered.

The criteria by which the suitability of the substratum is assessed vary widely according to the particular species concerned and have been reviewed by Williams (1964). He pointed out that such factors include physical features such as (i) the texture of the surface (Pomerat and Weiss, 1946; Barnes and Powell, 1950a; Crisp and Ryland, 1960; Williams, 1965a), (ii) the contour of the surface (Knight-Jones 1951b; Crisp and Barnes, 1954; Ryland, 1959; Wisely, 1960; Crisp and Austin, 1960), (iii) the angle of the surface (Hopkins, 1935a, Cole and Knight-Jones, 1939; Korringa, 1941; Pomerat and Reiner, 1942), (iv) the light-reflecting properties or colour of the surface (Visscher, 1928; Neu, 1933; Cole and Knight-Jones, 1939; Korringa, 1941; Pomerat and Gregg, 1942; Gregg, 1945; Walton Smith, 1948; Daniel, 1957), (v) the size of particles and interstices between the grains comprising the deposit (Wilson, 1952; Wieser, 1956, 1959,), and (vi) current strength (Pyefinch, 1948; Knight-Jones and Crisp, 1953; Crisp, 1955; Crisp and Meadows, 1963; Wilson, 1968). Biotic factors, however, play an important if not a dominant part in site selection; thus the presence or absence of a surface layer of bacteria or of material derived from bacterial action determines the degree of attractiveness of sand grains or other substrata for the settlement and metamorphosis of *Nassarius obsoletus* (Scheltema, 1961), *Ophelia bicornis* (Wilson 1948, 1951, 1952, 1953a,b, 1954, 1955), *Protodrilus* sp (Jägersten, 1940; Gray, 1966) *Spirorbis borealis* (Meadows and Williams, 1963), *Bugula* (Crisp and Ryland, 1960) and of many other species.

Now although these features of the habitat must be of considerable value in an assessment of the suitability of the environment for adult life, they are by no means a certain index. For example, a larva of a sublittoral animal might select a substratum which possessed the appropriate physico-chemical and biological features for adult life but that zone might then be emersed and the metamorphosed animal die from the effects of desiccation, thermal stress or reduced feeding time. Equally, larvae of an intertidal organism may locate an apparently suitable site for settlement but which is at the wrong tidal level for the maintenance of the adult. Clearly, the most important single feature of the environment which would indicate the suitability of the environment for adult life is the presence of adults of the same species. If they have survived, then the habitat is likely to be suitable not only at any one moment in time but over a long period. It is therefore perhaps not entirely surprising to find that the larvae of a great many intertidal organisms are capable of responding to and settling in the vicinity of adults of the same species, although in the absence of such adults, metamorphosis may be in response to other more general features of the environment. Such gregarious settlement occurs in oysters (Cole and Knight-Jones, 1949; Knight-Jones, 1951a; Bayne, 1969), several coelenterates (Duerden, 1902; Edmondson, 1929; Matthews, 1916; for review, Williams, 1965a, c), in barnacles (Knight-Jones and Stevenson, 1950; Knight-Jones, 1953b; Knight-Jones, 1955; Crisp and Meadows, 1962, 1963), in the polychaete *Sabellaria alveolata* (Wilson, 1968) and in the serpulid *Spirorbis* (Knight-Jones, 1951b, 1953a; Williams, 1964; Gee, 1965);

An equally reliable criterion of suitability is the presence of other organisms which commonly occupy a similar zone on the shore to that of the adults of the species concerned. The presence or absence of particular algae, for example, is sufficiently diagnostic of tidal level to serve as a stimulus for settlement. The serpulid *Spirorbis borealis* has been shown to settle preferentially on *Fucus serratus* or its extracts; *Spirorbis corallinae* on the calcareous alga *Corallina officinalis; Spirorbis tridentatus* on rocks and in dimly-lit situations (de Silva, 1962; Williams, 1964), whilst *Spirorbis rupestris* settles on *Lithothamnion* (de Silva and Knight-Jones, 1962; Gee and Knight-Jones, 1962; Gee, 1965). The use of either the presence of adults of the same species or of suitable algal substrates involves the recognition of chemical features such as quinone-tanned protein (Knight-Jones and Crisp, 1953; Knight-Jones, 1953b) or adsorbed arthropodin (Crisp and Meadows, 1962, 1963) in barnacles, or of compounds distinguishing the particular algal substrata (Crisp and Williams, 1960; Williams, 1964, 1965a; Gee, 1965). The orientation by *Protodrilus symbioticus* to surfaces which have been filmed with micro-organisms appears also to be primarily a response to adsorbed layers of organic material rather than to the bacteria themselves, since removal of bacteria from filmed surfaces by shaking does not alter the attractiveness of the sand (Meadows, 1964; Gray, 1966). Chemosensory mechanisms thus appear to play an important part in the detection of suitable substrata and, together with physical factors, also play a part in the final spacing-out of *Spirorbis borealis* (Wisely, 1960) and *Balanus balanoides* (Knight-Jones and Moyse, 1961; Crisp, 1961) which occurs immediately prior to settlement. In colonial forms such as the polychaete *Sabellaria alveolata,* however, no such spacing-out occurs (Wilson, 1968).

In short, there appears to be a hierarchy of factors, ranging from generalised responses to light to more specific orientation to the properties of the substratum, which serve to bring the larvae into the vicinity of suitable substrata. Temporary attachment followed by settlement may then be made in response to a variety of physical and chemical features of the substratum as well as to more precise stimuli such as the presence of adults of the same species. This hierarchy has been viewed in terms of an instinctive sequence (also Chapter 4) with a gradation of physical and chemical factors acting as 'releasers' for behavioural responses which culminate in settlement (Crisp and Meadows, 1963; Williams, 1965a). The appropriate sequence, together with the relevant releasers in *Spirorbis borealis,* is shown in Fig. 3.1. The gregarious response of

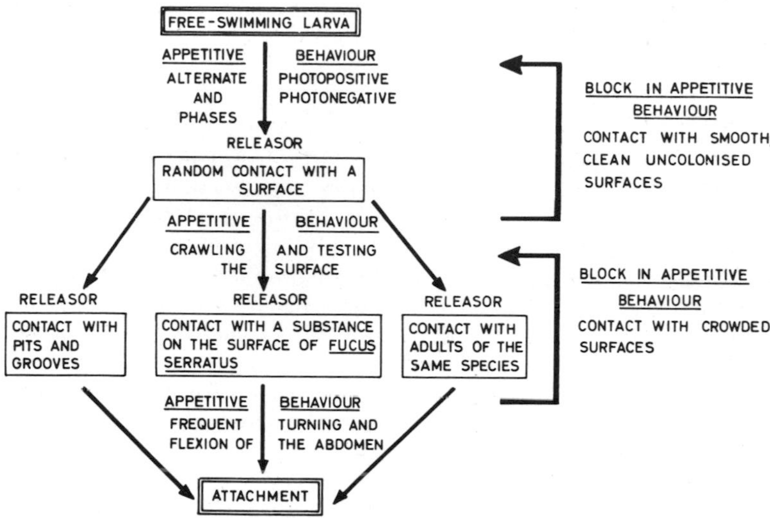

Fig. 3.1. Diagrammatic analysis of the settlement behaviour of *Spirorbis borealis.* (Modified after Williams, 1965a.)

many organisms ensures that the larvae initially colonise a suitable habitat for adult life, intraspecific competition for food and space being minimised by the tendency of the organisms to space themselves out (Wisely, 1960; Knight-Jones and Moyse, 1961; Crisp, 1961). In the absence of suitable substrata, settlement can be postponed (for review, see Thorson, 1950; Wilson, 1952, 1968; Knight-Jones and Moyse, 1961) although a long delay under experimental conditions results in abnormal metamorphosis (Wilson, 1932; Knight-Jones, 1953a) or even in the inability to metamorphose at all (Knight-Jones, 1953b, c; Williams, 1964; Wilson, 1968). After prolonged postponement of settlement, the threshold of sensitivity to the substratum appears to be diminished in some, but not all, larvae (see Crisp and Meadows, 1962; Williams, 1965a), and this would tend to lead to the colonisation of new environments once the parent populations had been replenished. For example, Cole and Knight-Jones (1949; also Knight-Jones 1951a, b) have shown that in oysters settlement of spat around the parent stock occurred most markedly when spatfall was light so that the parent population tended to be augmented; but when larvae were abundant settlement occurred over a wide area. Gregarious behaviour therefore ensures the settlement of organisms in positions which are suitable for adult life, and also allows the spread of the species into surrounding areas particularly when high densities occur in the parent locality. It would also tend to facilitate fertilisation of the gametes in those species which shed them freely into the water and is clearly necessary in sessile species such as barnacles where copulation occurs.

B. SETTLEMENT ON PARTICULATE SUBSTRATA

One of the earliest examples of the influence of the substratum in controlling the onset of metamorphosis in invertebrate larvae is shown by the echinoid *Mellita sexiesperforata.* Mortensen (1921) found that when larvae which were ready to metamorphose were placed in a vessel containing seawater and natural sand metamorphosis occurred within a few days, whereas a sample of the same batch of larvae placed in a vessel of seawater without sand did not metamorphose. Similar results were obtained on the larvae of three other species of echinoderm (Mortensen, 1938). Wilson (1928) similarly found that the larva of the polychaete *Polydora ciliata* could delay metamorphosis until a suitable crevice for adult life had been found. Again, Day and Wilson (1934; also Wilson, 1932, 1937) found that in *Scolecolepis fuliginosa* the larvae could postpone metamorphosis for as much as several weeks until a suitable substratum was found after which metamorphosis was complete in 12 hr or less. They concluded that the suitability of the substratum depended at least in part upon the grade of deposit, but that the organic content was also of importance. Later work by Wilson (1948, 1951, 1952, 1953a,b, 1954, 1955) who studied the factors influencing metamorphosis in the polychaete *Ophelia bicornis* has greatly extended and amplified the conclusions based on the species mentioned above.

Wilson (1948) studied the population of *Ophelia bicornis* which lives in restricted areas of the Exe estuary on the south coast of the U.K. and found that the larvae metamorphose most readily in response to physical contact with the natural sand from the Bullhill sandbank where the adults live. The majority of the particles of this sand passed a 40 mesh per inch sieve but were retained on a 86 mesh per inch grade and consisted of well-rounded smooth grains with little organic debris. He found that sands of smaller or more angular grains were less favourable to settlement and that in general the greater the dissimilarity from the sand of the Bullhill sandbank, the less likely was settlement and metamorphosis by the larvae. He concluded that it was the physical nature of the sand rather than any chemical substances diffusing from it which promoted settlement. Subsequent work (Wilson, 1953a,b, 1954, 1955) showed that one of the important attractive features of natural sand was the presence of a film of micro-organisms on the surface of the sand grains. He found that acid-cleaned sand lost its attractiveness and that such sand when stored in stoppered bottles in distilled water did not increase its attractiveness even after a year. On the other hand when the acid-cleaned sand was soaked in filtered seawater, then the longer it was in contact with the water, the more its attractiveness to *Ophelia* larvae increased. An even more effective way of rendering the acid-cleaned sand attractive was either to shake fresh Bullhill Bank sand with seawater and then place the filtered seawater in contact with the acid-cleaned sand, or else to place a few grains of natural sand in with the cleaned sand. In each case there was a marked increase in attractiveness especially after a few days of storage at room temperature. Similar results were obtained both in the light and in the dark, and suggested that the development of a growth of heterotrophic micro-organisms was responsible for the increase in attractiveness. Equally, when acid-cleaned sand was enclosed within a bolting-silk envelope and buried in fresh Bullhill Bank sand, there was a marked increase in attractiveness, although this effect was destroyed if the natural sand was first heated to 100°C. Thus any treatment which would be expected to preclude the development of a film of micro-organisms rendered the sands unattractive to *Ophelia* larvae.

Other sands in which an adult population of *Ophelia* does not occur appear to be actually repellent to the larvae. This may be partly due to the presence of dead organic material on the sand grains, and also to the presence of either too many micro-organisms or micro-organisms of the wrong kind. Thus, the factors stimulating settlement appear to include not only the grade of sand, for clean sand of the natural grade neither encourages nor discourages larvae to settle, but also the presence of micro-organisms of the correct type and in the correct quantity (Wilson, 1955). In much the same way, Nyholm (1950) found that larvae of the polychaete *Melinna cristata* metamorphosed only in response to a substratum suitable for adult life which in this case was one rich in organic debris. Again, Smidt (1951) found that in the absence of a suitable substratum, settlement and subsequent metamorphosis of the larvae of *Pygospio elegans* was inhibited and that sterile sand or pebbles impeded settlement. It is obvious, therefore, that choice of substratum, and inhibition of metamorphosis until a suitable one has been found, occurs in a number of polychaetes characteristic of particular types of deposits, and that settlement may be in response either to sands relatively free from organic debris, as in *Ophelia bicornis,* or in ones containing an abundance of such material as in *Melinna cristata.* In each case, however, the behaviour promotes settlement in substrata suitable for adult life.

Other organisms such as *Actinotrocha branchiata* (the larvae of *Phoronis mulleri*) also test the substratum and metamorphose in response to clay-mixed sand in which the adults are found (Silèn, 1954). Veligers of the gastropod *Nassarius obsoletus* also display a distinct preference for substrata suitable for adult life. Scheltema (1961) has shown that *Nassarius obsoletus* is a deposit-feeding species and that the larvae metamorphose in response to the properties of the bottom sediment. The closely-related *N. vibex,* however, is a scavenger and is widely distributed; this species has no distinct habitat preferences, and out of a series of eleven experiments 67·4 ± 11·8% of the larvae metamorphosed when sediment was present on the bottom of a vessel compared with 62·9 ± 11·5% metamorphosis in the absence of any sediment in the experimental containers. The ability to metamorphose in response to the properties of the sediment thus appears to correspond with the adaptive value of such a response. Scheltema (1961) showed that in *N. obsoletus* the presence of a suitable substratum greatly enhanced the rate of settlement of the larvae, and that in the absence of such a substratum metamorphosis could be delayed for as long as 20 days. Fig. 3.2 shows the percentage metamorphosis of 50

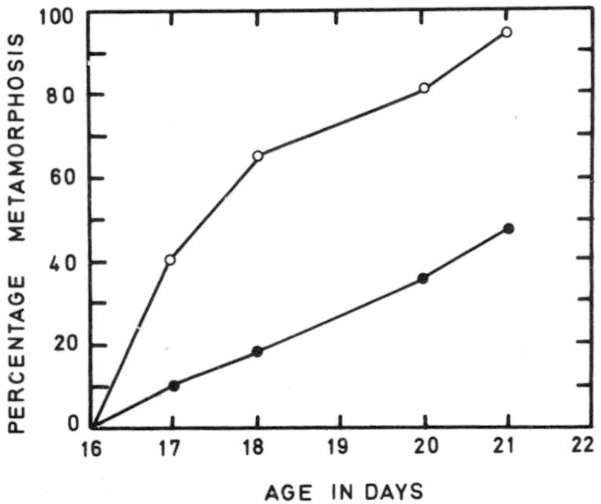

Fig. 3.2. Figure showing the percentage metamorphosis of veliger larvae of *Nassarius obsoletus* as a function of age in the presence of natural substratum (open circles) and in the absence of substratum (solid circles). (After Scheltema, 1961)

veligers which were placed in each of two 2-litre containers of seawater 16 days after emergence from the egg capsules. One of the vessels had freshly-collected substratum in the bottom whilst the control container had none. In the absence of a natural substratum metamorphosis was a linear function of age, but even after 21 days less that 50% of the larvae had metamorphosed. On the other hand, metamorphosis was rapid in the presence of a natural substratum and approached 100% after 21 days. The rate of settlement is, however, largely a function of age, for when contact with the substratum is delayed, metamorphosis is greatly accelerated when eventually the larvae are placed in contact with a suitable deposit.

Provided that the substrata were those which supported a natural population of *Nassarius obsoletus,* the mean grade and degree of sorting (p. 30) of the deposit had little effect on metamorphosis. However washing, followed by incineration for 1 hr to remove all trace of organic matter, oven heating at 210°C for 15 – 30 min, boiling in seawater and then in distilled water to remove part of the organic material and most of the natural microflora, heating in a water bath to 70 – 90°C for 5 – 15 min, and irradiation with ultra-violet light for 18 hours each resulted in a significant reduction in the attractiveness of the deposits. Unfortunately, as Scheltema (1961) pointed out, the differing percentages of larvae metamorphosing in the deposits could not be directly compared, for the experiments were conducted with different samples of sediment and with larvae of different ages. Subsequent experiments under more controlled conditions using the same batch of the larvae and substratum showed that 71·2 ± 5·3% metamorphosis occurred in the presence of untreated substratum, and 33·3 ± 4·8% in the presence of substratum heated in a water bath to 90°C for 15 min. By comparison, 7·2 ± 1·5% metamorphosed in the presence of incinerated substratum and 13·9 ± 2·6% in the absence of any substratum at all. It was concluded that since all these treatments affected the biological and chemical attributes of the substratum without greatly affecting the physical properties, it was the biological and chemical properties which were mainly responsible for the ability of the substratum to induce metamorphosis.

The attractive settlement-inducing factor appeared to be transferred to seawater, since when seawater was allowed to stand in contact with a natural substratum and was then carefully siphoned off, 48·5 ± 4·9% metamorphosis occurred in the 'conditioned' water without any added substratum, compared with only 11·2 ± 2·5% in a vessel containing normal seawater. Further, filtration of such conditioned seawater through a molecular filter with a pore size of 0·8 μ did not significantly reduce the attractiveness, so that the attractive factor appeared to be soluble in water. However, other experiments suggested that it might also become adsorbed onto sediments, since metamorphosis in the presence of sediments was always consistently higher than in conditioned seawater alone. There was also a slight reduction in attractiveness following filtration which might imply adsorption onto the filter itself. In substrate-conditioned seawater the mean percentage metamorphosis after 48 hours was 46·9 ± 6·5%, in molecular (0·8μ) filtered substrate-conditioned seawater 33·0 ± 5·6%, and in untreated seawater 12·3 ± 3·0%. The mean age of the larvae at the beginning of the experiment was 28·8 days. Scheltema (1961) concluded that chemoreception of the biologically active substance may have occurred even before contact with the substratum, but as has been subsequently shown by Crisp and Meadows (1962, 1963) and by Williams (1964) response to settlement factors in barnacles and in several polychaetes does not occur until contact is made with the substratum where adsorption of the water soluble substance has occurred. It seems likely, therefore, that in this animal, too, 'contact chemoreception' (see p. 102) with the substratum or with adsorbed layers on the walls of the experimental vessel, rather than olfaction, may be the stimulus to settlement in a particular deposit.

Other workers too, have found that a settlement-inducing substance is liberated from naturally occurring deposits into the surrounding water. Jägersten (1940), for example, found that metamorphosis of the archiannelid *Protodrilus rubropharyngeus* could be induced by allowing the animal to come into contact with a shell or pebble from the natural habitat, and that the settlement factor was liberated into the surrounding water. This substance proved to be resistant to heat, acid and alkalis and appeared to be in suspension since it could be removed by centrifugation or filtration. Jägersten (1940) concluded that the settlement factor was probably inorganic because of its resistance to destruction. He also inferred that it was produced by micro-organisms since natural gravel, or seawater in which such gravel had been in contact, could transfer its attractive properties to previously inert shell gravel.

More recently Gray (1966) has studied the responses of *Protodrilus symbioticus* to sand grains, the surface of which had been modified by various experimental procedures, and has made some important inferences on the nature of the attractive substances in these animals. He found that treatment with concentrated nitric or sulphuric acids, drying at above 70 – 80°C for 12 hr, heating the wet substratum above 50°C, and autoclaving at 15 lb/in^2 all nearly eliminated the attractiveness of the deposit to *Protodrilus*. Even drying at 16°C reduced the attractiveness index $\left(\frac{\text{no. animals aggregated in treated sand}}{\text{no. animals aggregated in natural sand}}\right)$ to 0·037. On the other hand when the sand was shaken for 10 min with either 5% formalin or with a solution of 90μg/l of the detergent CTAB (cetyltrimethylammonium bromide) only just over half of the attractiveness was destroyed. Assuming that these substances kill or remove the bacteria from the surface of the sand grains, it follows that only some 50% of the attractiveness is due to the presence of living micro-organisms. In fact the attractiveness index was 0·55 after treatment with formalin and 0·57 after treatment with CTAB; after the latter had been subsequently stored in freshly-boiled distilled water for 2 weeks the attractiveness fell still further to 0·27. It seems, therefore, that approximately two thirds of the attractiveness of the untreated sand is attributable to the presence of living micro-organisms, one third remaining even after the death or removal of the micro-organisms. There was little loss of attractiveness when the wet natural sand was heated to 40° C, although this attractiveness was destroyed on heating to 50°C. Many of the natural sand bacteria were found to survive heating up to 40° C, but none survived 50° C so that it is likely that the associated bacteria play an important part in making the sand attractive to *Protodrilus symbioticus.*

Gray (1966) then obtained direct evidence of the importance of bacteria. He placed equal volumes of acid-cleaned natural sand in each of eleven flasks together with 50 ml of seawater, and autoclaved them at 15 lb/in^2 for 30 min. Eight of the vessels were inoculated with a culture of sand bacteria and three control vessels were inoculated with peptone in sterile seawater. The attractiveness of the inoculated and control sand was then compared with that of natural sand at daily intervals. The number of bacteria in the inoculated sand was also estimated at the same time. The results are shown in Fig. 3.3. As Gray

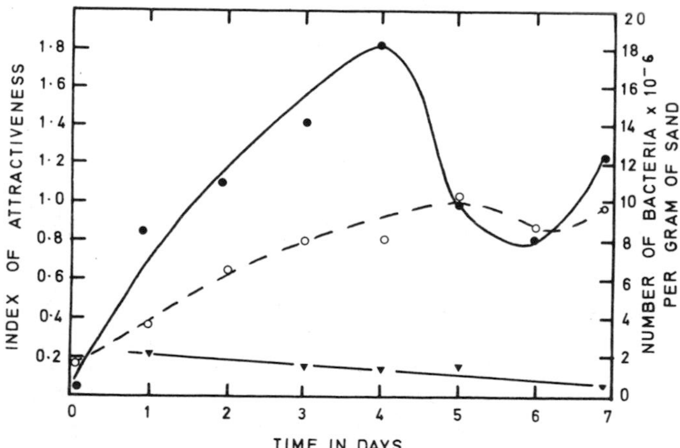

Fig. 3.3. Graphs showing the increase in bacterial numbers (solid circles) and index of attractiveness for *Protodrilus symbioticus* (open circles) of autoclaved sand which had been inoculated with natural sand bacteria. Triangles denote the attractiveness of sterile controls. Based on the mean of two experiments. (Data from Gray, 1966.)

(1966) pointed out, an increase in the number of bacteria from 1·4 x 10^3/gm of sand to 18 x 10^6/gm was associated with an increase in attractiveness of the inoculated sand from 0·2 to 1·0. Again, a decline in the number of bacteria after 4 days from 18 x 10^6/gm of sand to 7 x 10^6/gm was associated with a decline in the attractiveness. In contrast, the sterile control sand became gradually less attractive during the course of the experiment. In much the same way, there was an increase in the attractiveness and in the bacterial numbers in autoclaved sand which had been merely covered with 50 ml of laboratory seawater without any inoculation of sand bacteria. It will be noticed from the results of such an experiment (Fig. 3.4) that neither the attractiveness

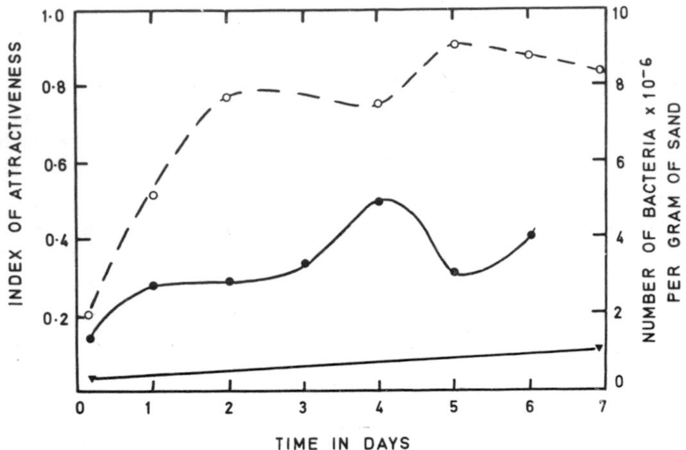

Fig. 3.4. Graphs showing the increase in bacterial numbers (solid circles), and index of attractiveness for *Protodrilus symbioticus* (open circles) of autoclaved sand to which laboratory seawater had been added. The attractiveness index = $\frac{\text{number of animals in soaked sand.}}{\text{number of animals in natural sand.}}$ The value for sterile controls is shown by triangles. (After Gray, 1966)

nor the total bacterial numbers reached as high values as in the inoculated sand. The increase in attractiveness is, however, clearly associated with an increase in the bacterial population; the attractiveness of the sterile control scarcely altered throughout the experiment.

Now although these experiments show a correlation between the numbers of bacteria and attractiveness of the sand, Gray (1966) made the interesting observation that in many sands the degree of attractiveness was due not so much to the numbers as to the kinds of bacteria. That is, in natural sands there was not necessarily any correlation between the total bacterial population and attractiveness. Furthermore, he showed that 90% of the bacteria could be removed from sand grains after 10 min treatment with a mechanical shaker. It was found that either autoclaved or acid-washed sand could be rendered attractive according to the species of bacteria used, and that the differences between the attractiveness before and after shaking were insignificant for both types of sand. It is thus evident that the attractiveness of the sands is due to the development of an organic film on the surfaces of the sand grains by bacteria and not to the presence of the bacteria themselves. The apparent correlation between bacterial population and attractiveness to *Protodrilus symbioticus* shown in Fig 3.3 and 3.4 is due to the more rapid development of such a film by a large population of bacteria. It would equally follow that a small population of favourable bacteria over a longer period of time would be able to produce an equally attractive organic film. Clearly, then, such results may be of general importance wherever bacteria have been shown to be implicated in the attractiveness of sands in site selection, and would provide a satisfactory explanation for the results obtained by Wilson (for his review see 1954; see also p. 89) on *Ophelia* as well as by Scheltema (1961) on *Nassarius obsoletus* and Jägersten (1940) on *Protodrilus rubropharyngeus.* Meadows (1964) has also shown that the removal of bacteria from sands by means of glycerol, sucrose and Na_2SO_4 did not alter the attractiveness of the deposit for *Corophium* (although distilled water did apparently eliminate the attractiveness). It therefore seems likely that in this animal too the response is primarily to the organic layer developed by the bacteria rather than to the micro-organisms themselves.

Meadows and Anderson (1968) have made a survey of the distribution, abundance and types of micro-organisms attached to the grains of intertidal deposits. They found that micro-organisms are not uniformly coated over the surface of the sand grains but tend to be confined to pits and grooves in the surface. Abrasion may therefore limit the abundance of bacteria on naturally-occurring sand grains, so that where little turbulance occurred in experimental vessels which were unstirred a more extensive film developed. As well as the increased surface area of fine-grained compared with coarser deposits, the less turbulant conditions in estuaries may thus facilitate increased populations of micro-organisms. Equally the population in coarse deposits would be less than expected from the surface area of the particles (p. 253). Zones of material which could be stained were also observed around the colonies of micro-organisms and it seems likely that these represent the attractive factor which remains after the detachment of the bacteria. Natural sand grains are thus likely to be strongly attractive not necessarily by virtue of dense bacterial populations but by the development of organic layers by small populations of the appropriate bacteria over a period of time. Adsorption of organic material from seawater may also play a significant part in the accumulation of such organic layers (Bader, Hood and Smith, 1960; Eisma *et al*, 1966). Many of the colonies of micro-organisms described by Meadows and Anderson (1968) consisted of one species and comprised some 5 – 150 cells per colony, and it is possible that the qualitive differences which occur under natural conditions may account for the specific nature of the settlement factor noted in many marine deposits.

C. SETTLEMENT ON ROCK SURFACES AND ON ALGAE

1. BARNACLES

The settlement of the larvae of intertidal animals on solid substrata has been studied in detail mainly in barnacles, serpulids, polyzoans, the hydroid *Clava,* and in the colonial polychaete *Sabellaria alveolata.* There are, however, large numbers of observations on other oganisms which emphasise the importance of behavioural responses in the settlement of such animals. Much of this earlier literature has been reviewed by Thorson (1946, 1950) and Wilson (1952, 1968). Apart from the work of Cole and Knight-Jones (1949) on the settlement of oyster larvae (p. 88) the first detailed analysis of the factors influencing settlement on solid substrata by the larvae of intertidal organisms was that of Knight-Jones and Stevenson (1950) on gregariousness in the barnacle *Elminius modestus.* They showed that the cyprids tended to settle in groups when colonising bare surfaces, and that settlement on bare test-plates was much greater when they were placed in the vicinity of settled barnacles than in areas where barnacles were scarce, even though the density of larvae in the plankton was probably similar in the two areas. The gregarious tendency was even more obvious when glass slides were exposed to settling barnacles for a few days. Some of the slides were wiped thoroughly to remove the settled barnacles whilst others were wiped gently so as to leave the barnacles attached to the slide. Then pairs of barnacled slides and pairs of cleaned slides were exposed for several days on the shore and it was found that consistently more barnacles settled on the plate which bore the initially barnacled slides than on the plate bearing the slides which had been wiped clean. Further, the density of settlement was not only greater on the slide bearing the attached barnacles, but was also greater on areas of the plate around the slide than on similar areas around slides from which the barnacles had been removed. This suggests that the gregarious effect influences the settlement of cyprids some distance from the parent individuals. Visscher (1928) had earlier observed cyprids of *Balanus improvisus* and

B. amphitrite moving over the substratum for distances exceeding 12 mm and for periods of as much as 1 hr; during this time testing movements were made. Thus contact with the settled barnacles followed by attachment and settlement even some distance away seems to provide the basis for gregariousness.

Knight-Jones (1953b) subsequently extended these observations to the barnacles *Balanus balanoides* and *B. crenatus* and also studied the mechanism of recognition in these species and in *Elminius modestus.* He showed that cyprids of *Balanus balanoides* could be collected by rinsing them from fronds of *Fucus* to which they had temporarily attached themselves. Cyprids are positively phototactic and could then be pipetted into clean vessels. Animals collected in this way were found to be able to postpone settling and metamorphosis for almost two weeks without ill-effects, but if a valve of *Mytilus* bearing the bases of freshly-detached barnacles was placed in the dish the majority of the cyprids settled within 24 hr. They were also capable of distinguishing between surfaces bearing bases of *B. balanoides* and those bearing the bases of even closely-related species; a similar ability was shown by the cyprids of *B. crenatus* and *Elminius modestus.* Once the cyprids had received the stimulus to settle by contact with a previously-settled member of their own species, they rarely swam away but embarked upon the turning movements which precede settling (p. 126). In some instances, however, where the adult barnacles or their bases were attached to an unfavourably smooth surface such as glass, movement to adjacent rougher surfaces occurred (Runnström, 1925; Barnes, Crisp, and Powell, 1951).

These results suggested that contact with a previously metamorphosed individual of the same species was a necessary prerequisite to settling by cyprids, and Knight-Jones (1953b) showed that contact with fragments of almost any part of the body including adult cirri and viscera, with which the cyprids do not normally come into contact, was effective. The settlement-stimulating substance thus appears to be generally distributed throughout the body. Since the isolated bases are composed mainly of cement which has similar properties to the cuticle, and are effective in promoting settlement, it is likely that the active substance is cuticular. That actual contact with such a substance is necessary for settlement was demonstrated by enclosing shells or stones bearing many settled barnacles in bags of bolting-silk. Each bag was placed in a dish containing cyprids together with a valve of *Mytilus.* In one series of experiments the *Mytilus* valves bore the bases of freshly-detached barnacles whilst in the other control series no bases were present. It was found that settlement readily occurred on the *Mytilus* valves which bore barnacles but that none occurred on those without barnacles even though settled barnacles were present in the dish but enclosed in a bolting-silk bag. Thus the settlement-stimulating substance present in the barnacle bases did not appear to be water-soluble; contact with the substance is clearly essential (p. 105).

Knight-Jones (1953b) showed that the attractiveness of the substance was retained even after heating for 30 min to 200°C, but not after heating to 275–300°C when charring occurred. The bases remained active after treatment with chloroform and toluene (fat solvents) as well as with protein solvents such as alchoholic solutions of phenol and aqueous solutions of urea and sodium sulphide, which also attacks substances with peptide linkages. Even boiling in dilute acids or treatment with cold concentrated acids or caustic alkalis had no effect, but boiling in concentrated acids or alkalis prevented settlement by cyprids. Settlement occurred after treatment with hydrogen peroxide, but not after treatment with sodium hypochlorite which dissolves quinone-tanned proteins and dissolved the bases of the barnacles. These chemical properties all suggest that the active settlement-stimulating substance in barnacles is a quinone-tanned protein which is known to be a component of the epicuticle and cement base of barnacles. Knight-Jones and Crisp (1953) also showed that the settlement-stimulating properties of the bases of barnacles are destroyed by substances which attack quinone-tanned proteins. In addition, the epicuticle of barnacles was shown to give a strong argentaffine reaction, local chromaffine staining, and a positive reaction to the diazo- and indophenol tests. These indicate the presence of a phenol and suggest that the epicuticle of barnacles does indeed contain a quinone-tanned protein.

Table VII. Table showing the total number of cyprids of *Balanus balanoides* settled on granite stones on top of which were placed various species of barnacles. (After Knight-Jones 1955).

Choice	*Species on stones*	*Family*	*Total no settled after 24 hr. Based on 25 experiments*
1	None	–	0
2	*Chthamalus stellatus*	Chthamalidae	0
3	*Verruca stroemia*	Verrucidae	2
4	*Elminius modestus*	Balanidae	15
5	*Balanus crenatus*	Balanidae	13
6	*Balanus balanoides*	Balanidae	71
7	Shell plates of *Balanus balanoides*	Balanidae	40

The settlement response is specific to particular barnacles, and

Knight-Jones (1955) has shown that the ability of adults to induce metamorphosis in other species of cyprids may be regarded as a measure of their systematic affinity. He counted the number of cyprids of *Balanus balanoides* settling in 24 hr in response to seven different substrata. In each experiment 20 freshly-collected cyprids were placed in a dish containing 250 ml seawater and a stone of rough granite. Each of the seven choices was repeated 25 times and the results expressed as the total number of cyprids settled after 24 hr. The first substratum was simply a stone of granite, while the others consisted of a recently detached and active adult specimen of *Chthamalus stellatus* (family Chthamalidae), *Verruca stroemia* (Verrucidae), *Elminius modestus* (Balanidae), *Balanus crenatus* (Balanidae) and *Balanus balanoides* (Balanidae). The final choice consisted of shell plates of *Balanus balanoides* placed flat on the stones so that no movements of the water occurred. The results are shown in Table VII, and indicate that many more cyprids of *B. balanoides* settled in response to the presence of members of the Balanidae than of the Chthamalidae or Verrucidae, neither of which showed any appreciable stimulatory effect. Both *Elminius modestus* and *Balanus crenatus* possessed comparable attractiveness for the settlement of the cyprids of *B. balanoides* and this, as has been pointed out by Knight-Jones (1955), agrees with their close systematic relationship with *B. balanoides* as assessed by the anatomy of the adults and larvae (for references see Knight-Jones, 1955). One further feature of interest is that in the absence of water currents, such as are set up by the movements of the cirri of the intact adult barnacles, settlement was greatly reduced. This implies that other physical factors play a part in controlling the attractiveness of the substratum to settling cyprids (p. 101).

We have already seen that smooth surfaces, such as are presented by glass plates, are unattractive to cyprids of *Balanus balanoides* (Runnström 1925; Barnes, Crisp and Powell, 1951). Knight-Jones and Crisp (1953) showed that barnacle cyprids crawl about on suitable surfaces, pulling themselves forward by alternate movements of the antennules. During this phase active testing of the substratum takes place, and changes in direction occur only infrequently. If an unfavourable surface, such as glass or loose particles or a sharp convex edge is encountered during this phase, the cyprids swim off almost at once. If on the other hand a favourable stimulus, such as the presence of a metamorphosed individual of the same species is encountered, then many random changes in direction occur and the cypris pivots on one antennule whilst testing the substratum with the other (also Doochin, 1951). It is during such a searching and testing phase, which precedes settlement and metamorphosis, that the cyprid may encounter both metamorphosed individuals of the same species and other physical features which may indicate conditions that are suitable for adult life. Crisp and Barnes (1954; also Barnes, Crisp and Powell, 1951), for example, have shown that cyprids of the barnacles *Balanus balanoides, B. crenatus* and *Elminius modestus* are capable of responding to grooves or other irregularities on surfaces; however when the surface of the substratum is a plane such as glass, or consists of randomly arranged irregularities as in ground glass, the major influence determining the orientation of the rostro-carinal axis of the adults is the direction of the incident light. Crisp and Barnes (1954) showed that the cyprid larvae of the three species of barnacles tended to settle in grooves and cavities on the substratum, and termed this response a "rugophilic" one (ruga = a wrinkle or groove). Further, the antero-posterior axis of the cyprid tends to be orientated along the axis of grooves, a response which they termed "rugotropism". The ability to orientate into grooves and cavities is probably made possible by the rotation of the animal about the antennules, which occurs immediately prior to settlement and results in the cyprids settling in regions which afford some measure of protection for the young barnacle. They found that there was a greater loss of settled specimens of *Balanus balanoides, B. crenatus* and *Elminius modestus* from smooth glass than from frosted glass, so that the rugophilic behaviour (settlement in a hollow) is an obvious advantage. On the other hand the rugotropic response is more difficult to interpret in ecological terms, but may be a reflection of the sensitivity to surface contour which is necessary for the cyprids to be able to settle in large grooves and hollows.

Crisp (1955) has demonstrated that as well as light (Barnes *et al.*, 1951) and surface contour (Crisp and Barnes, 1954) the direction and velocity of water currents play a part in the orientation of barnacles. He studied the behaviour of cyprids of *Balanus balanoides* and *Elminius modestus* at different rates of shear by placing groups of about ten larvae in a long tube through which seawater was allowed to flow at a measured rate. Larvae were introduced separately into the tube and the percentage attachment between a pair of marks 1 metre apart was recorded for each rate of flow. The results are shown in Fig. 3.5 from which the interesting fact emerges that the maximum percentage attachment by cyprids of *Balanus balanoides* was at a velocity gradient of 60 – 80 sec^{-1} whereas the maximum attachment in *Elminius modestus* was achieved at the lower velocity of $<50\ sec^{-1}$. In both cases the absence of current resulted in a lack of attachment, and often the cyprids were passively rolled along the bottom of the tube at low current velocities. This result may suggest an explanation for the results shown in Table VII in which the presence of an actively-beating *Balanus balanoides* was more effective in promoting settlement of cyprids of the same species than the shell plates of the adult barnacle (see Knight-

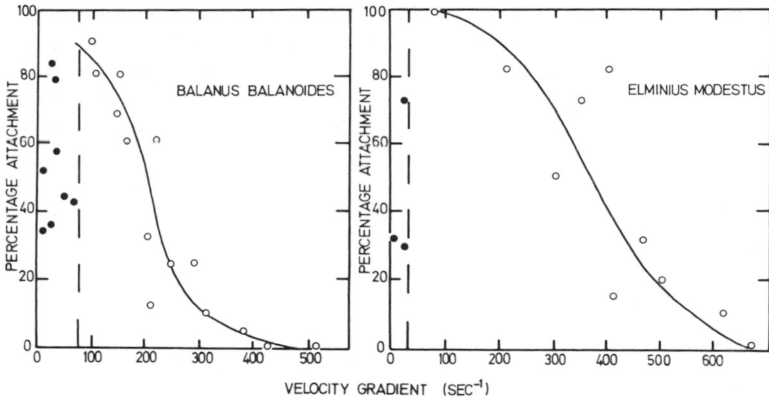

Fig. 3.5. Graphs showing the percentage attachment of cyprids of *Balanus balanoides* and *Elminius modestus* in tubes of 1m length plotted against the velocity gradient at the surface of the tube. Solid symbols indicate little or no reaction to current and desultory attachment. The broken line indicates the minimum current necessary to induce a strong attachment reaction. (Based on Crisp, 1955)

Jones, 1955). In *Balanus balanoides* an increase of the velocity gradient resulted in active swimming against the current, and 50% or more of the cyprids could still attach at a gradient of 100 sec^{-1} which correspond under conditions of laminar flow to a water velocity of 1 m/sec at only 1 cm from the surface. Beyond 100 sec^{-1} the percentage attachment became progressively lower up to a velocity gradient of 400 sec^{-1} when attachment never occurred. However, once attachment had occurred even high velocities did not displace the cyprids. Crisp (1955) has suggested that the difference in the shear at which attachment occurs corresponds with the observed ecological differences between the two barnacles, since *B. balanoides* tends to occur in more exposed situations whilst *E. modestus* settles more commonly in sheltered conditions (Moore, 1944). The adaptive value of the lack of attachment in the absence of a current is that the response would prevent settlement in regions where the water was stagnant, and facilitate settlement under conditions of moderate exposure (Barnes and Powell, 1953). Pyefinch (1948) also noticed that cyprids of *Balanus balanoides* attached in response to water currents, but that this effect was not apparent in cyprids of *B.crenatus.* As Crisp (1955) pointed out, the different threshold of response to shear might account for the known distribution of other barnacles in relation to exposure to wave action, *Chthamalus stellatus* being found in very exposed habitats (Hatton and Fischer-Piette, 1932; Moore and Kitching, 1939; also Chapter 1). A tendency for the posterior end of the barnacle to face the water current after metamorphosis was also noted in the cyprids of *Balanus balanoides* and *Elminius modestus.* This would result in the current being directed into the cirral net, and may therefore facilitate food capture under such conditions.

Although the physical features of the environment thus play a part in the settlement of cyprids and are perhaps especially important in the absence of adults of the same species, it is now generally recognised that the chemical characteristics of the substratum are of overriding importance. Crisp and Meadows (1962, 1963) have investigated the nature of the settlement-promoting substance in barnacles. They compared the number of cyprids settling on plates which had been soaked in seawater with those settling on plates which had been in contact with aqueous extracts of barnacles. An overwhelmingly greater number of cyprids settled on the plate which had been in contact with barnacle extract and this attractiveness was conferred in the relatively short time of 15 min. Since solid material had been centrifuged from the extract, it was inferred that the attractive principle present in barnacle tissues (as opposed to that present in the barnacle bases, p. 98) was water soluble. This "settling factor" was found to be resistant to boiling and did not pass through a dialysis membrane and must therefore be of a high molecular weight. It was also resistant to treatment with hydrogen peroxide and formalin but was destroyed by sodium hypochlorite.

Since the settlement-stimulating substance appeared to be transferred to solid surfaces, it was of interest to determine how long activity would withstand washing. Fig. 3.6 shows the result of an ex-

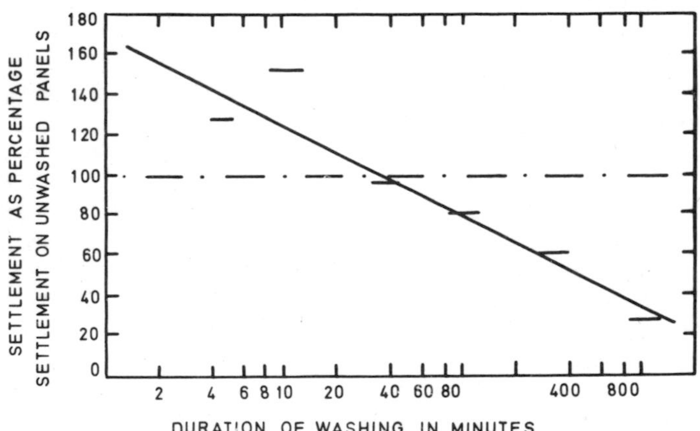

Fig. 3.6. Graph showing the fall in settlement of barnacle cyprids on "Tufnol" plates which had been treated with barnacle extract and then washed continuously prior to presentation to the cyprids. Plates were removed at various times (indicated by the horizontal lines) and the number of barnacles settling on them expressed as a percentage of the settlement on unwashed panels. (After Crisp and Meadows, 1962.)

periment in which "Tufnol" panels were continuously washed in seawater after having been first exposed to aqueous barnacle extract and were then presented at intervals to cyprids. The activity clearly fell off exponentially and was negligible after 24 hr. This loss of activity with time obviously imparts a complicating factor to the choice experiments since the activity of a treated panel at the beginning and end of an experiment might not be the same. However, the evidence suggests that the choice by cyprids is made very rapidly, attachment to the treated panels occurring almost immediately. They also found that dilution of the extract to which the experimental panels were exposed resulted in a corresponding reduction in attractiveness. Fig. 3.7 shows the number of

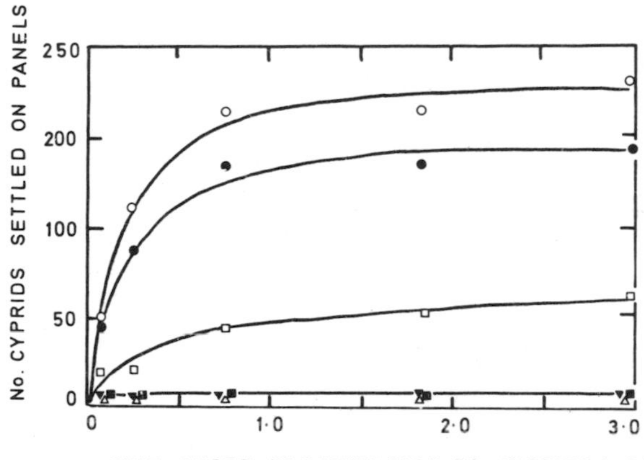

Fig. 3.7. Graphs showing the number of cyprids attached to panels treated with various concentrations of barnacle extract. The settlement on the plates was counted at several times after the initial presentation of the treated panels to the cyprids. Open circles – initial concentration of extract (2g of whole barnacle/ml seawater); solid circles – extract diluted x 100; open squares – extract diluted x 1000; open triangles – extract diluted x 10,000; solid triangles extract diluted x 100,000; solid squares – untreated (control) panels. (After Crisp and Meadows 1962.)

cyprids attached to panels which had been exposed to various concentrations of aqueous extract, the apparent fall in rate of attachment probably partly reflecting the decline in available free-swimming larvae. Aqueous extracts of other organisms including arthropods, the fish *Blennius pholis* and sponges such as *Ophlitaspongia seriata* all contained a settlement-stimulating substance, but the one from adult *Balanus balanoides* was more successful in promoting settlement of that species than extracts of other species. Cyprids of *Elminius modestus* show a similar sensitivity to panels treated with aqueous extracts of cirripedes and, as would be expected from the results of Knight-Jones (1955;

p. 99), settled preferentially in response to extracts of their own species. Such a response was found to occur over a wide range of concentrations of extract as is shown in Fig. 3.8.

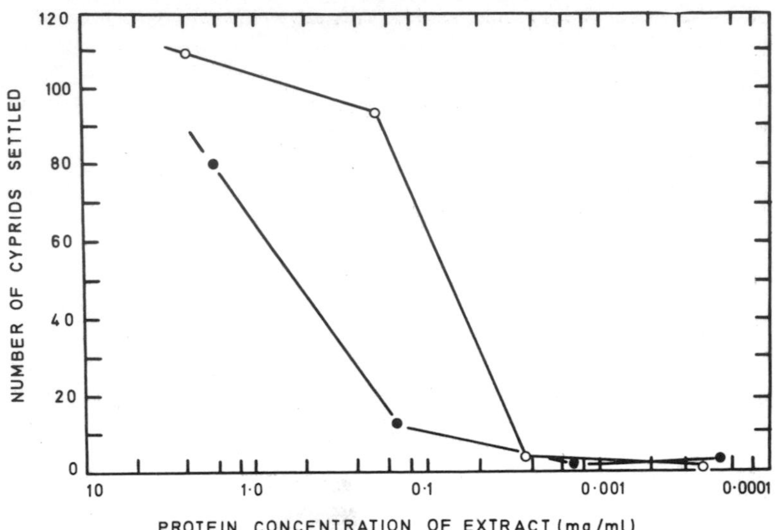

Fig. 3.8. Graphs showing the number of cyprids of *Elminius modestus* settling on slate panels treated with various concentrations of extracts of *Elminius modestus* (open circles) and of *Balanus balanoides* (solid circles). (After Crisp and Meadows, 1962.)

The adsorbed layer of settlement-stimulating substance proved to be just as resistant to treatment with acids, alkalis and other reagents as the barnacle fragments studied by Knight-Jones (1953b; p. 98). It seems likely that the substance responsible for the attractiveness of plates exposed to aqueous solutions of barnacle tissues may be the water soluble protein arthropodin which, as the cuticle hardens, becomes progressively less soluble. As Crisp and Meadows (1962) pointed out, arthropodin, unlike most proteins, is not denatured by boiling. The proportion of glycine and alanine residues varies in different species (Trim, 1941; Duchâteau and Florkin, 1954), and it is probable that such variation could account for the specific properties noted in the extracts of the different barnacles. They also suggested that once a cyprid has alighted on a surface bearing an adsorbed layer of the appropriate settlement factor, the simultaneous presentation of suitable tactile and chemical stimuli would initiate settlement.

Crisp and Meadows (1963) then demonstrated that all the physical and chemical properties of extracts of the settlement factor from barnacles were similar to those of the arthropodin extracted from insect cuticle, and went on to demonstrate the importance of adsorption in the response of cyprids to barnacle extracts. They showed conclusively that cyprids do not respond to the factor when in solution but only when adsorbed onto a surface. They placed ten slate panels which had previously been treated with a dilute seawater extract of barnacles (0·05 mg protein/ml) and ten untreated slate panels around the periphery of a rotating dish of seawater into which cyprids were introduced. A similar experiment was then made using another ten treated and ten untreated slate panels except that this time the seawater contained a barnacle extract of the same concentration as that used to treat the panels (0·05 mg protein/ml). The results are shown in Fig. 3.9. It will be seen that in the first experiment, as would be expected, many

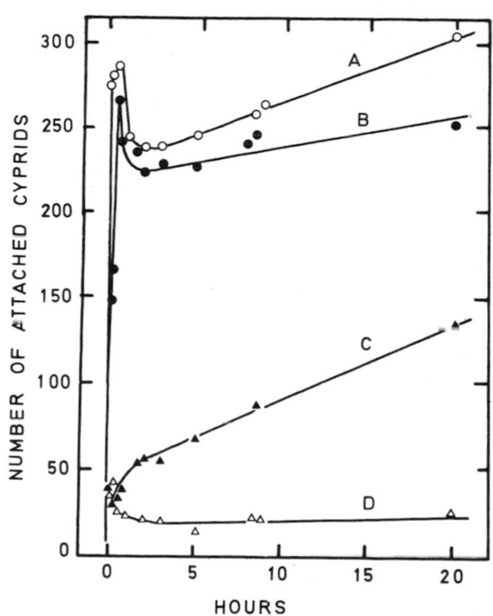

Fig. 3.9. Graphs showing the number of cyprids of *Balanus balanoides* attached to untreated panels and to panels treated with barnacle extract in two dishes. One dish contained seawater and the other contained the barnacle extract which had been used to treat half the panels. A = number attached to treated panels in seawater; B = number attached to treated panels in extract; C = number attached to untreated panels in extract; D = number attached to untreated panels in seawater.

more cyprids explored and attached to the treated panels than to the untreated panels in seawater (Fig. 3.9 A and D). However, a very meaningful result was obtained with the second experiment (Fig. 3.9 B and C) which showed that even in seawater solutions of the extract the treated panels were chosen in preference to the untreated ones. Such a response could only have been made by contact with the adsorbed layer of settlement factor on the treated panels. During the course of the experiment, the untreated panels in extract became more favourable presumably because of the adsorption of the settlement factor, and may then have competed for larvae with the treated panel which therefore attracted rather less cyprids than a comparable panel in sea-water (Fig. 3.9 A).

The settlement which occurred in response to adsorbed layers of barnacle extract appeared to result in a much stronger adhesion to the treated panels than to untreated panels. Further, this stimulus to settle lasted only whilst contact was made with the adsorbed layer. Removal of settled cyprids from a treated surface followed by presentation of an untreated surface resulted in an average of only 7% settlement contrasting with a 2% settlement when cyprids were removed from an untreated panel (and even these animals were only temporarily attached). The presence of adsorbed layers appeared to be of more importance than the nature of the surface in promoting settlement, although the presentation of both pits and barnacle extract greatly enhanced settlement. Currents appeared to be primarily of importance in bringing swimming cyprids into contact with a variety of different substrata prior to settlement. Should contact with a suitable substratum be delayed, then metamorphosis may be delayed too and the rate of settlement once contact has been made with a suitable substratum is correspondingly more rapid. For example, Crisp and Meadows (1963) found that the percentage of cyprids which had not settled after 24 hr on a treated panel was 69% in cyprids stored for only 1 hr, but was 29% in cyprids stored for 51·5 days, 19% in cyprids stored for 3·5 days and only 16% in cyprids stored for 5·5 days. Clearly if the numbers settling on treated panels were plotted as a function of time, then the rate would decline owing to the reduction in the number of free-swimming cyprids. This may be overcome by plotting $\log_{10} (N/N-x)$ against time t, where N is the total number of cyprids introduced and x is the number which had settled in time t (also Knight-Jones, 1953d; Williams, 1965c). The total numbers of cyprids stored for various times and settling on treated panels over a period of 12 hr are shown in Fig. 3.10 in which a constant rate of settlement should give a straight line. The rate of settlement clearly declined somewhat during the period of the experiment but nevertheless aging cyprids do not appear to lose the ability to distinguish treated surfaces (cf *Spirorbis borealis*, Knight-Jones, 1953d and p. 111). This work, then, shows the importance of adsorbed layers of settlement–stimulating substance in rendering surfaces attractive to cyprids. It is likely that, as Crisp and Meadows

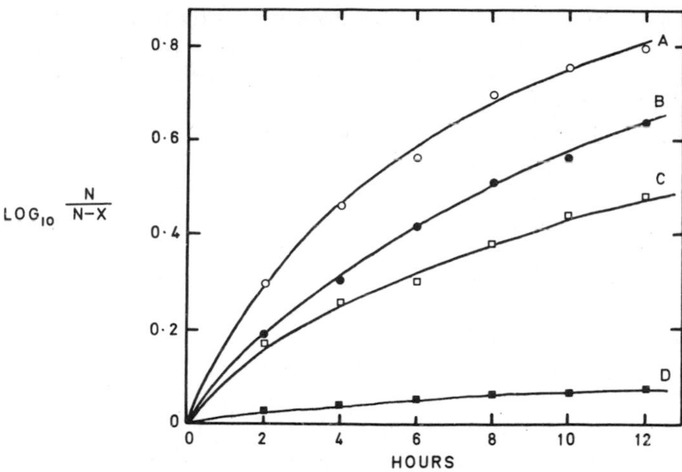

Fig. 3.10. Graphs showing the settlement of cyprids on a panel treated with barnacle extract. Settlement had been prevented (A) for 5·5 days by storage in a clean glass vessel, (B) for 3·5 days, (C) for 1·5 days, (D) for only 1 hr after collection from the shore. For explanation see text. (After Crisp and Meadows, 1963.)

(1963) concluded, the molecular configuration of soluble arthropodin adsorbed by its polar groups onto a solid surface presents a similar molecular configuration to the bound arthropodin of the cuticle which had previously been suggested to be the attractive factor in barnacle bases (Knight-Jones, 1953b; Knight-Jones and Crisp, 1953). Chemoreception occurs only after contact with the appropriate surface and is regarded by Crisp and Meadows (1963) as a true 'tactile chemical sense' which may possibly stimulate the receptors without the intervention of aqueous diffusion.

2. SERPULIDS

The settlement responses of serpulid polychaetes of the genus *Spirobis* bear remarkable similarities to those shown by barnacles, both the physical and chemical properties of the substratum as well as the presence of adults of the same species being involved in settlement. Knight-Jones (1951b) first described the settlement behaviour of *Spirorbis borealis,* which typically occurs on *Fucus,* and also compared the response with that of *S. pagenstecheri* which typically settles on stones and shells in the lower part of the intertidal zone. Both of these species are viviparous and the larvae are liberated mainly at the moon's quarters. After liberation the larvae, which measure approximately 4 mm in length, were found to be strongly positively phototactic and swam at about 3 mm/sec by means of the cilia of the prototroch. This behaviour lasted for 15 min to 2 hr, after which swimming became more random; for a period of from 1 – 2 hr the larvae then became photo-negative at times and visited a large number of different surfaces, exploring them and swimming off again. Once a suitable substratum for settlement had been encountered, and immediately prior to attachment, the larvae frequently changed the direction of crawling and flexed the abodmen sideways through an angle of approximately 120° (p. 126). Crawling then became slower and turning more frequent, and finally a milky fluid was secreted from an 'attachment gland' on the mid-dorsal surface of the posterior part of the thorax. This fluid formed the initial tube and the larvae then rolled onto the dorsal surface and underwent a cataclysmic metamorphosis. The whole process from initial fixation to rolling ventral side uppermost took only 1 – 2 min, after which a series of morphological changes, including the development of an operculum, took a further 5 – 10 min only.

Knight-Jones (1951b) then showed that a bacterial film is necessary to induce settling on glass. When larvae were placed in clean glass beakers settling and metamorphosis occurred only after a considerable delay and was often abnormal, whereas settlement was much more rapid in glass beakers which had been allowed to develop a bacterial film on the surface. After 24 hr, for example, only 15% of *S. borealis* larvae had settled in the clean beakers whilst 91% had metamorphosed in the filmed beakers. A piece of *Fucus* was, however, suitable for settlement whether its surface had been wiped or not. Further, the surface of *Fucus* was rendered much more attractive by the presence of previously settled *Spirorbis* on its surface, the ratio of larvae settling on colonised compared with uncolonised *Fucus* being at least 2:1 in favour of the previously colonised thallus.

As might be expected from the settlement behaviour of other animals, the larvae of *Spirorbis borealis* are able to postpone settlement until a suitable substratum has been encountered. Knight-Jones (1953d) studied the effect of storage of larvae in clean vessels on the rate of settlement when presented with *Fucus*. He carried out two series of experiments; in one of these each beaker contained a single larva so that the results represented the percentage settlement of isolated larvae, whilst in the second series each beaker contained five larvae. After the addition of *Fucus*, the number of metamorphosed larvae was counted each hour. In all, a total of 100 larvae was used in each of the isolated and each of the grouped experiments. One set of larvae was presented with *Fucus* immediately on liberation, then another after storage in a clean vessel for 3 hr, 6 hr and 12 hr respectively. The results are shown in Fig. 3.11 from which it will be noticed that in all cases the

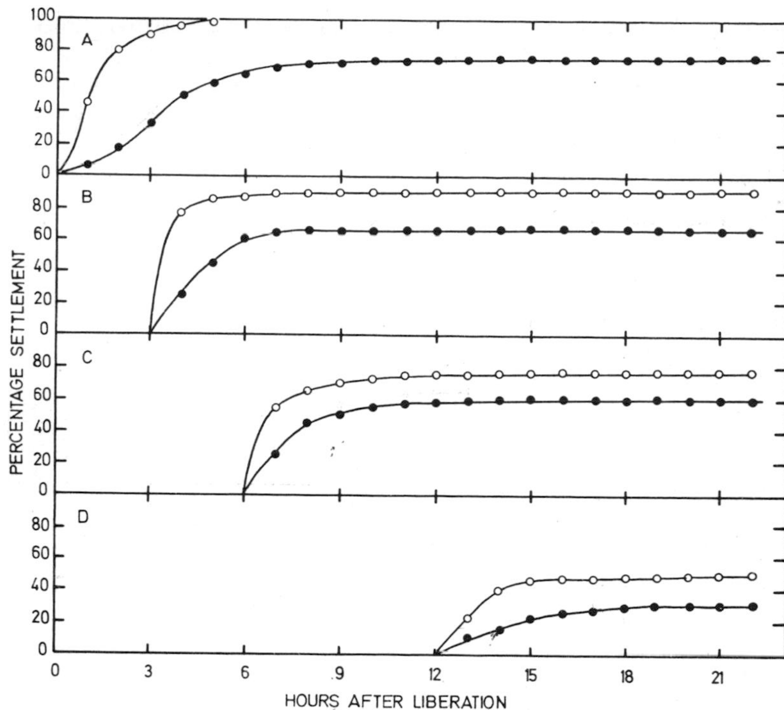

Fig. 3.11. Graphs showing the percentage settlement as a function of time in *Spirorbis borealis* which were presented with pieces of *Fucus* (A) immediately after liberation, (B) after 3 hr in a clean vessel, (C) after 6 hr in a clean vessel, (D) after 12 hr in a clean vessel. The open circles indicate the results obtained with groups of five larvae, the solid circles are those obtained with isolated larvae. (After Knight-Jones, 1953.)

grouped larvae settled at a faster rate than the isolated larvae. This is undoubtedly due to the settlement of one facilitating the settlement of others due to the gregarious tendency. As Knight-Jones (1953d) pointed out, the proportion of larvae which had settled in isolation compared with those tested in groups is very similar to the ratio between the settlement on bare *Fucus* and colonised *Fucus* in the gregariousness experiments. A further point of interest is that in the freshly-liberated larvae settlement was initially slow because of the photopositive behaviour of the larvae which prevented contact with the *Fucus.* But after 3hr storage the settlement occurred without delay and at a higher rate than in the freshly-liberated larvae. Storage for 6hr and 12 hr resulted in a lower proportion of settlement and the rate also appeared to be slower. As Knight-Jones (1953d) pointed out, the reduction in available larvae results in a decline in the percentage settlement with time. In order to take into account the reduction in the number of larvae available for settlement, the data can be plotted as $\log_{10}$ $(N/N{-}x)$ against time t (also p.106). The slope of this curve, which is shown in Fig. 3.12, is obviously given by d.Log_{10} $(N/N{-}x)$ dt and indicates the rate of settlement. From inspection of Figure 3.12, it is

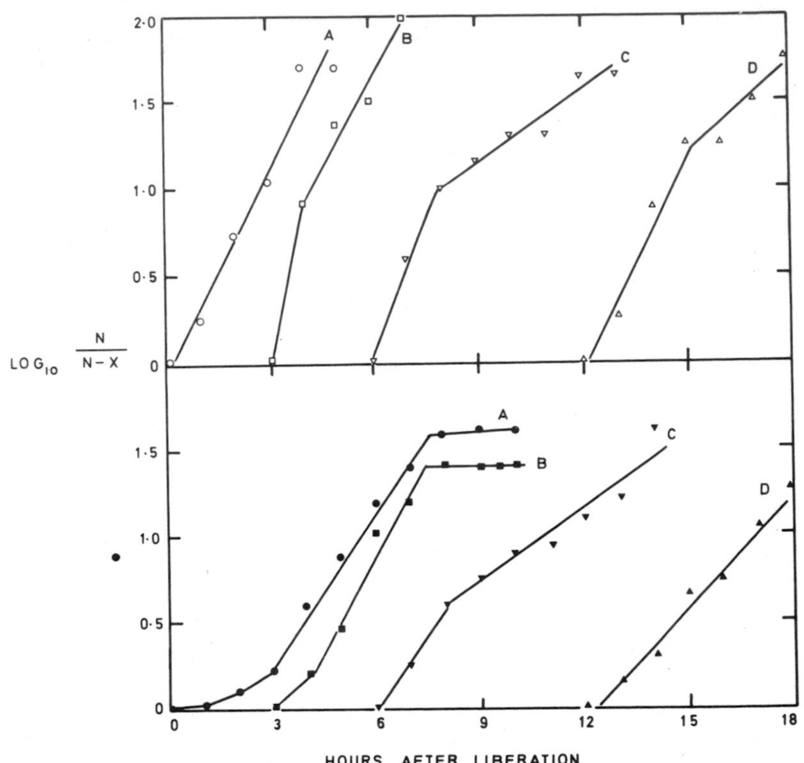

Fig. 3.12. Graphs showing the settlement of larvae of *Spirorbis borealis* when presented with *Fucus* (A) immediately after liberation, (B) after a delay of 3 hr (C) after a delay of 6 hr (D) after a delay of 12 hr. Data as in Fig. 3.11 after Knight-Jones (1953). Open symbols indicate grouped larvae, solid symbols indicate isolated larvae. For explanation see text.

apparent that the rate of settlement of the grouped larvae was not only greater (the line is steeper) but also proceeded earlier (the line is displaced to the left) compared with the isolated larvae. Amongst the grouped larvae themselves, the rate of settlement was greater in the animals stored for 3 hr than in the freshly-liberated larvae, whilst the rate of settlement of those stored for 6 hr was also initially high but then declined. After settlement had been prevented for 12 hr the rate of settlement was also slow, probably because of the exhaustion of the larvae (Knight-Jones, 1953d). In the isolated larvae there was a low rate of settlement during the first 3 hr, after which the rate increased. However after the larvae had spent 8 hr swimming, the rate of settlement again declined, perhaps due to the depletion of food reserves although a considerable variation was noted by Knight-Jones (1953d). As he pointed out, under natural conditions the danger of such exhaustion would be slight because the photonegative behaviour of the larvae would ensure contact with a suitable substratum in the neighbourhood of the parent population within a few hours of liberation.

We have seen that in barnacle cyprids (p.99) the gregarious tendency involves the recognition of previously settled members of the same species. This is true also of *Spirorbis borealis*. Knight-Jones (1951b) allowed larvae of *S. borealis* to settle on one series of filmed slate plates and *S. pagenstecheri* to settle on a similar series of plates. The plates bearing settled animals were then exposed in pairs (one of each pair bearing *S. borealis* and the other *S. pagenstecheri*) in 250 ml dishes containing seawater to which larvae of both species were added. After 24 hr the numbers of each species of larva settling on the two types of plates were counted. It was found that out of a large number of such experiments about 90% of the larvae of each of the species settled on the plate bearing individuals of their own species. As Knight-Jones (1951b) pointed out, it seems likely that chemical recognition occurs as in barnacle cyprids (p. 107). The gregarious tendency in *Spirorbis borealis* overrides even their preference for *Fucus*, since in the dark the larvae chose filmed plates bearing settled individuals in preference to bare *Fucus*. Such a response would prevent the settlement of *Spirorbis borealis* on *Fucus* in the upper part of the intertidal zone where excessive desiccation would tend to occur.

Garbarini (1936; see also Gross and Knight-Jones, 1957) had earlier reported that the larvae of *Spirorbis borealis* attached to *Fucus* but that few settled on *Laminaria, Himanthalia, Ascophyllum, Rhodymenia* or stones. The preference of *S. borealis* for *Fucus* compared with *Ascophyllum* was also noticed by Knight-Jones (1951b). De Silva (1962) subsequently made a thorough investigation of the substrate preferences of the larvae of *Spirorbis borealis, S. corallinae* and *S. tridentatus. S. borealis* occurs, as we have seen, primarily on *Fucus serratus, S. corallinae* on *Corallina officinalis* and *S. tridentatus* on rocks and stones. De Silva (1962) showed in a series of choice experiments that the occurrence of these three species on their typical substrata was due to the discrimination shown by the larvae during settlement since the larvae of each of the species tended to select the substratum upon which the species normally occurs. Other substrata were either less favourable or unfavourable. Other responses, too, tended to result in the selection of the appropriate substratum, since the larvae of *S. borealis* tended to be photopositive whilst those of *S. tridentatus* were photonegative during the settlement phase. This would tend to prevent settlement of *S. tridentatus* on algae and facilitate its settlement in dim situations where algae are scarce, whereas the reverse would be true for *S. borealis.* Gregariousness is also implicated in the settlement process of *S. corallinae* and *S. tridentatus* so that in both of these species, as well as in *S. borealis* and *S. pagenstecheri,* the selection of a suitable substratum for adult life is brought about not only by responses to the nature of the surface but also the presence of adults of the same species.

In much the same way, Gee and Knight-Jones (1962) have shown that the behaviour of the larvae of a new species of serpulid, *Spirorbis rupestris,* was appropriate to bring about settlement on the normal substratum occupied by the adult. In this case the adult serpulid lives on rock surfaces in well-illuminated situations throughout the lower part of the shore and is normally associated with the calcareous alga *Lithothamnion polymorphum* (in fact Gee and Knight-Jones, 1962, identified the alga as *Lithophyllum incrustans* but Gee, 1965 states that it is actually *Lithothamnion polymorphum*). They showed that in contrast to *S. borealis,* the larvae of *S. rupestris* never became pelagic even when first liberated provided that a suitable substratum was presented to them. The larvae thus tend to become attached to the same piece of rock as the parents, whereas a much smaller proportion of *S. borealis* larvae become attached to the parent frond. As in the other species of *Spirorbis,* the larvae appear to distinguish readily between suitable and unsuitable substrata – in this case between rocks bearing *Lithothamnion* and similar rocks without this coralline alga. Fig. 3.13 shows the percentage settlement out of 66 larvae in each of two dishes, one of which contained a rock bearing *Lithothamnion* and the other a rock only. It was also found that no settlement occurred on fronds of *Fucus serratus,* so that the choice of particular substrata by the different species of *Spirorbis* would tend to reduce competition between them. The difference in larval behaviour of *Spirorbis borealis* and *S. rupestris* can also be related to the substratum which they inhabit. In *S. rupestris,* for example, the virtual elimination of the pelagic larval phase in the presence of a suitable rock surface reduces the possibility of the larvae being washed away and subsequently failing to find a suitable substratum. On the other hand the fronds of *Fucus* inhabited by *S. borealis* are much more likely to be torn away and it would be disadvantageous for the population to be concentrated on such an ephemeral substratum. In this case a brief pelagic phase would allow

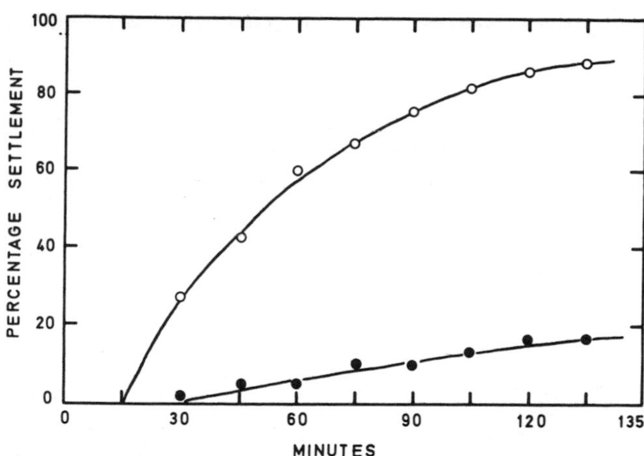

Fig. 3.13. Graphs showing the percentage settlement of larvae of *Spirorbis rupestris* on rock with *Lithothamnion* attached (open circles) and on bare rock (solid circles). (After Gee and Knight-Jones, 1962.)

colonisation of other algal fronds so that the population of adults could be distributed and the risk of destruction minimised.

The chemical basis of such discrimination in *Spirorbis borealis* has been studied by Williams (1964) and in *S. rupestris* by Gee (1965), although Crisp and Ryland (1960) had earlier demonstrated the importance of both surface texture and presence of a film of micro-organisms in promoting the settlement of *S. borealis* (also Meadows and Williams, 1963). They showed that larvae of this species avoided rough surfaces but were attracted to glossy or smooth filmed surfaces compared with cleaned ones. Williams (1964) offered three kinds of substrata to larvae of *Spirorbis borealis.* These were surfaces filmed with micro-organisms, unfilmed surfaces, and filmed surfaces treated with aqueous extracts of *Fucus serratus.* In all cases the presence of filmed surfaces treated with algal extract greatly enhanced the percentage settlement compared with filmed surfaces alone, although these in turn promoted greater settlement than the cleaned surfaces. Furthermore the rate of settlement on filmed plates which had been treated with algal extract was greater than on the other plates. The percentage of metamorphosed larvae on filmed panels treated with *Fucus* extract, on filmed surfaces alone and on unfilmed surfaces, is shown as a function of time in Fig. 3.14. The larvae were presented with the three types of surfaces immediately after release, after 6 hr and after 12 hr. The results in general agree with those of Knight-Jones (1953d) insofar as the maximum rate of settlement occurred some 3–6 hr after liberation, larvae more than 12 hr old becoming debilitated. Nevertheless in each case the marked preference

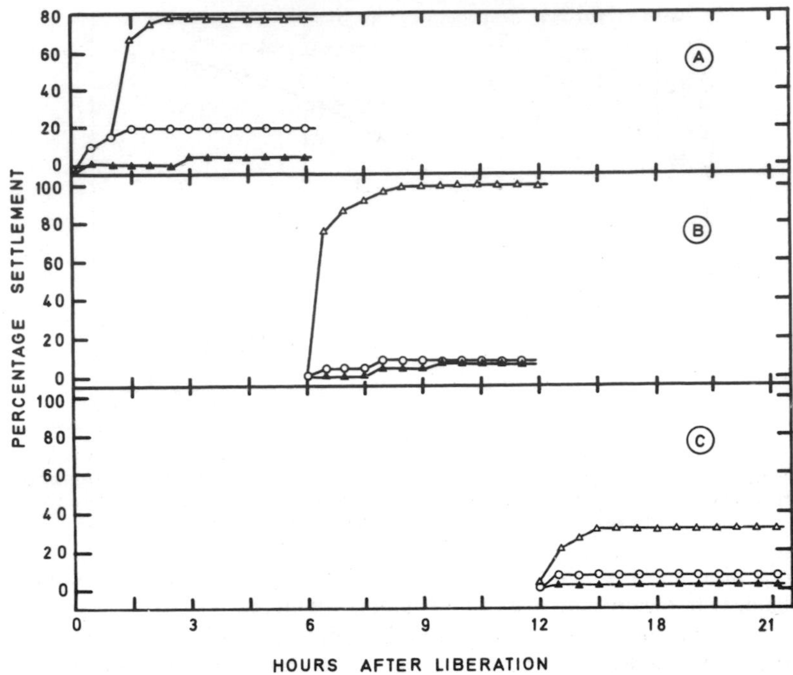

Fig. 3.14. Graphs showing the percentage settlement of larvae of *Spirorbis borealis* on filmed panels treated with *Fucus* extract (open triangles), on filmed panels (circles), and on unfilmed panels (solid triangles). The panels were presented to the larvae (A) immediately after release, (B) after a delay of 6 hr (C) after a delay of 12 hrs. (Based on Williams, 1964.)

for the filmed surfaces treated with *Fucus* extract is apparent. The larvae proved capable of responding to filmed surfaces treated with quite low concentrations of fucoid extract. An extract of *Fucus serratus* was prepared from 120 g of alga in 500 ml of distilled water. This extract was then diluted so that the final concentrations corresponded to 1/2, 1/5, 1/10 and 1/100 of the original extract. Williams (1964) found that the extract was active in promoting settlement at dilutions as low as 1/100 of the original concentration and appeared to contain little protein. Subsequent work (Williams, 1965a,b) suggested that the active factor was a mucopolysaccharide. Further, actual contact with an adsorbed layer of the settlement-stimulating substance appeared to be essential, since larvae failed to respond when placed in solutions of *Fucus* extract; in many respects therefore, the behaviour appears similar to the chemo-tactile response described above for barnacle cyprids.

Gee (1965) studied the settlement behaviour of *Spirorbis rupestris* to determine whether settlement was promoted by an aqueous extract of its characteristic substratum *Lithothamnion polymorphum,* as has been demonstrated for aqueous extracts of barnacles (Crisp and Meadows, 1962, 1963) in the settlement of cyprids and for *Spirorbis borealis* in response to extracts of *Fucus serratus* (Williams, 1964). He found that when four different batches of identical larvae of *Spirorbis rupestris* were presented with different coralline algae of a comparable form to that of *Lithothamnion polymorphum,* settlement occurred in overwhelmingly greater numbers on the *Lithothamnion* although thick *Lithophyllum incrustans,* which has a lobate surface, was rather more effective than the smoother species. It seems possible that the physical nature of the surface might provide some indication of substratum, but the evidence points to a chemical factor being of prime importance. Seawater extracts of *Lithothamnion* were then prepared by grinding weighed quantities with a pestle and mortar and removing debris in a centrifuge at 8,000 rpm for 10 min. Slate panels were then stored for 18 hr in contact with the extract and transferred to seawater. The proportion of larvae settling on them was then compared with that on panels which had been stored for a similar length of time in seawater and with that on untreated panels. About 120 larvae were used in each experiment and it was found that settlement was almost exclusively on the panels treated with *Lithothamnion* extract. Furthermore, panels were rendered overwhelmingly more attractive than controls in as short a period as 3 hr storage in *Lithothamnion* extract. It seems therefore that the settlement-stimulating substance is water-soluble and transferred to the surfaces by adsorption. Unlike the results for barnacles, however, the substance appeared to be destroyed by boiling, for panels soaked in boiled extract promoted only slightly heavier settlement than panels soaked in seawater and much less than those soaked in fresh extract.

Apart from being unstable at 100° C, the settlement-stimulating substance differs from that present in barnacles (Crisp and Meadows, 1962; 1963) and in fucoids (Williams, 1964, 1965a; p. 119) in being of small molecular weight. Equal numbers of larvae settled on panels treated with fluid from both inside and outside a cellulose dialysis tube of pore diameter 24 Å after 24 hr of dialysis. This suggested that a molecular equilibrium between the two fluids might have been attained. Unlike the barnacle extract, treatment with 10% seawater formalin destroyed the attractiveness of the substance. It is therefore evident that although chemical factors are effective in promoting settlement on the characteristic substrata occupied by the adults, such factors are of diverse physical and chemical properties in the different organisms concerned.

3. POLYZOANS

Ryland (1959), Crisp and Ryland (1960), Crisp and Williams (1960) and Williams (1965a,b) have investigated the substratum-specificity of polyzoan larvae. Ryland (1959) studied the settlement behaviour of the larvae of *Celleporella hyalina, Alcyonidium polyoum, A. hirsutum* and *Flustrellidra hispida* each of which has a lecithotrophic larva which settles shortly after liberation in contrast to the cyphonautes larvae of *Electra pilosa* and *Membranipora membranacea* which have a planktotrophic stage of several weeks. The settlement preferences of the four species of Polyzoa were tested by offering larvae choices between different species of alga and in total the larvae of each species of polyzoan were offered a choice of 14–15 different species of alga on which to settle. In *Alcyonidium hirsutum, A. polyoum* and *Flustrellidra hispida* each of which is a truly littoral species, settlement was greatest (*c.* 35%) on *Fucus serratus* (particularly on the non-fruiting tips). This choice corresponds with their normal occurrence on the shore which is also on *F. serratus.* In *Alcyonidium hirsutum,* algae including *Fucus spiralis, Gigartina stellata* and *Chondrus crispus* were also moderately favourable for settlement. Ryland (1959) noted that on exposed coasts small colonies of *A. hirsutum* are in fact moderately common on *Gigartina* and *Chondrus* as well as on *Fucus serratus* although they do not occur on *Fucus spiralis.* Although *A. hirsutum* favours *F. spiralis* in the laboratory its absence on this alga under natural conditions might be explained in terms of the desiccation or the reduced feeding time available at that level on the shore (Ryland, 1959).

In *Alcyonidium polyoum, Fucus vesiculosus* was very favourable for settlement in one of the series of experiments and the absence of the polyzoan from this alga under natural conditions might also be attributable to the fact that *F. vesiculosus* occurs too high on the shore for successful colonisation. On the other hand, *Gigartina* was favourable and is heavily colonised on the shore. *Dictyota* proved favourable but probably is never sufficiently abundant on the shore to provide an important substratum for polyzoans. The larvae of *Flustrellidra hispida* show similar preferences to those of *Alcyonidium polyoum,* and Ryland (1959) offered similar reasons for its absence from the apparently suitable *F. spiralis, F. vesiculosus* and *Dictyota dichotoma* under normal conditions. It is virtually confined to *F. serratus* but where exposure to wave action occurs, and *Gigartina* becomes common, colonies may be found on both species. Finally in *Celleporella hyalina* there was a less distinct preference for particular algal substrata, even the most favourable species (*Chondrus crispus*) inducing only about 20% settlement. Other algae such as *Laminaria saccharina* were nearly as favourable, and where this alga is abundant on the shore the colonies are found predominantly on it, rather than on neighbouring *Halidrys*. No larvae were found to settle on *Halidrys* under experimental conditions so that it is clear that despite the lack of distinct preferences in the laboratory, the larvae of *Celleporella hyalina* make a definite choice of alga under natural conditions.

The distribution of the three intertidal polyzoans *Alcyonidium hirsutum*, *A. polyoum* and *Flustrellidra hispida* on the shore can thus be explained in terms of the preferences of the larvae for mid-shore fucoids, especially *F. serratus*; the upper shore fucoids probably prove unsuitable even though the larvae under experimental conditions show a distinct preference for them compared with *Laminaria* and other lower shore algae. Larvae of the lower shore polyzoan *Celleporella hyalina* also appear to choose an alga appropriate to the normal habitat on the shore, although in this case the specificity of the selection is not so apparent as in the other species of polyzoan. As in barnacle cyprids (p. 100), the larvae of the polyzoans mentioned above actively settle in concavities or in the channelled grooves of a frond, and in this way receive protection from abrasion with other algae, rocks and sand.

Crisp and Ryland (1960) showed that the texture of the surface and the absence of a film of micro-organisms was of importance in rendering it attractive to *Bugula flabellata*. The larvae of *Spirorbis borealis* were attracted by a smooth filmed surface (p. 113), but *Bugula* was found to be indifferent to surface texture and repelled by the presence of films on the surface; on the other hand the larvae did not settle very well on unfilmed surfaces either. Crisp and Williams (1960) then showed that filmed surfaces were rendered highly attractive to the larvae of *Alcyonidium*, and the closely-related polyzoan *Flustrellidra hispida*, by treatment with aqueous extracts of fucoid algae; extracts of other algae were tested but found to be unattractive. The filmed surfaces rendered attractive included those treated with extracts of *Fucus serratus*, *F. vesiculosus*, *Ascophyllum nodosum* and *Pelvetia canaliculata*, but of these only *F. serratus* normally bears *Alcyonidium*. Thus, as Ryland (1959) suggested, other factors such as the high level on the shore at which the other fucoids occur must be invoked to explain the absence of colonies from these apparently suitable substrata.

Williams (1965a,b) has studied in more detail the chemical basis of substrate selection by polyzoan larvae. He studied *Alcyonidium polyoum*, *A. hirsutum* and *Flustrellidra hispida*, and showed that the larvae of all these species settle in significantly greater numbers on filmed surfaces treated with *Fucus* extract than on filmed surfaces and unfilmed surfaces. Further, there was no loss of activity when the *Fucus* extract was filtered through a cellulose ester membrane of $0{\cdot}5\mu$ pore diameter, so that the active factor must have been either in solution or colloidal suspension. One of the problems posed was whether the active factor itself was responsible for stimulating settlement, or whether it served as a substrate for micro-organisms which themselves induced settlement. Williams (1965a) investigated this by comparing the percentage settlement of larvae of *Alcyonidium polyoum* and *Flustrellidra hispida* on filmed and unfilmed panels, and filmed panels which had been treated with *Fucus serratus* extract. A final choice consisted of filmed panels which had been treated with *F. serratus* extract and then sterilised. The extract was known to be stable to boiling (also p. 120), and even when the micro-organisms had been destroyed the filmed plates which had been treated with extract still promoted as dense settlement as the unsterilised *F. serratus*-treated filmed panels. Williams (1965a) thus concluded that the active factor in the extract is almost certainly a product of the alga rather than a by-product of the modified metabolism of bacteria in the extract or on the surface of the filmed panels.

As described on p.116, Ryland (1959) had found that the larvae of various polyzoans exhibited a marked preference for members of the Fucacea. Since the water-soluble active factor mentioned above was extracted from *Fucus serratus,* it was clearly of importance to determine whether such a substance occurred in the other members of the Fucacea, and whether it occurred also in other types of algae. Williams (1965a) used larvae of *Alcyonidium polyoum* to investigate the activity of extracts of *Fucus serratus, F. vesiculosus, Ascophyllum nodosum, Pelvetia canaliculata, F. spiralis, Laminaria digitata, Ulva lactuca* and *Gigartina stellata.* The results are expressed in Table VIII as the ratio of the numbers of larvae settling on filmed plates treated with the particular algal extract, compared with the number settling on a similar plate treated with extract of *Fucus serratus.*

Table VIII. Ratio of the number of larvae of *Alcyonidium polyoum* settled on films treated with algal extract (NA) compared with settlement on films treated with extract of *Fucus serratus* (NF). (From Williams, 1965a)

Extract	*Ratio of NA/NF*
Fucus serratus	1·0
Fucus vesiculosus	1·04
Ascophyllum nodosum	1·72
Pelvetia canaliculata	0·63
Fucus spiralis	0·09
Laminaria digitata	0·25
Ulva lactuca	0·04
Gigartina stellata	0·00
Control (filmed surface)	0·22

It is obvious that the algae which stimulated heavier settlement than filmed surfaces alone are all members of the Fucacea, so that the settlement-stimulating substance for polyzoans appears to be restricted in its occurrence to this group of algae. These results also agree with those of Ryland (1959; also p. 116), who found that the fucoids were the most effective in stimulating settlement. For example, in the results of both Ryland (1959) and Williams (1965a), *F. serratus* and *F. vesiculosus* were found to stimulate heavy settlement and their extracts were also effective. There are, however, certain discrepancies, such as the fact that extracts of *Ascophyllum nodosum* on filmed surfaces are the most attractive of all the extracts, whereas settlement on pieces of the thallus of this alga is poor. Similarly, whereas settlement on pieces of *Gigartina stellata* is moderately heavy, the aqueous extract is even less effective than the filmed surface alone. The reason for these discrepancies is not altogether clear, but as Williams (1965a) pointed out, may be due to other properties, such as the physical features of the algal surface. Thus, although the presence of a settlement-stimulating substance in *Ascophyllum nodosum* is obvious from Table VIII, the convex surfaces presented by the thallus of this alga might by highly repellent. On the other hand although *Gigartina* possesses no settlement-stimulating substance, the concavities and smooth texture of the thallus might in themselves be sufficiently attractive features to induce settlement.

The physical and chemical properties of the settlement-stimulating substances were then investigated using the larvae of *Alcyonidium polyoum* and *Flustrellidra hispida.* Even after boiling or storage in a deep-freeze for 12 months, the extract of *Fucus serratus* proved as effective in promoting settlement as the freshly-prepared extract. After evaporation to dryness, however, the previously active extract no longer stimulated settlement when redissolved in water and used to treat filmed panels. The molecular weight of the substance appears to be high, since when the extract was placed in a dialysis sac of pore diameter 24Å the activity was retained inside the sac after 24 hr dialysis against distilled water or against running tap water. The distilled water into which substances could have dialysed from the extract did not induce greater settlement than the control panels. Treatment with iso-butanol resulted in a considerable purification of the extract since photosynthetic pigments and associated carotenoids were soluble in the iso-butanol layer, whereas the water-soluble active substance remained in the aqueous phase. The active substance could be precipitated completely from solution by treatment with a saturated solution of ammonium sulphate; the redissolved precipitate then proved as effective in promoting settlement as the crude extract. These physical and chemical properties show that the active substance (or substances) which stimulates the settlement of epiphytic polyzoa on fucoid algae are large molecules which are stable at both high and low temperatures; they are insoluble in fat solvents, and form stable solutions in seawater and may be organic polymers (Williams, 1965a,b). The great majority of proteins would be excluded because of the stability of the substance to boiling, although some proteins such as the water-soluble arthropodin will withstand such treatment (Crisp and Meadows, 1962, 1963; p. 104). The fact that the substance remained in the aqueous phase after treatment with iso-butanol excludes a lipoid or complex aromatic material, whilst both the biuret method and ultra-violet absorption between 280–260mμ have indicated that protein is scarce in solutions of the active factor. On the basis of these properties, Williams (1965a,b) suggested that the active settlement-stimulating substance present in the fucoids is likely to be a polysaccharide or mucopolysaccharide.

Selection of surfaces bearing films treated with a fucoid extract appears to be brought about, as in barnacles, by random settlement followed by renewed searching if the substratum appears unfavourable. On contact with a suitable film, however, the mucus which the larvae secrete serves to form a temporary attachment and so prevent them being swept away. Williams (1965a) made the interesting observation that on the addition of fucoid extract to water containing larvae attached by mucus to the substratum, the mucus appeared to alter in its physical properties and fibres were observed to form. Such a change would prevent dispersal and account for the attachment of larvae to surfaces bearing the settlement-stimulating substance.

4. HYDROIDS

The responses noted above for barnacles, serpulids and polyzoans, are by no means confined to organisms showing a complex level of organisation. Schijfsma (1939), for example, showed that the larvae of *Hydractinia echinata* are capable of site-selection. Pyefinch and Downing (1949) placed a microscope slide bearing a colony of *Tubularia larynx* on a wooden raft bearing very few *Tubularia* whilst a control slide with no *Tubularia* was placed nearby. A much heavier settlement occurred on the panel bearing the *Tubularia* colony than on the control, although it is not clear from this result whether the colonisation was from the plankton or due to larvae from the colony settling very soon after liberation. Williams (1965c) has recently made a detailed study of the behaviour of planula larvae of the hydroid *Clava squamata,* which is restricted in its distribution to two species of alga, *Ascophyllum nodosum* and *Fucus vesiculosus.* He showed that the responses of the planulae would assist them to locate substrata suitable for colonisation.

Light appeared to be one of the principal factors affecting the release of the larvae from the adult, liberation occurring on exposure to light. For example, when four ripe colonies were maintained in the dark (from midnight until dawn, except for short periods for examination), only three planulae were discovered in the vessels; but during exposure to bright light for 30 min at dawn, 245 planulae were liberated within this time. A similar response occurred at other periods of the day, so that the response appears to be to light, rather than in accordance with a diurnal rhythm. The larvae too, tended to crawl towards light, moving the anterior end from one side to the other as in a klinotactic response (Fraenkel and Gunn, 1940). Younger larvae, in particular, crawled actively towards light, but older larvae crawled more slowly; indeed when the latter were more than 30 hr old, and were allowed to come into contact with the surface of *Ascophyllum,* they crawled away from light.

The larvae of *Clava squamata* do not actively swim, but instead crawl over the surface of the substratum or are dispersed by currents. Williams (1965c) showed that the physical properties of the surface markedly influence the rate at which such crawling occurs. He counted the rate of dispersal of larvae from a centre circle in a dish; the loss of animals from the central arena could not, however, be merely plotted against time, as this would indicate a decreased rate of dispersal with increase in time, due to the progressive loss of larvae available for dispersal at successive time intervals. He therefore plotted $\log_{10} \frac{N}{N-x}$ against time t, where N was the original number of larvae and x the number dispersed outside the circle at time t. The results obtained for glossy surfaces, smooth surfaces, rough surfaces and very rough surfaces, are shown in Fig. 3.15. The rate of dispersal is given by the slope of the line $d. \log_{10} \frac{N}{N-x}\, dt$. The rate of activity of planulae is obviously greatly affected by the texture of the surface of the substratum, the rate of movement being greater on glossy and smooth surfaces than on rough surfaces.

Williams (1965c) showed that the adult colonies of *Clava squamata* occur in greatest numbers in pits and grooves on the surface of *Ascophyllum* fronds, and that this distribution could be explained in terms of the behaviour of the planulae. Larvae were presented with a plate of "Tufnol" in which a large number of shallow pits of 2 – 5 mm radius had been drilled, such that the surface area of the pits approximately equalled that of the plane surface. The distribution of the larvae was examined at 15 min intervals for 3 hr; the results are shown in Fig. 3.16. This indicates that, as in barnacle cyprids (p.100), an increasing proportion of the larvae came to occupy the pits. This was not merely the result of the pits acting as mechanical traps, however, for the

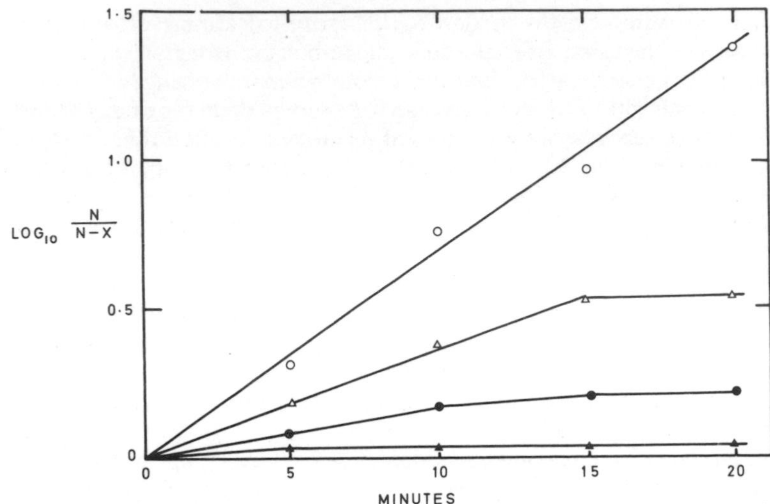

Fig. 3.15. Graph showing the rate of dispersal of planulae of *Clava squamata* from an area of 0·25 cm^2 over a distance of 1 cm on a glossy surface (open circles), a smooth surface (open triangles), a rough surface (solid circles) and a very rough surface (solid triangles). For explanation see text. (After Williams, 1965c.)

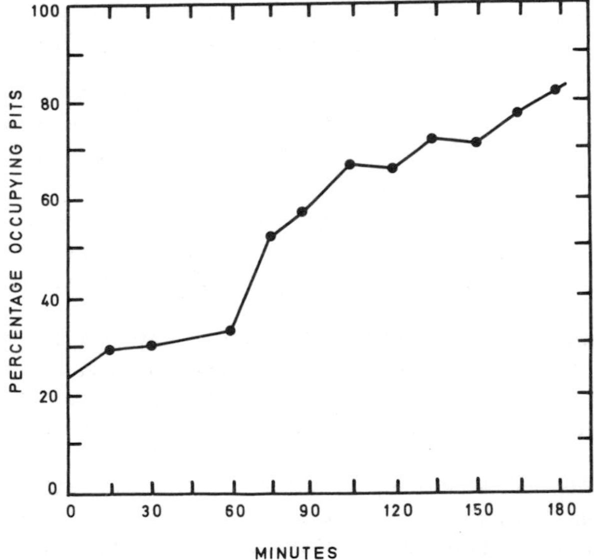

Fig. 3.16. Graph showing the percentage of the planulae of *Clava squamata* occupying pits of 2mm and 5mm radius when placed on a surface where the total area of pits was equal to the area of plane surface. (After Williams, 1965c.)

planulae were observed to leave the pits on occasions so that a thigmokinesis appears to be involved; the planulae also tended to crawl in the direction of water currents (positive rheotaxis).

Williams (1965c) has suggested that each of these responses may be of importance in site-selection by the planulae. The release of planulae in response to light (p. 121) would mean that under natural conditions the larvae would be liberated during daytime, and perhaps especially under conditions of increasing light intensity when uncovered by the tide. After release, the positive phototaxis of the larvae would assist movement up the littoral zone and facilitate contact with algal fronds. Having reached a suitable alga, the photonegative behaviour and the tendency to remain in irregularities on the surface of the thallus would enhance colonisation of regions protected from excessive desiccation. The positive rheotaxis may assist movements up the littoral zone for, since the release of planulae would occur primarily at low tide when the algae are illuminated by sunlight, movement with the currents of the flowing tide may help to prevent the loss of larvae in deeper waters where suitable algal substrata are absent.

5. SABELLARIA ALVEOLATA

A final example of the role of the settlement behaviour of larvae in the selection of suitable sites for adult life, is that of the colonial polychaete *Sabellaria alveolata.* Wilson (1968) made a detailed analysis of the behaviour of larvae reared from artificial fertilisations in the laboratory, and showed that the selection of substrata by them bore some remarkable similarities to the behaviour of barnacle cyprids (p. 96). The adult colonies are typically to be found on rocks near low water mark, where there is an abundance of calcareous shell sand suspended in the water by wave action. Wilson (1968) found that one of the factors responsible for inducing settlement was, indeed, the presence of turbulance. He showed that settlement in dishes containing sand and vigorously stirred seawater was greater than in dishes containing sand and unstirred seawater. There was, in addition, a tendency for older larvae, which had postponed settlement for some time, to settle at a faster rate than those larvae which had only just reached a stage at which they could settle and metamorphose (cf *Spirorbis* p. 110).

Although the swirling of the water evidently stimulated the larvae of *Sabellaria alveolata* to settle, especially when sand was in suspension, Wilson (1968) found that settlement was further enhanced when recently settled young were present. The rapid metamorphosis of larvae in stirred vessels was at least partly due to the influence of previously-settled individuals, much as has been found in serpulids and barnacles.

However, removal of the recently settled larvae as soon as they had metamorphosed, resulted in a slowing of the settlement in the stirred vessels; further experiments suggested that the larvae were responding, not to contact with the form of the tube of a metamorphosed individual, but to some property of the material comprising the tube. It was shown that the stimulus to metamorphose is received by the larva when it makes contact with complete adult tubes, or fragments of these, or primary mucoid tubes secreted by the larva at metamorphosis, or the first sandy tubes secreted by the metamorphosed larva. No metamorphosis occurred on contact with worms which had been removed from their tubes. The cementing substance which is used in tube formation and which adheres even to loose sand grains from fragmented tubes thus appears to be the most likely stimulating substance.

The chemical nature of this substance does not seem to be so specific as that demonstrated by Knight-Jones (1955) in barnacles, since sand from tubes of *Sabellaria spinulosa* was only a little less effective in promoting settlement than sand from tubes of *Sabellaria alveolata*. Although the chemical features of the substratum are thus of great importance in determining the suitability of the substratum for adult life, physical features other than turbulance, such as the stability of the tubes, are also important. The settlement-inducing properties of tubes or of sand derived from tubes were rendered ineffective if they were loose and moveable when the larva touched them. On the other hand, when such fragments were attached to the bottom of a glass vessel, they were rendered extremely attractive. This response is clearly of importance in allowing the settling larvae to distinguish between dead or disintegrating colonies, and those in a viable condition.

The nature of the settlement-stimulating substance contained in the cement and primary mucoid tubes of *Sabellaria alveolata* is rather different to that described earlier for other organisms. It is insoluble in water and its attractiveness is not destroyed by drying at low temperatures. On the other hand its attractiveness is destroyed by treatment with cold concentrated hydrochloric acid in contrast with the settlement-stimulating substance in barnacles (p. 98). The stimulus appears to be received by the larvae, however, by the same contact or "tactile chemical sense" described by Crisp and Meadows (1962, 1963) for barnacle cyprids. Bayne (1969) has recently demonstrated a very similar phenomenon in the settlement behaviour of larvae of the oyster *Ostrea edulis*. He found that the gregarious response of the larvae to the presence of previously-settled individuals could be simulated by the presence of surfaces which had been treated with a tissue extract of adult oysters. The extract was found to be species-specific and the active substance was of large molecular weight. It could be precipitated with ammonium sulphate and cold trichloracetic acid and appeared to be detected only after the larvae had alighted on the treated surface. Indeed, the presence of a tactile chemical sense appears to be a common factor in nearly all examples of chemically-induced metamorphosis, despite the variety of physical and chemical properties of the substances and of the types of animals concerned.

D. THE AVOIDANCE OF CROWDING

Although it is now apparent that gregariousness is of widespread occurrence, it is obvious that such behaviour might lead to intense intraspecific competition not only for space in which to grow, but also for food. Thus, whilst the presence of an adult of the same species is a good index of the suitability of the environment, other factors must operate to prevent excessive crowding. In the colonial *Sabellaria alveolata* no such spacing mechanism need occur for, as Wilson (1968) pointed out, the more massive the colony the greater its surface area and ability to collect food; spacing behaviour leading to the avoidance of crowding would, however, be expected in barnacles and serpulids. This aspect of settlement behaviour has been studied in *Spirorbis borealis* by Wisely (1960; also Knight-Jones and Moyse, 1961), and in barnacle cyprids by Knight-Jones and Moyse (1961) and Crisp (1961).

Wisely (1960) showed that the distribution of *Spirorbis borealis* on the fronds of *Fucus serratus* was remarkably even, and that despite the fact that the larvae are less than 0·5 mm long, and only about 0·25 mm broad, settlement did not occur closer than 0·5 mm. In fact, in their natural environment the distance between the centres of the adults was mainly about 1 mm even under crowded conditions whilst this distance in less crowded populations was mainly about 2 mm. (Fig. 3.17). Thus, although there is abundant room for the settlement of further larvae, such settlement does not occur under natural conditions, and in this way room is left for the settled individuals to grow to their full size of some 2 mm diameter. Although the densities which occur under natural conditions are only from 12–30/cm^2, artificially high settlement densities can be induced under laboratory conditions by delaying settlement and then presenting only small fragments of *Fucus*. Even under these conditions many of the larvae fail to settle; to produce a settlement density of 322/cm^2, for example, sufficient larvae had to be used to give a density of 500/cm^2 if all had settled. If, on the other hand, larvae are given the choice between bare *Fucus* and a piece of *Fucus* upon which small *Spirorbis* of 1 mm diameter, or even small sand grains, or pieces of shell, had been placed to a density of 10/cm^2, settlement occurs on the bare frond. Any raised object, including pieces of molluscan shells, precluded settlement; the spacing response in

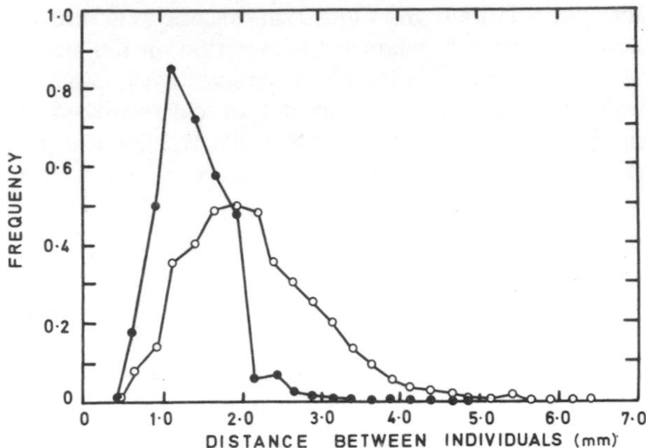

Fig. 3.17. Graphs showing the frequencies of occurrence in relation to the distance apart of adjacent *Spirorbis borealis* arranged in a linear series in the grooves on each side of the midrib of fronds of *Fucus serratus*. The open circles are for a sparse population and the solid circles for a crowded population. (After Wisely, 1960.)

Spirorbis borealis thus appears to depend, at least in part, upon mechanical contact with other adults.

Wisely (1960) then suggested an explanation for the spacing effects, and for the fact that crowding can be induced in the absence of suitable substrata, in terms of the behaviour of the larva. When a larva makes contact with a suitable substratum it makes exploratory movements, the rate of turning increasing immediately prior to settlement. It is during such a phase that contact would be made with previously settled individuals, and this would presumably inhibit settlement in the immediate vicinity. If a suitable substratum is witheld for some time, however, the larva makes only brief exploratory movements and changes direction frequently before settling. Since extensive excursions are not made, settlement is likely to be possible in a much smaller space than where the exploratory phase is more prolonged. Finally, if the larvae are kept swimming for as long as 8 hr before presentation of *Fucus* they settle almost where they alight, and the exploratory movements are restricted to swinging motions of the abdomen. In this case, settlement could occur wherever there was sufficient space for the larva. Normally, therefore, the settlement behaviour allows sufficient space between individuals to allow them to reach their maximum size but in the absence of suitable substrata, population densities of more than ten times the normal value may be attained by a modification of the exploratory phase of behaviour.

Crisp (1961) has shown that a similar mechanism exists in the settlement behaviour of cyprids of *Balanus balanoides.* He compared the distribution of distances between neighbours of a linear series in a random settlement of 3·05 individuals/cm, with the actual frequencies of distribution observed in a population of *Balanus balanoides* settled at a mean density of 3·05/cm along shallow grooves. The results are shown in Fig. 3.18; it is apparent that the settlement is quite different from

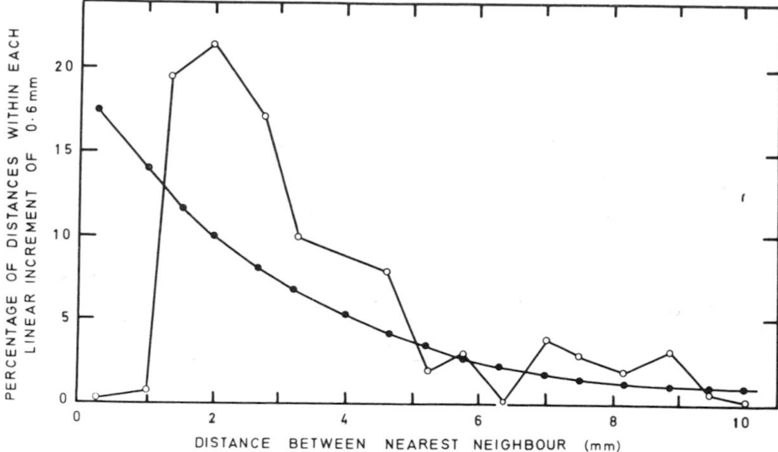

Fig. 3.18. Graphs showing the percentage distribution of distances between nearest neighbours in a linear series. Open circles show the observed distribution in a population of *Balanus balanoides* settled in narrow grooves. Solid circles indicate the distribution in a random series at the same population density. (After Crisp, 1961.)

the random distribution, insofar as there is a deficiency of spacings at less than 1·5 mm and an excess in the range of 1·5 to 5 mm. This is very similar to the results obtained for the spacings between linear series of *Spirorbis borealis* along grooves in the thallus of *Fucus serratus* (Wisely, 1960). In much the same way, Crisp (1961) showed that the distribution of barnacles on a planar surface is not random. In Fig. 3.19 the distribution of distances between the nearest neighbours on a planar surface for a randomly-distributed population at a mean density of 10·4 individuals/cm^2 is compared with the distances between neighbours in an actual *Balanus crenatus* population of the same density. It is clear that very few neighbours occurred within 1 mm of the reference individual, but that between 1·5 and 3 mm the frequency of spacings is very high. Later settling individuals both on planar surfaces as well as in grooves tend to maintain a separation of approximately 2 mm from the nearest individual.

Now it is known that extremely high densities of barnacles may

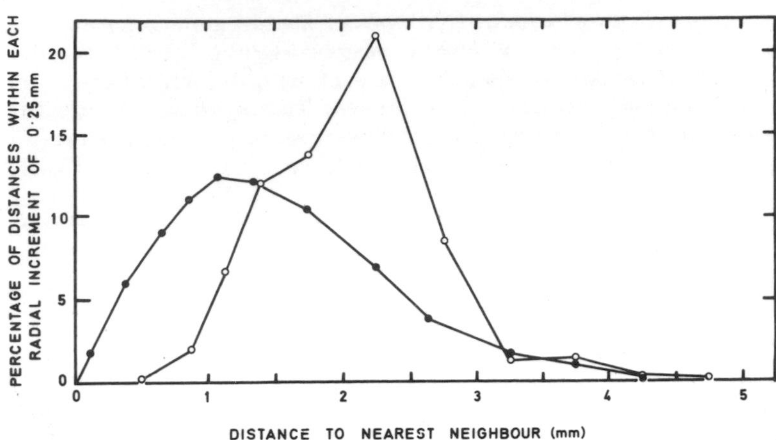

Fig. 3.19. Graphs showing the percentage distribution of distances between nearest neighbours in a two-dimensional array. Open circles show the observed distribution of the members of a population of young *Balanus crenatus*. Solid circles show the distribution in a random array at the same population density. (After Crisp, 1961.)

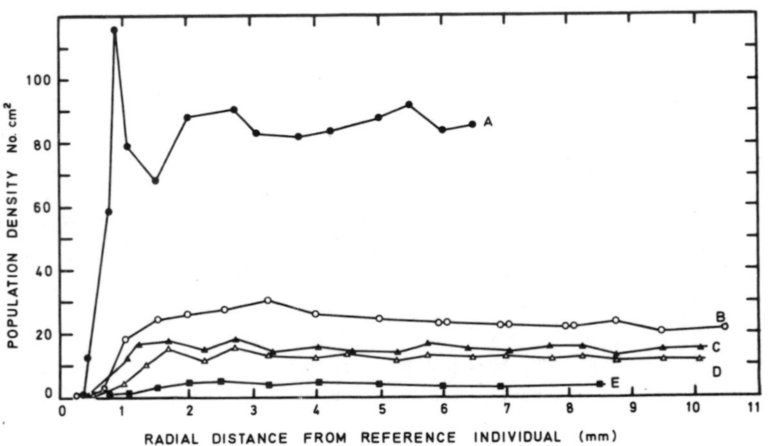

Fig. 3.20. Graphs showing the population density of *Elminius modestus* as a function of radial distance (mm) from a reference individual at five mean population densities of (A) 85 individuals/cm^2, (B) 22·5 individuals/cm^2, (C) 15·0 individuals/cm^2, (D) 12·5 individuals/cm^2, (E) 3·5 individuals/cm^2. (After Crisp, 1961.)

occur under some circumstances. Crisp (1961) studied the influence of high population densities on the settlement pattern of *Elminius modestus;* he found that even at high population densities of 85/cm^2, the territory within a radius of 1 mm of the reference individual tended to be avoided but that outside this zone the population density rose steeply (especially under very crowded conditions). Fig. 3.20 shows the population density of *Elminius modestus* plotted against the distance from a reference individual for mean densities ranging from 85 individuals/cm^2 to 3·5/cm^2. Crisp (1961) termed the distance between the reference individual and the point at which the curve cuts the 50% value for the population density, as the 'territorial separation'. It is evident that at low population densities of 3·5/cm^2 the territorial separation was 1·9 mm; at 12·5/cm^2 was 1·42 mm; at 15·0/cm^2 was 1·25 mm; at 22·5/cm^2 was 1·25 mm and at 85/cm^2 was only 0·86 mm. Thus the territorial separation becomes reduced as the population density increases. This behaviour appears to be the result of physical contact between the cyprids and neighbouring settled barnacles, the larger cyprids of *Balanus balanoides* and *B. crenatus* having a greater territorial separation than the small *Elminius* cyprids. Furthermore, settlement around the large spat of *B. crenatus* of mean diameter 1·7 mm had a territorial separation which exceeded that around cyprids and recently metamorphosed individuals of 0·44 mm diameter. It is significant that the magnitude of this difference in territorial separation is equal to the difference in the radius of the settled individuals. Thus, physical contact, as in *Spirorbis* (Wisely, 1960), determines the settlement site. In fact the value of the territorial separation turns out to be only a little greater than the sum of the length of the cyprid plus the radius of the previously settled individual; it is smaller in species with smaller cyprids and spat. As in *Spirorbis,* the repeated turning at the end of the exploratory phase and immediately prior to settlement (Knight-Jones and Crisp, 1953) would ensure that the cyprid kept at least its own length from previously settled individuals.

As Crisp (1961) pointed out, it is probable that contact with an individual of the same species stimulates further walking in cyprids. Thus early in the searching phase, contact with an adult would eliminate swimming and initiate crawling so that the cyprid would settle in the vicinity of the adult; towards the end of the settlement phase movement is slower, and contact with an adult of the same species at this stage would cause further searching. Because of the slower progress, settlement would occur nearby as long as no further contact was made with a previously settled individual. Since barnacles may reach over 1 cm diameter in one season, Crisp (1961) has suggested that the spacing behaviour is insufficient to allow space for optimal growth. Instead it is possible that the spacing allows time for the newly-settled spat to develop strong plates, after which tall columnar individuals may form when all the growing spat have covered the existing substratum (Barnes and Powell, 1950b; also Fig. 1.24). If spacing was sufficient to allow the unimpeded growth of all the spat, the spaces might become colonised with other competing organisms, with the consequent danger to the barnacles of being crushed or overgrown. The behaviour adopted thus represents a compromise between the space requirements of the young individuals and the prevention of colonisation by other organisms.

Knight-Jones and Moyse (1961) have given some interesting examples of how the specific recognition inherent in the gregarious tendencies of barnacle cyprids has become modified for spacing behaviour. They suggested that the spacing tendency in cyprids of *Balanus balanoides* is due partly to the chemical recognition response and partly to physical contact, as has been discussed above. Fig. 3.21

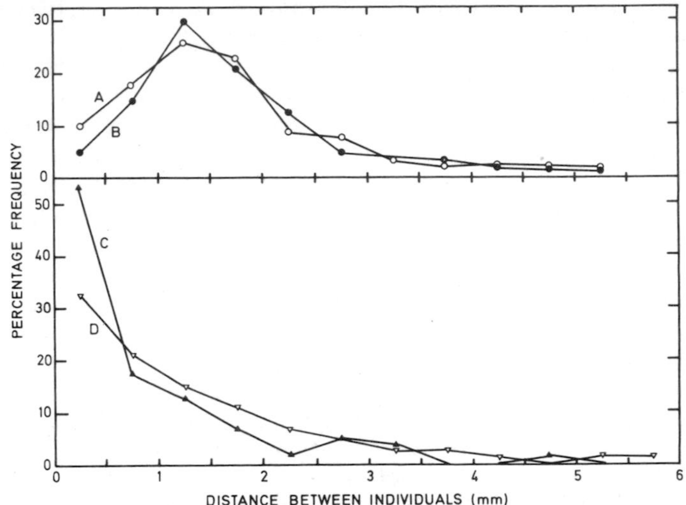

Fig. 3.21. Graphs showing the percentage frequency of distances between the cyprids of *Balanus balanoides* and the nearest previously settled barnacle. The nearest barnacle was (A) adult *Balanus balanoides,* (B) spat of *B. balanoides,* (C) adult *Chthamalus stellatus* and (D) adult *Elminius modestus.* Note that the cyprids spaced themselves out from their own species (A and B) but not from other species. (After Knight-Jones and Moyse, 1961.)

shows the percentage frequency of distances between cyprids of *Balanus balanoides* just settled on limestone and the nearest settled barnacle. It will be noticed that the cyprids spaced themselves out from adults or spat of their own species but that a high proportion settled close to, or even in contact with, other barnacles such as adult *Chthamalus stellatus* and *Elminius modestus.* As Knight-Jones and

Moyse (1961) pointed out, settlement in contact with *Elminius modestus* (Balanidae) was rather less than that in contact with *Chthamalus stellatus* (Chthamalidae), and this may be because the mechanism for species recognition is not perfect, so that a degree of spacing-out occurs in response to other genera of the same family (Balanidae), but not to other families (also Knight-Jones, 1955 and Table VII). Knight-Jones (1953b) had earlier described the settlement of barnacles in contact with the bases of detached adults of the same species (p. 97), and it follows that the spacing does not occur in response to them. Knight-Jones and Moyse (1961) therefore suggested that spacing behaviour as well as gregariousness depends firstly on contact with the cuticle of their own species. Secondly, for spacing to occur this stimulus must be coupled with the presence of a surface protuberance (cf Crisp, 1961; and p. 129).

The tendency of barnacle cyprids to space out from individuals of their own species, but not to other species, may account for the observed sharp boundaries between the barnacle zones (Chapter 1). As Knight-Jones and Moyse (1961) pointed out, if a barnacle species spaced out to all other species of barnacle, it would soon be pushed into extinction. The tendency to settle and space-out in zones where there are adults of the same species, and to settle on alien survivors in that zone, tends not only to make the barnacle zones distinct but to make interspecific competition between overlapping species of barnacles more severe than intraspecific competition. Such exclusion of one species by another is well-illustrated by the intertidal distribution of *Balanus balanoides* and *Chthamalus stellatus.* In the south of the British Isles, *Chthamalus stellatus* extends down even towards the lower parts of the intertidal zone, but towards the northern parts of its distribution occurs only at the highest tidal levels, being crowded from below by *Balanus balanoides* (Southward and Crisp, 1956). Interspecific competition for space is thus probably intense in overlapping species and illustrates the importance of biological factors as well as physical ones in controlling the zonational level of a particular species on the shore.

CHAPTER 4

The maintenance of zonation patterns

A. Introduction
B. Behaviour patterns in certain intertidal animals
   1. The behaviour of *Convoluta roscoffensis*
   2. The behaviour of chitons
   3. The behaviour of bivalves of muddy shores
   4. The behaviour of littorinids
      (a) *Malaraphe* (=*Littorina*) *neritoides*
      (b) *Littorina littorea*
      (c) Other littorinids
   5. The behaviour of *Lasaea rubra*
   6. The behaviour of *Hydrobia ulvae*
      (a) Responses to light
      (b) Responses to gravity
      (c) Interpretation of shore behaviour in terms of the responses to light and gravity
   7. The behaviour of the sand-hopper *Talitrus saltator*
      (a) Seaward migration
      (b) Landward migration
C. The role of directional environmental stimuli in the behaviour of intertidal animals
D. Learning in intertidal animals
   1. Learning in nereids
      (a) Habituation in *Nereis*
      (b) The effects of punishment on reflex behaviour patterns
      (c) The results of "reward-punishment" experiments
      (d) "Sensitisation' and non-associative learning in nereids
      (e) The biological role of sensitisation in nereids
   2. Learning in crustaceans
      (a) Learning of releasing components
      (b) Learning of orientational components
E. Conclusion

A. INTRODUCTION

We have seen that the distribution of animals on the shore is determined primarily by the ability of larvae or juveniles to select an environment which will be suitable for adult life. In sessile animals, such as barnacles and serpulids, the choice is an irreversible one; in others such as the mussel *Mytilus edulis,* settlement of larvae in areas of general suitability, followed by more specific site selection by the juveniles, occurs. The larvae of this animal initially tend to settle in shallow water where a high light intensity occurs (Verwey, 1952, 1954, 1957), but later, after a period of growth, detach themselves and are carried by inshore currents to other areas where resettlement takes place (de Blok and Geelen, 1958). Repeated floating and resettlement may take place, until finally the young mussels aggregate in areas where adult communities occur. Another example of site selection by the juvenile rather than the larva is illustrated by the winkle *Littorina littorea.* Smith and Newell (1955) have shown that the young littorinids at Whitstable, Kent, U.K., settle from the plankton onto the sea floor below the low water mark and are subsequently transported shorewards over the mud flats by wave action until they reach the lower margin of the stony beach. The young winkles then crawl shorewards and achieve their adult zonation only at the end of their first year.

Graham and Fretter (1947) have shown that migration of post-metamorphic juveniles also occurs in the blue-rayed limpet *Patina pellucida.* This animal occurs in two varieties which are found on different parts of the alga *Laminaria.* One variety, *pellucida,* lives on the fronds of *Laminaria,* while the second variety, *laevis,* lives in caves in the holdfasts. The animals breed maximally in winter and spring, the planktonic larvae settling during May as spat approximately 2 mm in length. A proportion of those which had settled on the fronds were shown to actively migrate during their first summer onto the holdfasts and then become indistinguishable from the *laevis* form. Thus the life-history of the variety *pellucida* involves a downward migration after spatfall; this migration prevents the limpets being cast off when the frond of *Laminaria digitata* disintegrates during the autumn.

These migrations may be regarded as extensions of the site-seeking phase in the life-history of the animal and finally result in the establishment of the adult zonational pattern. Once this is attained, however, such site-seeking behaviour is eliminated and is replaced in many intertidal animals by a series of behavioural responses appropriate to maintaining the adults in their optimal zone. The following sections deal with some of the ways in which a variety of adult intertidal animals remain in their optimal zone on the shore, despite occasional displacement and regular extensive feeding excursions.

B. BEHAVIOUR PATTERNS IN CERTAIN INTERTIDAL ANIMALS

1. THE BEHAVIOUR OF *CONVOLUTA ROSCOFFENSIS*

The behaviour of the intertidal turbellarian *Convoluta roscoffensis* has been known for a long time (Gamble and Keeble, 1904; Fraenkel, 1929), and represents a very simple form of tidal rhythm. The animals live in vast numbers in coarse sands and are green in colour owing to the presence of zoochlorellae in the tissues. When exposed by the tide, *Convoluta* migrates to the surface of the sand and the zoochlorellae are able to photosynthesise; but immediately before the incoming tide reaches them, the flatworms burrow and so avoid being dispersed by the waves. This behaviour is very similar to that described by Aleem (1949) for certain naviculoid diatoms; it has been shown in *Convoluta* to comprise a positive geotaxis when turbulance disturbs the worms, coupled with a negative geotaxis when the sand is undisturbed by wave action (Fraenkel, 1929). It has been suggested, however, that there is an inherent tidal rhythm in these animals, since Gamble and Keeble (1904) have shown the presence of a tidal rhythm of vertical migration in *Convoluta* which persists for up to one week in the laboratory in the absence of mechanical disturbance. It would be difficult to envisage an "internal clock" which made allowance for the progressive alteration of the time of ebb and flow of the tides, yet one which was not compensated in this way would be of little value to an intertidal animal. It seems likely, therefore, that external triggering stimuli such as mechanical disturbance are the immediate, if not the only factors controlling downward migration in *Convoluta.*

2. THE BEHAVIOUR OF CHITONS

A somewhat more complex pattern of behaviour is that described for the intertidal chiton *Lepidochitona cinereus* by Evans (1951). She has shown that this animal cannot tolerate the conditions of desiccation which occur on the top of the stones in the intertidal zone and that chitons normally live under stones when the tide has ebbed. When covered by the tide, the chitons browse on material adhering to the upper surfaces of the stones and migrate back to the moist microhabitat under the stones again when the tide recedes. This behaviour was shown to result from the combined responses of chitons to the stimuli of gravity and light.

Evans (1951) found that when chitons were placed on an inclined moist glass plate or a roughened perspex plate, the animals were positively geotactic (i.e. they crawled downwards); when the plate was immersed in seawater, however, this response was abolished. The chitons also tended to aggregate in regions of low light intensity because of the slower rate of crawling of the animals in low light intensities than in high ones. The mean distance moved in 10 min by 15 chitons in sunlight was 8·17 cm (± 0·64); under artificial light of approximately 7 lumens/cm^2/sec, 5·10 cm (± 0·43); and in the dark 1·38 cm (± 0·41). This response, in which the speed of crawling is directly related to the intensity of the stimulus (light), is termed an (photo-) ortho-kinesis (Ullyott, 1936a,b: Fraenkel and Gunn, 1940).

Clearly, the observed behaviour of *Lepidochitona cinereus* could then be explained in terms of its responses to the stimulus of light and of gravity on the shore. Evans (1951) suggested that as the tide recedes the stones with the chitons on them become exposed and the chitons, then out of water, react to gravity by moving downwards; since the surface of the stone is moist and the light intensity high the chitons travel at the maximum rate. This behaviour leads them to regions of low light intensity below the stone where, responding orthokinetically to the diminished light, they slow down and become aggregated. Thus a high light intensity has become a token stimulus for unfavourable conditions, and a low one indicates the moist conditions required by the animal. When the tide returns and covers the stones, the positive response to gravity is abolished and by random browsing movements the chitons become distributed over the surface of the stones again.

We have seen how there is a simple vertical migration in *Convoluta roscoffensis* in response to gravity at different stages of the tidal cycle, and that this simple response to gravity is enhanced in *Lepidochitona* by the imposition of an orthokinesis so that movement away from the adverse conditions is made faster. We now come to animals which are capable of making directed responses to light, and also in some cases to gravity.

3. THE BEHAVIOUR OF BIVALVES OF MUDDY SHORES

Macoma balthica is a small bivalve which is abundant on intertidal mudflats and is normally found at about mid-tide level (Fraser, 1932; Beanland, 1940; Spooner and Moore, 1940; Brady, 1943; Holme, 1949; Smidt, 1951; Newell, 1965). The animal lies just below the surface of the deposits in which it lives and is able to feed either as a suspension-feeder or, on occasions, at low water as a deposit-feeder (Brafield and Newell, 1961). It has also been shown by Brafield and Newell (1961) that *Macoma* rapidly clears the area around it of superficial material when deposit-feeding; they argue that it would be of advantage for the animal to move in order to tap fresh food resources. This movement was indeed shown to occur not only in *Macoma balthica,* but also in the

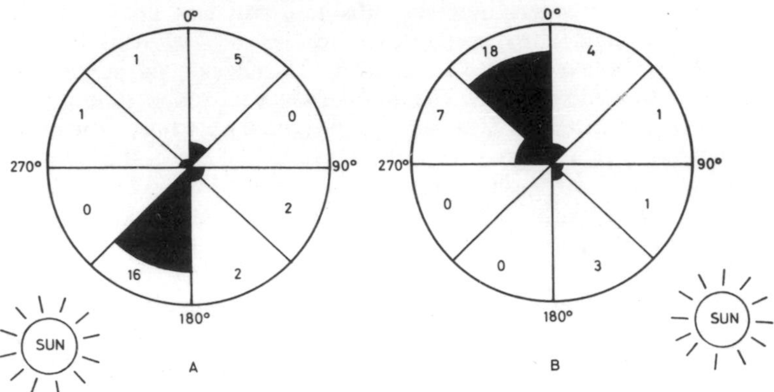

Fig. 4.1. Polar diagrams compiled by grouping the directions of straight furrows made by *Macoma balthica* into octants. Figures indicate the numbers of tracks lying in each octant. (A) The *Macoma* zone had been exposed for one hour only. (B) Orientation of tracks after 5 hr exposure. (Modified after Brafield and Newell, 1961.)

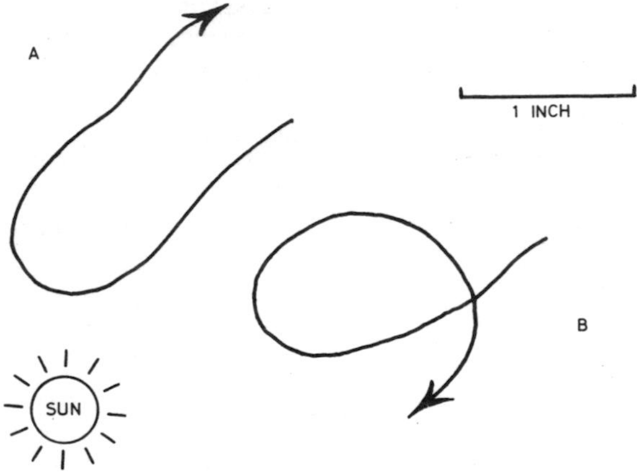

Fig. 4.2. Diagram showing a typical U-shaped track of *Macoma balthica* (A) and a closed looped track (B). (Redrawn after Brafield and Newell, 1961.)

cockle *Cardium edule* and in *Venus striatula,* all of which leave characteristic tracks in the deposits.

Shore observations of the tracks left by *Macoma* showed that when the animals had been exposed by the tide for one hour only, the vast majority of the animals were photopositive (Fig. 4.1A) but later, after five hours exposure, the majority were photonegative (Fig. 4.1B). Thus in *Macoma,* and also in other bivalves, the majority of the tracks were found to be U-shaped, the first-formed limb recording a movement towards the sun (Fig. 4.2). In this way the animals were able to undertake feeding excursions without becoming dispersed from their optimal zone on the shore. This behaviour pattern occurs in many other intertidal organisms which may also show additional types of behaviour appropriate to the differing environmental conditions.

4. THE BEHAVIOUR OF LITTORINIDS

Many littorinids such as *Melaraphe* (= *Littorina*) *neritoides* are found exclusively on rocky shores, whereas others such as *L. littoralis* are restricted to areas where fucoids occur. Finally there are those such as *L. saxatilis* and *L. littorea* which are widespread around the coasts of the British Isles and may occur on both muddy and rocky shores. Bearing in mind the widely different habitats in which littorinids occur, it is not surprising to find that the behavioural responses differ from species to species and also within any one species, according to the habitat from which the experimental animals were taken (p. 139); such variation in habitat and behaviour may in part explain the apparent discrepancies in the literature on the behaviour of littorinids (also Newell, 1958a,b).

(a) Melaraphe (= Littorina) neritoides

This littorinid lives high in the splash zone on rocky shores and inhabits crevices as well as dead barnacle shells. Fraenkel (1927) has shown that the characteristic zonational position of these animals can, like that of *Lepidochitona cinereus* (p. 134), be interpreted in terms of the responses to light and gravity.

When immersed in seawater, the animals are negatively geotactic and crawl towards the axis of rotation when placed in a horizontal rotating dish. They are also photonegative when immersed and the right way up; but this response changes when immersed upside-down. Under these conditions the animal becomes photopositive although this response does not occur when the animal is moist but not immersed.

We may now describe a hypothetical situation based upon the responses of *M. neritoides* to light and gravity. Suppose an animal is displaced by wave action so that it is immersed in water some way down in the intertidal zone. Being negatively geotactic, it would crawl upwards and in so doing it might encounter an immersed crevice. Being negatively phototactic it would crawl into the crevice and up the walls and then, being upside-down and still immersed, would become positively phototactic and crawl out of the crevice and up the rock face again. This would continue until the animal encountered a crevice which was not immersed in water. The photopositive response would therefore not occur when the animal was upside-down and the littorinid would remain in the depths of the crevice above the water (Fig. 4.3).

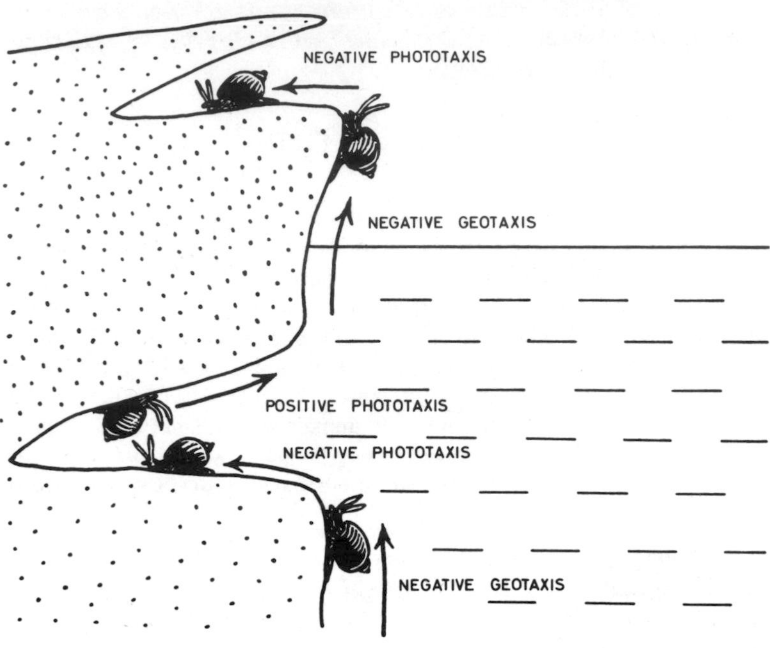

Fig. 4.3. Diagram illustrating the role of behavioural responses of *Melaraphe* (=*Littorina*) *neritoides* in the maintenance of its zonational position on the shore.

It is obvious that limited feeding excursions must occur from time to time, and it seems likely that the situation is rather more complex than that described by Fraenkel (1927); the behaviour pattern may, for example, alter according to the physiological state of the animal. Nevertheless his work is of great interest in demonstrating how the zonation pattern could be attained by means of relatively simple responses to light and gravity. Certain phases of this behaviour pattern have been recently confirmed by Charles (1961), although Evans (1965) has suggested that the behaviour of *M. neritoides* in a tidal tank cannot adequately be explained in terms of simple taxes (also p. 141).

(b) Littorina littorea

As mentioned above, *Littorina littorea* occurs in a wide variety of habitats and its behaviour may vary considerably according to locality. Newell (1958a,b) has shown that specimens of *L. littorea* from horizontal surfaces at Whitstable, Kent, orientate by means of a light-compass reaction, and can be made to reverse the direction of their crawling by shielding the eye facing the sun and then shining an image of the sun from the opposite side by means of a plane mirror. The winkles are normally either photopositive or photonegative (Fig. 4.4),

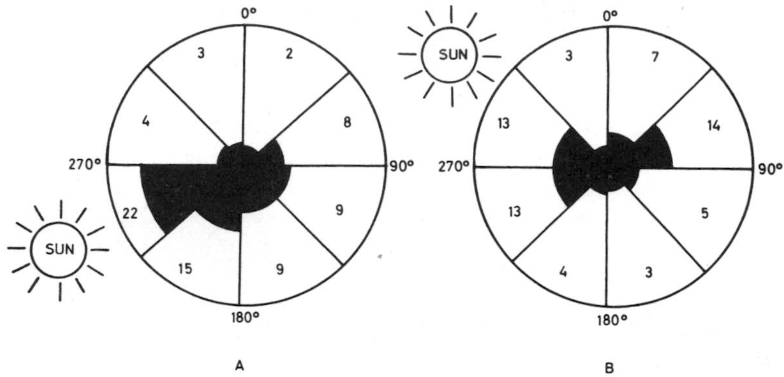

Fig. 4.4. Polar diagrams showing the direction and numbers of tracks of *Littorina littorea.* (A) 1 hr after the tide had receded. (B) At the time of low water. (After Newell, 1958b.)

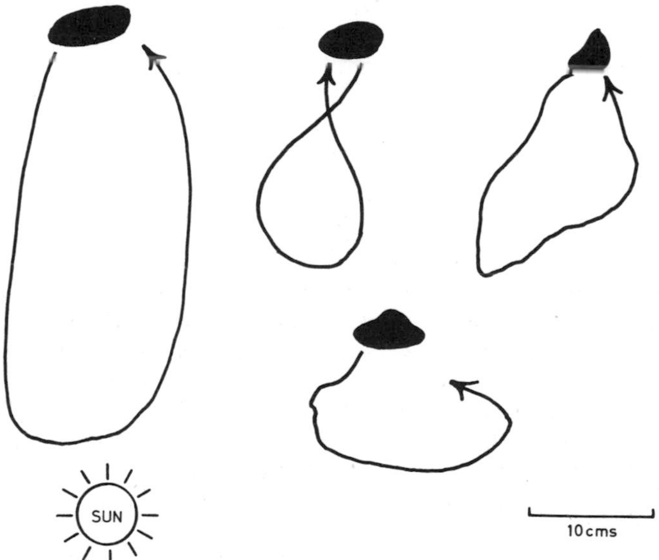

Fig. 4.5 Diagram illustrating a selection of tracks made by winkles crawling on sand in bright sunlight. (Redrawn after Newell, 1958a.)

and under natural conditions each animal reverses the direction of crawling so that the track is U-shaped as in the bivalves described earlier (p. 136)(Fig. 4.5).

Laboratory analysis showed that the animals collected from horizontal surfaces were always photopositive at first and later after 15–20 min, became photonegative; in this respect the behaviour is very similar to that of the bivalve *Macoma balthica* (Brafield and Newell, 1961). There are, however, important differences between these and animals collected from vertical surfaces; winkles from vertical surfaces tend to be at first photonegative and also positively geotactic. That is, when stimulated to make feeding excursions by the ebbing tide, the winkles on vertical surfaces crawl downwards, and only later become positively phototactic and negatively geotactic so that they return to their normal zonation position; the responses to light and gravity thus normally reinforce one another. The situation is further complicated by the fact that after a period which varies from a few hours to 10 days in the laboratory, winkles from horizontal surfaces climb up the walls of an aquarium and thereafter behave as animals from naturally-ocurring vertical surfaces. Equally animals from vertical surfaces may adapt to life on horizontal surfaces so that many gradations between the use of purely a light-compass reaction with reversal of response, and a geotaxis with reversal of response may be expected to occur in specimens collected from shores where a great variety of slopes are colonised.

(c) Other littorinids

The reactions of *Littorina littoralis* to light and gravity have been studied by Barkman (1955). He found that this species is always negatively phototactic and negatively geotactic. When placed in a tank with a tidal rise and fall of some 33 cm the animals aggregated at about 6 cm above the waterline. Barkman (1955) suggested that the winkles were negatively geotactic until halted by the effects of desiccation; he did not notice any reversal of response such as have been described for *Littorina littorea.* van Dongen (1956) showed that at least dim light is necessary for *L. littoralis* to respond to fucoids and Evans (1965) has to some extent confirmed this by showing that *L. littoralis* will not zone in darkness, even in the presence of fucoids. In view of the possibility of form-vision in other littorinids (Evans, 1961, 1965; Newell, 1965), it would be of great interest to determine the precise effect of light intensity in controlling zonation in *Littorina littoralis.*

Evans (1961) has shown that the littorinid *L. punctata* on the Ghana coast orientates not only to sun and gravity as described by Newell (1958a,b) for *Littorina littorea,* but also to the silhouette of the shore line. This important observation implies the existence of form-vision in littorinid eyes, a suggestion which is supported by a study of the optics of the eye of *L. littorea* by Newell (1965). Evans (1965) also suggested that certain other littorinids, notably *L. saxatilis* and *M. neritoides*, may use a degree of form vision in their orientation. He reported that when placed in a lighted tidal tank these species adopted a zonational position similar to that which they occupy on the shore, irrespective of the direction of the incident light. No such zonation occurred in the dark, even though the animals crawled about leaving randomly orientated trails. Further, if an upward slope such as is presented by a boulder did not extend up to high water, the animals were capable of altering their direction to descend the slope a little before the crest was reached, and then striking a new path to high water. Evans pointed out that although taxes may play a part in determining zonational position in these species, the behaviour described above would imply the presence of a negative geotaxis which reverses as a barrier is reached or even approached. This he regarded as highly improbable, and suggested that it is more satisfactory to suppose that the water surface is seen from below as a goal and involves image formation by the eye of these species (also p. 165). This suggestion would then account for the lack of response in the dark and the unimportance of light direction to the establishment of the characteristic zonation pattern in a tidal tank.

There is now a considerable body of evidence supporting the view that many species of littorinids are able to orientate to polarised light (Burdon-Jones and Charles, 1958a,b; Charles, 1961). Charles (1961) has summarised the evidence for orientation to polarised light in the four British littorinid species *Littorina littorea, L. littoralis, L. saxatilis,* and *M. neritoides.* He showed that all four species responded when crawling in air or in seawater to plane polarised light incident from above. Photonegative winkles crawl parallel to the plane of vibration and photopositive ones at right angles to the plane of vibration. The importance of these results, together with some general aspects of orientation to polarised light, is discussed in more detail on p. 153.

5. THE BEHAVIOUR OF *LASAEA RUBRA*

Lasaea rubra is a small bivalve which characteristically lives attached by temporary byssus threads in crevices, empty barnacle shells and in tufts of the lichen *Lichina pygmaea* (Morton, Boney and Corner, 1957). Morton (1960) has recently described the behaviour of this bivalve in some detail; he showed that on a flat surface in the dark or in uniform light, *Lasaea rubra* crawls freely with meandering trails; in directional light, even if this is of low intensity, darker crevices are found fairly directly. Orientation at higher levels of illumination is brought about by means of a klino-taxis in which the animal successively compares the light intensity on each side of its path. Morton (1960) suggested that in

Lasaea rubra the tissues of the foot are photosensitive. The foot is protruded successively on each side of the opaque shell during locomotion and in this way animals might be able to orientate down a light gradient and into a suitable crevice.

This negative phototaxis is superceded by a negative geotaxis which apparently dominates the responses to light to such an extent that the bivalve will crawl upwards even against a strong light gradient. As soon as contact is made with a small hole or crevice, or any situation affording lateral contact, the reactions to both light and gravity are superceded by positive thigmotaxis, or response to contact. Thus Morton suggested that in the natural habitat there is a hierarchy of responses, negative geotaxis providing a coarse adjustment to the securing and maintenance of position. The reaction to light (photo-klino-taxis) and lateral contact (thigmotaxis) then give an increasing precision in securing shelter from wave action and desiccation.

Each of these responses is a simple one to the directional environmental stimulus of light or gravity. The main difference from some of the species considered earlier is that in *Lasaea rubra* they have been combined into a definite hierarchy of importance. Yet such a complex behavioural pattern is by no means unusual in intertidal animals; both *Littorina littorea* and *M. neritoides,* for example, show the ability to use both light and gravity as directional stimuli (so too does the chiton *Lepidochitona cinereus*) Perhaps one of the more striking behaviour patterns using a combination of responses to light and gravity is that shown by the small estuarine prosobranch *Hydrobia ulvae.*

6. THE BEHAVIOUR OF *HYDROBIA ULVAE*

Hydrobia ulvae is a small prosobranch whose adult size rarely exceeds 4 mm and which occurs in vast numbers in estuarine muds. Thamdrup (1935) has recorded as many as 60,000/m^2 at Skalling, Denmark, but rather smaller numbers are commonly found in the British Isles; Holme (1949) observed 4,000/m^2 at the mouth of the Exe estuary, and Nicol (1935) recorded 8,000/m^2 on bare estuarine mud at Aberlady, Firth of Forth.

Newell (1960, 1962) has described the behavioural cycle of *Hydrobia ulvae* at Whitstable where the population does not exceed 10,000/m^2. When the tide was out and the mud-flats exposed, the animals crawled about leaving looped tracks much as described earlier for *Littorina littorea* (p. 139) and *Macoma balthica* (p. 135). After a time, the animals burrowed and remained beneath the surface of the substratum until just before the incoming tide reached them. Then they resurfaced and, crawling up the ripple-marks, launched themselves upside down afloat on the surface film of the water. The animals were then swept shorewards by the incoming tide and were later redeposited on the mud by the ebbing tide; they then began browsing again and the cycle of behaviour was completed. This sequence is illustrated in Fig. 4.6. Laboratory analysis of this rather complex behaviour cycle may be subdivided into (a) responses to light and (b) responses to gravity.

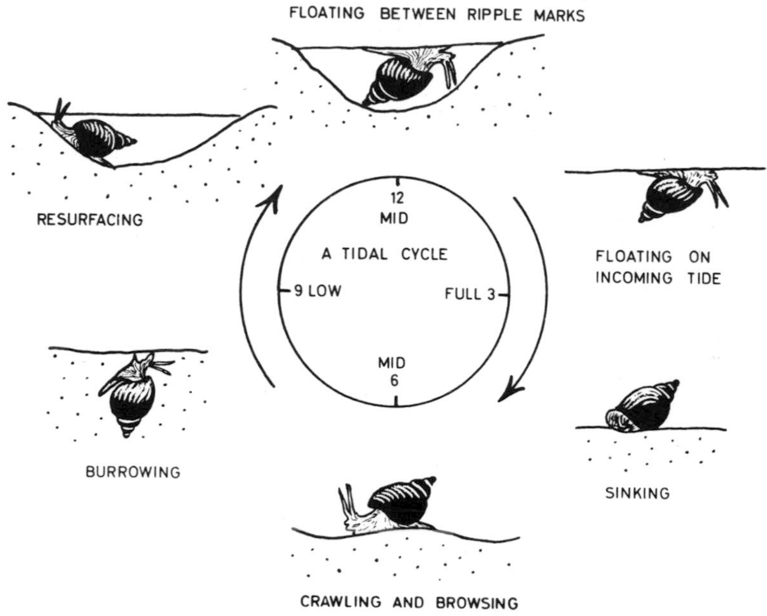

Fig. 4.6. Diagram showing the behaviour of *Hydrobia ulvae* as observed on the shore during one tidal cycle. (After Newell, 1962.)

(a) Responses to light

When specimens of *Hydrobia ulvae* were dark-adapted for 12 hr and then placed under unilateral illumination on a moist glass plate, it could be shown that the snails made a directional response to light. Fig. 4.7 shows the track made by an animal when lights A and B were switched on and off alternately; clearly the direction of crawling altered by approximately 180° each time the position of the light source was altered by this amount. Further, it was shown that the animal responded to the brighter of two lights when both were switched on at once (Fig. 4.8). Such behaviour, in which reaction to a second light source is inhibited rather than the animal following a course between the two lights, has been termed a telo-taxis (Fraenkel and Gunn, 1940). A special form of telo-taxis is a light-compass reaction, in which the

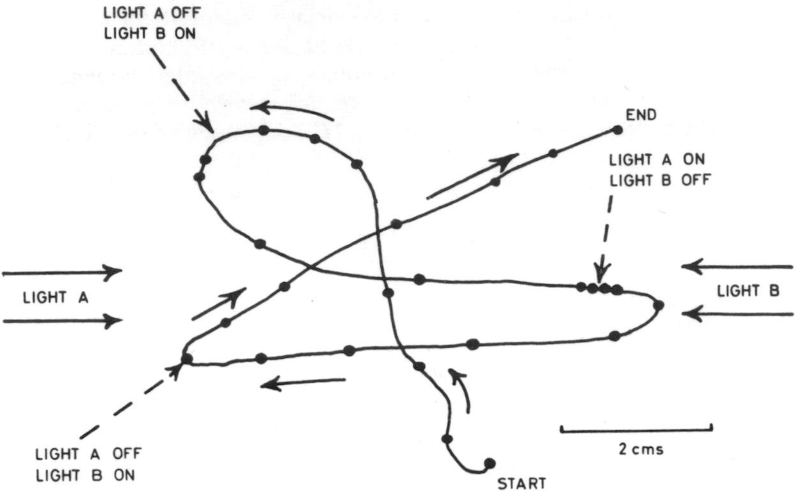

Fig. 4.7. Track made by a dark-adapted specimen of *H. ulvae* (12hr) orientating to light A when light B is off and light B when light A is off. Dots indicate half-minute intervals. (After Newell, 1962.)

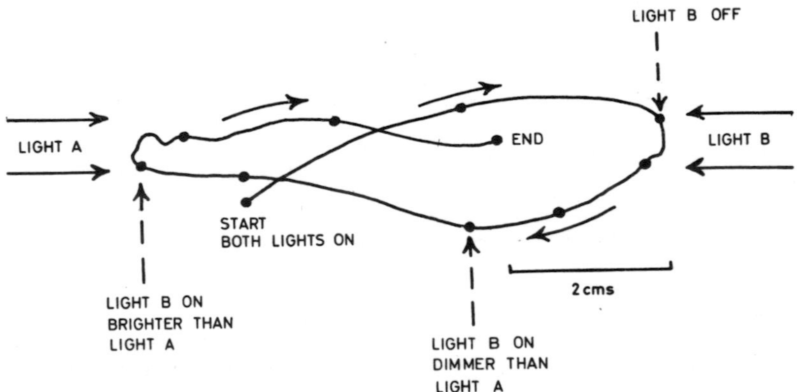

Fig. 4.8. Track showing the orientation of a dark-adapted specimen of *H. ulvae* (12hr) positively with respect to light A and inhibition of light B unless it is brighter than light A. Dots indicate half-minute intervals. (After Newell, 1962.)

animals do not necessarily orientate towards or away from the light. Into this category falls the photo-orientation of *Hydrobia ulvae* since not all animals were directly photopositive or photonegative.

After crawling for some time, which varied from individual to individual and with light intensity, the animals reversed the sign of their response to light so that the track was U-shaped just as in *Littorina littorea* and *Macoma balthica* (p. 135) (Fig. 4.9). The burrowing phase could be initiated by allowing light-adapted animals to come into contact with moist sand. Dark adapted animals, however, could not be induced to burrow and continually made movements normally preceding the floating phase. This implies that there is also some internal component in the behaviour of these animals.

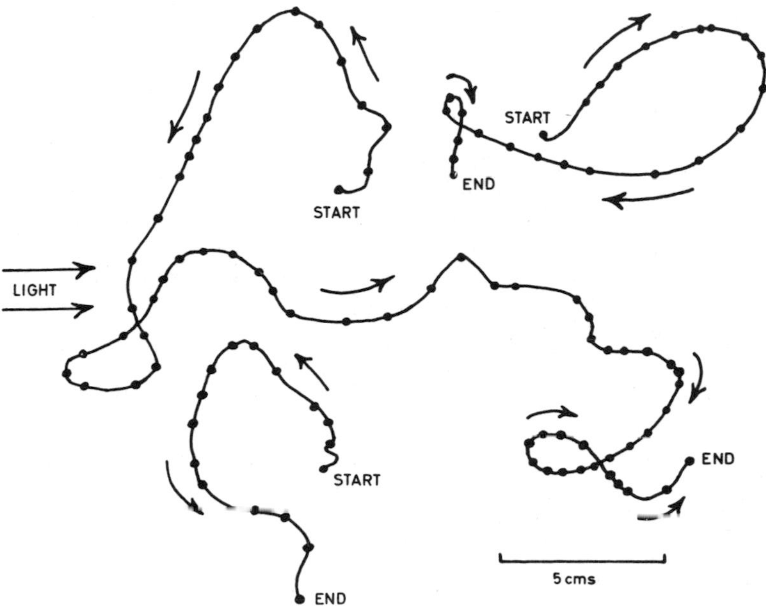

Fig. 4.9. Selection of tracks made by dark-adapted *H. ulvae* (12 hr) placed on a moist glass plate under unilateral illumination of 600 ft. candles. (After Newell, 1962.)

(b) Responses to gravity

The animals respond to gravity in such a way that they are normally brought to the surface film of the water between the ripple marks. Animals placed on a moist inclined glass plate partially immersed in water crawled downwards until they were immersed, and then crawled upwards and launched themselves on the surface film. Animals placed on the plate below the water crawled directly upwards and then floated on the water surface.

(c) Interpretation of shore behaviour in terms of the responses to light and gravity

It was suggested by Newell (1962) that when crawling on the mud-flats the animal uses a light-compass reaction with reversal of response as in *Littorina littorea* and *Macoma balthica.* After crawling has ceased (termed "light-adapted") the animals are in a state to burrow into the substratum. The burrowing phase allows the eyes to become "dark-adapted", and on resurfacing in response to an unknown stimulus possibly set up by the incoming tide, the eyes are subjected to an increase in light intensity. This increase is the immediate stimulus for floating movements, and the animals then respond to gravity in a way which brings them to the surface film. Animals resurfacing on the crests of ripple marks crawl down to the water surface and those resurfacing in the troughs crawl upwards. After a period afloat, the animals are passively stranded where there is a concave break of slope, which corresponds with a change from coarse to fine deposits (see Newell, 1964). Such settlement is enhanced by the increased tendency of the animals to attach themselves to any solid object after a long period afloat; the cycle of behaviour is then repeated.

Thus the behavioural cycle of *H. ulvae* may be interpreted in terms of relatively simple responses to the stimuli of light and gravity, although here the behaviour in response to a particular stimulus is not invariable, the animals altering in their response to light according to the events immediately preceding the stimulus. It is difficult, for example, to induce floating behaviour when the animals have already spent a considerable time on the surface film. Again, crawling and browsing stops after a time and cannot be induced again until the animal has completed burrowing and floating phases; similar arguments may apply to some of the other intertidal animals whose behaviour has been described. As was pointed out earlier, *Melaraphe neritoides* must alter its apparently inflexible behaviour to light and gravity in order to make feeding excursions; again, *Littorina littorea* does not crawl and browse during the whole of the intertidal period, although there is no apparent external factor which prevents it from doing so. It is obvious, therefore, that however adequately the distribution of intertidal animals can be explained in terms of simple responses to light and gravity, other more complex factors, such as the amount of feeding or locomotion which has occurred, play a part in controlling the behaviour patterns.

7. THE BEHAVIOUR OF THE SAND-HOPPER *TALITRUS SALTATOR*

The sand-hopper *Talitrus saltator* is an amphipod which is common on sandy beaches and normally lives buried in the sand during the daytime at, or near, the high water mark where optimal humidities prevail. During the night, however, or when humidity conditions are unsuitable, it wanders over much of the intertidal zone, as well as above the high water mark, scavenging for food or searching for optimal humidity conditions. A considerable amount of attention has therefore been focussed on this animal to determine the mechanisms which enable it to find its way about the shore (Williamson, 1951; Pardi and Papi, 1952, 1953, 1961; Papi and Pardi, 1953; Williamson, 1953; Pardi and Grassi, 1955; Pardi, 1960).

(a) Seaward migration

Pardi and Papi (1952), working on *Talitrus* from Pisa on the west coast of Italy, have found that sand-hoppers released onto dry sand above the high water mark made their way towards the sea. Further, the animals clustered to the west (seaward) side of a bowl with opaque sides, provided that the sun was allowed to shine on the hoppers. This suggested that the animals were responding to the sun and not orientating visually to the form of the landscape. The existence of a sun-compass reaction was subsequently confirmed by shading the hoppers from direct sunlight and presenting them with reflected sunlight. It was found that under these conditions the hoppers orientated in the same way to the reflected light as they had to the incident sunlight; they could also orientate at night using the moon as a light source (Papi and Pardi, 1953).

Pardi and Papi (1953) also showed that the hoppers were capable of orientating seawards even if the sun was obscured, provided that the sky was not cloudy. This suggested that the animals were also able to orientate to polarised light; when polaroid filters were rotated above the dish the hoppers were seen to change their direction of movement according to the rotation of the plane of polarisation. The suggestion that hoppers can orientate to polarised light is supported by the observation that disorientation occurs under cloudy conditions, which completely depolarises the light from the sky.

Animals from the west coast of Italy hopped westwards towards the sea when released above the high water mark on the west coast. Pardi and Papi (1952) then transferred one group of such hoppers out over the sea and another group across Italy to the Adriatic coast, where the sea lies to the east. It was found that the sand-hoppers still continued to face west in the bowl, even though this would then have taken them away from optimal conditions. Williamson (1953) suggested, however, that a sand hopper does not have a fixed "seaward" direction which is unalterable; specimens which show orientation in a fixed direction at the time of collection usually lost this ability within two days and subsequently orientated towards the sun irrespective of its position. Under such conditions it might be supposed that they may then acquire a new seaward sun-compass reaction relevant to the particular shore upon which they find themselves, although there is as yet no experimental verification of this. Indeed, from the results presented by Pardi (1960) the reverse would appear to be true. He took three groups of egg-carrying females from different areas and found that they orientated to the sun in a way which was relevant to the particular beach from which they were collected (Fig. 4.10). The eggs were then

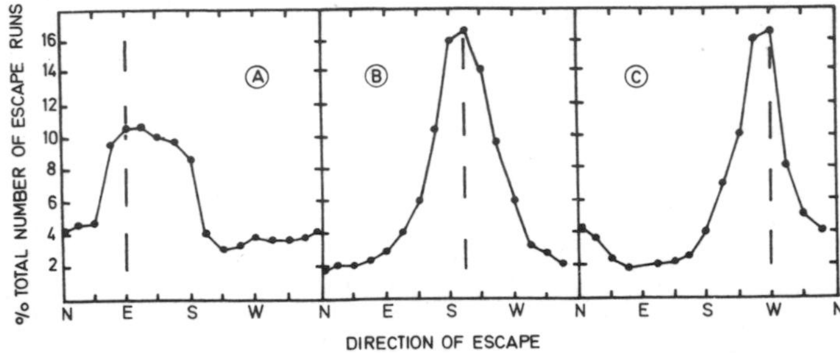

Fig. 4.10. Graphs showing the escape direction of three different populations of adult female *Talitrus saltator.* The line perpendicular to the home beach is shown (broken line). Ordinate, percentage of total number of escape runs; abcissa direction of escape run. (A) Line perpendicular to home beach at 81°. 2716 escape runs measured. (B) Line perpendicular to home beach at 201°. 6467 escape runs measured. (C) Line perpendicular to home beach at 267°. 5630 escape runs measured (Redrawn after Pardi, 1960.)

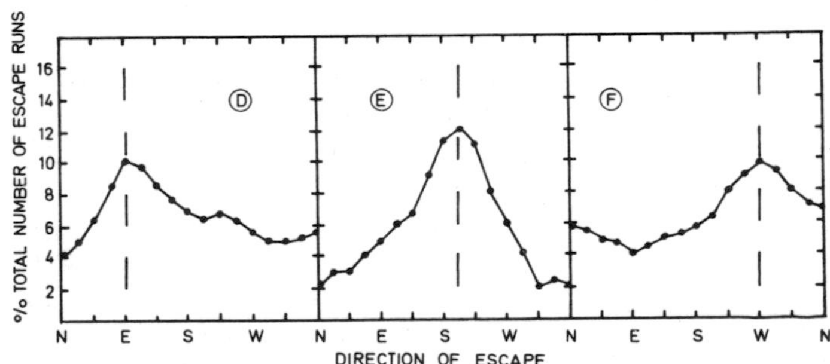

Fig. 4.11. Graphs showing the escape directions of young *Talitrus* reared in captivity. (D) Animals reared from adult females of population A (Fig. 4.10) 4337 escape runs measured. (E) Animals reared from adult females of population B (Fig. 4.10) 5988 escape runs measured. (F) Animals reared from adult females of population C (Fig. 4.10). 5363 escape runs measured. (Redrawn after Pardi, 1960.)

removed and reared in isolation from the females or any sight of the sun or sky. They were then tested in sunlight and showed a similar orientation to that of the adult females from which they were reared.

(Fig. 4.11). It seems difficult to avoid the conclusion from these experiments that the response to light is inherited; nevertheless the experiment does not rule out the possibility of light influencing the young while still attached to the female (p. 164). A further implication of this orientation to the sun is that *Talitrus* must make an internal correction for the movement of the sun across the sky; at any particular time of the day, the sand hoppers must orientate at a different angle to the sun in order to hop westwards. (Pardi and Papi, 1961). Hoppers

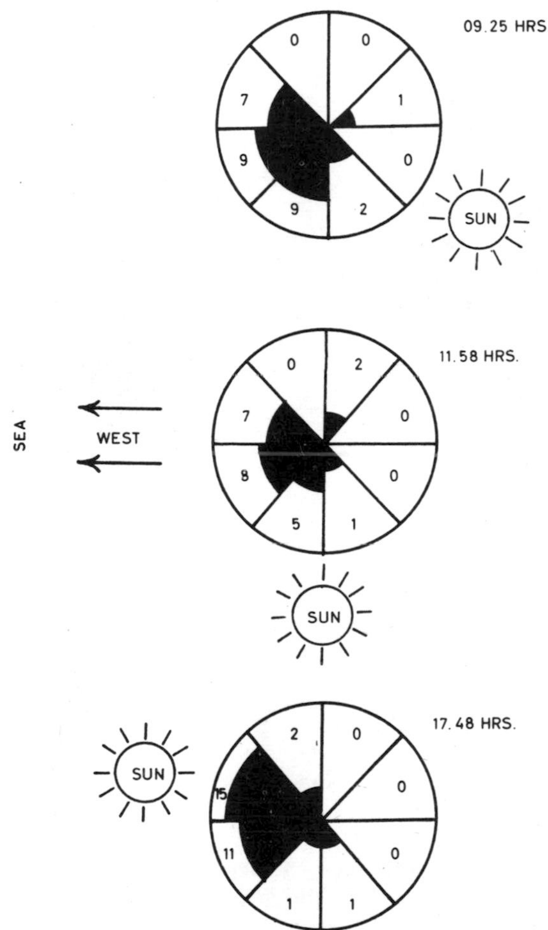

Fig. 4.12. Polar diagrams illustrating the results of three experiments at different times with *Talitrus saltator* from a beach with the sea to the west. (Modified after Pardi and Papi, 1961.)

released above high water mark on dry sand at dawn on the west coast of Italy would have to orientate at 180° to the sun, whereas if released towards noon the navigational angle would need to be 90°; towards sunset, with the sun in the west, the navigational angle would be 0°. (Fig. 4.12).

Such an internal correction in *Talitrus* involves a sense of time (Papi, 1955), the course of which is controlled by the daily alternation of light and dark. Pardi and Grassi (1955) have subjected *Talitrus* to artificial day/night conditions which were out of phase with the solar time and found that the sand hoppers migrated in a predictable direction according to the amount by which their "internal clock" was at variance with solar time. When 12 hr out of phase with solar time, for example, the hoppers orientated 180° away from the correct direction (Fig. 4.13). Again, when subjected to artificial illumination

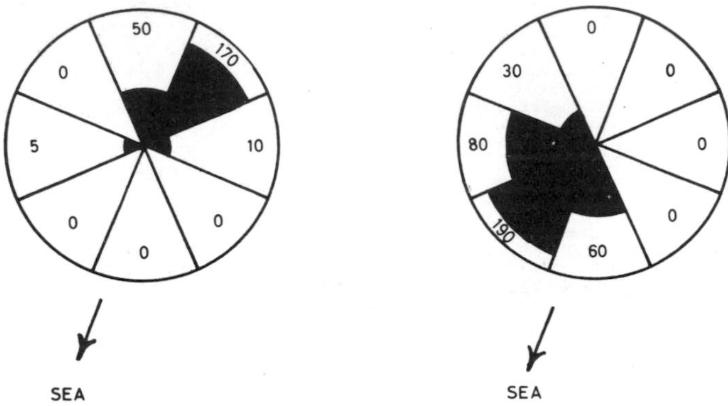

Fig. 4.13. Polar diagrams showing the number of experimental animals hopping in various directions when placed on the shore at 12 noon. (A) Animals had been kept 12 hr out of phase with solar time. (B) Animals in phase with solar time. (Modified after Pardi and Grassi, 1955.)

from noon to midnight and then placed on the shore above high water mark at noon, the animals adopted an orientation to the sun which they would normally have made at 06.00 hr.

(b) Landward migration

The response to light when the animals were placed on dry sand above high water resulted in the animals migrating into the moister conditions of the intertidal zone. If they were placed on moist sand, however, a very different response occurred; instead of a seaward migration, the animals hopped landwards. Pardi and Papi (1961) state that the direction taken is independent of the distance from sea or land

and conclude, therefore, that this response, like seaward migration depends on a sun-compass or on polarised light. Williamson (1951, 1953) on the other hand, has shown that specimens of *Talitrus* from the British Isles do not require either the sun or blue sky to be visible for landward migration. Further, when transferred to beaches facing another direction, such sand hoppers were not disorientated and navigated correctly towards the shore. He postulated, therefore, that the orientation of sand hoppers involves two different forms of navigation; (1) Seaward orientation involving a response to the sun or to polarised light; orientation to landmarks would not be possible under these circumstances since the eyes of *Talitrus* face forward, and suitably distinctive outlines would not occur in the seaward direction and (2) landward migration involving a response to coastal topography on certain shores.

It is clear that the sand hopper *Talitrus saltator*, like littorinids, is capable of using not only the sun but also polarised light and perhaps also the silhouette of the shore line as navigational aids. The next section deals, therefore, with some more general aspects of navigation by intertidal animals with particular reference to the ways in which a variety of directional environmental stimuli may be used in the behaviour patterns.

C. THE ROLE OF DIRECTIONAL ENVIRONMENTAL STIMULI IN THE BEHAVIOUR OF INTERTIDAL ANIMALS

From our consideration of some of the mechanisms by which intertidal animals maintain their characteristic position on the shore, it is apparent that the observed behaviour of these invertebrates is controlled by at least three components (also Schöne, 1964). (a) An internal one such as food requirement in browsing animals, the satiation of which may perhaps lead to inactivity in littorinids (p. 146). (b) An external releasing component or "token stimulus" which indicates favourable environmental conditions for the onset of a particular behaviour pattern. For example, although *Hydrobia ulvae* might be in a suitable state to undertake floating movements, it must first be subjected to an increase in light intensity before these will occur (p. 146). Again, *Littorina littorea* requires the stimulation of increased wave action such as occurs during the flowing or ebbing tide, to initiate browsing excursions (Newell, 1958b), and *Convoluta roscoffensis* also requires mechanical agitation before burrowing occurs (Fraenkel, 1929). Mechanical agitation in both of these animals serves to release a particular behaviour pattern; again in *Talitrus* the degree of moistness of the substratum initiates either seaward or landward migration. (c) There is an orientational component which is normally a directional stimulus set up by the environment and which serves to guide the animal during its behaviour.

It is with these latter more easily observable factors that most behavioural studies have been concerned. Such orientational stimuli include gravity, light, polarisation patterns and in certain instances shapes. One or more of these may be used by an animal which may, in addition, respond to non-directional stimuli which indicate the onset of particular phases of the tidal cycle and act as "releasers" (Tinbergen, 1951). Amongst these may be listed responses to mechanical agitation as in *Convoluta roscoffensis* (Fraenkel, 1929), brightness of light as in *Lepidochitona cinereus* (Evans, 1951), touch as in *Hydrobia ulvae* (Newell, 1962), and in some instances pressure as in the mud-dwelling amphipod *Corophium volutator* (Morgan, 1965). In all of the examples which have been given of the behaviour of intertidal animals, with the possible exception of *Convoluta roscoffensis* (Fraenkel, 1929), light is implicated to a greater or lesser extent. In the simpler behaviour patterns as shown by the chiton *Lepidochitona cinereus* (Evans, 1951), light has simply an orthokinetic effect stimulating the animals to crawl rapidly in bright light and more slowly in dim light. This is a non-directional response and its main effect is to enhance the speed of response of the animal to gravity (p. 135). Again, in the minute bivalve *Lasaea rubra* the dominant response is a negative geotaxis (Morton, 1960) but with a negative response to light playing an important part in the final selection of a suitable crevice.

In many other intertidal animals, particularly those such as *Littorina littorea, Hydrobia ulvae, Macoma balthica,* and *Talitrus saltator,* which inhabit featureless sandy or muddy shores, light-compass reactions play a dominant part in the behaviour of the animal. A common feature of the light-compass reaction of *Littorina littorea, Hydrobia ulvae, Macoma balthica,* and certain other bivalves, is the reversal of response to light both on the shore and under laboratory conditions, giving the characteristic looped tracks (Newell, 1958a,b; Newell, 1962; Brafield and Newell, 1961; for review Newell, 1966). Many littorinids can also orientate to polarised light (Burdon-Jones and Charles, 1958a,b; Charles, 1961); it seems likely, however, that although winkles, *Talitrus,* as well as certain other crustaceans including the mysid *Mysidium gracile* (Bainbridge and Waterman, 1957, 1958), have been observed to respond to changes in the plane of polarisation of the incident light, they may not be sensitive to the plane of polarisation itself but rather to changes in light intensity entering the eye (Stephens, Fingerman and Brown, 1953; Baylor and Smith, 1953; Waterman, 1954; Kalmus 1959; for reviews Carthy, 1957a,b; Waterman, 1961).

Such changes in light intensity would be expected to occur on a number of physical grounds as defined by Fresnel's laws of refraction of polarised light; these show that the amount of light refracted at the interphase of two non-birefringent media, such as the air/lens interface, is governed by the plane of polarisation of the incident light, the refractive indices of the two materials, and the angle of incidence of the light.

Charles (1961) showed that photonegative winkles orientate parallel to the plane of vibration and photopositive ones at right angles to it; this response is readily explicable in terms of Fresnel's laws and the structure of the littorinid eye. When the head of a littorinid is parallel to the plane of polarisation (*e* vector) of polarised light from above the animal, the plane of vibration is perpendicular to the plane of both the incident rays and the rays reflected from the surface of the eye. Charles pointed out that by Fresnel's laws, the minimum amount of light is then refracted into the eyes (Fig. 4.14A); however, when the head is at right angles to the plane of vibration, as in Figure 4.14B, the maximum amount of light is refracted into the eyes. Thus the responses of

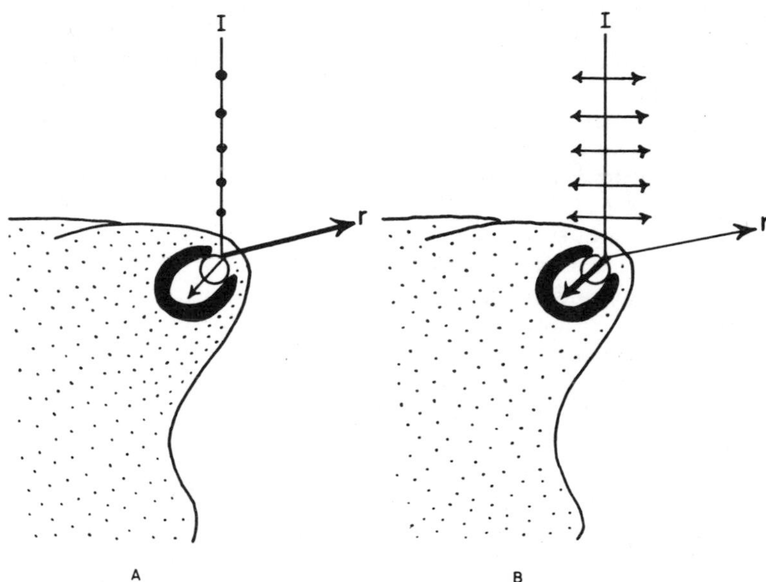

Fig. 4.14. Anterior views of transverse sections of the left side of a winkle head showing the left optic cup. *I*, vertical ray of plane polarised light incident upon the lens aperture. *r*, reflected ray. (A) Animal orientated parallel with the plane of vibration of the incident light, less light being refracted onto the retina. (B) Animal orientated at right angles to the plane of vibration of the incident light, more light being refracted onto the retina. (After Charles, 1961.)

littorinids to polarised light may be another aspect of the photopositive and photonegative phases described by Newell (1958a,b) for *Littorina littorea.* When the animals are in the photopositive phase they would crawl at right angles to the plane of polarisation whereas when the sign of the response is reversed, and the animals become photonegative, they align themselves parallel with the plane of vibration, less light being refracted onto the retina under these conditions. It has similarly been suggested that the orientation of *Drosophila* (Stephens, Fingerman and Brown, 1953); *Mysidium gracile* (Bainbridge and Waterman, 1958) *Daphnia* and other Cladocera (Baylor and Smith, 1953), and the discharges from the optic nerve of the eye of *Limulus* (Waterman, 1954) depend not on the sensitivity of the animal to polarised light itself, but on variations in light intensity set up by some external analyser refracting light in accordance with Fresnel's laws. It seems likely, therefore, that the responses of *Talitrus* to polarised light may also fall into the same category, namely of responses to different strengths of illumination of the eye, rather than to the plane of polarisation of the incident light. These and other aspects of orientation to polarised light are fully reviewed by Waterman (1959, 1961).

From the evidence summarised above, it seems that littorinids, and by inference other animals such as *Macoma balthica* and *Hydrobia ulvae* which make looped trails, orientate to the brightness of the light entering the eye, being at first photopositive and then photonegative. There is little information, however, on the factors controlling the reversal of response to light in these organisms, or of that controlling the reversal of response of *L. littorea* to gravity (p. 140) although Gowanloch and Hayes (1926) suggested that negative geotaxis in this animal is checked by the increasing desiccation encountered at higher levels on the shore.

In littorinids and in *Talitrus,* the situation is vastly complicated by the possibility of orientation to the topography of the shoreline (p. 151). Direct orientation to the shapes of features of the environment is an obvious method of maintaining a zonational position in intertidal organisms, and it seems likely that further investigation of the responses of such animals to shapes will show that this phenomenon is more widespread than has hitherto been supposed. Thorpe (1956), for example, reviewed the homing abilities of the limpet *Patella* and quoted the case of a limpet which he observed to return from an excursion of about 20 cm. He commented that the limpet showed some appreciation of the topography of the environment, and was not dependent upon a single orientational stimulus. This appears to be true also of the pulmonate mollusc *Onchidium* (Arey and Crozier, 1918; Piéron, 1909a,b).

Thus many intertidal animals are capable of orientating to a variety of directional stimuli according to the ecological situation in which they find themselves. The animal may respond simply to one stimulus such as light, as in *Littorina littorea, Macoma balthica,* or *Talitrus saltator* on a flat featureless shore, or they may respond to a combination of two directional stimuli which reinforce one another, as in the initial positive geotaxis and negative phototaxis of *Littorina littorea* from vertical surfaces, and in the negative photo-klino-taxis coupled with a positive thigmotaxis in *Lasaea rubra* crawling into a crevice (p. 141). Again, when the sun is obscured orientation to polarised light, or even to the topography of the environment, may occur in *Talitrus* and in littorinids. Moreover the animals behave in no way automatically to directional stimuli; their response varies according to the state of the animal and the situation is further complicated by the suggestion of Evans (1965) that in *Littorina saxatilis* and *Melaraphe neritoides* the water surface is seen from below as a goal. Apart from invoking the existence of form-vision in these animals, this suggestion implies the possibility of purposive behaviour in littorinids. Orientation to the topography of the environment in *Patella, Littorina punctata* and in *Talitrus* also involves a knowledge of the immediate environment which goes beyond the simple forms of behaviour we have discussed above. This leads us to consider the evidence for the occurrence of learning and associated processes in intertidal animals, and the role it plays in their mode of life.

D. LEARNING IN INTERTIDAL ANIMALS

We have seen how there is a considerable amount of evidence to support the view that the orientational components of behaviour can be learned by intertidal animals; this may also apply to the releasing components of their behaviour (p. 151). The experimental analysis of the mechanisms by which such releasing, as well as orientational components may be learned in intertidal animals has been studied primarily in nereids (for review Clark, 1964) and in certain Crustacea (for review Schöne, 1961a,b; 1964).

1. LEARNING IN NEREIDS

Much of the work on nereid behaviour has been concerned with the demonstration of learning processes in a variety of species of nereids, and with the role of the supra-oesophageal ganglion in such behaviour. Thorpe (1956) defined learning as the organisation of behaviour as a result of individual experience and went on to suggest that the simplest form of learning is represented by the process of habituation which is characterised by a relatively persistent waning of response resulting from repeated stimulation which is not followed by reinforcement.

(a) Habituation in *Nereis*

Such a process of habituation has been shown to occur in *Nereis* (Clark, 1960a,b) and also occurs in the sabellid worm *Branchiomma vesiculosum* (Nicol, 1950). Such worms normally respond to sudden stimulation by a contraction of the longitudinal muscles which results in the characteristic withdrawal response. Habituation of such behaviour can occur in *Nereis pelagica* in response to a variety of stimuli, including mechanical shock, sudden variations in light intensity (Clark, 1960a), and tactile stimulation of the posterior end; other species such as *N. diversicolor* and *Platynereis dumerilii* do not respond to the different stimuli in the same way although in these too, habituation has been demonstrated (Evans, quoted by Clark, 1964). Fig. 4.15 shows a

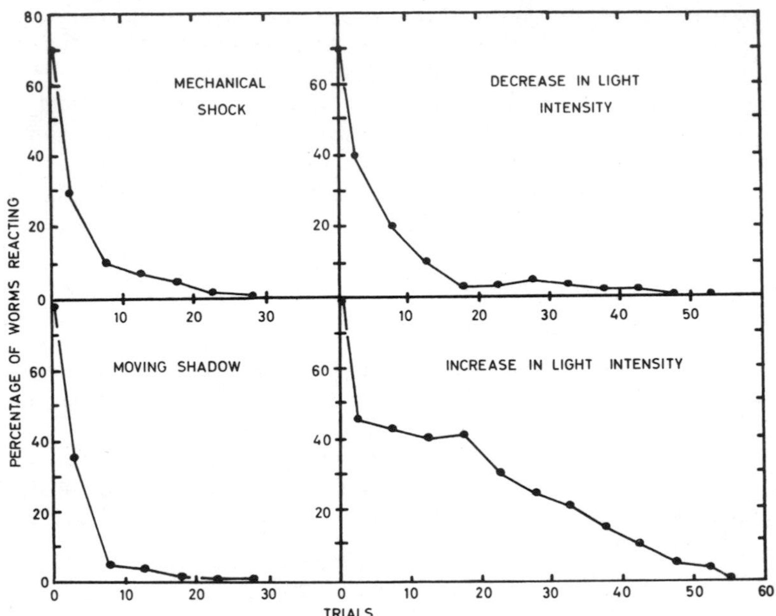

Fig. 4.15. Graphs showing habituation of the withdrawal reflex of *Nereis pelagica* to stimulation by mechanical shock, decrease in light intensity, moving shadows and increase in light intensity. (After Clark, 1960a,b.)

series of graphs illustrating the characteristic habituation of the withdrawal reflex in *Nereis pelagica* in response to a variety of stimuli. It also shows that habituation of withdrawal occurs more rapidly in response to mechanical shocks and moving shadows, than to changes in light intensity.

One might suppose that removal of the supra-oesophageal ganglion would profoundly affect the process of habituation. However, Clark (1964), quoting data of Evans, has shown that removal of the supra-oesophageal ganglion has little effect on habituation, and that decerebrate worms may sometimes habituate even more rapidly than intact ones (Fig. 4.16). Horridge (1959) had shown earlier that rapid

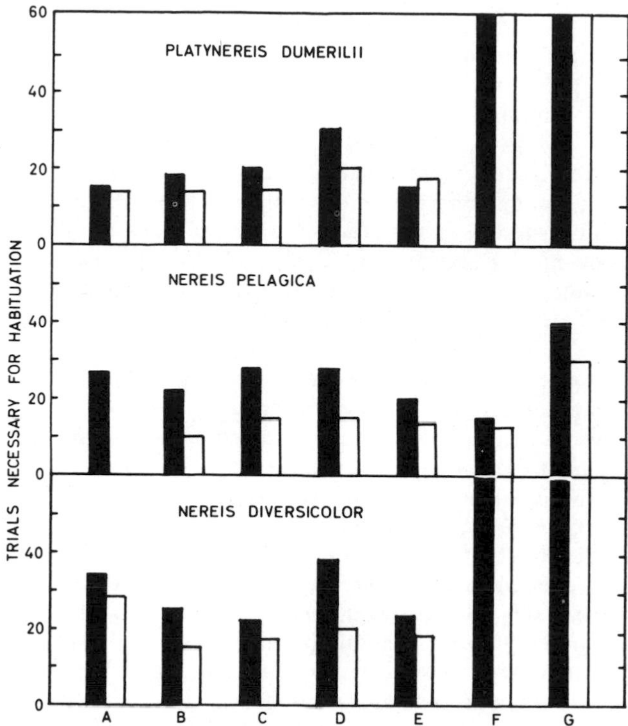

Fig. 4.16. Histograms showing the habituation rate of the withdrawal reflex of *Platynereis dumerilii, Nereis pelagica* and *N. diversicolor* to stimulation by (A) Increase in light intensity, (B) Decrease in light intensity, (C) Shadows, (D) Decrease in light intensity coupled with mechanical shock, (E) Mechanical shock, (F) Tactile stimulation of the anterior end, and (G) Tactile stimulation of the posterior end. Filled blocks, intact animals; open blocks decerebrate animals. (After Clark, 1964.)

habituation of the contraction reflex occurred in *Harmothoë* and *Nereis virens* in response to stimulation of the anal cirri. This was due to the properties of accommodation of the junction between the lateral giant axon and the motor neurone innervating the longitudinal muscles. Habituation in these worms may therefore be largely brought about by the properties of the segmental ganglia, although the supra-oesophageal ganglion may also be implicated to some extent in the habituation of complex behavioural responses.

(b) The effects of punishment on reflex behaviour patterns

Evans (1963a) has carried out a series of experiments on *Nereis diversicolor, N. virens* and *Perinereis cultrifera* to determine the role of the supra-oesophageal ganglion in the acquisition of new behaviour patterns. He found that if nereids were placed in the end of a glass tube they normally crawled straight to the other end. If they were then given an electric shock of 6 volts when they reached the end, after a number of trials they refused to approach the far end of the tube or even to go into the tube at all. Three successive refusals to enter the tube, or failure to reach the far end of the tube in 40 sec, was then regarded as a criterion of learning and allowed a comparison of the learning abilities of the three species. Evans (1963a) found that 25 individuals of *N. virens* took an average of only 21·5 trials to reach the criterion of learning whereas 5 specimens of *N. diversicolor* took an average of 47·8 trials and 5 specimens of *Perinereis cultrifera* took an average of 47 trials.

Nereis virens was then tested for its ability to retain or "remember" the training it received in this experiment and the results from such animals were compared with those obtained with an animal which was decerebrated immediately after training. It was found that the retention in intact and decerebrate animals was very similar, so that the presence of the supra-oesophageal ganglion appears to be quite unnecessary for the retention of the training pattern. Finally untrained "naive" worms were decerebrated and trained. Here it was found that although the animals were able to learn as quickly as intact ones, retention was so poor that one hour after training the animals required as many training trials as did untrained ones (Evans, 1963a). Thus, as Clark (1964) pointed out, although the supra-oesophageal ganglion is apparently unnecessary for the retention of previously learned behaviour, it does seem to play some part in the acquisition and later retention of new behaviour patterns.

(c) The results of "reward-punishment" experiments

Evans (1963b, 1966a,b) and Flint (1965) have studied the behaviour of nereids in "T-mazes". Evans (1963b) found that *Nereis virens, N. diversicolor* and *Perinereis cultrifera* all learn to avoid the arm of a T-maze associated with an electric shock, and moved into the arm in which they were allowed to remain undisturbed; this latter situation constituted a "reward". The learning process involved an initial random selection of arms varying greatly with individuals but the average length of this period was similar in *N. virens* (47·6 trials), *P. cultrifera* (52·3 trials) and *N. diversicolor* (49·0 trials). This phase was followed by a period of improved performance which differed in the three species; in *N. virens* it was only 10·6 trials, in *P. cultrifera* 22·7 trials and in *N. diversicolor* 26·5 trials. This phase was then followed by consistent avoidance of punishment. Evans (1963b) then found that decerebrated worms reverted to random selection of the arms of the maze; decerebrate untrained worms also failed to learn the maze (Fig. 4.17).

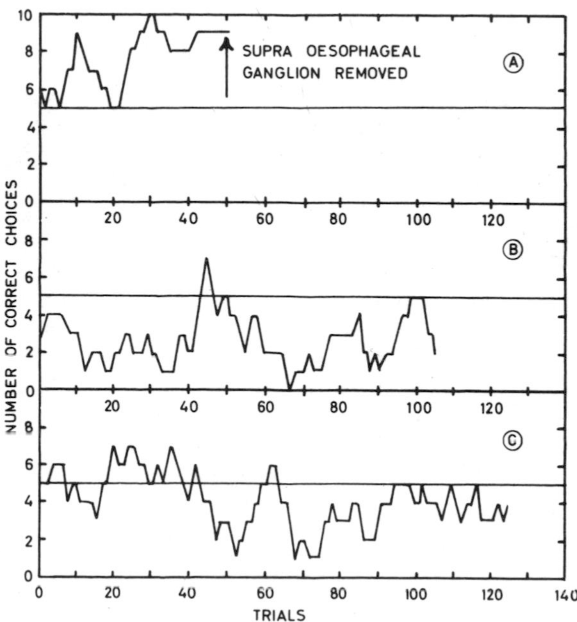

Fig. 4.17. Graphs showing the performance of an individual *Nereis virens* in a T-maze (A) before decerebration (B) after decerebration. (C) Naive worm decerebrated and then subjected to training. (After Evans, 1963b.)

These results suggested that the supra-oesophageal ganglion was of importance in the retention of such training behaviour. Flint (1965), however, has shown that the decerebrate worms fail the maze, not because the supra-oesophageal ganglion as such has been removed, but because removal of the ganglion involves the destruction of much of the sensory input from the anterior sense organs. She then showed that it was possible to sever the circum-oesophageal connectives, thus isolating the supra-oesophageal ganglion, in such a way that the afferent nerves of the cirri retained their connection *via* the connective ganglia with the rest of the nervous system. When this operation was performed on trained worms, a proportion of them were still able to perform in the maze correctly, thus indicating that it is the sensory input and not the supra-oesophageal ganglion which is of importance in the retention of learned behaviour. In this respect, as Clark (1964) pointed out, the "T-maze" experiments confirm the conclusions based on habituation and on the abolition of reflex behaviour patterns; retention of the acquired behaviour does not depend on the presence of the supra-oesophageal ganglion although it is of importance in the acquisition of new behaviour patterns.

(d) "Sensitisation" and non-associative learning in nereids

Evans (1966a,b) has shown that it is possible to "sensitise" *Nereis diversicolor* to the presence of food by pairing food presentation with changes in light intensity. Food presentation was shown to result in more worms crawling along their tubes to the exit at a significantly faster rate than did control animals. Initially 21·1% of nineteen *N. diversicolor* reached the end of the tube in a mean time of 105·5 sec in response to a sudden increase in illumination; after feeding with wheat-germ extract, however, 63·2% of the worms reacted and reached the end of the tube in only 80·8 sec. In the control animals which were not fed, 42·9% reached the end of the tube in 104·3 sec in response to the first light stimulus while 28·6% took 110·4 sec to reach the end in response to the second increase in light intensity. This difference in speed of response of the controls was shown to be insignificant (Evans, 1966b). Sensitisation has also been demonstrated in the withdrawal response of *Nereis pelagica* which becomes more reactive to shadows when mechanical shocks are interpolated into the series of stimuli (Clark, 1960b).

Evans (1966b) suggested that examples of withdrawal in response to a conditioning stimulus in other invertebrates may, in fact, be due to sensitisation effects, and not to a learned association between the stimulus and punishment. Similarly he pointed out that there is no evidence that *Nereis diversicolor* can learn to associate a sudden decrease in illumination with food presentation; the increased reactivity under such conditions appears to be an example of non-associative learning, achieved by the sensitisation effects of food presentation mentioned above. This conclusion is emphasised by his demonstration that although worms which were fed when subjected to an increase in light intensity subsequently became more reactive to light, they were no more reactive to light than worms which had been fed between trials; both groups were more reactive than the control animals which were subjected to an increase in light intensity only.

(e) The biological role of sensitisation in nereids

Evans (1966a,b) made the important suggestion that the sensitisation of nereid responses may serve a useful function in the life of the animal by rapidly achieving similar effects to associative learning. Under natural conditions withdrawal responses may thus be rapidly sensitised by a second stimulus, such as that set up by a predator, while feeding responses may be similarly sensitised by stimuli associated with the presence of food. In this way the animal may 'learn' the necessary releasing stimuli to initiate a particular behaviour pattern. Such a mechanism endows the individual with great flexibility since a wide variety of stimuli, or complexes of stimuli, may be used to sensitise the responses of the organism. Rather than a stereotyped system of releasers being used to initiate a sequence of behaviour patterns, the process of sensitisation would allow the acquisition of new releasing stimuli to be incorporated into the behaviour pattern.

2. LEARNING IN CRUSTACEANS

The extensive literature describing the behaviour of Crustacea has been recently reviewed by Schöne (1961a) who has based his terminology on that of Tinbergen (1951), Lorenz (1953), Baerends (1956) and Thorpe (1956). His review should be consulted for a detailed description of the many complex patterns of behaviour shown by crustaceans.

(a) Learning of releasing components

One of the earlier demonstrations of the learning of releasing stimuli by crustaceans is that of the prawn *Palaemonetes.* Mikhailoff (1923) fed prawns for some time in red surroundings and subsequently found that the animals would take food only when offered in red surroundings. Again Luther (1930), in a study of chemoreception in Brachyura, found that if the crabs *Eriocheir* and *Carcinus* were fed meat which had been soaked in coumarin, the presence of coumarin subsequently elicited feeding behaviour even in the absence of food. Whether these responses represent examples of true associative learning, or whether they are non-associative as has been described above for nereids, is unknown.

Many crustaceans have an elaborate courtship behaviour, examples of which have been described by Schöne (1961a), and many also have a complex social hierarchy in which the learning of releasing stimuli is of immense importance. Bovbjerg (1953) and Lowe (1956) investigated the influence of a number of factors including age, sex and weight, on the formation of a social hierarchy in the American crayfishes *Orconectes* and *Cambarellus shufeldtii.* It was found that animals which lost a fight came to associate the particular distinguishing features of the victor with submission resulting from conquest; by this process of association, the proportion of social contacts ending in an actual fight declined from the first to the twentieth day of communal life (Fig. 4.18). The learning of the distinctive features of the victor, and subsequently using these as releasing stimuli for submissive behaviour, is of obvious importance in the development of a social hierarchy and persists even after isolation for several days.

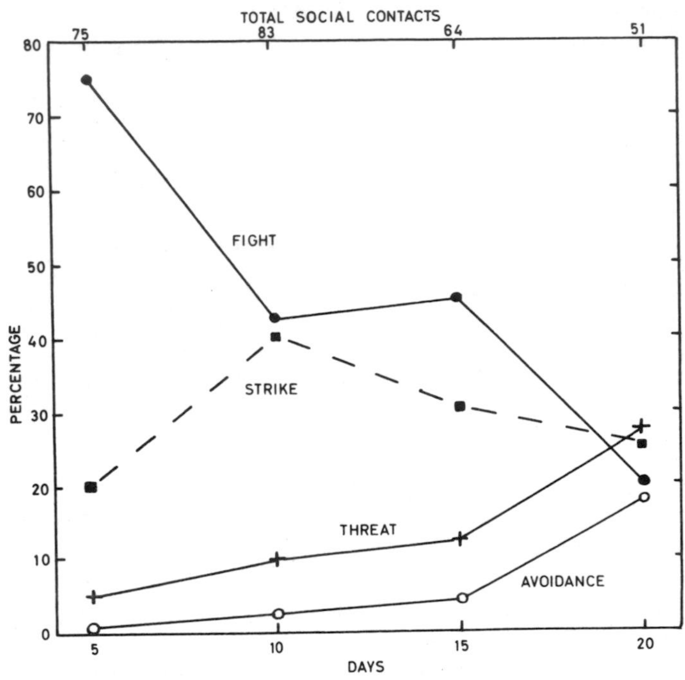

Fig. 4.18. Graphs showing the percentage of social contacts spent in fight, strike, threat and avoidance in *Cambarellus shufeldtii*. (After Lowe, 1956.)

(b) Learning of orientational components

Schöne (1961b) described an experiment in which a spiny lobster *Panulirus argus* was made to acquire both new releasing stimuli and also new orientational stimuli. The lobsters were confined in a box with two exits and were rewarded for choosing the exit through which the brightest light was shining. Fig. 4.19 shows the number of erroneous runs plotted against the number of trials; the number of 'non-spontaneous' runs, in which the animals remained near the starting point for more than 5 min is also shown. It is clear that the number of such 'non-spontaneous' runs declined rapidly over the period of the trials, and that the number of erroneous runs declined rather more slowly. From this, Schöne (1961b) concluded that the lobsters had learned two things. (a) They had to associate being placed on the starting line with the reward situation. That is, they had to learn the necessary releasing situation; it can be seen from Fig. 4.19 that this is learned after approximately six days (20 trials). (b) They had to learn to find the correct exit i.e. to learn the orientational stimulus. This process took some time in the case of light stimuli; in other experiments, however, Schöne (1961b) showed that the animals rapidly learned to find an exit in the dark if the reward situation was always on the same side. From this it may be concluded that orientational mechanisms other than visual ones plan an important part in the life of the animal.

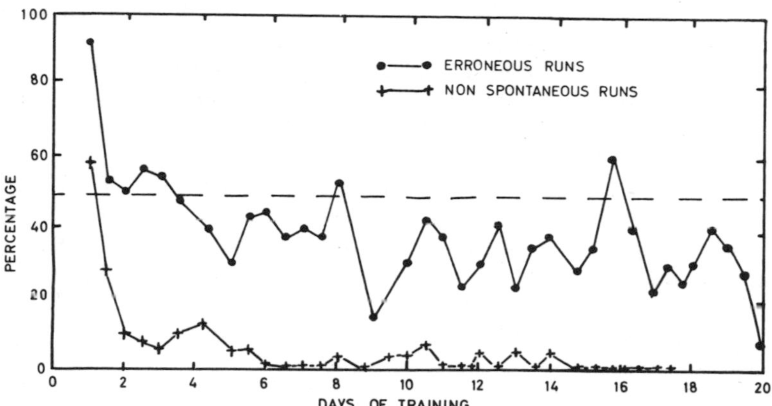

Fig. 4.19. Graphs showing the percentage of erroneous runs and of non-spontaneous runs plotted against the number of days training in Panulirus argus. (After Schöne, 1961b.)

E. CONCLUSION

The behaviour of intertidal organisms thus comprises three components (a) an internal one constituting a 'drive' (after the nomenclature of Tinbergen, 1951). (b) A pattern of external releasing stimuli which may be learned in an associative manner, as perhaps in the social behaviour of certain Crustacea (Lowe, 1956; for review Schöne, 1961a, b), or in a non-associative way, as in the sensitisation phenomena described by Evans (1966a,b) in nereids. Finally, there is (c) an orientational component. This is a directional stimulus of the environment and may aid either indirectly in the achievement of a particular behaviour pattern (e.g. sun-compass reactions), or directly, as in the orientation of many animals to the topography of the environment. Examples of such direct orientation include the shoreward migration of *Talitrus saltator* (Williamson, 1951, 1953), the orientation of *Patella* (Piéron 1909a,b; Thorpe, 1956) and that of certain littorinids (Evans, 1965); in each of these, such goal-seeking involves a knowledge of the immediate environment which can only have been acquired by a process of learning.

Learning, therefore, appears to be implicated in both the releasing and direct orientational phases of the behaviour of intertidal animals. Indeed, learning has not been specifically excluded from indirect orientational phases such as sun-compass reactions. Despite the results of Pardi (1960), obtained from studies of young *Talitrus* reared in the laboratory (p. 149), the influence of light on the offspring in the brood pouch has not been excluded. Again, in *Hydrobia ulvae* (p. 142) populations from different areas have been found to vary in the relative times spent afloat and browsing according to the particular tidal level from which they were collected (Newell, 1962). Here also, the eggs are carried by the adult for approximately 10 days before the veligers are released for a short pelagic life, and it seems not unlikely that the young of both *Talitrus* and *Hydrobia* may be affected by the directional stimuli impinging upon the parent. After release the young animals may then respond to directional environmental stimuli in a manner appropriate to the environment in which the adult population resides.

The properties of the sense-organs play a dominant part in all of such learning processes. Indeed, in nereids the presence of the supra-oesophageal ganglion appears to be unnecessary for the retention of certain learned behaviour patterns (p. 155, Clark, 1964). Clearly, orientation at a distance to topographical features of the environment is possible only if the eyes are capable of a degree of form-vision; geotaxis does not occur in individuals of a *Convoluta* population which are devoid of statocysts (Fraenkel, 1929). The behaviour of such organisms is thus intimately linked with the properties of their sense organs (for example Clark, 1956). The structure and functioning of the sense-organs of crustaceans have been reviewed by Waterman (1961), Cohen and Dijkgraaf (1961) and Barber (1961), whilst those of the molluscs are reviewed by Charles (1966) and Wells (1966). Pumphrey (1961) has presented a very clear review of the properties of visual organs in aquatic animals, whilst the occurrence and nature of pressure receptors in marine organisms has been described by Digby (1967). More recently Land (1968) has reviewed the evidence which suggests that in bivalve and gastropod molluscs the receptors involved in orientation responses are distinct from those involved in the responses to shadow; similar shadow-reflexes occur in a wide variety of other marine organisms (for such responses in echinoids see Millott, 1954; Millott and Yoshida 1959, 1960a,b; Yoshida and Millott, 1960; Yoshida, 1962).

In a detailed study of the structure and physiology of the eye of the scallop *Pecten maximus,* Land (1965, 1966a,b) showed that the focal point of the lens lies considerably behind the retina, so that the image does not fall near the retina cells. The back of the eye, however, is lined by an argentea composed of layers of guanine crystals whose function is to reflect the image back and produce a real, inverted image in the region of the retinal receptors. Such receptors are arranged in two anatomically distinct layers; the distal cells contain membranes formed from modified cilia, whilst the proximal receptors have a structure similar to that of other invertebrate photoreceptors and contain a large number of microvilli surrounding a conical extension of the cell body (also Dakin, 1928; Miller, 1960; Eakin, 1963, 1965). The distal retinal cells, representing modified cilia, respond to decreases in illumination caused by shadows whilst the proximal retinal cells containing microvilli, are concerned with orientation behaviour. The firing of the distal retinal cells is thus in response to the cessation of stimulation by light (an 'off-response') whilst the firing of the proximal retinal cells as in response to light (an 'on-response'). It seems likely that these differences may reside in hyperpolarisation phenomena which are known to occur in some inhibitory cells. The firing of the distal retinal cells is thus inhibited during stimulation by light but when the inhibitory stimulus is removed, as when a shadow passes across the field of vision, firing begins. Inhibitory, as well as excitatory processes in response to illumination, have been described in the bivalve *Spisula sollidissima* (Kennedy, 1960). In the nudibranch *Hermissenda crassicornis* the initial response of the cells is to become depolarised after stimulation by light, but inhibitory post-synaptic potentials may be developed in adjacent cells, and the summation of such hyperpolarisation may result in an inhibition of depolarisation (Dennis, 1965, 1967). Such hyperpolarisation is thus secondary in *Hermissenda* insofar as the first response of the receptors is depolarisation; in *Pecten,* however, primary light-induced hyperpolarisation seems likely to play an important part in the functioning of the cells of the distal retina involved in the shadow response (Land, 1968).

Land (1968) has also reviewed the evidence for form vision in mollusc eyes and pointed out that even in *Littorina littorea,* where the structure of the eye is such that form vision might occur during emersion (Newell, 1965), the acuity is probably poor. The presence of special receptor cells which respond to shadow appears to be common amongst the mollusca and, as Land (1968) pointed out, the cells of the compound eyes of the sabellid polychaete *Branchiomma vesiculosum,* which are thought to mediate the withdrawal response to shadow (Nicol, 1950), also have a ciliated structure very much as in the distal retinal cells of *Pecten* (Krasne and Lawrence, 1966). However in *Limulus* (Wilska and Hartline, 1941), insect dorsal ocelli (Ruck, 1961), and barnacles (Gwilliam, 1963), responses to a decrease in light intensity are brought about not by means of special receptors, as in the molluscs mentioned above, but by synaptic inhibition by late order neurones of 'on-responding' primary receptors. Whether the structural and electrophysiological differences in the receptors concerned in orientation and defensive reponses to shadow in molluscs occur in many other intertidal invertebrates is not yet known but offers a most promising field for future investigation.

CHAPTER 5

Mechanisms of feeding

A. INTRODUCTION

Although there is a great variety of feeding mechanisms in intertidal animals, it is possible to attempt a system of classification based upon the source of food. Such a system has been used by Yonge (1956) who proposed that the bottom fauna could be divided into three major categories: (a) suspension feeders, (b) deposit feeders and (c) carnivores. Other workers have also recognised the presence of three major categories, but have based their classification upon the size of the particles of material ingested (Jordan and Hirsch, 1927; Yonge, 1928; Prosser *et al,* 1950; Fretter and Graham, 1962). Jordan and Hirsch (1927) suggested a division into (a) Microphagous animals (b) Macrophagous animals and (c) animals absorbing food in solution. Whilst the latter system and that proposed by Yonge (1928) are useful schemes for comparison of the feeding mechanisms themselves, they are not perhaps as satisfactory from an ecological point of view. In this context the scheme proposed by Hunt (1925) also Yonge, (1956), based upon the three main sources of food available to marine animals, is preferable and forms the basis of the classification used below.

B. BROWSING

By far the most common examples of browsing organisms in the intertidal zone are to be found amongst the Mollusca, especially the chitons and gastropods. In these two classes the toothed radula and the buccal mass form a complex and highly specialised means of rasping food from the substratum. Typically, the buccal complex in gastropods consists of a toothed, ribbon-like strip of chitinised material, the radula, which lies in a posterior diverticulum of the buccal cavity, the radular sac. The radula is added to continuously at its posterior end and stretches forward over the dorsal surface of the buccal mass or odontophore (Fig. 5.1.) The cuticle-covered buccal mass contains one or more

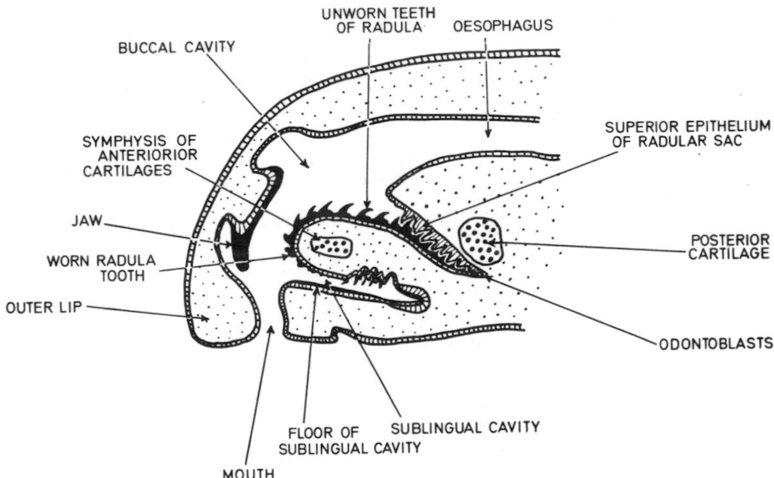

Fig. 5.1. Diagrammatic vertical longitudinal section through the head of *Patella vulgata*. (After Runham and Thornton, 1967.)

cartilagenous skeletal elements and is supplied with a complex musculature which serves to manipulate the radula during the feeding movements. Such muscles include protractors and retractors, together with intrinsic muscles altering the shape of the buccal mass, as well as tensor muscles in certain forms; the tensor muscles serve to maintain the buccal mass in any particular position.

The radula of prosobranch gastropods bears a series of teeth arranged in transverse and longitudinal series, whose number and form are characteristic of a particular species. The dentition consists typically of a large central, or rachidian tooth in the middle of each row, with a number of lateral teeth on each side (the largest of which may be termed the dominant tooth) followed by a series of marginal teeth. The number of radular teeth is, however, related to the diet and feeding mechanism employed, and it is customary to represent the dentition of a particlar species by means of a radular formula some of which are shown in Table IX (also Fig. 5.3).

The radula possessed by the simpler prosobranchs is rhipidoglossan

Table IX. Table showing the six main types of radula found among the Prosobranchia. D = dominant tooth; α = large, uncountable number of marginal teeth; n = large but variable number of teeth; R = ranchidian tooth. (Compiled from Fretter and Graham, 1962.)

Type of Radula	*Radula Formula*	*Examples of Species*
Rhipidoglossan	$\alpha + 1 + 4 + R + 4 + 1 + \alpha$ or $\alpha + 1 + D + 3 + R + 3 + D + 1 + \alpha$	Zeugobranchia eg, *Scissurella, Emarginula, Puncturella, Diodora.* Trochacea eg, *Margarites* spp, *Monodonta, Gibbula* spp, *Cantharidus* spp, *Calliostoma* spp, *Skenea* spp.
Docoglossan	$3 + D + 2 + R^* + 2 + D + 3$ * absent or small	*Acmaea* spp, *Patella* spp.
Taenioglossan	$3 + R + 3$ or $2 + 1 + R + 1 + 2$	All intertidal mesogastropods except *Clathrus* spp.
Ptenoglossan	$n + O + n$	Specialised forms include *Clathrus* spp.
Rachiglossan	$1 + R + 1$	Muricacea eg, *Trophon* spp, *Nucella, Urosalpinx, Ocenebra,* Buccinacea eg, *Buccinum, Nassarius* spp.
Toxoglossan	$D + O + D$ or $1 + O + O + O + 1$	*Mangelia* spp, *Philbertia* spp.

in form and is similar to the unspecialised teeth found in the chitons; from this the taenioglossan, rachiglossan and toxoglossan types may have evolved. The docoglossan type is regarded as a separate line by Fretter and Graham (1962) whilst the Ptenoglossan is thought to represent a regression towards the ancestral rhipidoglossan type of radula. Some examples of the feeding mechanism in different gastropods are given below.

1. FEEDING MOVEMENTS IN THE RHIPIDOGLOSSA

The complex movements associated with feeding in certain trochids have been described in considerable detail by Nisbet (1953) for *Monodonta lineata,* and by Ankel (1938) and Eigenbrodt (1941) for *Gibbula cineraria.* This work has been fully reviewed by Fretter and Graham (1962) and by Owen (1966). Nisbet (1953) distinguished the following six phases of a continuous sequence of movements involved in feeding in *Monodonta.*

Firstly the slit-like mouth is opened and exposes the paired chitinous jaws which lie on the anterior wall of the buccal cavity and which represent local thickenings of the cuticle that covers much of the buccal cavity.

The second phase involves the protraction of the buccal mass which carries the radula forwards and downwards into contact with the substratum. At the same time the groove in which the radula normally lies when the buccal mass is retracted, is obliterated, and the radula then lies level with the rest of the surface of the buccal mass. This is effected partly by the action of the tensor muscles, partly as an indirect effect of the protraction of the buccal mass on the subradular membrane and partly by hydrostatic pressure in the blood spaces of the buccal mass. (Fig. 5.2A)

The third phase involves a downward movement of the subradular membrane which carries the radula with it. The teeth lie flat on the radula within the buccal cavity, but as they move down and over the anterior cartilages of the buccal mass they become erected and come to stand nearly vertically on the surface of the buccal mass; the lateral teeth move forward so as to form a radiating series on each side of the central teeth. The region at which such movement occurs has been termed the "bending plane" by Fretter and Graham (1962) after the nomenclature of Ankel (1936b). During the forward movement of the radula, each row of teeth is in turn affected, rotating forward so as to stand vertically at the tip of the buccal mass (Fig. 5.2B). At the end of this phase, therefore, the teeth are in the forward position both on the tip and underside of the buccal mass.

The next phase is the retraction of the radula. This is brought about not by a movement of the radula over the stationary cartilages of the

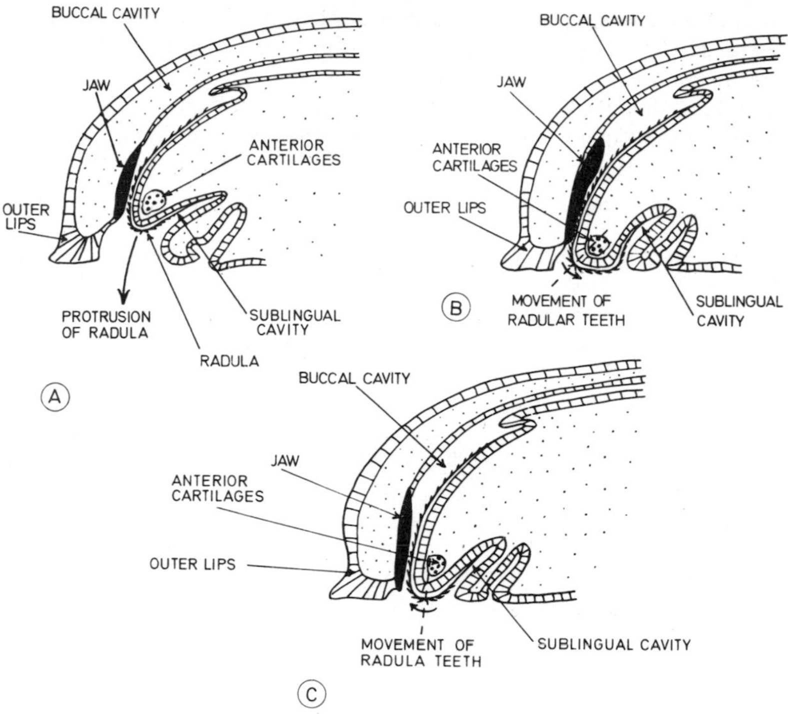

Fig. 5.2. Diagrams showing the feeding mechanism in *Monodonta lineata.* (A) Initial protrusion of buccal mass and radula. (B) Downward movement of the subradular membrane causing rotation of the radular teeth at the bending plane (broken line). (C) Posterior movement of the anterior cartilages and bending plane with the return of the radular teeth to their original position. (Based on Nisbet, 1953; Fretter and Graham, 1962.)

buccal mass, as in the protraction process, but by a downward and backward movement of the cartilages underneath the stationary radula. This results in a movement of the bending plane of the radula from an antero-dorsal to a postero-ventral position with a consequent movement of the radular teeth back to their original position. The lateral teeth collapse inwards carrying with them any particles into the midline and holding them there (Fig. 5.2C).

The fifth stage is the withdrawal of the buccal mass together with the material which has been rasped from the substratum, and merges with the next phase in which the lips are closed.

As Fretter and Graham (1962) pointed out, it is clear from this work that the marginal teeth of the radula are brushed lightly against the substratum with the median and lateral teeth acting mainly to withdraw the collected food into the buccal cavity. The jaws are weak and could not be used for the removal of large pieces of material from the sub-stratum. Hence *Monodonta,* in common with other molluscs such as *Gibbula* and *Haliotis* possessing this type of radula, is microphagous, probably feeding off a variety of micro-organisms as well as organic debris.

2. FEEDING MOVEMENTS IN THE TAENIOGLOSSA

The taenioglossan radula (Table IX and Fig. 5.3) is found in most of the Mesogastropoda, which includes the vast majority of intertidal gastropods. The principal difference from the rhipidoglossan type is the

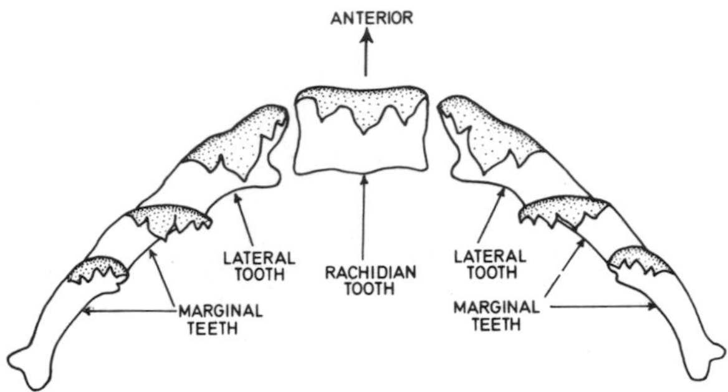

Fig. 5.3. Diagram showing a single transverse row of radular teeth of *Littorina littorea.* (After Fretter and Graham, 1962.)

reduction in the size of the marginal teeth and corresponding increase in the size and importance of the rachidian and lateral teeth. The marginal teeth still brush in particles as in the rhipidoglossan radula, but in addition the rachidian and lateral teeth actively rasp the substratum, and either gather particles directly or pass them to the marginal teeth for collection.

The form of the buccal mass in *Littorina* and in certain rissoids has been described by Johansson (1939) and by Nisbet (1953); that of rissoids has been described by Bregenzer (1916) and Krull (1935) (for reviews Fretter and Graham, 1962; Owen, 1966). Essentially, the musculature comprises paired lateral buccal protractors and ventral retractors whose effect is to pull the buccal mass forward and to stretch the subradular membrane laterally in such a way as to open the radular teeth. The buccal mass is then applied firmly to the substratum and is, in fact, rotated so that its tip is applied at right angles to the substratum.

The second phase of the feeding action is for the subradular membrane to be moved out of the buccal cavity and over the bending plane of the odontophore cartilages in a laterally stretched state; this results in the erection and outspreading of the radular teeth just as described above for the rhipidoglossan types. The third phase involves the ventral movement of the odontophore cartilages, and moves the bending plane in such a way that the teeth in contact with the substratum execute powerful rasping movements (Eigenbrodt, 1941). Nisbet (1953) stated that several of such to and fro movements of the bending plane may occur so that the radular teeth are opened and closed several times while pressed against the substratum. Finally, as in the rhipidoglossan mechanism, the odontophore cartilages move back and the radular teeth fold back onto the sub-radular membrane, carrying with them the particles rasped from the substratum.

Thus the principal difference between the rhipidoglossan and taenioglossan types is that the latter system involves the increasing mobility of the odontophore cartilages which are used both in the retraction process and in the erection of the radular teeth against the substratum. There is a corresponding simplification of the musculature of the odontophore in the taenioglossans, since the radula is pressed firmly against the substratum rather than being held over it as in the rhipidoglossans; this latter process calls for delicate adjustment and the use of a system of tensor muscles (Fretter and Graham, 1962). The modification of the radular formula from the rhipidoglossan type has allowed a great variety of food materials to be utilised, including encrusting algae as in the littorinids, the adoption of deposit-feeding as in the estuarine mesogastropod *Hydrobia ulvae* (p. 259), and predation as in the sponge-eating forms *Cerithiopsis* and *Triphora* (p. 184).

3. FEEDING MOVEMENTS IN THE DOCOGLOSSA

The docoglossan type of radula is possessed by the limpets *Patella* and *Patina,* the feeding movements involved differing in certain important respects from those described above for the rhipidoglossan and taenioglossan types. As pointed out by Fretter and Graham (1962), the first difference is that the radula does not move over the bending plane, but is applied firmly to the surface of the substratum by the protracting buccal mass. Numerous rows of teeth are in contact with the substratum and are not rotated but are drawn across the substratum like a rasp, although from an examination of the grooving of the jaw, Runham and Thornton (1967) have suggested that some erection of the dominant teeth occurs immediately prior to the radula passing over the bending plane. This difference in the feeding movements is correlated with an increase in the musculature concerned with the protraction of the buccal mass; such a system must involve a considerable increase in friction with the substratum compared with the rhipidoglossan and taenioglossan systems, in which only the row of radular teeth passing over the bending plane at the tip of the buccal mass is applied to the substratum. There is also an increase in the width of attachment of the radular teeth to the membrane, the teeth being no longer required to rotate during the feeding process; in addition the teeth have an increased hardness, no doubt to accommodate the wear and tear of the rasping movements. Runham and Thornton (1967) have recently shown that in *Patella vulgata* the marginal teeth have a surface coating which differs from the rest of the tooth, and this may be true also of the lateral teeth (Wayte, unpublished, quoted by Runham and Thornton, 1967). The increase in the hardness may, therefore, reside in the properties of the outer layer of the radular teeth, the structure of which is shown in detail by Runham and Thornton (1967).

The second major difference from the types previously described lies in the means by which food material is conveyed into the buccal cavity. In rhipidoglossan and taenioglossan types the backward movement of the odontophore cartilages and retraction of the buccal mass causes a rotation of the marginal teeth, this movement carrying food material onto the lateral teeth and back into the buccal cavity (p. 171). In the docoglossans, however, the protraction process is the one which forces the radula forwards, down against the substratum, and then forwards towards the dorsal lip, scraping the substratum with numerous rows of teeth as it does so (Ankel, 1938; Eigenbrodt, 1941). The subradular membrane is then pulled inwards and the material withdrawn into the buccal cavity. Rotation of the teeth on the radular membrane thus plays no part in the transport of food into the buccal cavity. The buccal mass is therefore not only well supplied with powerful protractor muscles but, as pointed out by Fretter and Graham (1962), is also stiffened by muscles against the mechanical shock set up by the protraction movements.

A similar process to that described above appears to be used by the intertidal chitons; the effective rasping stroke is a forward one, brought about by the protractor muscles of the buccal mass. The majority of the Polyplacophora are thus herbivores, grazing on small algae which they rasp from the substratum when the intertidal zone is immersed by the tide (p. 134).

C. PREDATION

1. COELENTERATES

Although the great majority of predators are active organisms whose powers of movement exceed those of their prey, this is by no means true of all intertidal predators. The coelenterates, for example, provide an interesting example of a captorial method of feeding in which the tentacles may be used either to entangle prey which is of large size relative to the coelenterate or, as in the anemone *Metridium,* to trap small particles which are then transferred to the mouth by means of cilia. Both macrophagous and microphagous coelenterates make use of their characteristic cnidae or nematocysts to entangle or pierce the prey. Weill (1934) was the first to classify systematically the forms of cnidae occurring in the coelenterates, although Carlgren (1940) and Cutress (1955) have also made significant contributions to the study of the nematocysts of the Anthozoa (the alcyonarians and zoantharians). Basically, two types of cnidae may be recognised. Firstly, there are spirocysts which are restricted to the zoantharian anthozoans (true corals and anemones), and in which the capsule is thin-walled and permeable to water. The second type of cnidae are the nematocysts proper which occur in all coelenterates. There is, however, some evidence which suggests that the spirocysts are really a form of holotrichous nematocyst (Cutress, 1955). Hand (1961) has pointed out that although Weill (1934) described 18 different types of nematocyst proper, they may be divided into two major groups. Firstly there are the astomocnidae which have closed tubes, and secondly there are the stomocnidae in which the tube is open at the tip. The further division of nematocysts is shown in Table X, the astomocnidae being subdivided into two categories, (a) desmonemes (volvents) which have a tube that when everted coils round projecting spines on the prey, and (b) rhopalonemes (including acrophores and anacrophores) which occur in the Siphonophora. The stomocnidae are subdivided firstly into (a) haplonemes in which there is no butt or swollen basal region to the tube, and (b) heteronemes in which a butt is present. Further subdivision is based upon the form of the everted tube or of the butt, and whether spines are present on either or both of these structures.

Cutress (1955) suggested that the amastigophores of Weill (1934) should be replaced by two other categories, q-mastigophores and macrobasic p-mastigophores, but Hand (1961) has argued convincingly for the retention of the scheme outlined by Weill (1934) and which is summarised in Table X. The different nematocyst types are by no means evenly represented amongst the coelenterates. The hydrozoans, for example, show a great variety of structural types, although the function of only a few of them is well known. On the other hand the anthozoans have only stomocnides and are the only group possessing spirocysts. Even amongst the anthozoans specialisations occur; the basitrichous haplonemes, for example, are the only nematocysts found in the Alcyonaria (soft corals), and are rare or even absent in the Actiniaria (sea anemones).

Table X. Table showing the major types of nematocysts proper. (After Hand (1961) and based on Weill, 1934; also Schulze, 1917.)

ASTOMOCNIDAE *Nematocyst tube closed at tip*	*STOMOCNIDAE* *Nematocyst tube open at tip*
A. *DESMONEMES* (Volvents) Unarmed tube which coils round projecting spines on the prey. B. *RHOPALONEMES* (Acrophores + Anacrophores) Characteristic of siphonophores. Tube simple and sac-like.	A. *HAPLONEMES.* No enlarged basal region or butt. Probably mainly adhesive in function. 1. ISORHIZAS. The tube is of even diameter or isorhizic. (a) *Holotrichs* (large glutinants) – armed haplonemes. (b) *Atrichs* (small glutinants) – unarmed haplonemes. (c) *Basitrichs* – tube armed at the base only. 2. ANISORHIZAS. The tube tapers or is slightly swollen at the base i.e. is anisorhizic. Characteristically found in siphonophores and in *Tubularia.* B. *HETERONEMES* (Penetrants). An enlarged basal portion or butt is present. Any particular heteroneme may be either armed (hoplotelic) or unarmed (anaplotelic). 1. RHABDOIDES. – The butt is of even diameter. (a) *Mastigophores* – a terminal thread is present. (b) *Amastigophores* – no terminal thread. 2. RHOPALOIDES. The butt is not of even diameter. (a) *Euryteles* – the butt is dilated at the distal end. (b) *Stenoteles* – the butt is dilated at the proximal (basal) end.

Although there are evidently many different types of nematocyst, the function of only a few of them is known in detail; the precise factors initiating discharge and the mechanism of eversion are also largely unknown. Pantin (1942) has shown that in the anemones *Metridium senile* and *Anemonia sulcata,* the nematocysts can be made to discharge in response to strong mechanical stimulation or to electrical stimulation, although he found that in both cases discharge occurred only in the vicinity of the stimulus. There was no indication of a spread of excitation such as would be expected to occur if the nematocysts were under nervous control. It thus appears that the nematocysts are independent effectors whose discharge is facilitated by stimuli set up by the presence of suitable prey in the vicinity. The mechanical stimuli required to initiate discharge was, however, much higher than that which would be set up by the prey, and it was found that both a mechanical and a chemical factor are involved in nematocyst discharge in response to suitable food. Extracts of food alone were not sufficient to initiate a discharge but lowered the threshold of the nematocysts to mechanical stimulation. In the presence of food such as molluscan mantle extract and human saliva, even light mechanical stimulation was sufficient to initiate a discharge. The sensitising substance appears to be a surface-active lipoid, agents such as saponin bringing about a vigorous discharge. Much the same result was obtained by Mackie (1960) who showed that the presence of food extract sensitised the cnidoblasts of *Physalia,* a greater discharge in response to mechanical stimulation occurring in the presence of food than in its absence.

Ewer (1947) has shown that in *Hydra* four different types of nematocysts have different thresholds of sensitivity to mechanical and chemical stimulation according to the part they play in the life of the animal, and it seems likely that a similar situation exists in marine coelenterates possessing a variety of structural types of cnidae. In the absence of a chemical stimulus the atrichous isorhizas (small glutinants) of *Hydra* have the lowest threshold to mechanical stimulation but are not discharged by stimuli of short duration as set up by particles of silt falling on the tentacles; such atrichous isorhizas are used in *Hydra* for the attachment of the tentacles during locomotion. A similar function might be ascribed to the spirocysts of anemones since Skaer and Picken (1965) have shown that those of the anemone *Corynactis viridis* are sticky and are known to be especially abundant in larval anemones (Stephenson, 1927). In *Hydra* the threshold of the atrichous isorhizas to mechanical stimulation is raised when a chemical stimulus occurs whilst that of the stenoteles (penetrants) and desmonemes (volvents) is lowered, the threshold of the desmonemes being higher than that of the stenoteles. The tube of the stenoteles is armed with spiral rows of spines (Fig. 5.6) and discharges a poison through the open tip as the prey is penetrated (Toppe, 1909; also Kline, 1961; Lane, 1961; Barnes, 1966). The desmonemes are discharged after the stenoteles and only when a bristly prey violently stimulates the tentacles. The conclusion drawn from the work of Pantin (1942) and Ewer (1947) was that the discharge of nematocysts is a peripheral phenomenon operating independently of the nervous system.

Lentz and Barrnett (1962) have recently confirmed and amplified these conclusions in *Hydra* and have, in addition, studied the effects of enzyme substrates and inhibitors, as well as neuro-pharmacological agents, on nematocyst discharge. They showed that the addition of organic phosphates enhanced discharge provided that mechanical stimulation also occurred, whilst enzymatic inhibitors eliminated a discharge produced in this way. Again, acetylcholine produced a massive discharge, this being augmented by eserine and inhibited by hexamethonium and tubocurare. From these and other experiments, Lentz and Barrnett (1962) concluded that mature nematocysts contain enzymes responding to chemicals. These serve as substrates for the nematocyst enzymes and therefore, provided mechanical stimulation also occurs, act as effector substances.

It was also shown that a combination of nervous and mechanical stimulation may play a part in discharge. Lentz and Barrnett (1962) therefore concluded that either a combination of mechanical and chemical, or mechanical and nervous stimulation is effective in promoting discharge of nematocysts. The enzymes of the nematocyst appear to be located in the surface of the capsule or in the cnidocil (Lentz and Barrnett, 1961a,b). Such a chemical facilitation of discharge is apparently independent of the nervous system since the response is unaffected by hexamethonium which would block a nervous mechanism. The nervous system, equally, appeared to facilitate discharge even in the absence of chemical substrates. Lentz and Barrnett (1962) concluded, therefore, that nematocysts are independent effectors insofar as a minimum of two stimuli, one of which is local and mechanical, are required. Nevertheless, in *Hydra* the nervous system, as well as external chemical substrates, appears to lower the threshold to mechanical stimulation. Other evidence, too, suggests that although nematocysts may be regarded as independent effectors, the magnitude of discharge and nature of the nematocysts involved may be influenced by a variety of factors. Burnett, Lentz and Warren (1960) found that in fully-fed *Hydra*, nematocysts were still discharged in response to stimulation by *Artemia,* but that the cnidoblasts were extruded from the epidermis so that the prey was not retained: seasonal variations in the response of *Hydra* to prey organisms have also been noted (Bouchet, 1961). Again, Ross (1960; also Davenport, Ross and Sutton, 1961) has shown that in the anemone *Calliactis parasitica,* which is normally attached to whelk shells containing the hermit crab *Pagurus bernhardus,* preliminary attachment is by means of nematocysts. If the anemone is already attached to a shell however, no attachment occurs to a second whelk shell presented to the anemone. The threshold of discharge thus appears to be higher in anemones attached to whelk shells, and this response appears to be mediated through the pedal disc. Evidence of possible nervous elements associated with cnidoblasts has been described not only in *Hydra* (Spangenberg and Ham, 1960; Lentz and Barrnett, 1961a,b) but also in the anemones *Metridium* and *Sagartia*

(Lentz and Wood, 1964), so that it seems possible on both behavioural and anatomical grounds that the discharge of nematocysts may be influenced, if not initiated, by nervous elements.

The mechanism of discharge of nematocysts has been investigated by Picken (1953) and Robson (1953) both of whom studied the large holotrichous isorhizas (glutinants) of the anemone *Corynactis viridis.* This work established that prior to discharge the barbs face inwards, occluding the lumen of the undischarged nematocyst thread. On discharge the nematocyst thread is everted so that the barbs come to lie facing outwards from the external surface of the tube. It might seem that there is a simple eversion of the tube on discharge, but close examination of the arrangement of the barbs showed that in the discharged thread the barbs were no longer arranged in a hexagonal packing, but in three right-handed helices running round the tube 120° out of phase with one another (Fig. 5.6) Since the spacing between the barbs of the helix was greater than that between the barbs of the undischarged thread, an increase in diameter must have occurred. Equally, the barb arrangement showed that the increase in diameter which occurred on eversion was exceeded by an increase in the length of the filament. In fact there is an almost three-fold increase in length compared with only a 50% increase in diameter. The changes in barb pattern as observed by a light-microscope are shown in Fig. 5.4. This anisometric expansion was thought to be possibly due to some intrinsic property of the walls of the tube (Picken, 1953; Robson, 1953).

More recently, the structure of nematocysts both in the discharged and undischarged state, as well as in various states of eversion, have been studied in a variety of coelenterates by means of an electron microscope. Notable advances have been made, especially with nematocysts of *Corynactis* (Skaer and Picken, 1965; Picken and Skaer, 1966), *Metridium* (Hand, 1961; Westfall and Hand, 1962; Westfall, 1964) and *Hydra* (for review Slautterback, 1961; Chapman, 1961; Lentz, 1966). Skaer and Picken (1965; also Picken and Skaer, 1966) have shown that in the spirocysts, microbasic mastigophores and holotrichous isorhizas of *Corynactis viridis* the walls of the undischarged thread are pleated, and it is the unfolding of these pleats during discharge that causes the expansion in width of the thread. Further, the anisometric increase in length may be attributed to the fact that the pleats are in a left-handed helical arrangement and join the core of the thread at angles between 30° and 90° to the long axis. The pleats are wrapped round the central core of the undischarged thread as is shown in Fig. 5.5B; their size diminishes and the pitch-angle increases from the base to the tip of the nematocyst thread. The barbs of this undischarged thread are closely packed into the lumen of the tube as is shown in Fig. 5.5A and as can be seen also in the transverse section. In fact there are normally some

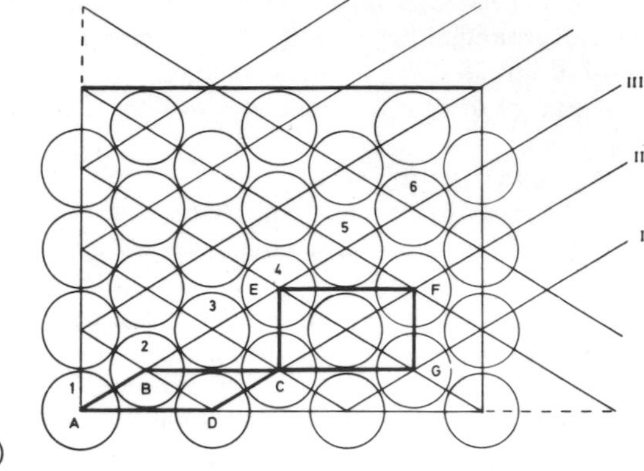

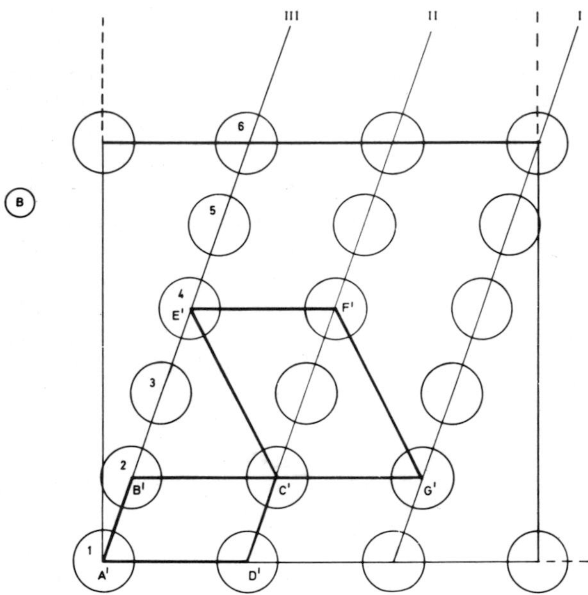

Fig. 5.4. Diagram showing the arrangement of barbs (open circles) as seen under a light-microscope in (A) an undischarged nematocyst thread regarded as a plane sheet, (B) a discharged and expanded thread regarded as a plane sheet. Note that the conversion of the close-packed barb arrangement of an undischarged thread (ABCDEFG in (A) above) into the expanded condition ($A^1B^1C^1D^1E^1F^1G^1$ in (B) below) involves and anisometric increase in area. The axes of the right-handed helices in which the barbs are arranged are labelled I, II and III respectively. (After Picken, 1953.)

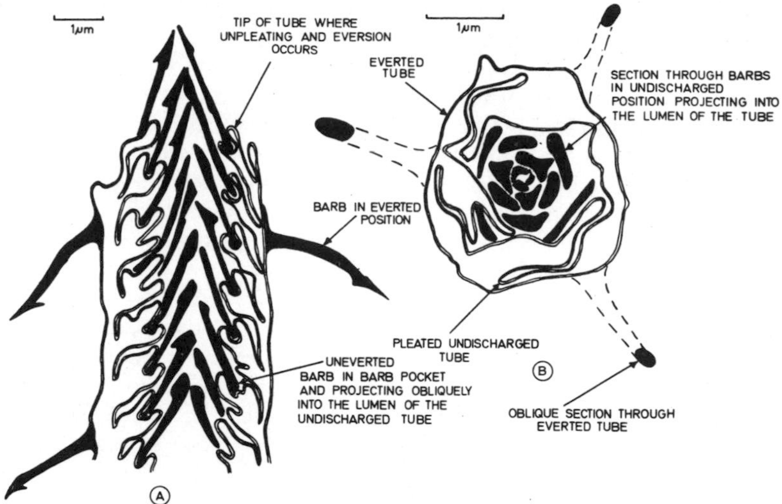

Fig. 5.5. Semi-diagrammatic sections through a partially everted holotrichous isorhiza of *Corynactis viridis*. (Based on electron micrographs of Skaer and Picken, 1965). (A) Longitudinal section through the tip of a partially everted isorhiza showing the zone of unpleating and transformation of the barb pattern from the uneverted to the everted position. (B) Transverse section through a partially everted isorhiza to show pleated arrangement of the walls of the uneverted tube. Note the triradial arrangement of the uneverted barbs.

12 – 15 barbs seen in a transverse section of an uneverted thread and this pattern was interpreted by Skaer and Picken (1965) as a transverse section of three helical rows of barbs. The distances between the bases of adjacent helices is more or less constant at 0·4 to 0·5 μm throughout the length of the uneverted tube and a transverse section shows portions of five barbs from each of the three barb helices. Now, although the whorl separation remains approximately constant, the diameter of the thread decreases towards the tip, and at the same time the length of the barbs decreases. This results in a decrease in the amount of overlap of the barbs near the tip of the undischarged filament so that the number of barbs seen in cross section decreases from 15 to 12 in that region.

Skaer and Picken (1965) then showed that the total surface area of the pleated uneverted tube was similar to that of the smooth cylindrical everted thread, so that eversion evidently involves unpleating at the tip of the advancing thread without any major expansion of the surface area of the walls. The helical arrangement of the pleats would also impart a rotation to the thread during eversion (Picken, 1953). This process not only gives stability to the discharging thread, but also gives a powerful cutting action to the anteriorly-directed barbs at the tip.

Such unpleating, and consequent anisometric expansion of the thread, occurs even when a dried partially everted thread is sectioned and then placed in water again (Robson, 1953; also Skaer and Picken, 1965). A segment of a partially everted thread consists of an outer everted region with the characteristic helical arrangement of spines pointing outwards; inside this tube is the uneverted tube of narrow diameter in which the spines point inwards towards the lumen. When such a portion of nematocyst thread was placed in water, the inner tube underwent anisometric expansion (unpleating), but no eversion occurred, so that the barbs faced inwards. It is thus evident that the eversion which occurs on expansion of an intact thread is a mechanical effect due to the inverted arrangement of the undischarged tube and the fact that it makes contact with water only at the advancing tip. The remaining undischarged portion of the tube is thus drawn up as eversion proceeds. The helical arrangement of barbs on the discharged thread is clearly seen in Fig. 5.6.

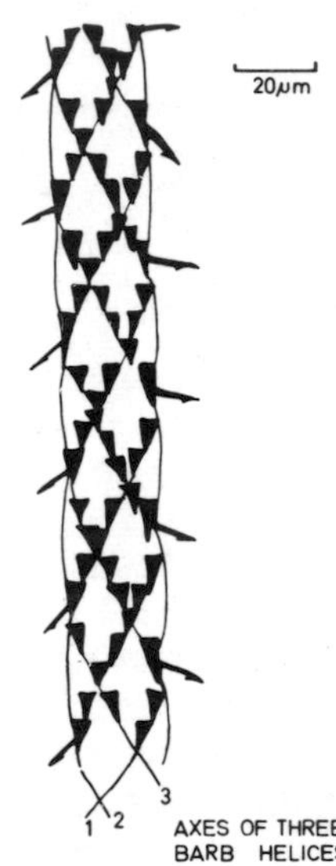

Fig. 5.6. Diagram showing the helical arrangement of everted barbs in a discharged holotrichous isorhiza of *Corynactis viridis.* (Based on Skaer and Picken, 1965.) Note that under an electron microscope the barb bases are triangular rather than circular as seen under a light microscope. The axes of the three helical rows of barbs are indicated.

As has been mentioned on p. 177, the factors controlling discharge are complex, but in many cases mechanical stimulation is of importance. In *Hydra* stimulation of the cnidocil appears to be a major controlling factor (Burnett, Lentz and Warren, 1960; also Lentz, 1966) but such projecting structures appear to be present only in the Hydrozoa (Hand, 1961). It is possible that structures such as the ciliary cones described in *Anemonia sulcata* by Pantin (1942) and in *Corynactis viridis* (Skaer and Picken, 1965; Picken and Skaer, 1966) may serve some excitatory function. There is indeed a close relationship between the trigger-like cnidocil of *Hydra* and the capsule of the cnidoblast. Chapman (1961), Slautterback (1961) and Lentz (1966) have reviewed the structure of the cnidocil of the stenoteles of *Hydra*, and have shown that the cnidocil consists of a central cilium which is surrounded by 20 – 21 dense rods, or supporting rods of the cnidoblast. The same number of larger rods arise from these supporting rods of the cnidoblast and support the cnidocil; microtubules also extend from the region of the cnidocil to the nematoblast capsule at some stages in the development of the nematocyst. It is clear, therefore, that there is some structural evidence for the role of the cnidocil in initiating discharge. Indeed, in *Hydra* the desmonemes, which have long cnidocils, are the first to be stimulated and to discharge on contact with *Artemia*, whilst much closer contact is necessary to stimulate the shorter cnidocils of the stenoteles which therefore tend to be discharged later (Burnett, Lentz and Warren, 1960).

The precise mechanism by which stimulation of the cnidocil initiates discharge is still largely unknown. It has been thought that the dense granules which surround the cnidocil, and which may be 5 – hydroxytryptamine, may play a part not only in exerting the toxic effects of the stenoteles (Welsh, 1960, 1961), but also in discharge (Lentz, 1966). Several authors have emphasised the possibility of a rise in the pressure of the contents of the nematoblast forcing eversion of the thread (Yanagita, 1943; Yanagita and Wada, 1954; for review Lentz, 1966). Other workers have suggested that the wall of the nematoblast may contract (Jones, 1947), but the capsular wall is possibly of a collagen-like substance (Lenhoff, Kline and Hurley, 1957; Johnson and Lenhoff, 1958; Chapman, 1961; Picken and Skaer, 1966) and no contractile elements have been detected either within, or outside the capsule. Mattern, Park and Daniel (1965) suggested that in *Hydra* the supporting rods of the cnidocil and of the cnidoblast may be mechanically disconnected on stimulation and would cause the operculum to open if the contents of the capsule were at a high hydrostatic pressure. Picken and Skaer (1966) have analysed the depression of freezing point of the contents of the holotrichous isorhizas of *Corynactis*, and have indeed found that the capsular contents are at a very high osmotic pressure prior to discharge. Thus values for the depression of freezing point of the contents of undischarged capsules were between $-4{\cdot}16$ and $-5{\cdot}94°$C. During discharge, absorption of water evidently occurred as the depression of freezing point of a sample of capsular fluid at the end of discharge was −3·83° C and fluid exuding from the tip of discharged threads gave freezing points of from −2·54 to −3·45° C. It is not yet known whether the high osmotic pressure of the capsular contents prior to discharge is sufficient to initiate the process, but since an increase in the volume of the tube occurs on discharge, it follows that absorption of water would be necessary during the process of eversion in order to maintain the turgidity of the advancing thread. The cavity of the undischarged thread and the space between the capsular wall and the thread is filled with a flocculent proteinaceous material. It is probably the absorption of water by this substance which results in an increase in the volume of the capsular fluid of at least 200% on discharge (Skaer and Picken, 1965). Such hygroscopic protein thus possesses the necessary properties both for the initiation of discharge following stimulation and initial entry of water into the capsule, and for the maintenance of the turgor of the advancing thread.

2. PREDATORY MOLLUSCS

The adoption of a predatory mode of life amongst the Mollusca is associated with profound modifications of the rhipidoglossan and docoglossan types of radula, and jaws are developed in some forms such as *Clathrus.* In the sponge-eating molluscs *Cerithiopsis* and *Triphora*, the jaws are also prominent and are used to separate off pieces of tissue which are then taken into the buccal cavity by the radula (Fretter, 1951). Members of the Lamellariacea and Cypraeacea are also highly adapted for such an existence, and often have specific food preferences. *Velutina plicatilis,* for example, feeds on *Tubularia indivisa* (Ankel, 1936a) while *V. velutina* eats the ascidians *Ascidia, Phallusia* and *Styela* (Ankel, 1936a; Diehl, 1956). Other genera also feed on ascidians; *Erato* feeds by thrusting the proboscis directly through the inhalant aperture of the colonial ascidians *Botrylloides* or *Botryllus,* while *Trivia* eats through the indigestible body wall of its prey before engulfing the inner tissue.

Other forms such as the ascidian-eating prosobranch *Pleurobranchus* possess a radula whose two halves are opposable as the radula is withdrawn, the gripping action of the radula being aided by the movements of the jaws which push food onto the radular teeth (Thompson and Slinn, 1959). A similar system is used by the herbivorous opisthobranch *Aplysia* which commonly feeds on *Ulva.* Howells (1936) showed that the food is first gripped by the foot and pushed into the mouth; the radular teeth are opposable and the withdrawal of the radula carries the alga into the buccal cavity. After successive movements have resulted in the ingestion of approximately 2 cm of *Ulva,* the jaws grip the weed which is then broken as the radula next withdraws into the mouth.

The aeolids feed on coelenterates (Graham, 1938b) and remove small pieces solely by means of jaws; other opisthobranchs, such as the ascidian-eating dorids, principally employ a suctorial system coupled with a modified radula which is used to penetrate the prey before the tissues are sucked out. *Adalaria proxima*, for example, feeds on the polyzoan *Electra pilosa* by means of the pumping action of its modified buccal mass (Thompson, 1958) whilst *Goniodoris nodosa* and *G. castanea*, which feed on *Dendrodoa* and *Botryllus* respectively, similarly cut into the prey by means of the modified radula before sucking out the tissues (Forrest, 1953).

So far we have considered principally the mechanisms by which intertidal molluscs feed upon relatively soft-tissued sessile organisms. Such habits are especially characteristic of the opisthobranchs although *Onchidoris fusca* preys upon barnacles (Barnes and Powell, 1954). Certain gastropods, however, particularly members of the Mesogastropoda and Neogastropoda, have developed a remarkable ability to penetrate the tissues of even thick-shelled bivalves, gastropods and barnacles. The two main groups of prosobranchs in which such boring is well-developed are the Muricacea [including the rough tingle *Ocenebra*, the American tingle *Urosalpinx*, and the dog-whelk *Thais* (= *Nucella*)] and the naticids which includes the two common sandy shore species *Natica catena* and *N. alderi*.

Members of the Muricacea attack a variety of organisms including barnacles, gastropods and lamellibranchs. *Thais* (= *Nucella*) *lapillus*, for example, feeds upon barnacles, mussels and limpets whilst *Urosalpinx* and *Ocenebra erinacea* eat young bivalves, especially *Venus* and oysters. The mechanism of boring in these animals has been recently described in detail by Carriker (1955, 1958, 1959, 1961). He has suggested that penetration is achieved by a combination of chemical and mechanical activity. In *Urosalpinx*, and also in *Thais*, the prey is first gripped by the foot and the proboscis is applied againt the periostracum of the bivalve. The radula is then rasped against the shell for a short while but during this time makes little impression on the shell. Then the animal creeps forward and applies the sucker-like "accessory boring organ", (Carriker, 1961) to the etched area. After a time, which varies from a few minutes up to one hour, the accessory boring organ is retracted into a sac lying in the pedal tissue and rasping begins again; this sequence persists until the shell of the prey is penetrated. The nature of the secretion of the accessory boring organ is not known. However, since chips of calcite are visible in the gut of *Nucella* (Fretter and Graham, 1962), it is clear that the calcareous matter is not dissolved, although it may be softened by a chelating agent or by an enzyme attacking the organic matrix of the shell (Carriker, 1959). Despite the uncertainty of the nature of the secretion of the accessory boring organ, Carriker has demonstrated that the presence of both accessory boring organ and proboscis are essential for the boring process. Removal of either proboscis or accessory boring organ prevented boring until both organs had regenerated.

The naticids feed mainly on a wide variety of bivalves including *Donax vittatus, Tellina, Mactra corallina, Macoma balthica* and *Nucula* (Piéron, 1933); the mechanism by which they bore into their prey has been fully described by Ziegelmeier (1954). The bivalve is first held by the propodium, the proboscis being twisted through 90° to the

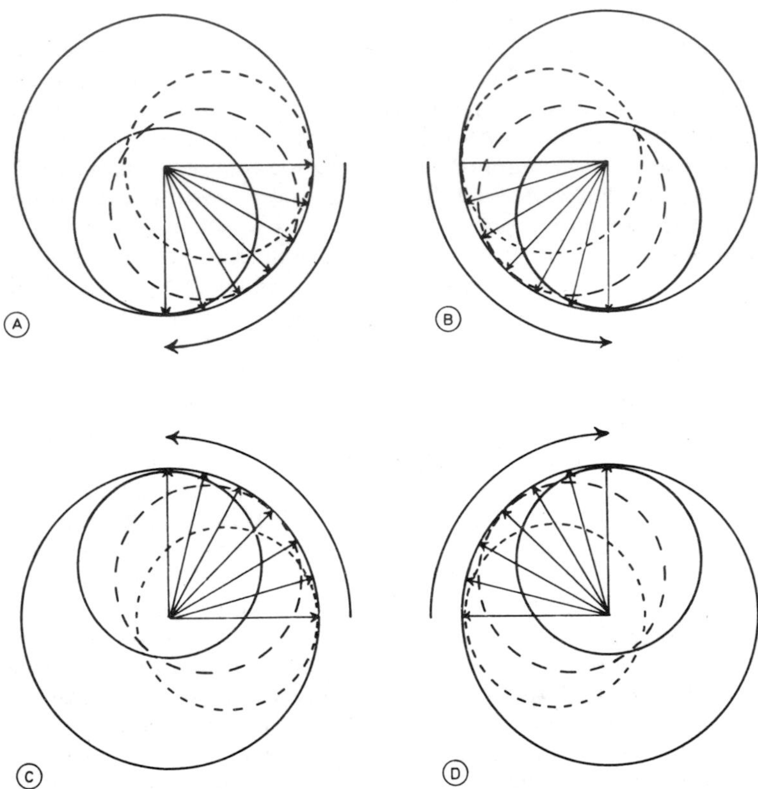

Fig. 5.7. Diagram illustrating four phases of the movement of the proboscis of *Natica* during boring. The overall borehole is indicated by the large circles, A, B, C and D. Boring is accomplished in sectors with a rest between each phase. The starting point of the radula at each phase is indicated by a dotted circle, an intermediate position by a dashed circle and the final position by a full circle. The proboscis therefore starts twisted laterally, the direction of its untwisting in the different phases being indicated by the curved arrows. The radial arrows indicate the direction of radular movement during each phase of rasping. (Modified from Ziegelmeier, 1954.)

right or left before being applied to the shell. As the radula makes its successive rasps, the proboscis is gradually rotated back to its normal position (Fig. 5.7). There is then a brief rest of a few minutes during which an accessory boring organ on the ventral lip is brought near to the hole but is not inserted into it as in the Muricacea. The proboscis is then applied in a twisted position to another quadrant of the drilling area and gradually untwists again in such a way that the final rasping movements are in an anterioposterior direction. This sequence is repeated and results in the production of a cavity with a characteristic central knob where the radular action is less effective than towards the margins of the drilling area (Fretter and Graham, 1962).

Ziegelmeier noted the presence of white shell material passing down the oesophagus which suggests that secretions from the accessory boring organ do not dissolve the shell. From a study of the structure of the organ, he suggested that it was a tactile sense organ necessary for boring while *Natica* is buried in the sand. Fretter and Graham (1962), however, have shown the presence of gland cells in the accessory boring organ of *Natica catena,* and concluded that the organ may be implicated in the boring process in much the same way as it is in the Muricacea. Similar periods of radular activity and quiescence, during which the accessory boring organ is brought near to the hole, have been noted by Ziegelmeier and it is reasonable to assume that here too, chemical attack of the shell takes place as in *Urosalpinx* (Carriker, 1955, 1958, 1959, 1961).

3. OTHER PREDATORY ORGANISMS

Other intertidal animals such as the regular echinoid *Psammechinus miliaris* also feed upon barnacles, cockles and occasionally adult oysters (Hancock, 1957). This species of sea urchin first erodes the hinge ligament and this possibly leads to a relaxation of the adductor muscles. The urchin may then insert its teeth into the hinge mechanism after which contraction of the adductor muscles of the bivalve causes the valves to gape so that the soft tissues can be ingested. *Psammechinus miliaris* may also eat as many as 21 barnacles per day during March (Hancock, 1957), although this is less than small *Asterias* which have been reported to eat as many as 49 barnacles per day (Hancock, 1955).

It is generally assumed that the errant forms of intertidal polychaetes are macrophagous carnivores, and most bear powerful jaws, which would appear to support this view. Nevertheless *Nereis diversicolor* with its powerful jaws and paragnaths, hitherto regarded as a typical predatory polychaete (Linke, 1939), has recently been shown to be a filter-feeder on occasions (Harley, 1950; also p. 191). Sanders (1956) has shown that the American species *Nephtys incisa* is also a deposit-feeder although Clark (1962) has suggested that this is not typical of the family as a whole. Gut analyses of *Nephtys cirrosa* showed the diet to consist largely of a variety of polychaetes including *Nephtys* as well as eggs and nauplii; *Nephtys hombergi* also appears to feed on a similar diet (Clark, 1962).

D. SCAVENGING

A variety of intertidal animals are scavengers. These include the shore crab *Carcinus maenas*, the sand-hopper *Talitrus saltator* (also p. 146), *Nassarius* as well as the whelk *Buccinum undatum* which although normally sublittoral, is occasionally to be found at extreme low water, and the isopod *Idotea* which is often abundant amongst intertidal seaweed.

Idotea is a good example of an omnivorous scavenger (Naylor, 1955), and although seaweeds provide the bulk of the food available to idoteids, most of the species will eat animal remains, or even living animals when these are available. The process of feeding in *Idotea* is similar to that described for the intertidal isopod *Ligia oceanica* by Nicholls (1931a,b). In *Idotea* spines on the maxillules and maxillipedes scrape the food which is then bitten by the incisor processes of the mandibles (Fig. 5.8). Toothed spines on the maxillules and mandibles

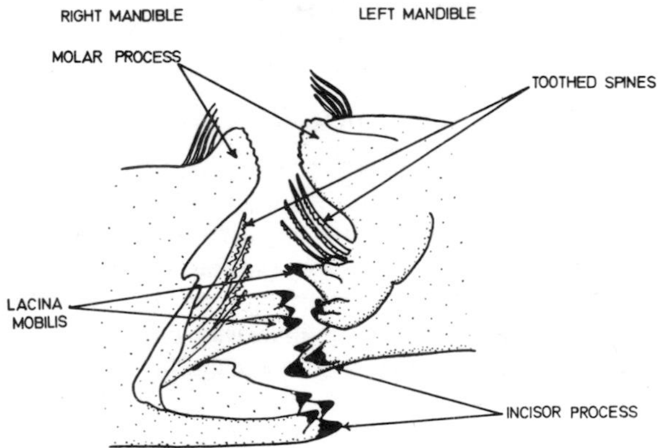

Fig. 5.8. Diagrammatic view of the opposed mandibles of *Idotea*. (After Naylor, 1955.)

then push food upwards towards the molar processes of the mandibles where it is broken into smaller particles. Escaping particles of food are brushed forward by a series of brush setae on the maxillules, maxillae and maxillipedes where they are transferred to the mandibles and thence move upwards to the molar processes by the lateral movements of the mandibles. Fine particles of food which tend to adhere to the mouthparts are cleaned off by means of spines on the maxillipedal palp and on the first leg. This scraping and biting method of feeding is clearly well-suited to a scavenging mode of life where a wide variety of materials is utilised for food.

E. SUSPENSION FEEDING

Suspension feeding is of widespread occurrence amongst intertidal invertebrates and examples may be found amongst all major classes of shore animals. Nevertheless the precise mechanism of food collection is known in detail for only a few representative forms, some of which are dealt with below. More general aspects of suspension-feeding, together with suspension feeding mechanisms in other organisms, are discussed in a series of excellent reviews by Yonge (1928), Jørgensen (1955, 1966), von Buddenbrock (1956) and van Gansen (1960). For comparative purposes, three main methods may be recognised by which intertidal organisms abstract suspended matter from the water. These include (1) mucous bag mechanisms, (2) ciliary mechanisms and (3) feeding by means of setose appendages.

1. MUCOUS BAG MECHANISMS

MacGinitie (1939a) has described the mode of suspension feeding adopted by *Chaetopterus variopedatus.* This animal has a highly specialised structure; the main organs concerned with feeding are the peristomial funnel and lips, the dorsal ciliated groove (which ends in a dorsal cupule on the 13th segment), the pair of aliform notopodia of the 12th segment, and the fans on the 14th, 15th and 16th segments (Fig. 5.9a).

The first part of the feeding sequence involves the spreading of the aliform notopodia out against the sides of the leathery U-shaped tube near one or other of the ends. The distal ends of the aliform notopodia then secrete mucus, this process gradually spreading inwards along the length of the notopodia. The cilia on the inner surfaces of the aliform notopodia then carry the mucus down to the dorsal surface of the worm; at this stage the mucus forms a sheet between the faces of the aliform notopodia. The sheet is then extended posteriorly as a bag by the cilia of the dorsal groove (5.9b), and ends in the concave part of the dorsal cupule. Thus a closed bag is formed whose anterior margins are the aliform notopodia, and whose posterior end is sealed and contained by the dorsal cupule. The aliform notopodia are meanwhile held in contact with the walls of the tube and the three fans beat in sequence, thus drawing water through the bag and along the dorsal surface of the worm. These activities result in suspended matter being trapped in the

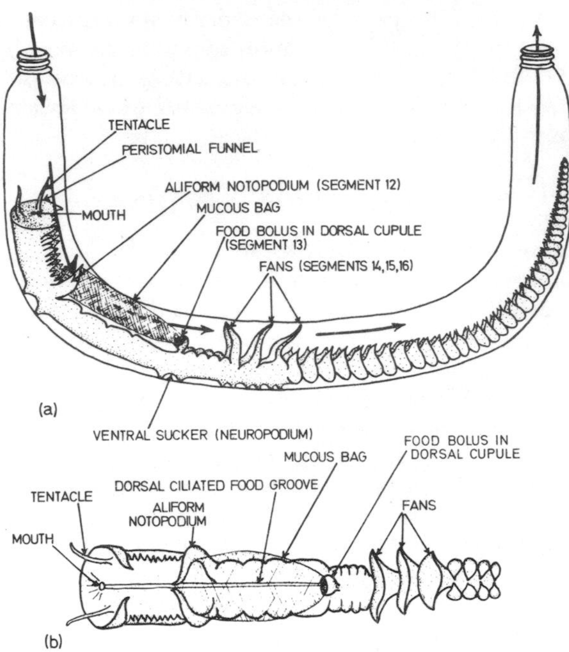

Fig. 5.9. Diagrams showing the feeding currents in *Chaetopterus variopedatus* (After MacGinitie, 1939). (a) Lateral view of the animal in its tube. (b) View of the dorsal surface showing position of food groove.

mucous bag which is continuously rolled up into a food bolus by the cilia of the dorsal cupule and replaced at a corresponding rate by the cells of the aliform notopodia. Large particles are partially excluded by the constricted aperture of the tube, but those which do enter are allowed to pass between the aliform notopodia and the tube thus being excluded from the mucous bag.

When the food bolus in the dorsal cupule has reached a certain size, secretion of mucus by the aliform notopodia ceases and the remainder of the bag is rolled up by the cilia of the dorsal cupule. This organ turns anteriorly and ejects the bolus into the dorsal food groove whose cilia then beat anteriorly and carry the bolus of mucus with entrapped particles forward towards the mouth where it is swallowed.

It is clear that such a feeding mechanism is well-adapted for entrapping not only organic debris, but also the nanoplankton, much of which would pass through all but the finest of sieving mechanisms. Such a food source may be especially abundant in estuaries where the nutrients are enriched by run-off from the land. It is therefore not altogether surprising to find that another polychaete, *Nereis diversi-*

color, which is characteristic of estuarine conditions, has adopted a mucous bag feeding mechanism. *Nereis diversicolor* has often been regarded as a typical errant polychaete, and the presence of its powerful jaws has led to the reasonable assumption that the animal is an active predator (for example Linke, 1939; MacGinitie and MacGinitie, 1949). However, Harley (1950) has demonstrated that *Nereis diversicolor* is a suspension-feeder on occasions.

When about to start feeding by this method, the animal moves to one end of the tube and expands the anterior region of the body from the 6th – 13th setigers. Mucus is then secreted by the parapodial glands and sticks to the wall of the tube. The animal retreats down the tube, rotating repeatedly from side to side through a semi-circle and moving the first 4 – 14 pairs of parapodia in circles. The setae are meanwhile alternately protruded and withdrawn, the whole sequence resulting in the construction of a feeding funnel from the threads secreted by the parapodial glands. Thus the completed funnel is attached anteriorly to the walls of the tube and tapers posteriorly to the body of the worm. A feeding current is created by wave-like undulations of the posterior segments, particle-laden water being drawn through the feeding cone. The worm then moves forward and swallows the mucous cone together with its entrapped particles. Such feeding cycles may follow one another intermittently for up to 2 hr with only short intervals of a few minutes between each cycle.

2. CILIARY MECHANISMS OF FOOD COLLECTION

(a) Polychaetes

Such mucous bag mechanisms are shown by rather few intertidal invertebrates, of which *Chaetopterus* and *Nereis diversicolor* are the most obvious examples. A much more common mechanism of suspension feeding is that adopted by the sabellid and serpulid fan worms, ascidians, certain gastropods, as well as by the vast majority of the bivalves. In all of these forms mucus plays an important part in the transport of suspended material to the mouth but the initiation of the feeding current, as well as the deflection of the suspended material into the food grooves, is carried out by cilia.

The most detailed description of the mechanism of feeding in sabellid polychaetes is that of *Sabella pavonina* given by Nicol (1930) who also referred to feeding processes in some other tubicolous polychaetes. During feeding in *Sabella pavonina* the two halves of the branchial crown are expanded in such a way as to form a flattened plate-like funnel (Fig. 5.10). The filaments of the crown are kept in position by means of a basal web and their pinnules stick out from each

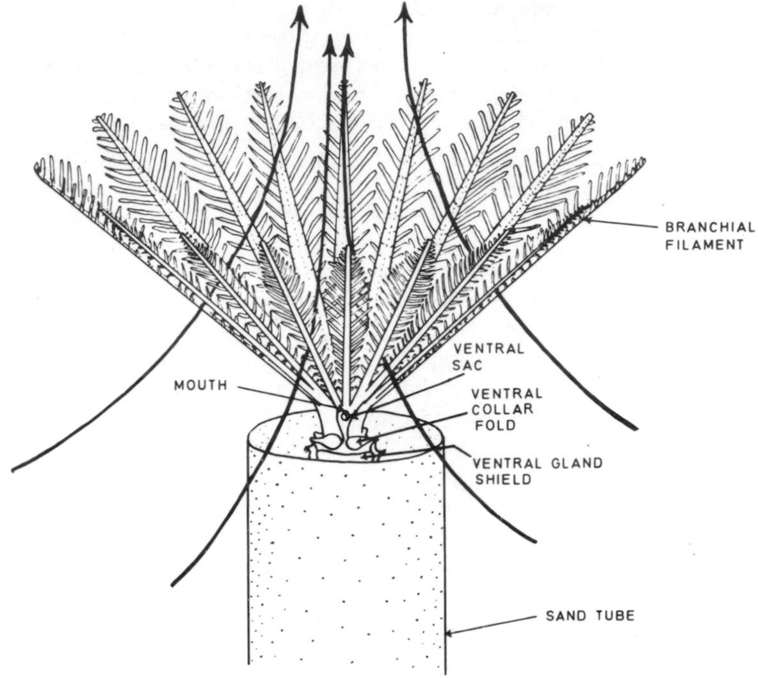

Fig. 5.10. Simplified diagram showing the direction of the feeding currents in *Sabella*. The longitudinal food grooves are indicated by a broken line.

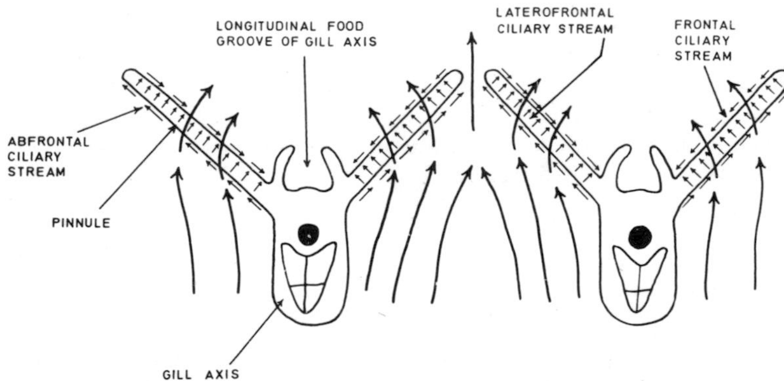

Fig. 5.11. Diagrammatic section of two branchial filaments of *Sabella* showing the direction of flow of the feeding currents (large arrows) and of the beating of the cilia of the pinnules (small arrows). (After Nicol, 1930.)

side at an angle of approximately 90° to one another forming an interlocking food net towards the base of the funnel. The feeding current is set up by the activity of abfrontal cilia (Fig. 5.11) which beat from the base to the tip of the pinnules. The water then passes between the pinnules owing to the activity of the latero-frontal cilia (Fig. 5.11 and 12) which beat at right angles to the abfrontal cilia, and thus enters the branchial funnel. Particles are meanwhile thrown into the frontal groove of the pinnules by the latero-frontal cilia and are carried down to the base of the pinnule by the frontal cilia (Fig. 5.11). Another factor which tends to throw particles into the frontal groove is the eddy formed in this region by the flow of water past the pinnule.

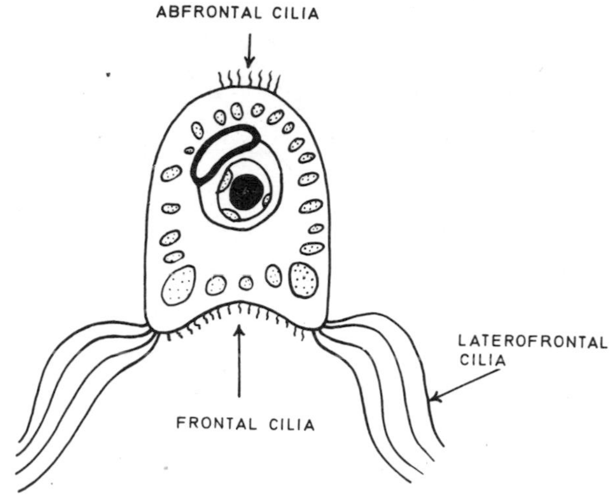

Fig. 5.12. Transverse section through a pinnule of *Sabella* showing the ciliation. (After Nicol, 1930.)

Once particles reach the base of the pinnule they enter the longitudinal groove of the gill axis and are there sorted into three grades. This food groove is ciliated on its inner and outer faces, the cilia of the inner face beating outwards from the base of the groove towards its free edges (Fig. 5.13). There is also a pair of longitudinal ridges on the inner face of the groove which effectively divides it into a basal region into which small particles can pass, a middle region where medium sized particles come to rest, and an outer region which excludes large particles from the food groove (Fig. 5.13). Along the base, and along the median longitudinal ridge as well as along the outer margin of the food groove the cilia are arranged in tracts which do not beat outwards but towards the base of the branchial filaments, and

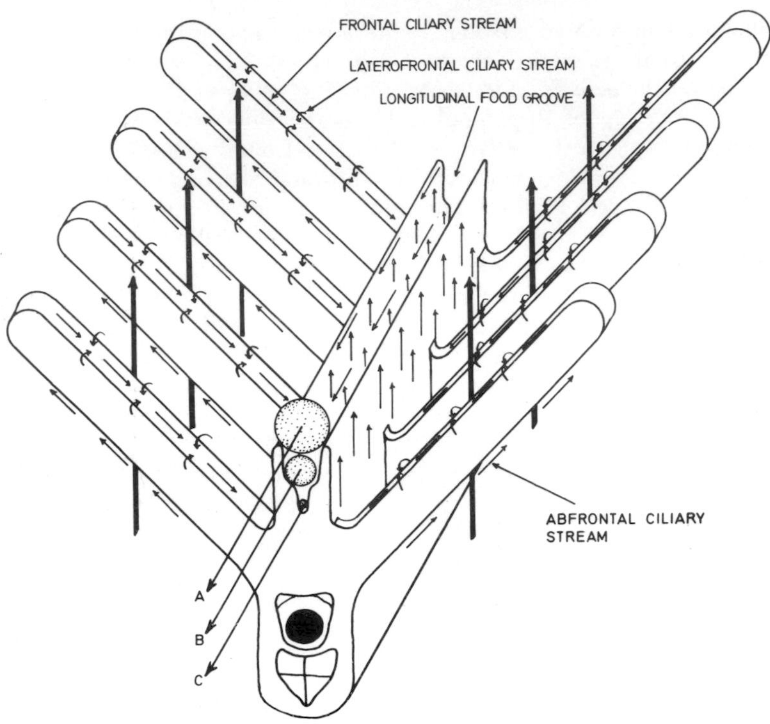

Fig. 5.13. Diagrammatic section through a branchial filament of *Sabella* showing the ciliary streams on the pinnules and the mode of functioning of the longitudinal food groove. A, B and C are three particle sizes sorted by the structure of the food groove. The feeding currents are indicated by heavy arrows. (Based on Nicol, 1930.)

hence transport particles towards the mouth. Here the groove abuts on the lateral and dorsal lips upon which three corresponding ciliary tracts are developed. Fine particles are carried in a basal groove to the mouth; medium particles pass along a ridge on the dorsal lip and then into a groove on the lateral lip where they are subsequently carried to the ventral sac to assist in tube formation. Coarse particles are rejected *via* the palps.

Thus the food collection mechanism in *Sabella* depends upon the interaction of four ciliary tracts comprising one abfrontal, one frontal, and two laterofrontal rows of cilia, the final suitability of the food material being assessed on a basis of paricles size (cf bivalves, p. 209). Much the same system operates in other sabellids such as *Branchiomma, Dasychone* and *Myxicola.* In serpulids such as *Filograna, Pomatoceros* and *Hydroides,* however, there are no abfrontal cilia the current entering the branchial funnel being created by the laterofrontal cilia. The water therefore enters the branchial funnel at right angles to the pinnules rather than obliquely (as in the sabellids).

(b) Prosobranchs

As mentioned earlier (p. 189) many intertidal organisms are suspension feeders. We have considered above only polychaete examples and may now review some of the devices used by those molluscs which use this system of feeding. Suspension feeding is found not only in the lamellibranchs, where it attains its highest development, but also in the prosobranch gastropods. It is of interest, therefore, to consider food collection the gastropods first and then examine the more specialised mechanisms which occur in the bivalves. In a typical prosobranch such as *Littorina littorea* or *Buccinum undatum,* the respiratory current is drawn into the mantle cavity at the left of the head by the activity of the lateral cilia of the gill filaments. This inhalant opening may be prolonged into a siphon as in *Buccinum,* or may be a simple aperture as

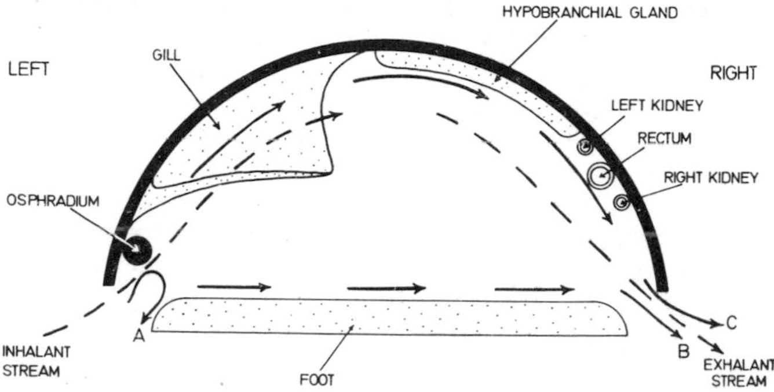

Fig. 5.14. Diagrammatic vertical section through a pectinibranch prosobranch (e.g. *Littorina*) showing the course of the three main particle-rejection tracts. A, current which rejects coarse particles; B, current which rejects medium particles and C, path taken by fine particles.

in the mesogastropods. The water then passes between the gill leaflets and passes out *via* the exhalant aperture on the right side of the head (Fig. 5.14). It is clear that sediment drawn in with this respiratory current must be eliminated to avoid interference with gas exchange. Yonge (1937b) has shown that in *Aporrhais* disposal of the suspended matter is brought about by three ciliary tracts which occur also in a wide variety of other pectinibranch prosobranchs. The first tract of cilia (A in Fig. 5.14) is located on the margin of the inhalant aperture and carries large particles to the exterior by way of the inhalant opening. Such particles tend to settle almost at once where the relatively narrow inhalant aperture opens into the wide mantle cavity and where a reduction in the velocity of the inhalant stream occurs (Orton, 1912). The second tract of cilia (B in Fig. 5.14) lies on the floor of the mantle cavity and carries medium-sized particles across to the right side where they are conveyed forward to the exhalant aperture. The third stream (C in Fig. 5.14) is set up by the frontal cilia on the gills and by cilia on the surface of the hypobranchial gland which lies in the roof of the mantle cavity (Fig. 5.14). Fine particles are carried by this stream between the gills and across the hypobranchial gland where they are entangled in mucus and carried to the exhalant aperture in a consolidated state. The osphradium lies in the inhalant stream and probably functions as a chemoreceptor in some forms or as a 'silt-detector (Hulbert and Yonge, 1937).

The adoption of suspension feeding has involved the modification of such rejection tracts so that particles initially voided through the exhalant aperture (currents B and C), as well as particles normally ejected through the inhalant aperture (current A), are utilised for food. Orton (1912, 1913a) has described the feeding mechanism in *Capulus ungaricus* and in two members of the Calyptraeidae, the American slipper limpet *Crepidula fornicata* (also Yonge 1928, 1938; Werner, 1952, 1953) and the Chinaman's hat limpet *Calyptraea chinensis* (Yonge 1928, 1938). In each of these genera, suspension feeding has been associated with a corresponding loss of mobility, although suspension feeding can also occur in the free-living *Turritella communis* (Graham 1938a; Yonge, 1946) and also in *Bithynia tentaculata* (Schäfer 1953a,b; Lilly, 1953; also Fretter and Graham, 1962). Since the food of *Turritella communis* consists of a large proportion of material swept up from the substratum (Hunt, 1925), this animal is perhaps best included with the deposit feeders. The feeding mechanism is therefore described in more detail on p. 230.

Orton (1912) and Yonge (1938) have described the feeding mechanism of *Capulus ungaricus,* which has adopted a sedentary suspension-feeding mode of life. Yonge (1938) showed that the gill in *Capulus* is large and extends obliquely across the roof of the mantle cavity which also contains a large osphradium (Fig. 5.15). Fig. 5.16 shows a dorsal view of the mantle cavity in which the courses of the ciliary tracts are indicated. The inhalant current is created by the activity of the lateral cilia and impinges on the osphradium. Dense particles settle just inside the inhalant aperture and are then carried anteriorly to an ingestion area on the dorsal surface of the foot; this feeding current represents the rejection current A of a typical prosobranch (Fig. 5.14). Medium particles are carried across the floor of the mantle cavity to the right hand side of the animal. The fine particles are carried by the frontal and abfrontal cilia to the tip of the gill filaments; here currents B and C unite near the exhalant aperture and are carried round to the ingestion

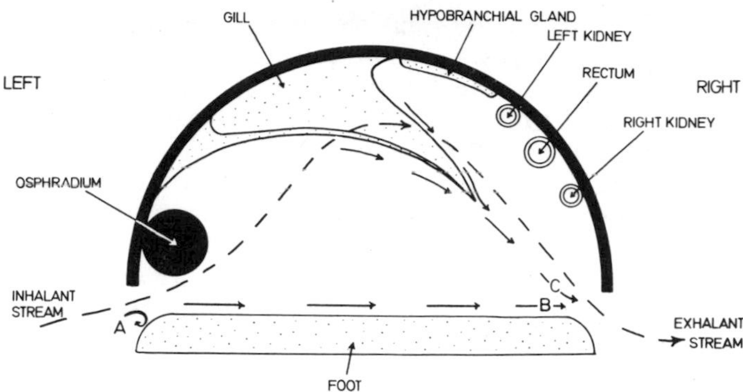

Fig. 5.15. Diagrammatic vertical section through *Capulus ungaricus* showing the ciliary currents in the mantle cavity. A, current carrying coarse particles; B, current carrying medium-sized particles; C, current carrying fine particles. Note the increase in size of the osphradium and in the length of the gill compared with a typical pectinibranch prosobranch (see Fig. 5.14). Note also that current A, turns anteriorly rather than to the outside whilst currents B and C unite on the right-hand side near the exhalant aperture of the mantle cavity.

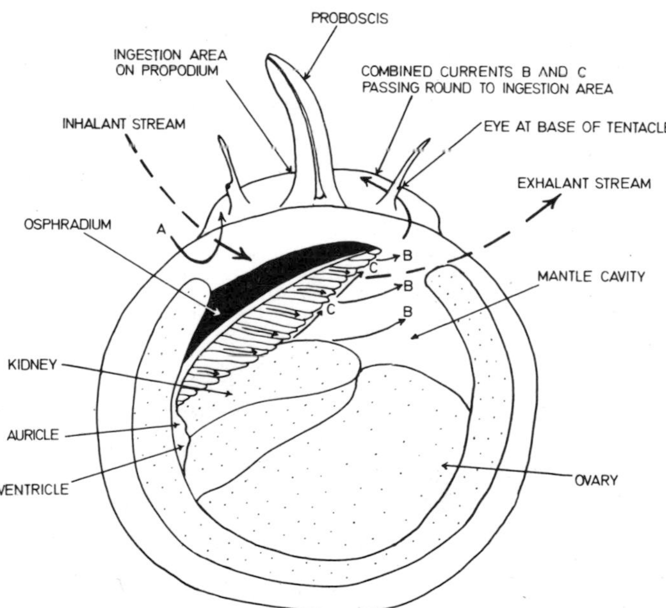

5.16. Dorsal view of *Capulus ungaricus* after removal from the shell. (After Fretter and Graham, 1962.) The courses of the feeding currents are indicated by arrows. A, current carrying coarse particles directly to ingestion area; B, current carrying medium-sized particles; C, current carrying fine particles.

area in front of the mouth. This area is well supplied with mucous glands (Yonge, 1938) whose secretions entangle the food particles which are then engulfed with the aid of the proboscis; the latter has a groove along its length in which food is conveyed to the mouth partly by ciliary and partly by muscular means.

In *Crepidula*, the elaboration of suspension feeding has been associated with an enormous elongation of the gill filaments coupled with an increase in the size of the mantle cavity and a reduction of the osphradium. An endostyle is developed on the inhalant side of the gills and produces mucus which is carried on the frontal surface of the gill filaments together with entrapped particles (Fig. 5.17). During feeding,

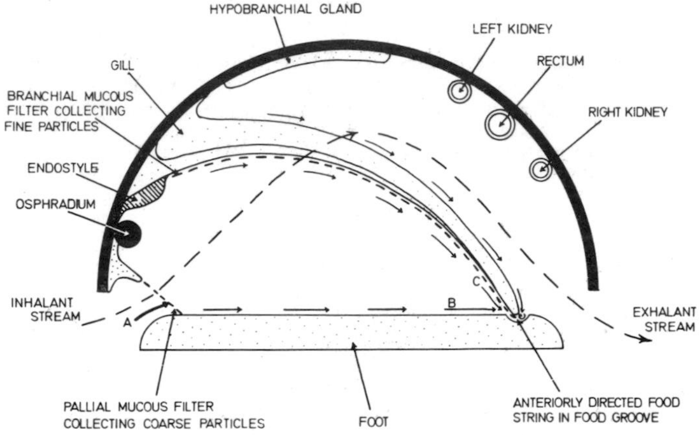

Fig. 5.17. Diagrammatic vertical section through *Crepidula fornicata* showing the ciliary currents in the mantle cavity. A, current carrying coarse particles; B, current carrying medium-sized particles; C, current carrying fine particles. Note the endostyle as well as pallial and branchial mucous filters. Note also the longitudinal food groove in which particle-laden mucus is wound into a food string.

large particles are strained off by means of a mucous filter which is secreted continuously from a filter gland and which covers the inhalant aperture (Werner, 1952, 1953; for review, Jørgensen, 1966). The particle-loaded mucus is then carried into a food pouch situated in the edge of the mantle just anterior to the mouth; food may then be either ingested or rejected as pseudofaeces. The path of the dense particles into the food pouch thus represents a modification of current A of Fig. 5.14. Some medium particles are carried across the floor of the mantle cavity to the right hand side where they enter a longitudinal food groove (Fig. 5.17); this is rejection current B of Fig. 5.14. Finally, most of the medium and fine particles are entangled by a second mucous filter secreted by the endostyle and lying against the frontal surface of the gill. Such particles are thus prevented from passing between the filaments and instead are carried to the tips of the gill filaments; the tip of the gill is applied to the longitudinal food groove and there the mucus is twisted into a rope. (Fig. 5.17). In this way medium and fine particles entangled in mucus are carried anteriorly towards the mouth. Such strings of food-laden mucus are grasped by the radula and drawn into the mouth where they are retained by the mandibles prior to swallowing.

As Yonge (1938) pointed out, the adoption of suspension feeding in these prosobranchs has resulted in profound modifications of the structure and size of the gill from its respiratory function in *Buccinum* and *Littorina.* Firstly there is an increase in the length of the filaments relative to their width at the base. In *Buccinum* and other prosobranchs the height is approximately equal to the width of the filament at its base, but in *Capulus* the ratio of height to width is approximately 3:1 and in *Crepidula* 26:1 (Fig. 5.18). The general arrangement of the

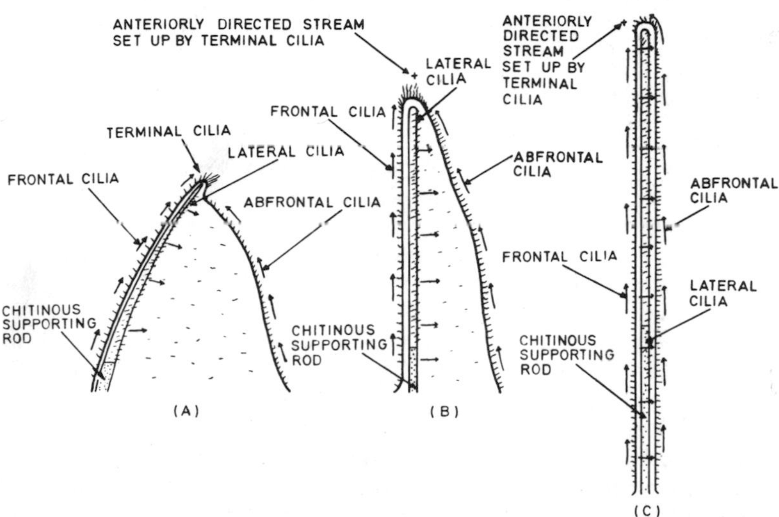

Fig. 5.18. Diagram of a gill filament of (A) *Buccinum undatum,* (B) *Capulus ungaricus* and (C) *Crepidula fornicata* showing the different proportion of the gill leaflets in the three genera. (After Yonge, 1938.)

ciliary tracts is similar in all three types, but whereas mucous glands occur in the lateral face of the leaflets in *Buccinum,* they are confined to the frontal surface in *Capulus,* thereby aiding in the trapping of particles on this surface. Mucous glands are rare in the filaments of *Crepidula;* the scarcity of such glands is associated with the presence of an endostyle which renders mucous glands on the filaments unnecessary.

Thus the adoption of suspension feeding in prosobranchs has been associated with the conversion of all three particle rejection tracts of a typical prosobranch into food streams. In *Capulus* there has been an enormous enlargement of the gill which is made possible by its oblique arrangement in the mantle cavity. Suspension feeding has also involved an elongation of the length of the filaments, and the development of mucous glands in their frontal surfaces. In *Crepidula* there is an extreme elongation of the gill filaments, coupled with the development of a food pouch and endostyle. It seems likely that, as Yonge (1938) suggested, the change in the shape of the filaments from triangular to linear has the effect of increasing the area carrying lateral cilia (thus increasing the flow of water through the mantle cavity). It is important to notice, however, that despite these adaptations to suspension feeding, such prosobranchs still grasp the food by means of a radula, and complex sorting devices such as occur in the bivalves are absent. It is this latter group, upon which the adoption of suspension feeding has imposed its most profound effects, that is considered next.

(c) Bivalves

Within the lamellibranchs, the Protobranchia including the genus *Nucula*, have been shown to be primarily deposit-feeders (for review, Yonge, 1928), the gills being mainly concerned with the production of a respiratory current. In this group the palp proboscides, which are tentacle-like extensions of the labial palps, are used for the collection of food particles from the substratum, although Stasek (1961, 1965) has recently shown that the gills may also play an important part in food collection (p. 227). The two main groups of lamellibranchs which have adopted suspension feeding are the Filibranchia and Eulamellibranchia. In both of these groups, as in the prosobranchs considered above, suspension feeding has involved the conversion of silt rejection tracts into food streams, with a corresponding elongation of the filaments and increase in complexity of their ciliation. In addition, although the sorting mechanisms are still on a quantitative basis according to the density of the particles, they are much more sophisticated than those of the Capulidae of Calyptraeidea. This increase in complexity is brought about not only by the presence of sorting mechanisms on the gills (Yonge, 1928; Atkins, 1936, 1937a,b, 1938, 1943; Jørgensen 1955, 1966), but by the appearance of ridged labial palps which sort out, and transfer particles to the mouth (Wallengren, 1905; Kellogg, 1915; Yonge, 1926; for review, Yonge, 1928; Jørgensen, 1955, 1966).

In the protobranchs such as *Nucula*, the gills represent the least

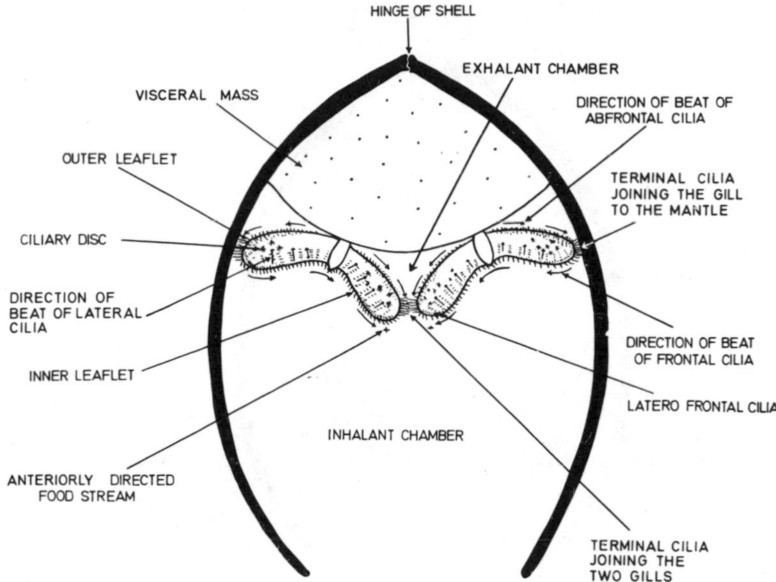

Fig. 5.19. Diagrammatic transverse section through *Nucula* (Protobranchiata) showing the direction of the ciliary streams set up by the gills. (Modified from Atkins, 1936.)

modified condition each consisting of a series of flat leaflets alternating on each side of a central axis (Fig. 5.19). Each leaflet bears a series of abfrontal cilia, frontal cilia, lateral cilia and, near the tip, laterofrontal cilia. In addition, there are lateral ciliated discs which join the leaflets to their neighbours, as well as terminal cilia which unite the gill with the mantle laterally and with the leaflets of the other gill in the midline (Orton, 1912; Yonge, 1928). The respiratory current is drawn in by the activity of the lateral cilia which beat across the surface of the leaflets. The frontal cilia in *Nucula* carry particles to the midline where they are carried anteriorly (Fig. 5.19) and normally rejected. Caspers (1940), however, has shown that *Nucula* is capable of ingesting carmine and other particles suspended in the inhalant stream. This protobranch is therefore capable of suspension feeding, although the complexity of the labial palps suggests that these play a major part in nutrition under natural conditions (also Stasek, 1961; 1965 and p. 227).

The Filibranchia, which include the scallop *Pecten*, the mussel *Mytilus*, and *Glycymeris* and *Arca*, show a more specialised gill structure. In these animals the gill leaflets have become elongated into filaments, the tips remaining in contact with one another medially and with the mantle laterally. This results in the formation of a W-shaped

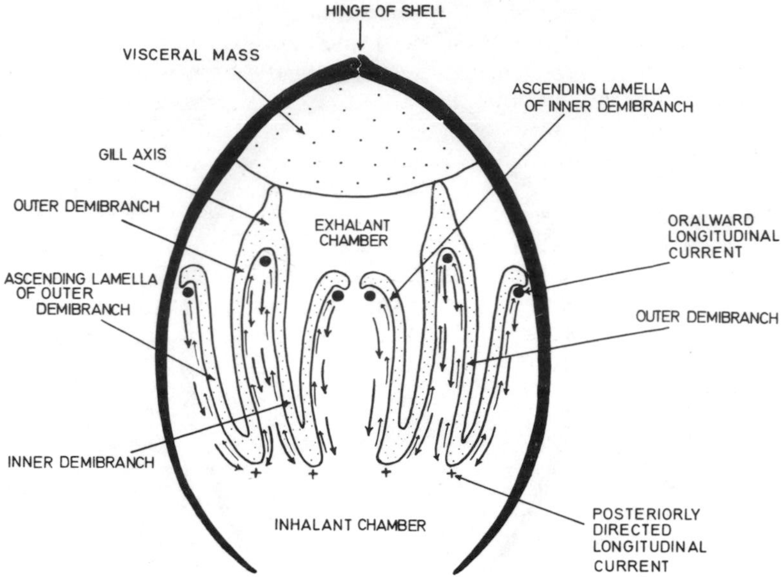

Fig. 5.20. Diagrammatic transverse section through *Arca* (Filibranchiata) showing the form of the gill and the direction of the ciliary currents. (Based on Atkins, 1936.) A large dot indicates the position of oralward longitudinal currents. A cross indicates posteriorly directed longitudinal currents.

gill on each side, since each filament is now U-shaped (Fig. 5.20) and is composed of ascending and descending regions. The ascending and descending arms of each filament are united to each other by tissue junctions whilst the filaments are united laterally by ciliary junctions as in the Protobranchia. The ciliation of the filaments is otherwise the same as described above for *Nucula.* There are, in addition, a number of anteriorly and posteriorly directed ciliary tracts, the location and direction of which have been described in detail by Atkins (1936, 1937a, b; also Jørgensen 1955, 1966 and p. 203).

The Eulamellibranchia includes the majority of suspension-feeding lamellibranchs; typical genera include the cockle *Cardium,* the venerids, the clam *Mya,* and the razor shells. In these animals the gill filaments are similar to those of the Filibranchia, except that the interfilamentar ciliated discs are replaced by tissue junctions. This renders the whole gill structure more solid than in the Filibranchia, and occludes the suprabranchial space to a series of chambers (Fig. 5.21 and 22). Particles are effectively retained on the frontal surface of this rigid structure and abfrontal cilia are absent. Many filibranchs and eulamellibranchs have not only anteriorly-directed food grooves, but also rejection tracts on the gills. This separation involves the use of sorting devices on the gills many of which are described by Atkins (1936, 1937a,b; for review Jørgensen 1955, 1966). Six different types of mechanism by which particles of different density are separated on the gills of intertidal lamellibranchs may be recognised, one or more of which may be present in a single animal.

1 In the filibranchs *Glycymeris* and *Arca* the labial palps are small, and there is a correspondingly efficient sorting mechanism on the gills themselves. Each filament has three tracts of cilia on its frontal surface; these are a median tract of coarse ventrally-beating cilia which become active only when stimulated by heavy particles on the gills. On each side of the coarse cilia is a tract of fine, dorsally-beating cilia which transport food into orally-directed currents along the dorsal edges of the gill lamellae (Fig. 5.20). Posteriorly directed rejection currents run along the ventral edges of the demibranchs and transport large particles away from the mouth.

2 In *Solen, Chlamys, Pecten* and *Ostrea,* as well as in certain sublittoral lamellibranchs, the surface of the gill lamella is thrown into a series of folds so that a corresponding series of dorso-ventral ridges and

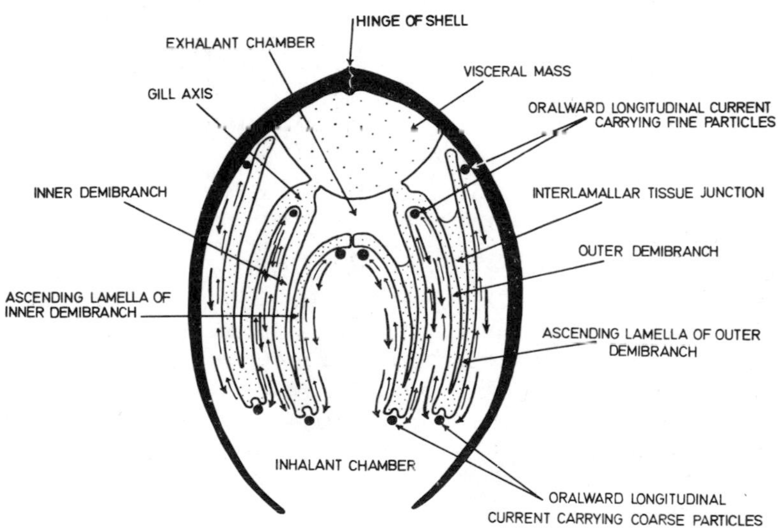

Fig. 5.21. Diagrammatic transverse section through *Solen* or *Ensis* (Eulamellibranchiata) showing the form of the gill and the direction of the ciliary currents. (Based on Atkins, 1936.) A large dot indicates the position of oralward currents. Note that fine particles are carried oralwards in the dorsal longitudinal grooves but coarse particles are carried orally along the ventral margins of the gills. On the left of the diagram is an ordinary gill filament but that on the right is a principal gill filament with interlamellar tissue junctions.

grooves are formed. Only small food particles can fall into the grooves, the cilia of which beat dorsally and transport this fine material into anteriorly-directed tracts between the bases of the demibranchs on each side of the body and at the dorsal edges of the ascending gill lamellae. The coarse particles are transported to the ventral margins of the demibranchs, and thence towards the mouth (Fig. 5.21). This route is more hazardous, and if excessive amounts of suspended material are drawn in with the inhalant stream, the whole demibranch contracts so that material is excluded from the grooves and passes to the ventral margins of the demibranch. Here the accumulated material may be thrown onto the mantle by twitching of the gill and thence carried posteriorly and rejected (Graham, 1931).

3 Plicate gills are a common feature of the sorting mechanisms of many lamellibranchs but in some, for example *Pinna fragilis, Thracia, Conchlodesma* and *Lyonsia,* all the frontal cilia, both on the plical

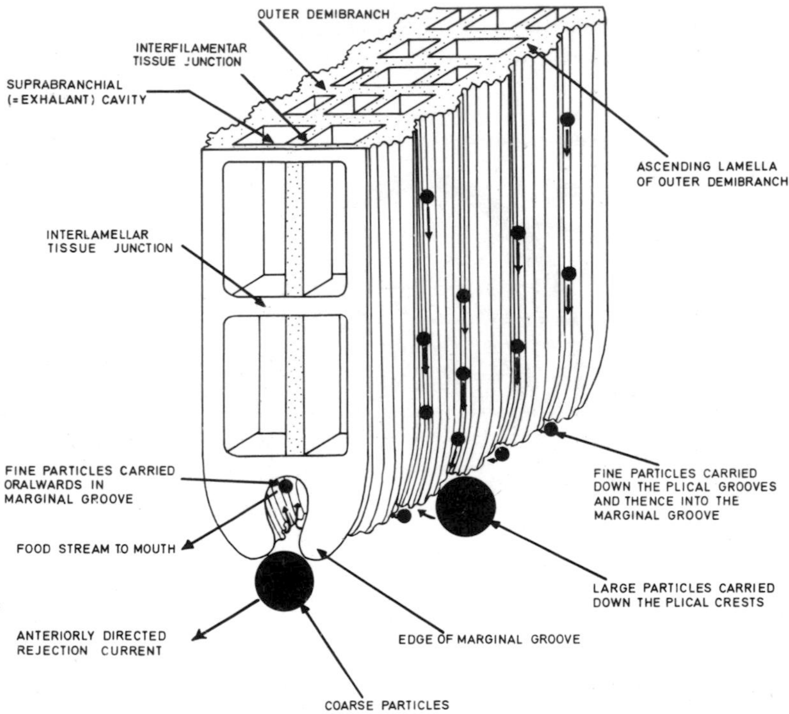

Fig. 5.22. Simplified block diagram illustrating the use of the plicae and marginal groove in sorting coarse particles from fine ones on a Eulamellibranch gill. (Modified from Atkins, 1937a.)

crests and in the grooves, beat towards the free ventral edge of the demibranchs. Fine particles pass down the troughs whilst coarse ones are carried by the frontal cilia on the crests of the plicae. A marginal groove, whose edges can be opposed, runs along the ventral edge of the gills (Fig. 5.22). The plical grooves lead into the depths of the marginal groove which is lined in this region with fine anteriorly-beating cilia. Coarse particles travelling down the crests of the plicae are normally excluded from the food stream in the marginal groove, although they may travel for some distance towards the mouth before falling off onto the mantle and being rejected. As in other forms with plicate gills, the plicae can be contracted so that even fine particles are excluded from the grooves and travel down with the coarse particles to be rejected. Thus the walls of the marginal groove of *Pinna* are deeply scalloped by the alternate ridges and grooves of the plicae. This structure plays an important part in the sorting process and provides the means by which fine particles enter the marginal groove.

4 In forms such as *Solecurtus, Lutraria* and *Cardium edule* the plicae do not extend as far as the marginal groove whose margins are, in consequence, of even height. Sorting by the marginal groove takes place in these animals by the apposition of the walls so that fine particles only can pass in, coarse particles being excluded by the presence of relatively coarse terminal cilia which beat obliquely anteriorly and which roof in the marginal groove.

5 In some lamellibranchs, notably those which live in a muddy habitat, a series of terminal guarding cilia is highly developed and forms the principal method of particle sorting. Typical genera with these structures include *Modiolus adriaticus, M. phaseolinus, Modiolaria discors, Musculas* (= *Modiolaria*), *marmoratus, Kellia suborbicularis, Montacuta ferruginosa,* many venerids, *Petricola pholadiformis, Solecurtus* and *Hiatella* (Atkins, 1937). On the tip of each filamentar lobe forming the edge of the marginal groove is set a fan-shaped group of long fine cilia which are intermittently active when stimulated by the presence of suspended particles. These cilia arch over the marginal groove and separate a channel lined with fine cilia at the depth of the groove from a more superficial one lined with coarse cilia (Fig. 5.23). Such a system is widely developed amongst the lamellibranchs, although the size and arrangement of the guarding cilia varies; in *Montacuta ferruginosa* and *Entovalva perrieri,* for example, the cilia are present on one side of the marginal groove only (that corresponding to the ascending lamellae), and arch over towards the opposite side of the groove.

6 The final method of sorting occurs in *Mactra corallina, Spisula subtruncata* and to a lesser extent in *Spisula elliptica* and *Donax vittatus,* the highest development occurring in *Spisula subtruncata.* Here there is a series of cirrus-like frontal cilia on the flat gill lamellae which also

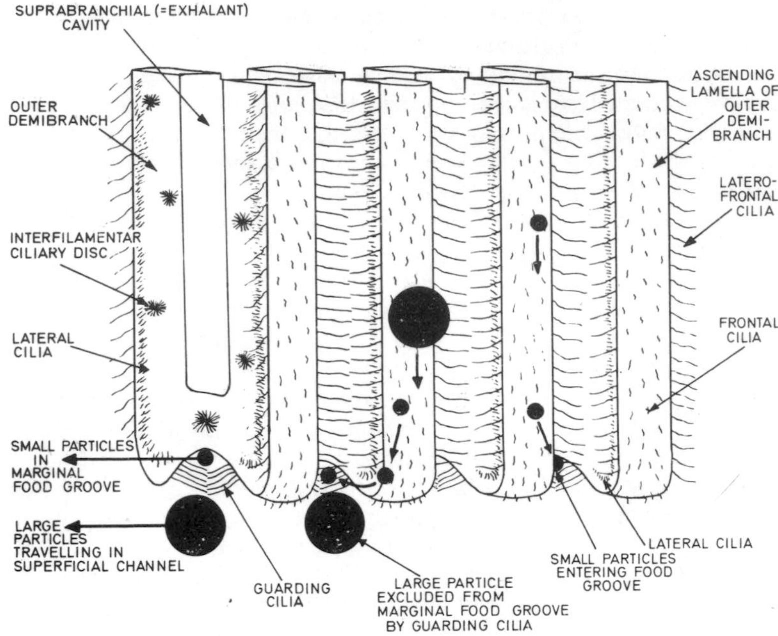

Fig. 5.23 Simplified block diagram illustrating the use of guarding cilia in the sorting of coarse particles from fine ones on a filibranch gill, e.g. *Musculus* (= *Modiolaria*) *marmoratus.* (Modified from Atkins, 1937a.)

bear fine frontal cilia. When stimulated by the presence of coarse particles, the stout frontal cirri beat towards the free margins of the demibranchs, throwing coarse particles off the gill and onto the mantle where they are rejected (Fig. 5.24). Small particles are transported by the fine frontal cilia, enter the wide marginal groove and travel orally along a tiny channel.

Having reviewed the various sorting mechanisms described by Atkins (1936, 1937) which occur on the gills of lamellibranchs, we may now summarise the mode of feeding of these animals, as inferred by observations on specimens from which one valve and the mantle of that side had been removed. A feeding current is drawn through the interfilamentary slits by the action of the lateral cilia, particles being thrown onto the frontal surfaces of the filaments by the laterofrontal cilia. When the particles touch the frontal surfaces of the filaments, mucus is secreted in proportion to the number and size of the particles striking the filaments (Kellogg, 1915; Nelson, 1960). The entrapped material is then subjected to a variety of ciliary sorting mechanisms, fine particles

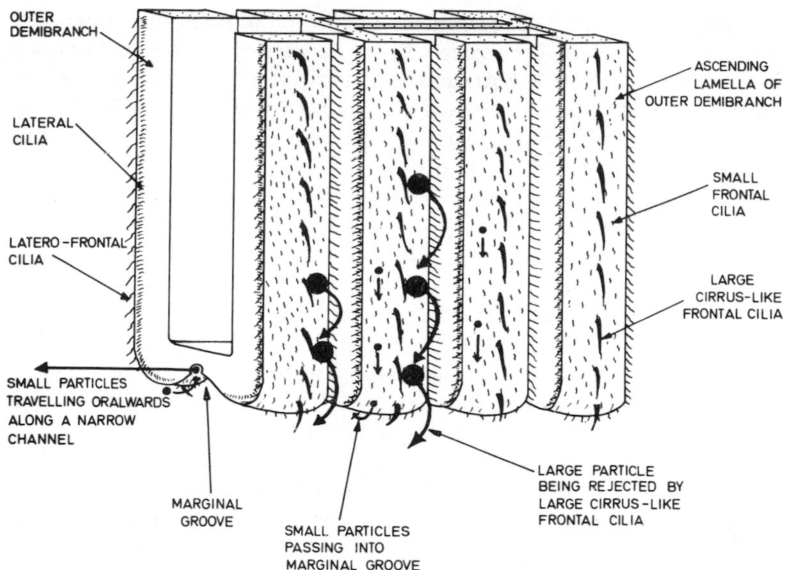

Fig. 5.24. Simplified block diagram illustrating the use of cirrus-like frontal cilia in sorting coarse from fine particles on the gill of a sand-dwelling Eulamellibranch, e.g. *Mactra*. (Modified from Atkins, 1937a.)

travelling to the mouth along protected food grooves and coarser particles passing along more superficial routes.

MacGinitie (1941) has questioned the validity of this interpretation of the feeding mechanism of lamellibranchs on the grounds that removal of one valve and part of the mantle seriously affects the normal functioning of the gill. He sealed glass windows into the valves of the bivalves *Schizothaerus nuttallii, Pecten circularis, Ostrea lurida* and *Mytilus californianus,* and allowed them to begin to regenerate the shell before making observations on the feeding behaviour of the undisturbed animals. Under such conditions, he found that the entire gill surface was covered with mucus which therefore acted as a filter in rather the same way as in *Crepidula* (Werner, 1959; also p. 198). Clearly, under these circumstances sorting of the entangled particles as described by Atkins (1936, 1937) could not occur, particles of all sizes being transported to the labial palps. These organs had the function of removing unsuitable particles and hence were the prime sorting organ before material was ingested. When disturbed, or when carborundum or carmine powder was added to the water, mucus secretion ceased, but particles could still be conveyed to the labial palps by cilia as described by Atkins (1936, 1937) although MacGinitie (1941) stated that they were generally rejected on arrival at the palps.

Jørgensen (1955, 1966) has attempted to unite these two apparently opposing views by suggesting that the complex sorting devices described by Atkins (1936, 1937) operate under turbid conditions, where the fine particles suitable for intracellular digestion (Yonge 1931; 1937a) must be separated from coarse particles. In clear water, complex sorting devices would be superfluous. Instead, all particles may be entrapped in a mucous filter (MacGinitie 1941), and passed by the frontal cilia to the marginal grooves and thence to the labial palps. It should be noted here that the rejection tracts described by Atkins (1936, 1937) would then become food-collection tracts in that system. As Jørgensen (1955, 1966) pointed out, the presence of elaborate ciliary sorting mechanisms on the gills would be surprising if no sorting took place in that region. It therefore seems likely that the feeding mechanism of bivalves is more versatile than had hitherto been supposed and can be modified according to the changing turbidity of the inhalant stream.

All particles, whether sorted on the gills or not, must pass to the mouth across the labial palps which may be placed either in contact with the margins of the demibranchs or spread transversely so that material from the gills cannot reach the inner surfaces of the palps and is rejected (MacGinitie, 1941). The inner surfaces of the palps are ridged in a complicated way (Fig. 5.25), and as many as eight different ciliary tracts may be present on the surface of a single ridge and groove (Allen, 1958). In general, however, the ciliary tracts can be grouped into three main types which are common to the palps of all lamellibranchs (Fig. 5.25; Purchon 1955). (a) Ventrally beating rejection tracts in the bases of the grooves into which dense particles fall. (b) Acceptance tracts which occur on the crests of the ridges and which beat across the direction of the folds; small suspended particles do not fall into the grooves and are swept by the acceptance tract across ridges and grooves towards the mouth. (c) Finally there is a variety of resorting tracts whose number and arrangement varies from species to species; these serve to redistribute the particles which fall into the grooves. As in the sorting mechanisms of the gill, muscular activity plays an important part in modifying the sorting activity of the palps (Ansell, 1961). In the presence of excess material, the ridges and grooves are deep so that many of the particles fall into the rejection tracts; when suspended material is scarce, however, the ridges are allowed to overlap one another so that more particles are carried to the mouth by the acceptance tracts on the crests.

Thus the bivalves have well-developed sorting mechanisms on both gills and labial palps, although in the Tellinacea there appears to be a reciprocal relationship between the degree of development of palps and gills (Yonge, 1949). Where the gills are large and capable of sorting material the palps are correspondingly small, and *vice versa.* The

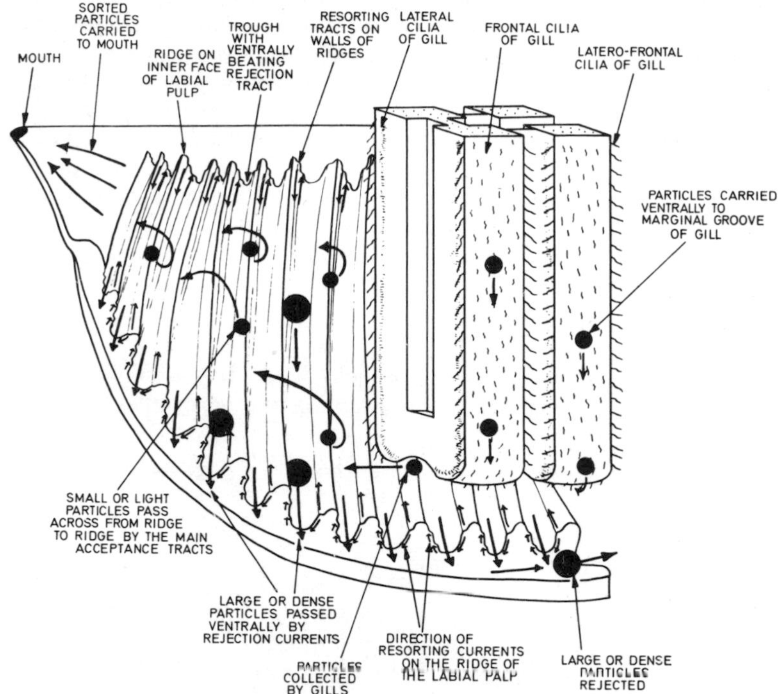

Fig. 5.25. Simplified diagram showing the relationship between the anterior end of the gill and the labial palps in a Eulamellibranch. For simplicity only the inner, ridged surface of the palp part of one side is shown. The course of the main resorting tracts is shown by arrows at the dorsal and ventral margins of the palp; acceptance and rejection tracts are shown by large arrows. (Based partly on Allen, 1958a.)

situation is further complicated by the ability of the labial and branchial sorting mechanisms to vary the amount and grade of particles entering the food grooves, and this may account for some of the discrepancies between accounts of the filtering efficiency of lamellibranchs (for review, Jφrgensen, 1966). Finally, there is some indication that a qualitative selection of particles may take place, although quantitative selection on the basis of particle size and density is the rule. Menzel (1955), for example, showed that young *Crassostrea virginica* were capable of abstracting organic plankton from a bolus of mixed carmine, charcoal and plankton particles which had accumulated on the palps, while Buley (1936) found that the average percentages of dinoflagellates and diatoms in the stomach of *Mytilus californianus* throughout a period of seven months were 97·4 and 2·6, whereas those in the surrounding water throughout this period were 2·4 and 97·6 respectively. Again, Loosanoff (1949), working on *Crassostrea virginica,* found that when this oyster is fed a mixture of diatoms and the purple bacterium *Cromatium* the pseudofaeces consist largely of the bacterium whereas the true faeces are composed largely of remains of diatoms; it was also able to reject yeast cells mixed in suspension with other micro-organisms. It seems possible that the labial palps play a part in such a qualitative separation (Menzel, 1955) although the precise mechanism is not clear. The significance of such qualitative sorting is also unknown, but it may serve to prevent the ingestion of toxic organisms such as *Phaeocysts* which at times occur in large numbers in the inshore plankton.

(d) Ascidians

The final example of ciliary suspension feeding is shown by the ascidians. These organisms are abundant in the lower intertidal zone,

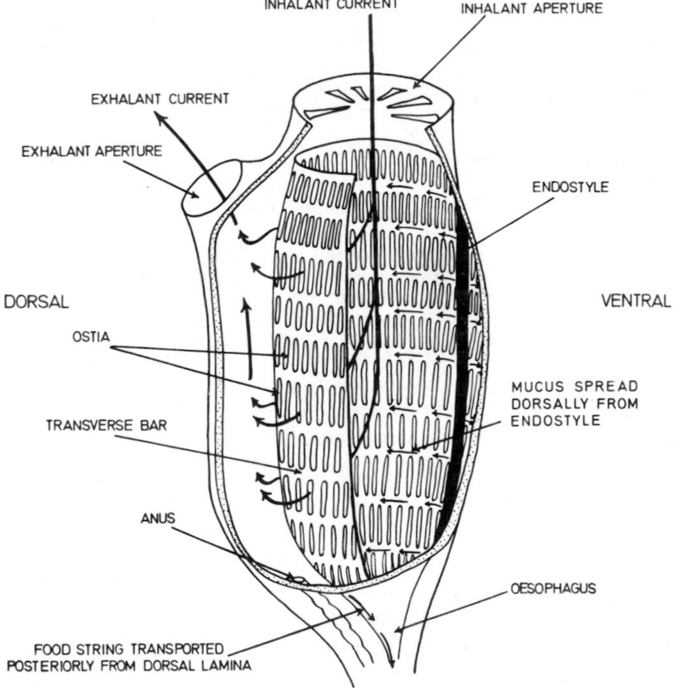

Fig. 5.26. Diagrammatic sagittal section of *Clavelina* showing the branchial basket and feeding currents. The side of the tunic and part of the branchial basket are shown cut away. (Based on Werner and Werner, 1954.)

and all have a basically similar feeding pattern. Water is drawn into the pharynx by the lateral cilia of the gill bars which, together with a large number of lateral branches, subdivide the pharyngeal wall into a branchial basket penetrated by numerous ostia (Fig. 5.26 and 27). The endostyle is a longitudinal strip of ciliated and mucus-secreting cells running along the ventral wall of the pharynx, while a longitudinal strip of posteriorly-beating cilia in the dorsal wall forms the dorsal lamina. The cilia of the endostyle lash laterally and spread mucus onto the faces of the gill bars, whose frontal cilia beat dorsally carrying mucus towards the dorsal lamina. According to most descriptions (for example Orton, 1913a,b; van Weel, 1940; Berrill, 1950) the particles are strained from the water passing through the pharynx by the ostia, whose size would therefore determine the size of particle retained by the animal, the particles then being entrapped in mucus, transported dorsally towards the lamina and thence posteriorly to be engulfed. Recently, however, it has become apparent that whereas the ostia could retain only the larger phytoplankton organisms (Werner and Werner, 1954), ascidians can in fact retain graphite particles of only 1μ in diameter from the inhalant stream (Jørgensen, 1949a, 1952, 1966). The answer, as in bivalves, appears to lie in the ability of ascidians to cover the filtering system with a sheet of mucus (MacGinitie, 1939b; Jørgensen and Goldberg,

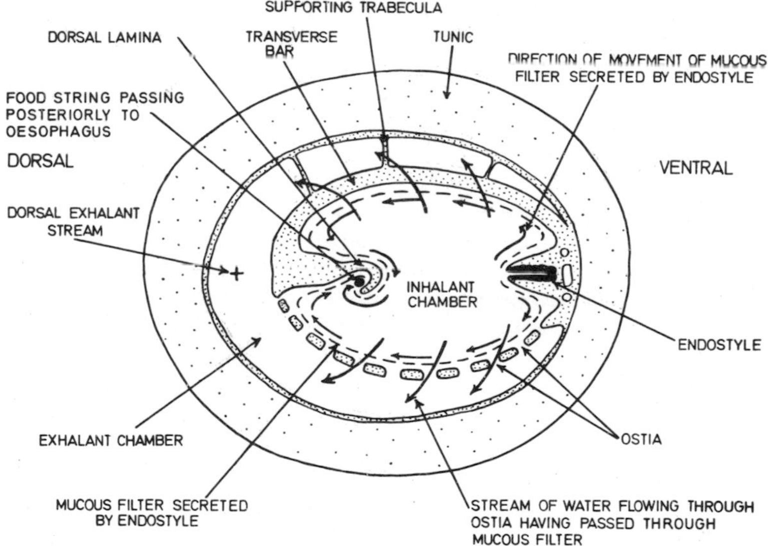

Fig. 5.27. Diagrammatic horizontal section through *Clavelina* showing a transverse bar (above) and the branchial basket penetrated by ostia (below). (After Werner and Werner, 1954.)

1953; Werner and Werner, 1954; Werner, 1959). In *Ciona intestinalis, Ascidia californica, Diplosoma pizone,* and *Clavelina lepadiformis,* the sheet of mucus, which acts as a fine filter, is cut off when the animals are disturbed, with the result that only rather coarse particles are retained and transported dorsally by the cilia of the gill bars. This effect is clearly similar to that described in bivalves by MacGinitie (1941) (also p. 206). Under normal conditions the particles entrapped in the mucous sheet are transported dorsally towards the dorsal lamina where a mucous string is formed and transported posteriorly for ingestion (Fig. 5.26 and 27). Ascidians are thus mainly non-selective suspension feeders, although MacGinitie (1939b) has described the occurrence of large cilia which reject coarse particles; these are subsequently expelled by squirting.

In the course of this review of ciliary mechanisms of food collection, it has become apparent that mucus may be used in feeding in two ways; firstly as a medium for the transport of particles which may have been to some extent sorted by gills, palps or other devices, and secondly, as a filter itself, most of the particles trapped in it being transported for ingestion. This second method is an unselective one suited, in particular, to the retention of very fine particles; it is used in *Crepidula* (Werner, 1959), on occasions in bivalves (MacGinitie, 1941) and in ascidians (MacGinitie, 1939b; Werner and Werner, 1954; Werner, 1959; Jørgensen and Goldberg, 1953; for review Jørgensen, 1966). Clearly this second use of mucus differs only in degree from the 'mucous bag' mechanisms described for *Chaetopterus* and *Nereis diversicolor* (pp. 189–191). At this point these two systems of food collection become distinguishable only by the absence of supporting rods or bars for the mucous sheet in the polychaetes. Werner (1959) has shown that the coarse mantle filter in *Crepidula* is built up as a net of mucous fibres, and it seems likely, as Jørgensen (1966) suggests, that the 'mucous sheets' covering the frontal surfaces of the gill of *Crepidula,* bivalves, and ascidians may, in fact, be a network of mucus fibres rather than homogeneous sheets. Werner (1959) has suggested that such net-like filters could be formed either by the secretion of threads of mucus which subsequently harden and adhere to one another or more regular apertures could be formed by the secretion of nets from specialised glands such as he has described for the coarse filter of *Crepidula.* Whatever the mode of secretion of such filters, it is clear that they are of widespread occurrence in suspension feeders and are admirably suited for the abstraction of fine particles from the inhalant stream.

3. SUSPENSION FEEDING BY MEANS OF SETOSE APPENDAGES

The use of setose appendages is characteristic of crustacean suspension feeders such as the barnacles which have been studied by Crisp and Southward (1961). Feeding is carried out by movements of pairs of setose cirri which are of differing form and function. In balanoids the first three pairs are short and stout (Fig. 5.28) with the

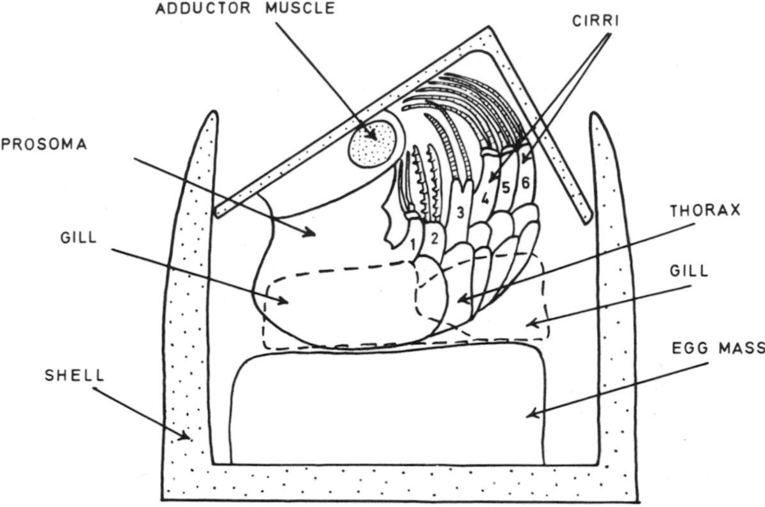

Fig. 5.28. Diagrammatic vertical longitudinal section of a balanoid barnacle with the depressor muscles and gills of one side removed. For simplicity the lateral setae of the 4th, 5th and 6th pairs of cirri are not shown. (After Crisp and Southward, 1961.)

two rami of each limb of different length, whereas cirri 4 – 6 are long and thin with rami of almost equal length. In chthamalids the first two pairs of cirri only are short, and the remaining four pairs are long and setose. The morphological differences correspond with their functions, the long cirri being used for the collection of coarse particles and the short ones for fine particles.

During captorial feeding on coarse particles the long cirri are extended relatively slowly when body fluid is forced into them by muscular activity. The cirri are then retracted and rolled up by means of a series of flexor muscles, entrapped particles being scraped off by means of a series of special setae on the short limbs closer to the head end, whence they are then passed to the mouth. Although this feeding method appears to be an unselective one, selection of particles can occur at or near the mouth and also in some balanoids, such as *Balanus balanoides* and *B. hameri,* on the cirri (Crisp and Southward, 1961). In this case, instead of the unwanted particles being wiped off onto the smaller cirri, the extended cirral net is rotated through 90° – 180° and the local water currents are allowed to sweep away the accumulated material. Indeed barnacles have a wide repertoire of movements which is fully described by Crisp and Southward (1961), and in still water several sweeps of the cirri may be made before retraction. Conversely, in a water current the cirri may be held stationary across the water flow, and under these circumstances thus act as a passive food net. This last method is the main feeding activity of stalked barnacles (Batham, 1945; Barnes and Reese, 1959).

Crisp and Southward (1961) have found that a diameter of 30μ is the smallest particle size which can be retained by the cirral net. Nevertheless, Southward (1955a,b) and Barnes (1959) have found that particles too small to be retained by the cirral net are ingested by barnacles, and the faecal pellets contain particles of a range of sizes down to a few microns in diameter. The capture of such fine material is made possible by the action of the small cirri (1 – 2 in chthamalids or 1 – 3 in balanoids) which are held across the inhalant aperture of the mantle cavity during the extrusion of the large cirri, and whose setae are so closely packed that spaces between the bristles are as small as 1 – 2μ. As the mantle cavity begins to expand prior to the extrusion of the cirri there is a slow inflow of water into it; this continues during the initial phases of the normal cirral beat and the slow flow facilitates the removal of fine particles from suspension. There is then a rapid transition from a slow inflow of water into the mantle cavity to a rapid outflow, during which the large cirri begin to retract and the inhalant aperture is closed partly by movements of the rostral plates and partly by the gills acting as simple flap-valves. At the same time an exhalant aperture is opened in front of the cirri, and water is expelled as a stream. In this way the filtered water is separated from the particle-laden water surrounding the barnacle (Fig. 5.29).

Haustorius arenarius is a common burrowing amphipod of sandy shores and this animal, too, has two different mechanisms to deal with either large or small particles. In filtering off small particles, the setose maxillae are rotated in such a way as to act as a combined suction pump and filter plate, while the maxillipeds remove accumulated food particles from the maxillae and pass them to the mouth (Dennell, 1933). In many ways the feeding movements are comparable with those described above for barnacles insofar as the inhalant and exhalant streams leave by different apertures created by the movements of the limbs. The maxillae are held almost horizontally beneath the head, and are moved downwards and outwards during the initial phase of feeding. This movement enlarges the food basin below the head into which water flows through the sieve of setae bordering the inner margins of the inner lobes of the maxillae. The food basin is now full of water and

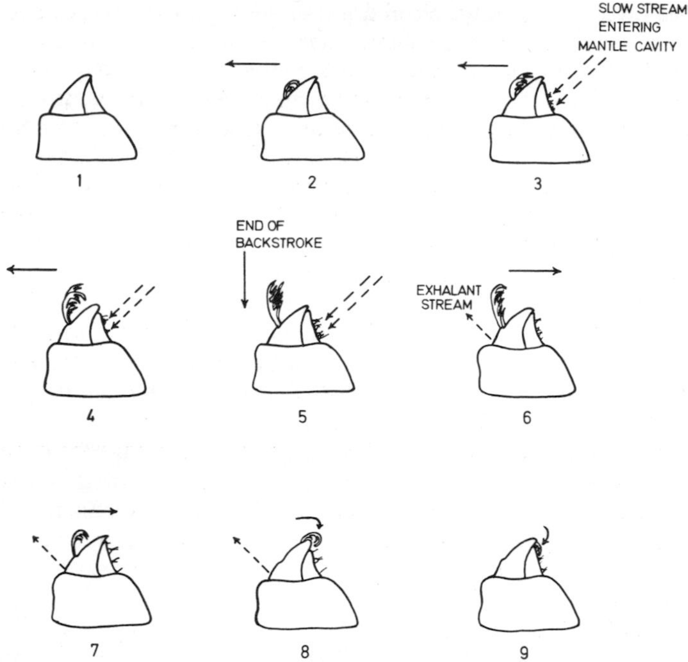

Fig. 5.29. Nine stages in the normal beat of *Balanus balanus.* (after Crisp and Southward, 1961.) The continuous arrow indicates the direction of motion of the cirri; broken arrows indicate the entry and exit of water to and from the mantle cavity during which fine particles are filtered.

the maxillae subsequently close the gap between their inner margins, so that the filtered water cannot leave by the route it entered. The maxillae are next moved upwards, constricting the food basin and forcing the water forwards through a second series of setae on the outer lobes of the maxillae which retain any particles which have escaped the marginal setae between the inner lobes of the maxillae during the inhalant phase. Filtered material on the inner lobes of the maxillae is subsequently transferred by way of the maxillipeds to the mouth. The movement of the maxillipeds is latero-medial; this allows a parallel comb-row of setae on the second joint of the palp to work over the sieve of setae bordering the inner margins of the maxillae, and to transfer the food particles to the comb-rows. The particles are then transferred to the serrated setae on the distal joint of the maxillipedal palp, from which they are removed by the maxillules and passed to the mandibles. *Haustorius* can also, like other amphipods, feed off large food masses.

The two examples considered above show how barnacles and the burrowing amphipod *Haustorius arenarius* filter particles suspended in the water. Several other crustaceans are filter feeders, but show a transition towards deposit feeding in that material from the substratum is initially stirred up into suspension before being filtered for food. Such a mechanism has been described in *Pagurus bernhardus* by Orton (1926), and in *Galathea dispersa* by Nicol (1932) although she has shown that the related *Porcellana longicornis* is a true suspension feeder. Nicol (1932) found that the food requirements and feeding mechanism of *Porcellana longicornis* and *P. platycheles* were identical; she selected that of *P. longicornis* as an example of the feeding mechanism of both species. Although Potts (1915) found relatively large pieces of alga in the stomach, and Dalyell (1853) recorded that *Porcellana* will eat mussels, Nicol (1932) found that the animal was principally a suspension feeder.

The respiratory current set up by the alternate beating of the flagella of the exopodites plays an important part in the feeding process of *Porcellana longicornis.* Water is drawn by the beating of the flagella of one side over the dorsal and lateral surfaces of the animal (Fig. 5.30);

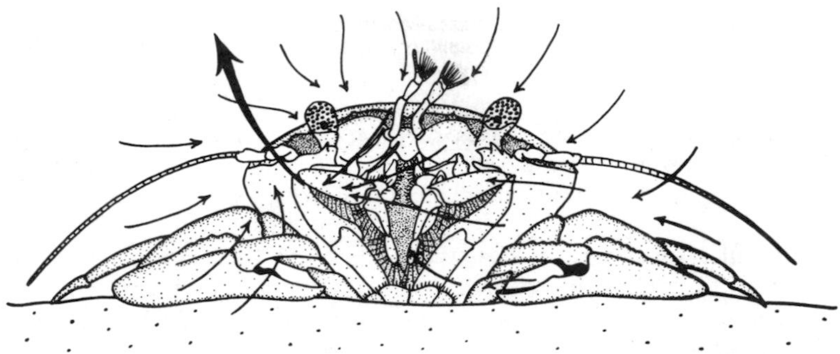

Fig. 5.30. Diagrammatic anterior view of *Porcellana longicornis* showing the direction of the respiratory current when the flagella of the right side are beating. (After Nicol, 1932.)

the beating of these flagella subsequently stops and the flagella of the opposite side start to beat, causing a complete reversal of the respiratory current. In this way a current is created about the animal which impinges on the antennules so that the animal is made aware of the presence of suitable food in the vicinity. When feeding begins, these movements of the flagella stop and are replaced by feeding movements of the 3rd maxillipeds which, like the flagella, are flexed and extended alternately on each side. These movements result in a flow of particle-laden water towards the mouth from all directions, except below where the animal is pressed to the substratum. This flow is illustrated in Fig. 5.31 where the inflow caused by the extension of the left maxilliped is stronger than on the right where the maxilliped is flexed.

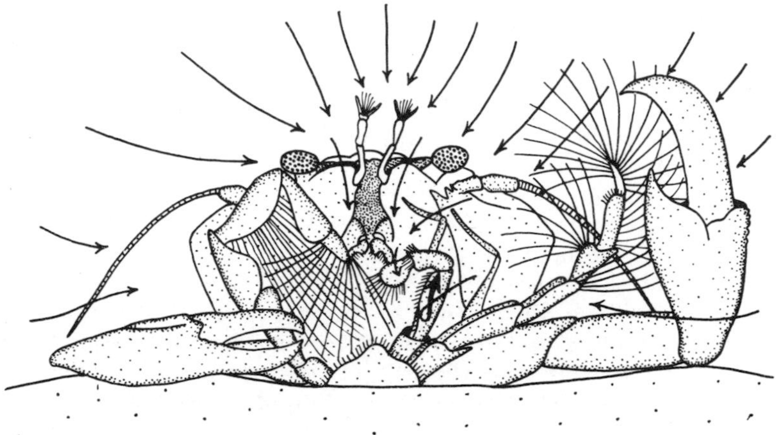

Fig. 5.31. Anterior view of *Porcellana longicornis* showing the direction of the feeding currents drawing food in suspension rapidly towards the left side and more slowly towards its right where the maxilliped is flexed. (After Nicol, 1932).

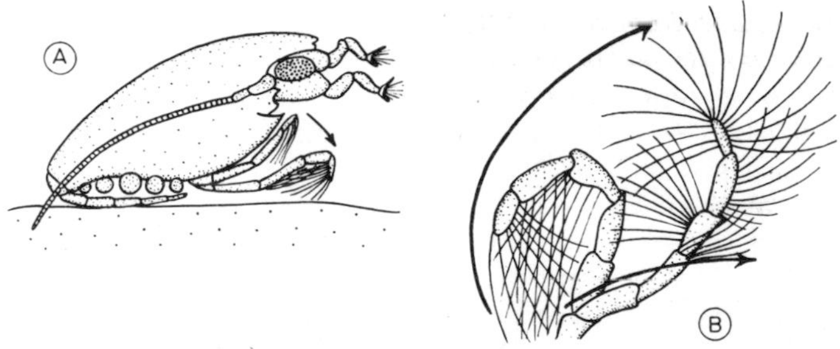

Fig. 5.32. Diagram showing the feeding movements in *Porcellana longicornis* (After Nicol, 1932.) (A) Lateral view showing the lowering of the 3rd maxillipeds prior to feeding. (B) Ventral view of a third maxilliped in the flexed and unflexed positions with arrows showing the movements of the setae of the filter net.

Food particles are trapped on the setae of the 3rd maxillipeds, each is spread out into a wide spoon-shaped net when the maxilliped is extended. As the maxilliped is flexed, the basal segments move towards the midline and a considerable volume of water is forced out between the setae which move closer together during flexion of the 3rd maxilliped and which thus retain particles from suspension;

material is removed from these filtering setae by a terminal brush on the 2nd maxilliped (Fig. 5.32). As soon as the 3rd maxilliped is flexed, the 2nd maxilliped of that side is moved in such a way that its terminal setae are inserted between the filtering setae of the 3rd maxilliped. Material is then removed, and the tip of the 2nd maxilliped inserted into the mouth. Cleaning movements then occur in which the 3rd maxilliped is turned laterally and unflexed so that the filtering setae are drawn through the terminal brush of the 2nd maxilliped. The feeding cycle is then repeated, although such cleaning movements may be made several times in succession if a large number of particles have been caught by the filtering setae. These feeding movements of *Porcellana longicornis* are summarised in Fig. 5.32A & B.

From this account, it is clear that barnacles and *Haustorius* filter particles during the inhalant phase of the feeding movement and prevent particles from being ejected by passing the exhalant stream out through a different aperture; that is, they maintain a one-way flow through the filtering setae. Porcellanids filter particles from the exhalant stream which is set up when the 3rd maxilliped begins to flex; under these circumstances the separation of inhalant and exhalant streams is unnecessary and does not occur.

Pagurus and *Galathea* are not true suspension feeders, although, as will be seen below, their feeding movements are similar to those of the Porcellanids. Orton (1926) showed that when placed on muddy gravel, specimens of *Pagurus bernhardus* would scrape the substratum with the smaller claw and pass the products to the 3rd maxillipeds. The material was then transferred by the brush-like setae of the 3rd maxillipeds to the other mouthparts, where sorting and rejection of unsuitable material took place. This general mechanism of feeding has been described in more detail by Nicol (1932) working on the Galatheids, of which *G. dispersa* was selected as an example.

Galathea dispersa, in common with other galatheids such as *G. squamifera, G. strigosa* and *Munida rondeletii,* can feed either on large pieces of organic material or on organic debris and micro-organisms resuspended from the substratum. In each of these species the latter mechanism is more common. When feeding on fine material, *G. dispersa* uses the setose and extremely mobile 3rd maxillipeds like brooms to sweep small particles towards the mouth. These limbs often work together but may also be used alternately (Fig. 5.33); they are also capable of sweeping material not only from in front of the animal but from underneath. When the brushing movement of each 3rd maxilliped is completed, it is retracted and folded in such a way that the third segment of the endopodite is at right angles to the other parts of the limb and the terminal segments lie in a parallel plane to the basal segments, but nearer the midline. Here the short 2nd maxillipeds, which

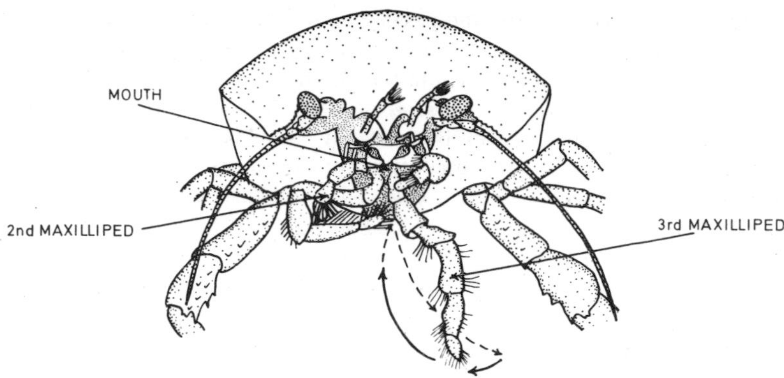

Fig. 5.33. Anterior view of *Galathea dispersa* feeding on finely-divided material. The left 3rd maxilliped is extended and collecting particles off the substratum; the right is flexed and the bristles are about to be cleaned by the 2nd maxilliped. The left 2nd maxilliped has completed the cleaning movement and is transferring collected particles to the mouth. (After Nicol, 1932.)

always work alternately, remove particles from the tufts of the 3rd maxillipeds by means of a series of stiff terminal setae. They are then bent ventrally so that their tips bearing the particles are combed by the inner mouthparts where sorting takes place. Unfortunately, the precise mechanism by which this occurs is not known either for Pagurids or Galatheids. Unsuitable material is rejected into the respiratory stream created by the flagella of the exopodites of the two pairs of maxillipeds.

We have just seen that the feeding mechanisms of the Galatheids and Pagurids, although showing certain resemblances to those of the Porcellanids, are in fact more closely allied to those of the true deposit feeders, which are considered next.

F. DEPOSIT FEEDING

A great variety of intertidal organisms feed directly upon the deposits in which they live. Such organisms include four main categories of deposit feeders. Firstly, there are animals such as terebellid polychaetes and the scaphopod *Dentalium* which transport particles to the mouth by means of complex ciliary tracts. A second category includes animals such as the bivalves, *Macoma balthica* and *Scrobicularia plana*, which feed by sucking up the deposits, and others which utilise the thin layer of material resuspended from the deposits by wave action. Among the latter we may include certain other members of the Tellinacea such as *Donax vittatus* and *Gari* spp. as well as the polychaete *Owenia fusiformis*, although it is obvious that at this point the distinction between deposit feeders and suspension feeders becomes arbitrary since most intertidal suspension feeders must also to some extent utilise re-suspended deposits as a food source. The third category includes those animals, such as the amphipod *Corophium volutator,* which select only the nutritive particles from the deposits in which they live. Finally there are animals such as the lugworm *Arenicola* and the maldanid *Clymenella torquata* as well as the enteropneust *Saccoglossus* which swallow the deposits more or less indiscriminately and pass out the indigestible component as faeces.

The feeding mechanism of terebellid polychaetes has been described in some detail for *Amphitrite johnstoni* by Dales (1955) who also gave a comparative account of the mechanisms used by the other principal terebellid genera. In terebellids, the head bears a crown of extensile hollow tentacles which are the principal organs of food collection. Each tentacle is extensively ciliated on its adoral surface and is supplied with a series of longitudinal, transverse and oblique muscles (Fig. 5.34). The

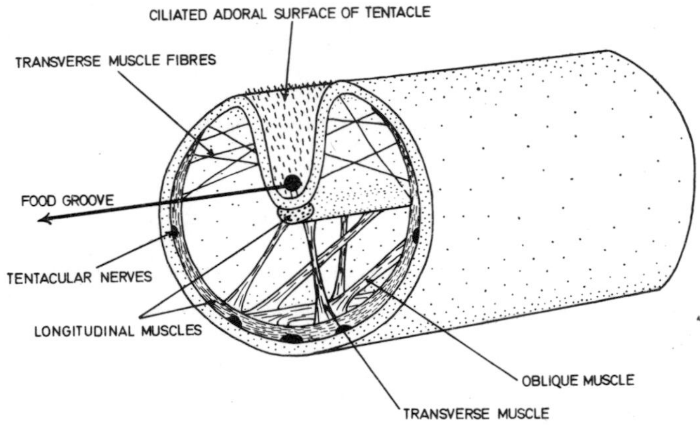

Fig. 5.34. Diagram showing the structure of a section of a tentacle of *Terebella lapidaria.* (After Dales, 1955.)

longitudinal muscles are responsible for the retraction of the tentacle whilst the transverse and oblique muscles allow a variety of movements including the rolling and curling of the tentacle over the surface of the substratum. During feeding, the tentacles are rolled over and the ciliary groove opened out so that each tentacle can be extended by ciliary creeping. After a certain degree of extension, a region near the tip of the tentacle becomes flattened by the contraction of the transverse muscles and secretes mucus so that the tentacle becomes anchored at this point, partly by the mucus and partly by means of a suction pad.

This point of attachment allows a further extension of the terminal part of the tentacle which is released and reattached continuously. As a result of such extension of all the tentacles, the surface of the deposits around the entrance of the burrow is covered with a radiating system of tentacles (Fig. 5.35). Particles are conveyed to the lips by one or more of three different methods (Dales, 1955). Fine particles are passed to the mouth along the ciliated groove of the tentacle, which may be either an open gutter as in Fig. 5.35 or closed to form a tube. Larger particles are carried to the lips partly by the action of the cilia lining the food groove and partly by the squeezing action of the sides of the groove. Finally, large particles which cannot be transported by either of these two methods are gripped by the walls of the food groove and the whole tentacle is then retracted towards the mouth (cf *Dentalium* p. 224).

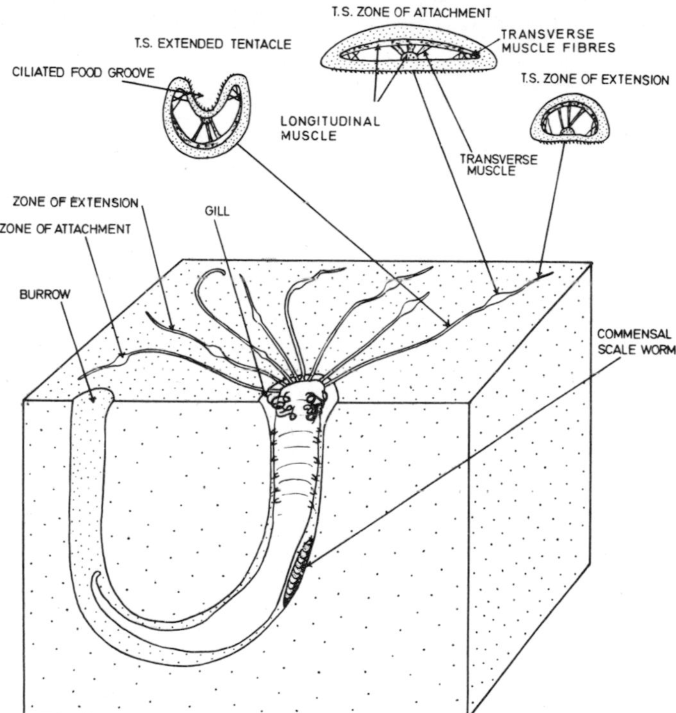

Fig. 5.35. Diagram showing the structure and use of the tentacles in a terebellid (*Amphitrite johnstoni*) during feeding. (Based on Dales, 1955.) *Amphitrite johnstoni* is often accompanied by the commensal scale worm *Gattyana cirrosa* which is shown in the diagram and which under natural conditions moves up the burrow intermittently to remove some of the food accumulated by the tentacles.

The cilia of the food groove do not reach to the base of the tentacles, and particles are transferred to the mouth *via* the lips where a degree of sorting occurs. The upper lip is formed from the prostomium and its ciliated oral surface is covered with shallow grooves running down towards the mouth. It is provided with a complex musculature which allows a wide variety of movements associated with the removal of particles entangled on the tentacles. Some of these movements are shown in Fig. 5.36. In addition to the dorsal lip, there are two pairs of ventral lips, the inner pair of which is derived from the stomodaeum and the outer pair from the ventral part of the first segment. The pairs of lips can be distinguished by their appearance, the outer pair being yellow or whitish and provided with abundant mucous glands and cilia, while the inner pair is grooved and reddish in colour with no mucous glands or cilia. Dales (1955) has referred to the outer fold of the outer lip pair as 'lip A', and its inner fold as 'lip B' while the outer fold of the inner lip is 'lip C' and its inner fold 'lip D' (Fig. 5.36).

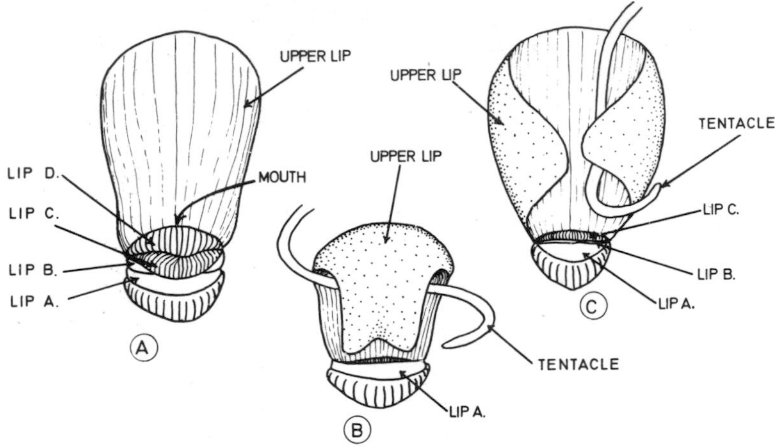

Fig. 5.36. Diagram illustrating lip movements in *Amphitrite gracilis* (after Dales, 1955.) (A) Diagram showing the arrangement of the lips; (B) and (C) two movements of the upper lip.

The longitudinal muscles of the outer lip pair (A and B) are derived from the segmental muscles of the ventral side and play an important part in the feeding process. When a food bolus arrives in the groove between lips A and B the longitudinal muscles of this region contract and deepen the groove (Fig. 5.37A). The longitudinal muscles are then serially contracted towards the mouth so that the bolus is transferred to the outer margin of lip C (Fig. 5.37B). The inner lips are provided with a thick band of transverse muscle under which lies a thinner band of longitudinal muscle running from the inner edge of lip B to the inner edge of lip C. Contraction of this longitudinal muscle causes the inner lip pair open and receive the food bolus which is then gripped by the apposition of lips C and D. The longitudinal muscle on the oral side of lip D then contracts so that the bolus is tilted towards the mouth and rolls in (Fig. 5.37C). As Dales (1955) points out, this is an oversimplified scheme and the food is often in the form of a mucous string running the length of the lip. Nevertheless, it does suggest the principle underlying the characteristic rippling movements of the ventral lips and the ways in which such movements results in the transfer of food to the mouth.

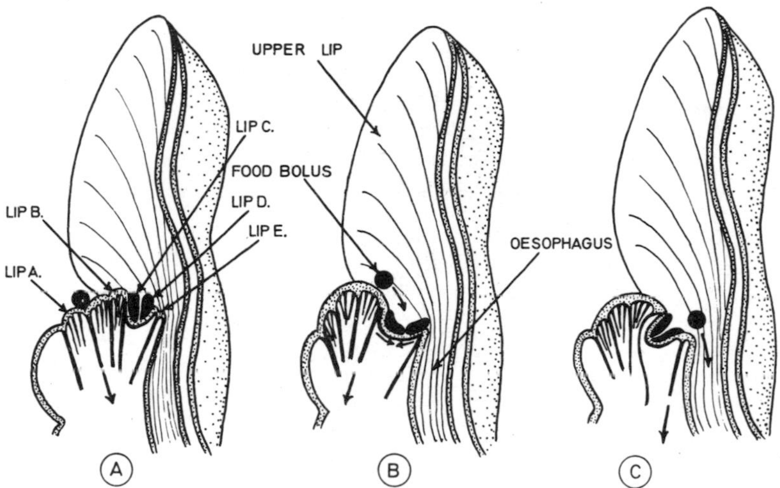

Fig. 5.37. Diagram illustrating the lip movements in a terebellid, e.g. *Amphitrite johnstoni.* (Based on Dales, 1955.) For explanation see text.

There is a considerable variety of other feeding movements in terebellids. In *Amphitrite johnstoni,* for example, the tentacles may be wiped across the outer lips whose cilia aggregate particles between lips A and B. The accumulated material is then transferred to the mouth as described for the food bolus above. Again, some particles may be wiped directly onto the upper lip, whence they are transported directly into the mouth. Larger particles which cannot be scooped up by the inner lip, pair C and D, are rejected to the outer lip groove where they become entangled in mucus before being either rejected in non-tubicolous forms or incorporated into the tube of genera such as the sand mason *Lanice conchilega*.

Feeding in the Scaphopoda shows some obvious similarities with the mechanism described for terebellid polychaetes by Dales (1955). The scaphopods, of which the tusk-shell *Dentalium* is an example, live in a slightly curved tapering shell and the anterior end of the animal bears a series of characteristic tentacle-like captacula. The tip of each captaculum is swollen to form the alveolus (Fig. 5.38) and it had been supposed that during feeding groups of captacula attach themselves to foraminifera with the alveoli and draw them in towards the mouth (Morton, 1959). In this respect the feeding mechanism is comparable to that of a terebellid which grips large particles with the walls of the food groove and subsequently retracts the tentacle towards the mouth (p. 221). Recently, however, Dinamani (1964) has shown that *Dentalium* normally feeds by means of ciliary tracts which extend along the length of each captaculum.

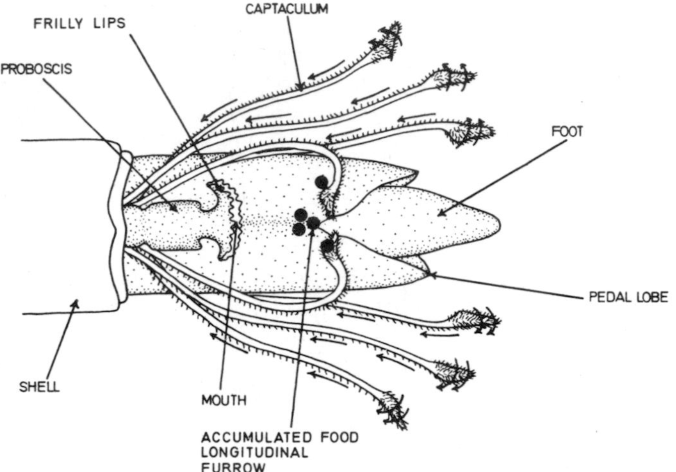

Fig. 5.38. Dorsal view of *Dentalium conspicuum* with the anterior portion of the shell and mantle removed to show the longitudinal furrow and the captacula entwining particles. (After Dinamani, 1964.) Arrows indicate the direction of the ciliary streams on the captacula.

When *Dentalium* is feeding, each captaculum is extended between the particles of sediment and is then held in position by the temporary attachment of the alveolus to the sand grains of the substratum. This is comparable to the "zone of attachment" noted above (p. 220) in terebellids. The captacula feel about in all directions, the long cilia of the alveolus whisking particles up onto the back of the enlarged tip. The particles are then transported along the length of the captaculum by a series of short cilia beating in a metachronal rhythm towards the dorsal surface of the foot. (Fig. 5.38). Here the matter collected by the captacula accumulates in a longitudinal furrow behind the pedal lobes of the foot (Fig. 5.39) which then periodically swings upwards, tipping particles in the groove towards the mouth. Food particles are then selected from the accumulated material by the frilly lips and shorter captacula. Material may also be accumulated in the dorsal groove of the foot by the action of the pedal lobes which are moved to and fro in the substratum in such a way that particles are dislodged and wafted towards the mouth. Dinamani (1964) did not observe *Dentalium* abstracting isolated foraminifernas from the deposits as had earlier been reported by Morton (1959), nor were any found in the gut. Nevertheless, it seems possible that, like terebellids, a variety of feeding mechanisms may be employed according to the availability of food in the substratum.

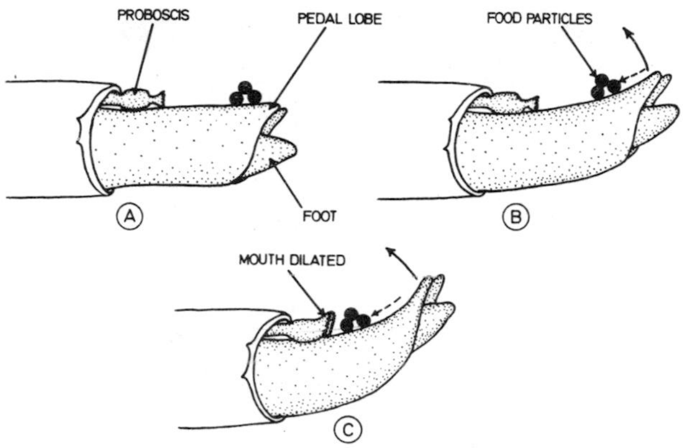

Fig. 5.39. Lateral view of the anterior end of *Dentalium conspicuum* showing three successive stages in the action of the foot in bringing particles nearer the frilly lips. (After Dinamani, 1964.) Solid arrows indicate the movement of the foot; broken arrows indicate movement of the food particles.

We come now to the second category of deposit feeders, namely those which either suck material directly from the substratum or utilise the thin layer of resuspended material at the surface of the deposits. The feeding mechanism of the small polychaete *Owenia fusiformis* may be included in this category although, as pointed out on p. 220 the distinction between this mechanism and that of the true suspension feeders is an arbitrary one. *Owenia fusiformis* is a widely distributed and locally common tubicolous polychaete of sandy shores. The tubes, made of light-coloured particles, are from 15 – 20 cm long and 2 – 5 mm across, their extreme flexibility allowing the animal to rebury itself when necessary. Feeding can be accomplished in two different ways; either in a ciliary manner or by picking up large particles with the lips (Dales, 1957).

The anterior end of the animal bears a crown of short branched structures developed from the prostomium and which are the main feeding organs. There are four main branches or units on each side, each unit being divided into 4 – 6 branches. Each branch ends in 2 – 4 small lobes which are usually arranged in pairs (Fig. 5.40). The margins of the

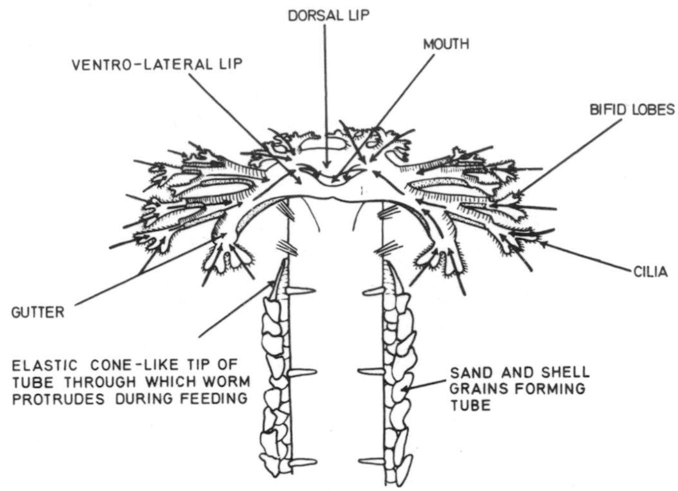

Fig. 5.40. Diagram of *Owenia fusiformis* from a ventral aspect showing feeding currents. (Based on Dales, 1957.)

lobes are raised so that the oral surface is trough-shaped, the edges of the trough bearing tracts of long cilia which beat inwards towards the axis of the stem and sweep particles into the crown. Small particles come to rest in the gutters and are transported orally by short cilia which beat towards the mouth. As in the terebellids, each food trough is capable of a variety of movements including inrolling which serves to free the crown of unwanted particles and to direct the particle-laden food strings to the mouth. Dales (1957) has also found that *Owenia fusiformis* can ingest large particles of food. Here the animal bends over until the crown is applied to the surface of the substratum, then the crown retracts by the action of a series of circular muscles which extend into all the branches except the bifid tips and which act against the hydrostatic pressure of the body fluid. Grains as large as 0·2 mm may be ingested in this way, larger ones being either rejected or used for tube-building.

Amongst the bivalves, two of the protobranch families, the Nuculidae and the Nuculanidae, have been traditionally regarded as exclusively selective deposit feeders (for example Yonge, 1939). In these families, there are well developed extensile palp proboscides which reach out from between the shell valves and gather material directly from the substratum. Although the Nuculidae and Nuculanidae undoubtedly do feed by means of the palp proboscides, Stasek (1961, 1965) has questioned the validity of the assumption that these families are exclusively deposit feeders. He suggested that the role of the ctenidia in food collection has been overlooked because it was thought that they were too small to play an important part in food collection. It was also assumed that there was no connection between the ctenidia and palps and that in any case in *Nucula* small particles pass between the ctenidial platelets into the suprabranchial chamber (for example Hirasaka, 1927; Atkins, 1936; Yonge, 1939). Nevertheless, Stasek (1961, 1965) has demonstrated that the ctenidia in certain nuculids and nuculanids not only gather particles but sort them according to size. Fine particles are passed to the anterior end of the ctenidia and coarser particles accumulate more posteriorly. The fine material is transferred to the palps at a point near the junction of the two palp lamellae in *Yoldia* (Stasek, 1965) and then pass towards the mouth along the dorsal part of the palp folds where little sorting takes place. Coarse particles are transferred to the inner lamellae and pass across a series of palp folds where rigorous sorting occurs.

As in other nuculanids, *Yoldia* draws water into the mantle cavity by means of intermittent pumping movements of the ctenidia. It thus appears that this stream is both a respiratory one and plays an important part in the nutrition of the animal. Stasek (1961, 1965) has demonstrated that suspension feeding may well occur in all three families of the Protobranchia, the Nuculanidae, Nuculidae and the Solemyidae (Yonge, 1939), although in the first two families deposit feeding by means of the palp proboscides may play an important part in nutrition; in the Solemyidae movements of the foot and mantle are known to aid in deposit feeding (Owen, 1961).

Of the eulamellibranchs, those most characteristically adapted to deposit feeding belong to the four families comprising the Tellinacea. Yonge (1950a) has described in detail such adaptations in *Tellina tenuis*, and *Macoma balthica* (Tellinidae), *Abra alba* and *Scrobicularia plana* (Semelidae), *Donax vittatus* (Donacidae) and *Gari tellinella, G. fervensis, Solecurtus chamosolen* and *S. scopula* (Asaphidae). The most obvious structures associated with deposit feeding are the siphons which are invariably separate. In the Tellinidae and Semelidae these are extremely long and mobile due to their intrinsic musculature which allows the inhalant siphon to actively suck material from the surface of the substratum (Fig. 5.41). In the other two families (Donacidae and Asaphidae) the siphons are relatively wider and shorter and are used for the intake of less consolidated material from the substratum

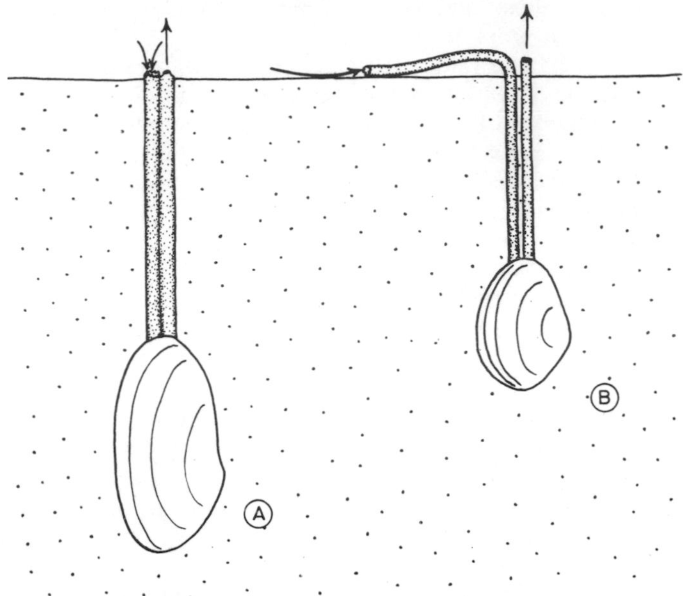

Fig. 5.41. Diagrams showing the use of the siphons in feeding in (A) a suspension feeding bivalve (*Mya arenaria*) and (B) a deposit feeding bivalve (*Scrobicularia plana*.) (After Yonge, 1950.) Note the fused siphons in *Mya* and the separate siphons in *Scrobicularia*. Arrows indicate the inhalant and exhalant streams.

(Fig. 5.41). The siphons are also devoid of filtering tentacles around the inhalant aperture such as occur in the siphonate suspension feeders *Mya, Spisula* and *Lutraria* (Yonge, 1948). Other modifications include an increase in the size of the foot for burrowing, and the development of a special sense organ associated with the cruciform muscle at the base of the siphons. It seems possible that, as suggested by Yonge (1950a), this sense organ is a proprioceptor providing information about the degree of extension of the siphons.

It is obvious that the branchial and labial sorting mechanisms of such deposit feeding genera must be adapted to deal with considerably greater quantities of particles than in suspension feeders (p. 220). Yonge (1950a) has shown that the greatest modifications occur in the Tellinidae where the outer demibranch is upturned and there is no food groove in the outer margin of the inner demibranch which bears the only oral stream (Fig. 5.42a). In the Semelidae there is a similar upturned outer demibranch but there are two oral streams, one along the gill axis and one along the tip of the inner demibranch (Fig. 5.42b). Finally, in the remaining Tellinacea the outer demibranch is reflected as in the suspension feeding eulamellibranchs described earlier (p. 202). In

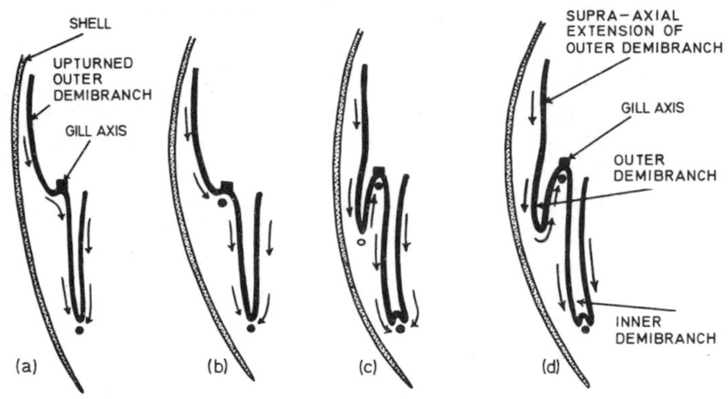

Fig. 5.42. Diagrams showing the form of the gill in vertical section in (a) *Tellina tenuis, T. fabula, T. donacina, Macoma balthica, Abra alba* and *A. nitida.* (b) *Tellina crassa* and *Scrobicularia plana.* (c) *Donax vittatus, Gari tellinella* and *G. fervensis.* (d) *Solecurtus chamosolen* and *S. scopula.* (After Yonge, 1950.) Arrows indicate major currents on the gill; solid circles, oralward currents and open circle, incipient oralward current.

such Tellinacea the oral streams are normally confined to the gill axis and the tip of the inner demibranch where there is a marginal groove, although in *Donax* and *Gari* spp. there is a tendency for particles to move anteriorly also along the tip of the outer demibranch (Fig. 5.42c and d) due to the presence of a series of coarse cilia (Atkins, 1937b). It has been suggested by Yonge (1950a) that the characteristic form of the outer demibranch in the Tellinidae and Semelidae (Fig. 5.42a and b) reduces the risk of choking the gill with the relatively immense quantities of bottom material which is swept in with the inhalant stream. There is also a series of coarse cirrus-like frontal cilia (also p. 205 and Fig. 5.24) which may facilitate the rapid forward passage of such food material. In addition to such modifications of the gills and siphons, the form of the stomach is highly characteristic of the group and has been described by Yonge (1950a).

A number of gastropods, too, are deposit-feeders, the best known are *Aporrhais pes-pelicani* and *Turritella communis* and deposit-feeding also occurs in the estuarine prosobranch *Hydrobia ulvae* which burrows in the deposits during the benthonic phase of its behavioural cycle (p. 142). Graham (1938a) and Yonge (1946) have shown that although *Turritella communis* is a burrowing form, it feeds by abstracting particles from the inhalant stream in much the same way as in *Bithynia tentaculata* (Schäfer, 1953a,b; Lilly 1953; Fretter and Graham, 1962) *Calyptraea chinensis* and *Capulus ungaricus* (p. 195). In this respect *Turritella communis* is perhaps best regarded as a ciliary suspension feeder; nevertheless, since feeding takes place while the animal is buried, it seems probable that the major constituents of the food are derived from the deposits themselves and hence justify the inclusion of this species in the category of deposit-feeders.

When placed on the surface of a muddy gravel, *Turritella* soon burrows by working its way down at an angle of approximately 10° making a series of side to side movements as it does so (Yonge, 1946). Eventually the animal is buried completely and remains in a stationary position with the head and anterior part of the foot at the surface of the substratum. An inhalant depression is made by means of the ciliated foot which moves the deposits to the right where they are consolidated by mucus from the pedal gland. Meanwhile a shallower depression is formed to the right of the mound by the expulsion of water in the exhalant stream. During feeding, coarse particles are prevented from entering the inhalant opening to the mantle cavity by a series of pinnate tentacles. Medium-sized particles pass across the floor of the mantle cavity towards the right hand side where there is a longitudinal food groove. Fine particles pass to the tips of the ctenidial filaments which are applied to the longitudinal food groove. Here the particles are entangled in mucus from the hypobranchial gland and carried anteriorly to a flattened feeding area in front of the mouth where the particle-laden food string is periodically engulfed by means of the radula.

Unlike *Turritella, Aporrhais* is a selective deposit feeder and thus is to be included in the third category of deposit feeders. Hunt (1925) recognised that *Aporrhais* and *Turritella* were deposit feeders by the presence of bottom material in their stomachs, and subsequently Yonge (1937b) described the feeding mechanism of *Aporrhais pes-pelicani* in some detail. This animal is occasionally found at extreme low water but normally lives in muddy gravel at between 5 and 100 fathoms. The lip of the shell is prolonged into five processes, four of which are conspicuous and one rather small. The terminal process is blade-like and plays an important part in burrowing (Yonge, 1937b). When placed in a vessel containing a mixture of mud and gravel *Aporrhais* begins to burrow by inserting the terminal process into the substratum and then crawling obliquely downwards. Mud is then thrown up over the dorsal surface of the shell and the animal gradually becomes buried. When the large terminal whorl of the shell is just covered by the deposits, the animal crawls horizontally with the spire still above the surface. During the forward progression the shell comes to lie in a shallow groove caused by the lateral displacement of the deposits, and finally even the apical whorls of the spire are covered by material falling in from the sides of the groove so that the animal is completely buried. At this stage the animal ceases to crawl and begins the construction of inhalant and exhalant channels in the deposit. The proboscis is extended anteriorly and upwards until it reaches the surface and then moves round to form

Fig. 5.43. Diagrams showing the form of the shell of *Aporrhais pes-pelicani* (A) from above, (B) from the side. (After Yonge, 1937.)

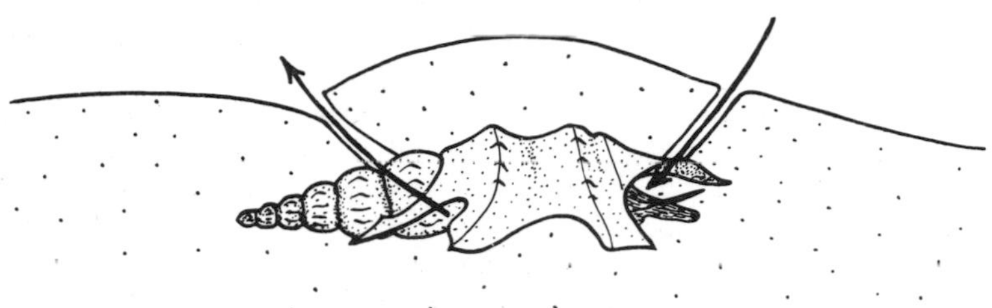

Fig. 5.44. Diagram showing the position of *Aporrhais pes-pelicani* when buried. Arrows indicate inhalant streams. (After Yonge, 1936.)

an inhalant opening of approximately 2mm in diameter (Fig. 5.43). The walls of this inhalant channel are then consolidated by the secretion of mucus from the walls of the proboscis which is then withdrawn and protruded upwards from the hinder end of the lip (Fig. 5.44), when an exhalant channel is constructed in the same way.

Although a considerable amount of suspended material is drawn into the mantle cavity with the respiratory stream, such particles are not used as a food source as they are in *Turritella, Bithynia, Crepidula, Capulus* and *Calyptraea* (p. 195). Coarse particles are rejected on entering the mantle cavity and medium sized particles pass across the floor of the mantle cavity and then move forward. Fine particles pass between the gill filaments and are then consolidated with mucus from the large hypobranchial gland before passing anteriorly to join a rejection tract running round the base of the foot (also p. 195 and Fig. 5.14). Feeding is carried out by means of the jaws and radula which is adapted for grasping plant material and bottom-living diatoms. When the local supply of food material is depleted, *Aporrhais* moves to another region where the process of inhalant and exhalant canal formation is repeated prior to feeding.

Hydrobia ulvae lies buried in the deposits during much of the time it is exposed by the tide (Newell, 1960; 1962; also p. 142) and its mode of burrowing is not unlike that of *Turritella communis*. Unlike *Turritella*, however, *Hydrobia ulvae* actively engulfs material from the surface of the deposits by means of the radula. In this respect its feeding-mechanism is similar to that of *Aporrhais pes-pelicani*, although at other times, as when *H. ulvae* is floating from the surface of the water by means of a mucous raft (p. 146), it feeds by periodically engulfing the particle laden raft which is drawn towards the mouth by means of the cilia on the foot (Newell, 1960, 1962).

The amphipod, *Corophium volutator*, is another burrowing animal which, like the prosobranchs mentioned above, is able to feed either as a suspension feeder or as a selective deposit feeder. Hart (1930) has described two distinct methods of feeding in this animal. The first is true selective deposit feeding in which material is picked up by the second gnathopods while the animal is moving over the surface of the substratum, and the second involves the filtration of particles drawn in with the respiratory stream. Meadows and Reid (1966) have recently shown that the setae on the first and second gnathopods play an important part in the process of deposit feeding, although they expressed some doubt concerning the importance of filter feeding in the adult. They showed that although the respiratory stream passes through the setal combs, fine particles were not retained on the filter. Nevertheless, as they point out, small animals do not leave their burrows and hence must abstract much of their food from the respiratory stream. It seems likely, therefore, that aggregates of suspended matter may be utilised for food and that this method of feeding may be of importance during certain phases of the life of *Corophium*.

Although Hart (1930) states that *Corophium* also feeds when on the surface of the deposit, Meadows and Reid (1966) were unable to confirm this. They found that the animal normally scrapes superficial organic debris into the burrow with the second antennae and release it just inside the entrance (Fig. 5.45). Semi-suspended aggregates of material are then drawn onto the setal filter by the current created by the beating of the pleopods. Material retained on the filtering setae of the gnathopods (Fig. 5.46) is then brushed forward onto the combs of the first gnathopods where sorting takes place by the movements of the first gnathopods against each other. The sorted material is then transferred to the maxillipeds whose outer surfaces bear a series of spines which comb material out of the filtering setae on the first gnathopod.

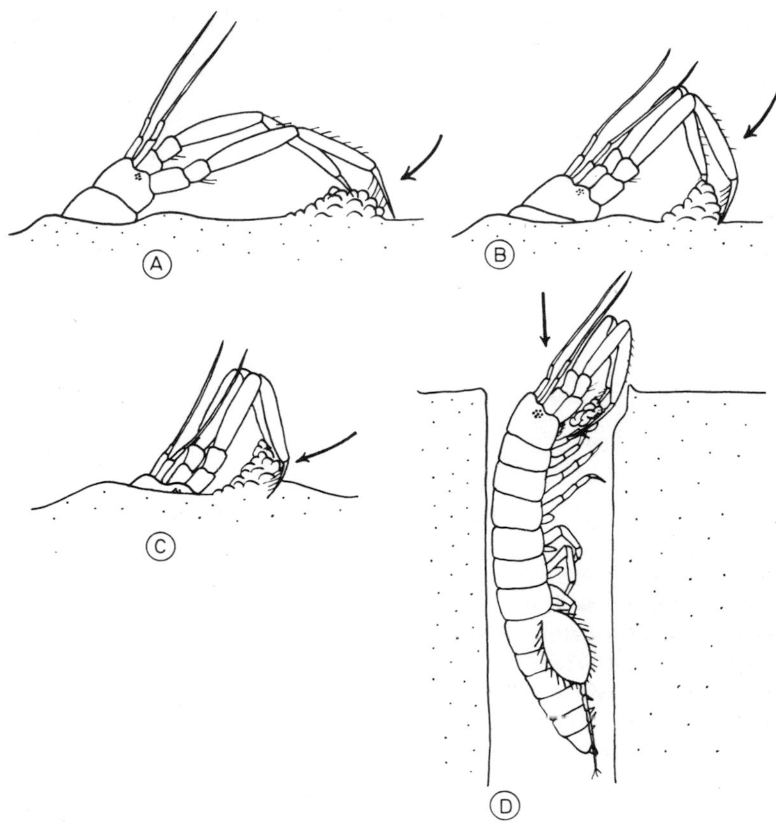

Fig. 5.45. Diagram illustrating how *Corophium volutator* removes organic debris from the surface of the substratum (A, B and C) and retreats into its burrow (D), (After Meadows and Reid, 1966).

The particles are finally passed to the first maxillae and thence to the mandibles which bear both outer incisor and inner molar processes. Thus *Corophium volutator,* which is extremely abundant in estuarine muds, is primarily a selective deposit feeder, either browsing on the surface of the substratum as described by Hart (1930), or drawing material into the burrow and filtering off suitable aggregates of organic material (Meadows and Reid, 1966). Filter feeding may also be of value to the juveniles which do not leave the burrow (Meadows and Reid, 1966), but the importance of this mechanism is difficult to assess since the setae appear to be unable to retain fine material drawn in with the respiratory stream.

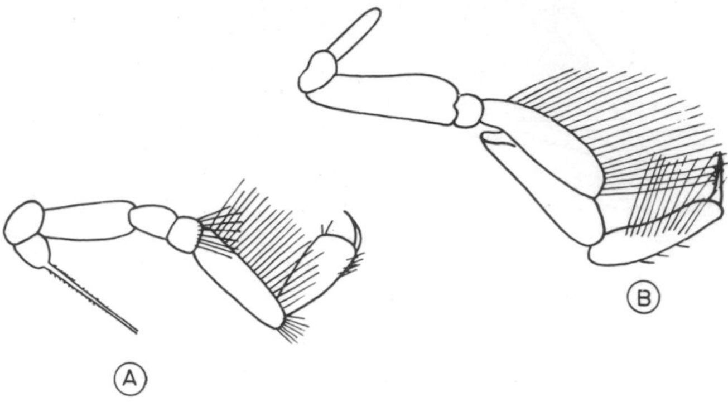

Fig. 5.46. Diagrams of the first gnathopod (A) and second gnathopod (B) on the left side of *Corophium volutator* showing the long setae which overlap with each other and with those of the opposite side to form the setal basket. (After Meadows and Reid, 1966.)

The final category of deposit feeders includes organisms such as the enteropneust *Saccoglossus*, the maldanid polychaete *Clymenella torquata* and the lugworm *Arenicola marina*, all of which feed indiscriminately upon the deposits in which they live. *Saccoglossus* presents an interesting transition between a ciliary deposit feeder (cf terebellids) and a mud swallower such as *Arenicola*. Although feeding has been observed to take place at the surface of the deposits (Burdon-Jones, 1952), *Saccoglossus* normally feeds in superficial temporary shafts which lead off from the main mucus-lined burrow (Knight-Jones, 1953a). The proboscis of the animal is extensively covered with mucous glands and cilia except in the region of the ciliary organ where mucous glands are absent. The cilia of the general surface beat towards the base of the proboscis, the dorsal cilia of the basal region beating latero-ventrally so that particles are driven down each side of the proboscis stalk towards the mouth (Fig. 5.47). Particles entrapped in mucus are carried into a groove of the ciliary organ by the beating of cilia on ridges which form the margins of the groove and then join a mucous stream leading to the mouth. The epithelium of the groove contains a number of sensory cells which lie above a special tract of nerve fibres (Knight-Jones, 1952) so that the ciliary organ not only carries particles to the mouth but probably also functions as an organ of taste. During feeding, the proboscis is moved about and particles become entrapped in the mucous sheath which surrounds the proboscis. This sheath is then moved posteriorly by the cilia on the surface of the proboscis and collects as a ring around the proboscis base. The anterior edge of the collar is usually dilated so that the mouth is held open and the particle-laden mucus is drawn into the mouth as a continuous string

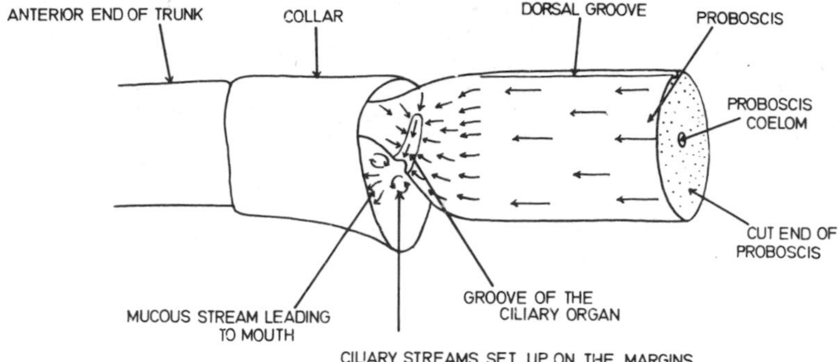

Fig. 5.47. Diagram showing the ciliary currents at the base of the proboscis of *Saccoglossus* viewed from the right side. (After Knight-Jones, 1953.)

by the cilia lining the buccal cavity. No sorting takes place prior to the particles being engulfed so that *Saccoglossus* may be regarded as an unselective deposit feeder.

In contrast to the ciliary mechanism of food collection in *Saccoglossus,* the lugworm *Arenicola marina* (Wells, 1945) and the maldanid polychaete *Clymenella torquata* (Ullman and Bookhout, 1949) are true mud swallowers which do not draw in particles from the substratum by ciliary activity. Instead, the eversible pharynx is used to engulf particles which are then swallowed indiscriminately, indigestible material being voided with the faeces. *Arenicola marina* is a common inhabitant of muddy sand and may be regarded as a typical example of a mud swallower. Thamdrup (1935), Linke (1939) and Wells (1944, 1945) have shown that the animal lives in a U-shaped burrow, and that burrowing is accomplished not by swallowing the substratum but by a combination of proboscis movements and mucus secretion (also Chapman and Newell, 1947). The worm burrows vertically to a depth of approximately 20 – 30 cm and then turns horizontally so that an L-shaped gallery is formed (Thamdrup, 1935; Wells, 1945; Jacobsen, 1967). Finally it moves anteriorly and then retreats, withdrawing sand; this aids in the formation of the "head shaft" or funnel. New material is constantly drawn down into the burrow by this means and, after being engulfed, indigestible material is voided as a cast at the exit of the tail shaft (Fig. 5.48).

Kruger (1959) has shown that the respiratory stream which *Arenicola marina* draws through the burrow may have the effect of causing suspended matter to be strained off by the sand grains in the head shaft. He showed that carmine particles are retained in the sand immediately anterior to the worm and are subsequently ingested. Further, since the levels of organic nitrogen in the oesophagus were

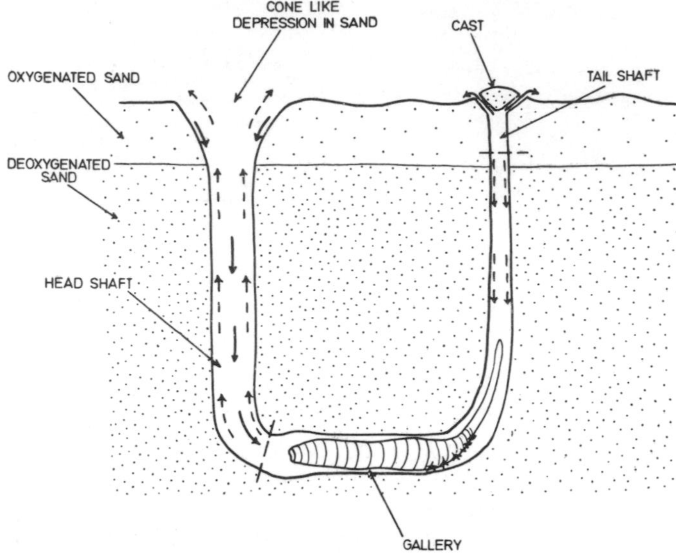

Fig. 5.48. Diagrammatic section through a lugworm burrow with the animal lying in the gallery. The broken arrows indicate water currents; solid arrows indicate sand movements. (After Wells, 1945.)

found to be from 5 – 10 times that found in the surrounding deposits, he suggested that *Arenicola marina* should be regarded as a suspension feeder utilising the plankton-enriched particles in the head shaft rather than the organic matter of the deposits themselves. However, these conclusions have been recently questioned by Jacobsen (1967) who found that the nitrogen values of the faeces and of the contents of the oesophagus of worms fed on sterilised sand were as high as that of animals taken straight from the shore. Jacobsen therefore concluded that the high organic nitrogen values recorded by Kruger (1959) were due to a mucous secretion by the worm and not to the presence of plankton-enriched food. He also pointed out that frequent reversal of the respiratory stream takes place, which would hardly aid the effective sieving of plankton in the head shaft. Further, as pointed out by van Dam (1938), suspension feeders normally pump rather a large volume of water compared with the oxygen utilisation, the normal oxygen utilisation in *Anodonta* being only 2 – 10%. *Arenicola* has a utilisation of 30 – 50%, unlike that occurring in most suspension feeders (Jacobsen, 1967). A final factor which tends to emphasise the importance of the organic matter of the deposits in the nutrition of *Arenicola marina* is that there is a good correlation between the total biomass of the lugworm and the organic carbon and nitrogen of the deposits in

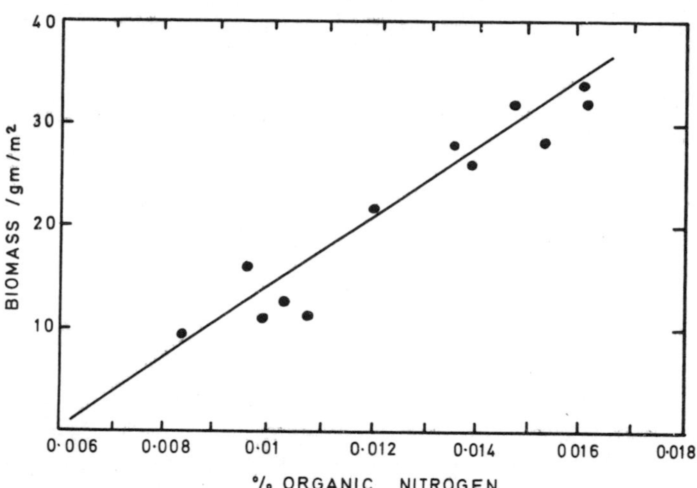

Fig. 5.49. Graph illustrating the relationship between the biomass (dry wt gm) of *Arenicola marina* and levels of organic nitrogen in the deposits at Whitstable, Kent, during 1965. (After Longbottom, 1968.)

which it lives (Fig. 5.49) (Longbottom, 1968 and p. 259). Nevertheless, Jacobsen (1967) has confirmed that, as found by Kruger (1959), the percentage organic nitrogen in the head shaft is high, although it is not clear whether such values also represent the presence of mucous contamination by the worm.

As pointed out by Jacobsen (1967), further information is necessary before the relative importance of the two mechanisms can be assessed. The possibility of organic nitrogen enrichment by the straining mechanism suggested by Kruger (1959) is an interesting one, which may well prove to be of importance in coarse deposits where the organic content of the substratum is low (Newell, 1965; Longbottom, 1968, and p. 249). Where the substratum is relatively fine, such a sieving mechanism may be of minor importance compared with the organic content of the deposits themselves. Thus here again, as in many of the other intertidal organisms whose feeding mechanisms have been reviewed above, it is possible that *Arenicola marina* may use different mechanisms of food collection to utilise such food material as may be locally available.

CHAPTER 6

Factors affecting feeding

A. INTRODUCTION

In Chapter 5 we have considered examples of the mechanisms by which intertidal animals are able to utilise the food sources available to them, whether these are in the form of encrusting algae, other intertidal animals, suspended plankton, or dead and decaying matter on the sea floor. Such sources of food are, however, not of equal abundance in all intertidal habitats. On rocky shores, for example, encrusting algae provide a rich food source for browsing organisms such as prosobranchs, opisthobranchs and chitons (p. 174), whilst the plankton is utilised by an abundant suspension feeding population including serpulid worms, barnacles (p. 213) and bivalves (p. 200). On sandy shores the unstable substratum precludes the development of encrusting algae, and the browsing component of the fauna is therefore correspondingly reduced. The plankton still affords a food supply for suspension feeding organisms which thus form the dominant component of the fauna. Where the substratum becomes finer, and where less turbulant conditions prevail, the organic content of the substratum is considerably higher than on coarse sandy shores. It is in such habitats that the deposit feeding population forms an important part of the faunal assemblage whilst browsing organisms, as on sandy shores, are scarce. It should be stressed, however, that the intertidal environment is normally complex with a variety of habitats impinging upon, or even grading into one another. Nevertheless, some areas are dominated by rocky, sandy or muddy shores, and it is by a study of organisms from these habitats that some indication of the biological and environmental factors controlling not only the rate of feeding but also the nature of the feeding mechanisms, may be gained.

B. FACTORS AFFECTING THE RATE OF FEEDING

A wide variety of environmental and biological factors influence the rate at which intertidal animals feed. Many suspension feeders, for example, are able to feed only when immersed by the tide or splashed by the waves, whilst a variety of browsing organisms are active mainly when recently exposed. Among the more important environmental factors which influence feeding are the degree of exposure by the tide, the quality and quantity of suspended material in the inhalant stream, and the temperature; biological factors include the size of the organism under consideration as well as the presence or absence of food in the gut. These, and a variety of other factors have recently been extensively reviewed by Jφrgensen (1966) and by Owen (1966).

1. ENVIRONMENTAL FACTORS

As has been mentioned above, perhaps the most obvious environmental factor which might influence the rate at which intertidal animals feed is the duration of available feeding time. Suspension feeders living high in the intertidal zone, for example, might be expected to increase their feeding rate in order to compensate for the reduced time of immersion. In fact, there are few examples in the literature of such compensatory processes. Indeed, Segal, Rao and James (1953) have shown that the reverse is true of the mussel *Mytilus californianus.* The rate of water propulsion of mussels from 10 m depth was found to be faster than of mussels collected from low in the intertidal zone, whilst the rate of water propulsion of the latter was faster than in mussels collected from the mid-intertidal zone. Segal, Rao and James (1953) suggested that such intraspecific differences in pumping rate were due to the different temperature conditions at the three shore levels: as had been suggested by Rao (1953a) to account for variations in the pumping rate of *Mytilus californianus* with latitude. Since the populations of *M. californianus* at different levels on the shore mix during reproduction, it was suggested that any differences which exist in the adults must have been imposed by the environmental conditions after the establishment of the adult zonation pattern (also Chapter 8). Such differences in filtration rate with tidal level apparently do not occur in the mussel *Mytilus edulis* (Jφrgensen, 1960), although the mid-tide specimens selected for the feeding experiments were collected from tidal pools and thus might be expected to resemble the permanently-immersed low tide animals with which they were compared. Southward (1964) has also demonstrated a suppression in the cirral activity of high level populations of the barnacle *Chthamalus dalli* compared with low level populations. He has also shown that animals taken from wave beaten localities have a lower rate of cirral activity than high level populations from sheltered localities. He therefore suggested that rather than representing an adaptation to temperature, such differences in the rate of cirral beating are to be correlated with the low rates of growth of high level and wave beaten populations of barnacles. Moore (1935a) has shown that *Balanus balanoides* from sheltered habitats grow slower than those from exposed situations, though whether this is due to a difference in the feeding rate or in the quantity of food available is not clear.

Morton, Boney and Corner (1957) have studied the filtration rate of the small intertidal crevice-dwelling bivalve *Lasaea rubra* in relation to tidal level. The upper limit of this animal at Wembury, Devon is mean high water of spring tides, and in this situation it is immersed for an average of only 1 hr in 12 and may not be covered at all for 12 consecutive days during the period of neap tides. Animals from the lower limit of distribution, however, are immersed for an average of 8 hr per 12-hourly tidal cycle and are not exposed continuously for more than 12 hr at neap tides. Since the animal is a suspension feeder (Ballantine and Morton, 1956) and appears to thrive at its upper limit of distribution, it might be expected to show interesting adaptations in its feeding behaviour during its short immersion period. Morton, Boney and Corner (1957) found that animals collected from a level corresponding with mean high water spring tides at Wembury showed two main adaptations to the reduced period of feeding time compared with *Lasaea rubra* collected from a level corresponding with mean low water of neap tides; the high level animals responded more quickly to wetting, and their filtration rate was initially higher than that of the low level animals.

A further factor which offsets the reduced period of submersion of high level *Lasaea rubra* is that the animals are small enough to be able to utilise the thin film of water covering the rocks from wave splash. As Colman (1933) showed, wave splash prolongs the effective submersion period of intertidal animals and *Lasaea rubra* is well adapted to feed even when the water film is scarcely deep enough to cover the shell (Fig. 6.1) (Morton, Boney and Corner, 1957). Under such conditions, the speed of response to wetting may play an important part in an extension of the feeding period of the animal. Fig. 6.2 shows the result of an experiment in which 60 animals were placed on a slide and covered with a thin film of water. The percentage emerging and filtering

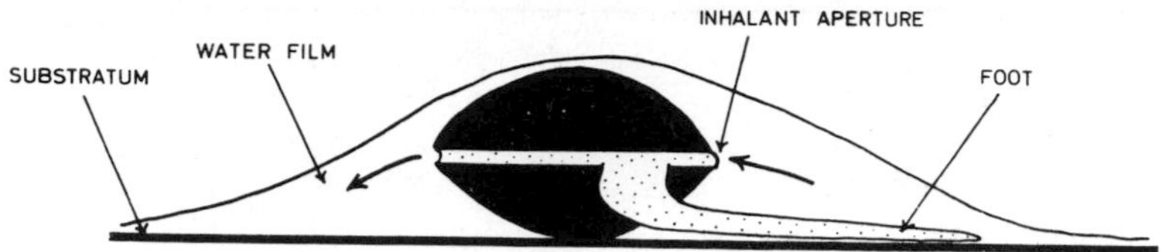

Fig. 6.1. Diagram illustrating the utilisation of the water film for feeding by *Lasaea rubra.* (After Morton, Boney and Corner, 1957.)

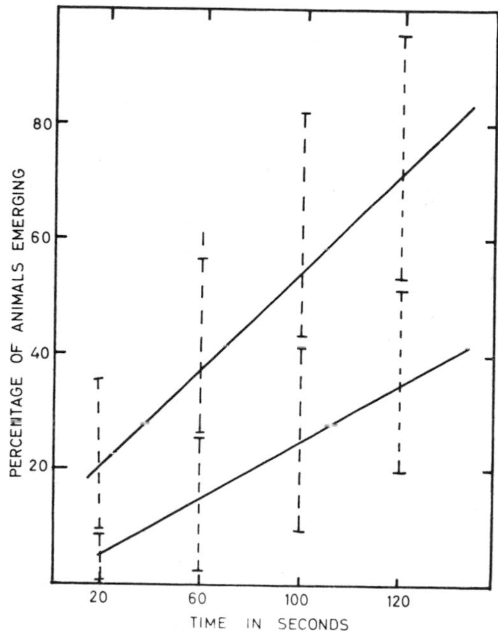

Fig. 6.2. Graph showing the percentage emergence of groups of high level (upper line) and low level (lower line) *Lasaea rubra* collected from Plymouth and measured after 20, 60, 100 and 140 secs under a film of seawater on a slide. The broken lines indicate the approximate scatter of the results. (Replotted after Morton, Boney and Corner, 1957.)

was then recorded after 20, 60, 100 and 140 seconds; it was clear that the high level animals responded significantly faster to wetting than the low level ones.

The second factor which to some extent offsets the reduced feeding time in high level populations of *Lasaea rubra,* is that the rate of feeding during the first two hours of immersion is approximately double that of animals from lower on the shore (Ballantine and Morton, 1956;

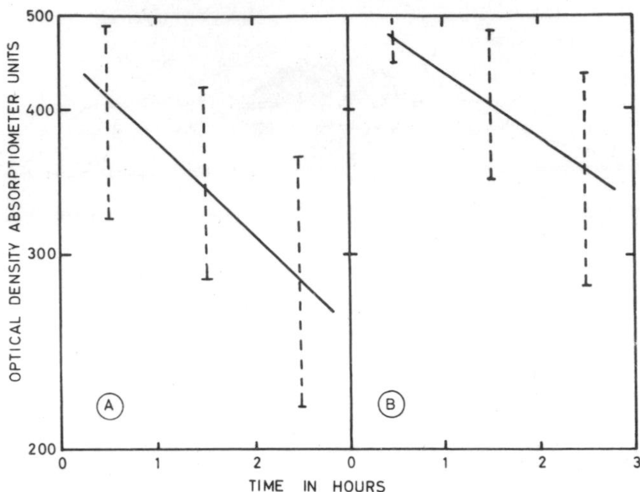

Fig. 6.3. Graphs showing the results of an experiment in which groups of 10 *Lasaea rubra* were allowed to filter for up to 1·5 hr. in 3·75 ml. of a culture of *Phaeodactylum tricornutum.* A, animals collected from Wembury high spring tide level; B, animals from Wembury low neap tide level. The approximate scatter is indicated by the broken lines. (Replotted after Morton, Boney and Corner, 1957.)

Morton, Boney and Corner, 1957). Fig. 6.3 shows the result of an experiment in which 10 *Lasaea rubra* of approximately 2 mm in length were allowed to filter in vessels containing 3·75 ml of a culture of the diatom *Phaeodactylum tricornutum,* the decrease in optical density being recorded after 0·5 hr, 1 hr and 1·5 hr; the filtration rate of the high level animals is obviously higher than that of the low level bivalves, and thus compensates for the difference in submersion time of the two populations. Such adaptive responses have been shown in few other intertidal suspension feeders. However barnacles respond almost immediately to wetting (Barnes and Barnes, 1958), which ensures that they are also able to filter feed for the full duration of the submersion period (see Southward and Crisp, 1965).

Just as the feeding time of suspension feeders is limited to the period when they are wetted or submersed by the tide, the duration of the feeding period of many other intertidal animals is restricted to certain phases of the tidal cycle. The winkle, *Littorina littorea,* for example, normally makes its browsing excursions when the tide has ebbed, when it uses the sun as a light-compass (Newell, 1958a,b; also p. 139). Many of the inhabitants of crevices, too, move out during the period of exposure to scavenge over the surface of the rocks; animals in this category include the isopod *Ligia oceanica* (Nicholls, 1931a,b), as well as *Eulalia viridis, Anurida bisetosa, Halotydeus hydrodromus, Bdella longicornis* and *Anuridella marina,* each of which has been observed to emerge from crevices when exposed by the tide (Glynne-Williams and Hobart, 1952; also p. 54). The sand-hopper, *Talitrus saltator,* also actively migrates down into the intertidal zone from above the high water mark when the tide ebbs (p. 146).

We see, therefore, that during the time the intertidal zone is submersed by the tide, the suspension feeding component of the fauna becomes active, utilising the plankton as a food source. The deposit feeding component, as well as certain aquatic predators, is also active at this time since many of the bivalve deposit feeders require to resuspend the deposits in a water stream before they can be ingested (Chapter 5). With the onset of the more terrestrial conditions which occur during the intertidal period, however, a great variety of other intertidal animals move out from their places of shelter to browse or scavenge in the intertidal zone. Only those suspension feeders which remain immersed in pools left by the receding tide are able to maintain continuous feeding throughout the tidal cycle. Nevertheless, certain suspension feeding animals living high in the intertidal zone, such as the bivalve *Lasaea rubra* and barnacles, do possess behavioural responses which ensure that the brief period of submersion is promptly utilised for feeding. Other intertidal suspension feeders apparently do not show such adaptations, but these results may in part be due to the difficulty of measuring feeding rates under conditions which are comparable with those on the shore. As Jørgensen (1966) pointed out, many laboratory measurements of the feeding rates of bivalves are lower than the rates of undisturbed specimens measured under natural conditions. It may be, therefore, that this factor has masked small quantitative differences between the feeding rates of high and low level populations of intertidal suspension feeding animals.

The quantity and nature of the suspended matter available to suspension feeding animals appears to have little effect on the rate of transport of water by a wide variety of these organisms (Jørgensen, 1966), although excessive quantities of silt or food may clog the filtration mechanism and thus reduce water transport. Lund (1957) found that the oyster *Crassostrea virginica* retained and engulfed suspensions containing 10 mg/l of Fuller's earth or 5 – 10 mg/l of a mixture of Fuller's earth and chloroplasts without any production of pseudofaeces. The animal was therefore not distinguishing between inert and organic suspended material (a possible reason for which is suggested on p. 263). However, high concentrations of Fuller's earth or other inert suspensions depress the filtration rate by 50% when present in concentrations of 100 mg/l, and by more than 80% when present in excess of 1000 mg/l (Loosanoff and Tommers, 1948). Other suspension-feeders, such as *Ostrea gigas, Mytilus edulis* and *Venerupis semidecussata,* are more tolerant of high concentrations of suspended matter such as bentonite (Chiba and Oshima, 1957); but, as Jørgensen (1966) pointed out, even in low bentonite concentrations the recorded values for filtration rates were rather low compared with values obtained by other workers. Despite this possible source of error, it seems likely that bivalves such as the clam *Mya arenaria,* the mussel *Mytilus edulis,* the cockle *Cardium edule,* and others which live in habitats where the silt concentration is high, would be more tolerant of high concentrations of suspended material than bivalves living on rocky shores or in coarse sandy deposits. Indeed, suspension feeding animals living on, or in, fine deposits may utilise micro-organisms adsorbed onto the fine particles in much the same way as has been shown to be true of certain deposit feeders (p. 259; Jørgensen, 1966). Barnacles too, grow faster under estuarine conditions, which suggests that, like deposit

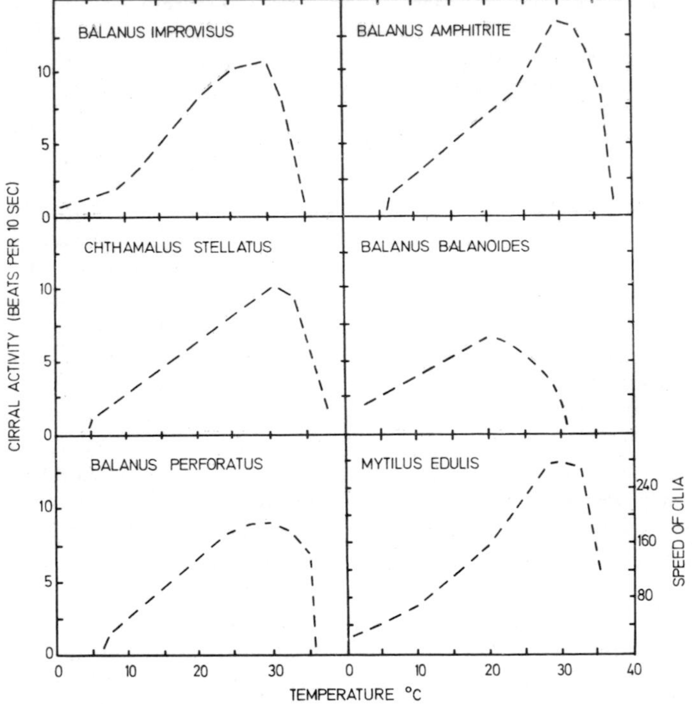

Fig. 6.4. Graphs illustrating the effect of temperature on the cirral activity of five species of barnacles, (after Southward, 1957, 1962, 1964), and on ciliary activity of the gill of *Mytilus edulis.* (After Gray, 1923.)

feeding organisms (p. 219), they utilise the rich food supplies occurring in estuaries where the deposits are fine (Moore, 1936).

Fluctuations in temperature have a profound effect on the rate of activity of invertebrates (Fig. 6.4). Gray (1923) showed that the rate of movement of the cilia on the gills of *Mytilus edulis* increased with temperature, the Q_{10} varying between 3·1 (1 – 10°C) and 1·92 (22·5 – 32·5°C) up to approximately 33°C after which decline occurred until at 37·5°C the cilia ceased to create any detectable current. Southward (1955b,c; 1957, 1962, 1964) has shown that the rate of cirral activity of barnacles, too, is strongly dependent upon temperature, the cirri being fully active only within a narrow range which is correlated with the geographical distribution of the species. Again, Galtsoff (1928) and Hopkins (1933, 1935a) found that the gills of *Crassostrea virginica* performed their maximum work at 30°C; the influence of temperature is discussed in more detail in Chapters 8 and 9.

The optimal temperatures for activity may, however, vary according to the thermal conditions in the habitat. Rao (1953a), for example, has shown that in *Mytilus californianus* the rate of water propulsion in animals taken from higher latitudes on the western coast of the United States (Friday Harbour, 48°27 N) was greater than in similar sized animals taken from lower latitudes (Fort Ross, 38°31 N) and measured at the same temperature. Again, these last animals transported water faster at comparable temperatures than similar sized animals taken from still lower latitudes (Los Angeles, 34°OO N). Mussels from Friday Harbour transported water at the same rate at 6·5°C as those from Fort Ross at 10°C and those from Los Angeles at 12°C. Rao (1953a) estimated that the mean annual temperature of Friday Harbour was 6·5°C, and Fort Ross 10°C and Los Angeles 16°C. That is, by a process of thermal acclimation, the mussels from a cold environment (Friday Harbour) had raised their level of activity, whilst those from a warm environment (Los Angeles) had a suppressed rate of pumping.

Temperature fluctuations of relatively short duration thus exert a profound effect on the rate of feeding of such suspension feeding organisms, but prolonged exposure to high or low temperatures results in the development of compensatory changes such as those cited above. Such changes have been reviewed by Bullock (1955), and are discussed in more detail in Chapters 8 and 9 together with more general aspects of the effect of temperature on intertidal organisms.

2. BIOLOGICAL FACTORS

The absolute rate at which suspension feeding animals pump water is often considerably greater in large animals than in small ones (for example Rao, 1953a). A more useful basis for comparison between different sized animals is, however, the weight-specific pumping rate. This expression represents the volume of water pumped per unit time, divided by the weight or size of the animal. When such weight-specific pumping rates are plotted against the weight of the soft parts, small animals are seen to pump faster than larger animals of the same species. (Fig. 6.5) The same relationship applies to the weight specific beating rate of barnacles (Newell and Northcroft, 1965). The curve in such filter

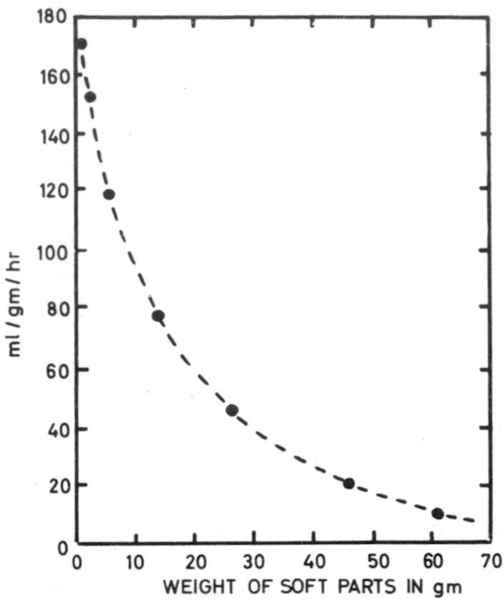

Fig. 6.5. Graph showing the weight-specific pumping rate of *Mytilus californianus* plotted against the weight of soft parts. (After Rao, 1953a.)

feeders is a logarithmic one, so that when plotted on a logarithmic scale as in Fig. 6.6. a straight line with a negative slope is obtained. Unfortunately, although such relationship between the weight-specific pumping rate and tissue weight appears to apply to a wide variety of suspension feeders (for review, Jørgensen, 1966), the differing experimental methods used by the investigators makes strict comparisons between pumping rates difficult. Jørgensen (1966) points out that members of the epifauna appear to be less susceptible to disturbance than burrowing bivalves. Thus the great difference in level between the values of the weight-specific pumping rates in *Crassostrea virginica* (Loosanoff and Nomejko, 1946) and *Ostrea edulis* (Allen, 1962) compared with *Venus mercenaria* (Rice and Smith, 1958), *Venus striatula* and *Mya arenaria* (Allen, 1962), may be partly attributable to the experimental conditions under which feeding was measured. However,

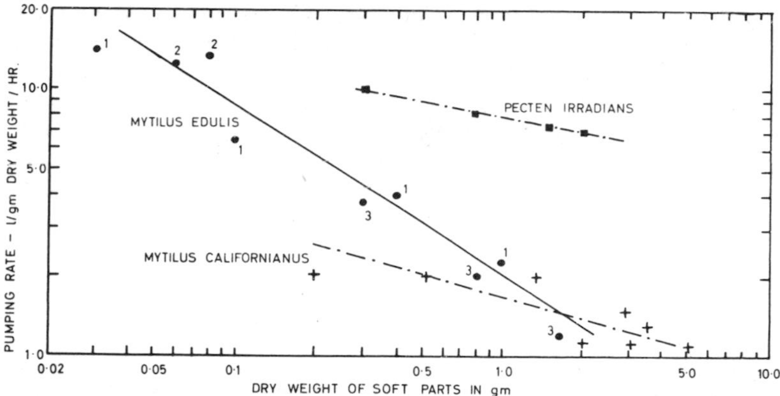

Fig. 6.6. Graphs illustrating the relationship between the weight-specific pumping rate in certain bivalves and the approximate dry weight of the body tissues. (Re-calculated from Jørgensen, 1966.) *Mytilus edulis,* dots and continuous line. (1, 2 and 3 indicates data compiled from different sources by Jørgensen, 1966). *Mytilus californianus*, crosses and broken line. (Based on Rao, 1953) *Pecteri irradians*, solid squares and broken line. (Based on Chipman and Hopkins, 1954. *Note:* The measurements on *Pecten irradians* were made at 22 – 26°C which may well account for the high level of pumping in this animal compared with *Mytilus* in which pumping rates were estimated at only 12 – 15°C. Assuming a Q_{10} of 2, the pumping rate of *Pecten irradians* would be similar to that of *Mytilus* spp. at 12 – 15°C.

the temperature conditions under which feeding was measured would also adequately account for such differences (see note to Figure 6.6). It is therefore not yet known whether the rate of feeding of members of the epifauna is, in general, higher than that of infaunal suspension feeding bivalves when measurements are made at similar temperatures under conditions as close as possible to those occurring in the habitat.

As we have seen, the different experimental methods used to measure the feeding rate of suspension feeding organisms has led to a considerable variation in the recorded rates. Such variation would necessarily obscure relatively small variations in feeding rate which might be caused by the presence or absence of food in the gut. Nevertheless, it is generally stated that the rate of water transport in undisturbed suspension feeding animals is maintained almost continuously, irrespective of the amount of material collected. Under natural conditions, feeding would therefore be expected to occur more or less continuously when the suspension feeding animals are covered by the tide, and cease when they are exposed, although a large number of workers have reported the occurrence of phases of activity and quiescence even in the pumping of fully immersed bivalves (for review van Dam, 1938, 1954; Verwey 1956); the extent to which this is due to the experimental conditions is, however, not yet known. Rao (1954) has shown that even in permanently immersed *Mytilus edulis* and *M. californianus* a tidal rhythm in feeding occurred, maximal rates of water transport coinciding with high water and minimal rates with low water. Such a rhythm was found to persist for several weeks in the laboratory; Jørgensen (1960), however, was later unable to confirm the presence of a tidal rhythm in *Mytilus edulis.* Jørgensen (1966) suggested that the animals used in the experiments of Rao (1954) were pumping at reduced rates, and that the variations noted were likely to be due to variations in the conditions responsible for the suppressed rates of water transport. As pointed out on p. 240, the contradictory results obtained by Rao (1954) and by Jørgensen (1960) may be partly due to the different intertidal situations from which the animals were selected. Although Jørgensen (1960) collected *Mytilus edulis* from mid-tide level, the bivalves were normally submersed in a tidal pool during the intertidal period; under these conditions their feeding responses would be expected to approximate more closely with low level populations than with high level ones.

The observed tidal variations in filtering rates noted for *Mytilus edulis* and *M. californianus* by Rao (1954), as well as the differences in rates noted at different tidal levels in *Mytilus californianus* (Segal, Rao and James, 1953) and in *Lasaea rubra* (Ballantine and Morton, 1956; Morton, Boney and Corner, 1957), have obvious ecological implications. Nevertheless, it is clear that the differing susceptibility of intertidal suspension feeders to experimental conditions may introduce a source of error into such estimations, and further evidence is required before adaptive variations in feeding rate with tidal level can be regarded as of general occurrence among suspension feeders.

C. FACTORS LEADING TO QUALITATIVE DIFFERENCES IN FEEDING

The differences between rocky shores and sandy or muddy shores are sufficiently obvious to account fully for the marked faunal differences which occur; these have been discussed in Chapter 2. Although the various feeding mechanisms of these animals have been discussed already (Chapter 5), the factors which account for the abundance of deposit feeding animals in fine substrata have not been thoroughly studied until recently. Yonge (1952), Southward (1952, 1953), Fox *et al.* (1952) and Rullier (1959), as well as a number of other workers, recognised that organic matter is more abundant in fine deposits than in coarse ones, and that it might represent a food source for deposit feeding organisms. Nevertheless, there has been little attempt other than the work cited below to determine the precise relationship between particle size and organic matter; neither have there been many attempts to determine the nutritive component of the organic matter in intertidal deposits.

1. ORGANIC MATTER IN RELATION TO GRADE OF DEPOSIT

The organic component of intertidal deposits may be conveniently analysed either in terms of the organic nitrogen comprising proteins, or as organic carbon, both of these factors being best related to the median particle size of the deposit (Morgans, 1956 and p. 30). Wood (1964, 1965), Newell (1965) and Longbottom (1968) have recently attempted to correlate the organic nitrogen of estuarine deposits with particle size. Wood (1964) studied the fine muds of estuaries in Australia, and although the coarsest particles he analysed were of the order of only 40 μ in diameter, it is possible to extrapolate his data to include coarser materials. Newell (1965) and Longbottom (1968) both analysed the organic nitrogen component of deposits from those characteristic of coarse sandy shores to fine estuarine muds. In all of these investigations it was found that the organic nitrogen of the deposits increased logarithmically as the deposits became finer. These results are summarised in Fig. 6.7 in which the percentage of organic nitrogen is plotted against the median particle diameter of the deposits; also shown are the approximate values calculated from the data of Brambell and Cole (1939) and Buchanan (1966). Other workers, for example Jacobsen (1967), Rullier (1959), Thamdrup (1935), and Callame (1961), have assessed the organic nitrogen of intertidal deposits, but since the details of the precise grade of deposit are not given their results cannot easily be compared with those shown in Fig. 6.7. From the results illustrated, it is clear that the organic nitrogen of particles of median particle diameter 0·15 mm is between 0·01% (Longbottom, 1968) and 0·02% (Buchanan, 1966). This compares with values of 0·012% found by Newell (1965) and 0·013% (extrapolated from Wood, 1964). Thamdrup (1935) obtained values of between 0·019% and 0·045% organic nitrogen at the surface of the deposit, and stated that this value was reduced to 0·01% at 8 cm depth below the surface of the deposits. (also Jacobsen, 1967).

It is apparent from Fig. 6.7 that there is a broad similarity in level between the values for organic nitrogen obtained by a variety of workers. The slopes of the regression lines which relate percentage of organic nitrogen to median particle diameter are also similar for the data of Wood (1964) and Longbottom (1968). The slope obtained by Newell (1965) is somewhat steeper, and Longbottom (1968) has suggested that the reason for this is that the points obtained from fine semi-fluid muds are artificially high because of the development of

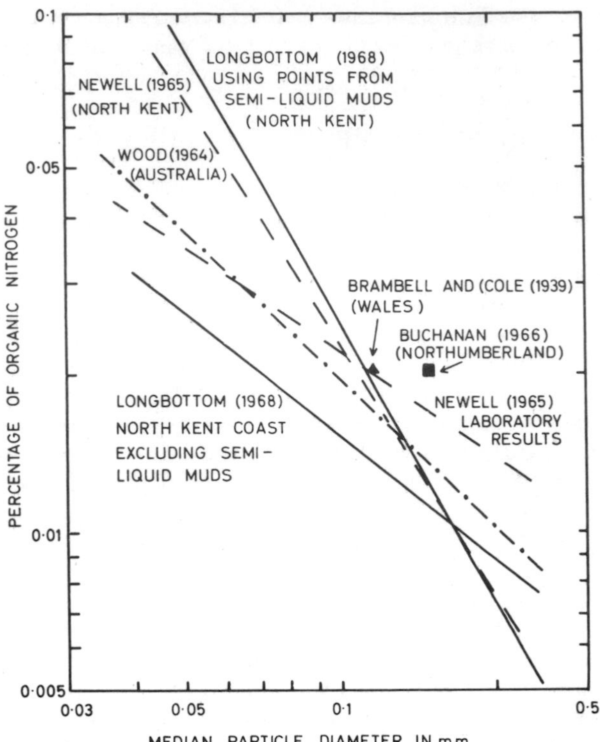

Fig. 6.7. Graphs summarising the relationship between particle size and percentage organic nitrogen in marine deposits as found by Wood (1964), Newell (1965) and Longbottom (1968). The results of Brambell and Cole (1939) and of Buchanan (1966) are also shown. (Based on Longbottom, 1968.)

micro-organisms in the interstitial spaces. Accordingly, he plotted his data from a range of deposits which did not include such muds; he pointed out that the line obtained by Newell (1965) for laboratory-cultured deposits was much more similar to his data and those of Wood (1964). This regression line for cultured deposits is also shown in Fig. 6.7, the high level being due to the artificially large amounts of organic carbon added to the deposits (0·65%) compared with the 0·1 – 0·3% which occurs on the shore in equivalent grades of deposit. Despite the difference in level, it is clear that the organic nitrogen of intertidal deposits increases logarithmically as the particles become finer, and also under natural conditions reaches similar levels in comparable grades of deposit.

Much the same relationship exists between the organic carbon component of intertidal deposits and particle size as has been described above for organic nitrogen. There is, however, more variation in the data of workers from different areas around the coasts due, in part, to the presence of coal in some intertidal deposits (Southward, 1952). The level of organic carbon should therefore be used in conjunction with organic nitrogen determinations where coal is likely to interfere with the measurements. Other factors which will greatly alter the value obtained from sample to sample are small fragments of organic debris. Longbottom (1968) has summarised the work of various authors, and has confirmed the results of Newell (1965) which showed that the percentage of organic carbon increased logarithmically with decreasing particle size. Once again, the semi-fluid muds which Newell (1965) analysed apparently contained an exceptionally high proportion of organic debris and micro-organisms; when less fluid muds were used the slope of the line was shallower, and resembles data obtained by Longbottom (1968). Fig. 6.8 shows these data together with an indication of the levels obtained by Brambell and Cole (1939) and by Buchanan (1966), although these workers used a very broad range of sieves so that the precise median particle size is difficult to calculate.

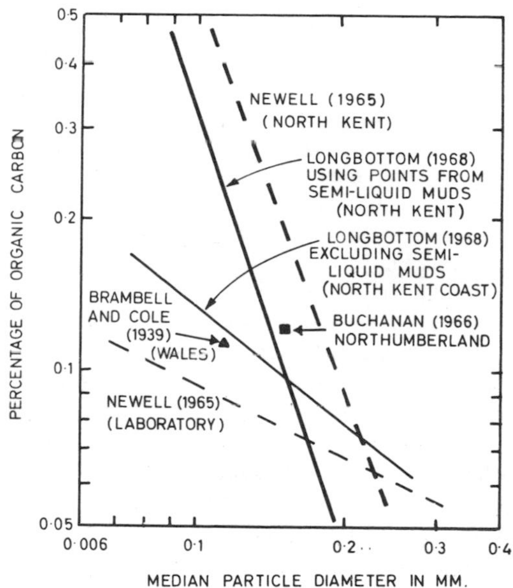

Fig. 6.8. Graphs summarising the relationship between particle size and percentage organic carbon in marine deposits, as found by Newell (1965) and Longbottom (1968). The results of Brambell and Cole (1939) and of Buchanan (1966) are also shown. (Based on Longbottom, 1968.)

2. THE SOURCE OF ORGANIC MATTER IN DEPOSITS

Having established the relationship between particle size and organic matter in the deposits, it is of interest to determine experimentally the cause of the logarithmic increase in organic matter with decrease in particle size. Newell (1965) sterilised a series of sand samples of widely differing grades with hydrogen peroxide and then cultured them in filtered seawater at 18° C. Analyses of organic carbon were then made on sub-samples at regular intervals and it was found that the deposits developed a higher organic carbon value in the fine grades than in the coarse ones. Since similar results were obtained in duplicate samples kept in the dark and under illumination, it was concluded that the development of an organic carbon component was due to the growth of heterotrophic micro-organisms. Further, the slope of the curve relating organic carbon to grade of deposit was similar to that obtained on shore deposits (Fig. 6.8). The implication is therefore that the organic carbon measured on the shore represents, at least in part, a population of micro-organisms. A similar experiment was carried out by Newell (1965) to determine the possible source of the organic nitrogen component of intertidal deposits. Deposits of various grades were sterilised with hydrogen peroxide and then 0·65% of powdered *Fucus* was added to each sample. Duplicate sets of the deposits were then cultured in filtered seawater at 19° C in the light and in the dark. Sub-samples were analysed at regular intervals for 27 days and it was found that in both illuminated samples and in those stored in the dark an organic nitrogen component developed which was highest in the fine deposits. The organic carbon value, mainly representing the powdered *Fucus*, was reduced as the nitrogen value increased. It was therefore inferred that the increase in organic nitrogen was also due to the presence of a population of heterotrophic micro-organisms which utilised the powdered *Fucus* to obtain the necessary energy to build protein. The curve relating such organic nitrogen to grade of deposit is shown in Fig. 6.7 and it is clear that the slope is very similar to that obtained on natural deposits by Wood (1964) and by Longbottom (1968).

We are now in a position to suggest that the organic nitrogen component of intertidal deposits represents the protein contained in a population of heterotrophic micro-organisms (probably mainly bacteria) whilst the organic carbon also represents such micro-organisms, although at least some of the organic carbon values must include that contained in organic debris of non-bacterial origin. The results of direct analyses of micro-organisms in intertidal deposits are therefore of interest. Reuszer (1933), Zobell (1936, 1938, 1943, 1946), Zobell and Anderson (1936), Fox *et al.* (1952) and Darnell (1967a,b) have emphasised the importance of surfaces in increasing the adsorption of particulate or sub-particulate nutrients, which enables large numbers of bacteria to thrive in fine deposits. Zobell found that bacteria will attach themselves to recently sterilised surfaces which probably quickly adsorb nutrients from seawater, whilst Zobell and Anderson (1936) suggested that exo-enzymes may be concentrated on surfaces such as sand grains; they subsequently found that there was a tenfold increase in the rate of destruction of hydrocarbons by bacteria when sand was available. Direct observations on bacteria by Pearse *et al.* (1942), Anderson and Meadows (1965), Meadows (1965) and Meadows and Anderson (1966, 1968) have shown that bacteria are firmly attached to the surface of sand grains and often remain attached even after fixation and staining. Indeed, Jansson (1966) suggested that beach sand, with its large surface area, represents ideal conditions for the culture of bacteria.

These direct observations on bacteria suggest that the surface area of the deposits largely controls the level of bacterial population. Newell (1965) suggested that this supported the conclusion that the organic carbon and nitrogen in marine deposits did, indeed, represent a population of bacteria, since the relation between organic carbon and nitrogen and grade of deposit is similar to that relating surface area to grade of deposit, assuming the deposit to be composed of a series of spheres whose diameter is equivalent to the median particle size. Longbottom (1968) has, in fact, demonstrated this clearly by calculating the surface area of a series of spheres of uniform diameter and

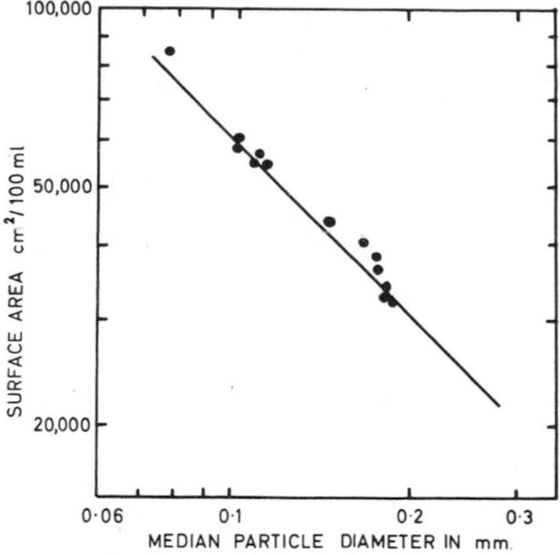

Fig. 6.9. Graph showing the calculated line for the surface area ($cm^2/100$ ml) of a system of uniform spheres plotted against particle diameter. Dots indicate experimentally-determined surface areas of a series of deposits of differing median particle diameter. (After Longbottom, 1968.)

plotting these values against the diameters of the particles. (Fig. 6.9). He then calculated the surface areas of a number of grades of natural deposits. As will be seen from Fig. 6.9, there is a close correspondence between the theoretical and calculated values, and the deposits may be conveniently regarded as a series of spheres of uniform diameter. The slope of the calculated line could then be compared with the slope of the line relating organic nitrogen to the grade of deposit. This is shown in Fig. 6.10 where the slopes of the two lines are similar. This is a strong indication that the organic nitrogen is related to the surface area of the deposits – as would be expected on the assumption that the nitrogen represents a population of bacteria attached to the surface of the sand grains.

Longbottom (personal communication) has also calculated the maximum possible amount of organic carbon and nitrogen which would be present in a series of deposits, assuming them to be composed of a

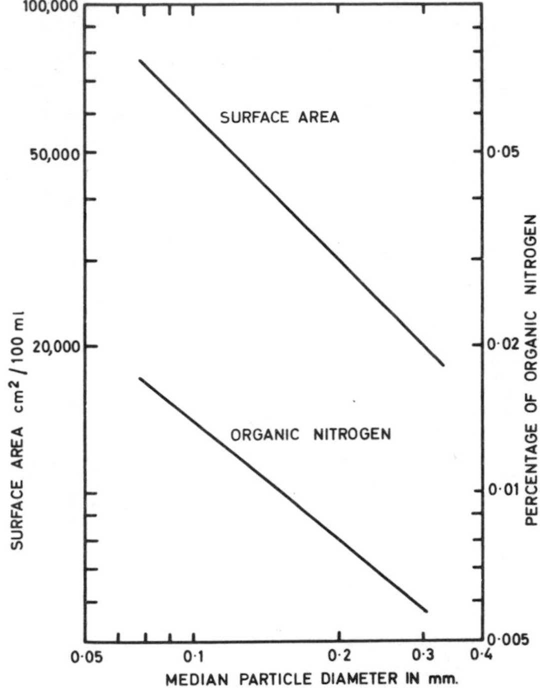

Fig. 6.10. Graphs showing the similarity between the slopes of the lines relating the percentage of organic nitrogen and the surface area (cm^2/100 ml) to the median particle diameter of intertidal deposits from the north Kent coast. (After Longbottom, 1968.)

series of uniform spheres, if the surface of each was coated with a single layer of bacteria. The basis of his calculations is as follows: Zobell and Feltham (1942) estimated that the weight of a single bacterium is $0{\cdot}5 \times 10^{-12}$ gm ($0{\cdot}5 \times 10^{-6} \mu g$). As approximately 80% of this is composed of water, the organic material represents $0{\cdot}1 \times 10^{-6}$ μg, which corresponds approximately with Jørgensen's (1966) estimation that 10^7 bacteria contain 1μg of organic matter. Organic nitrogen may be expected to comprise approximately 7·5% of the organic matter present (Nicolle and Alilare, 1909). Finally, the average size of a bacterium must be estimated. Jørgensen (1966) takes a diameter of 1μ as that of an average spherical bacterium whilst the dimensions of rods are of the order of 2·5μ x 0·5μ; For simplicity, it seems safe to assume that one bacterium occupies an area of $1\mu^2$.

Knowing the calculated surface area of the deposits (Fig. 6.9) and the space which one bacterium occupies ($1\ \mu^2$), it is possible to calculate the number of bacteria which would be present if a layer one bacterium thick was present on the surface of the grains. Then, using Jørgensen's (1966) estimation that 10^7 bacteria are equivalent to 1 μg of organic matter, the percentage organic matter per 100 ml of spheres (= per 256 gm of deposit, since the specific gravity of silicious sands is approximately 2·56) can be calculated. Finally, the percentage organic nitrogen of such a system can be calculated assuming that it comprises 7·5% of the organic matter; similarly, the percentage organic carbon can

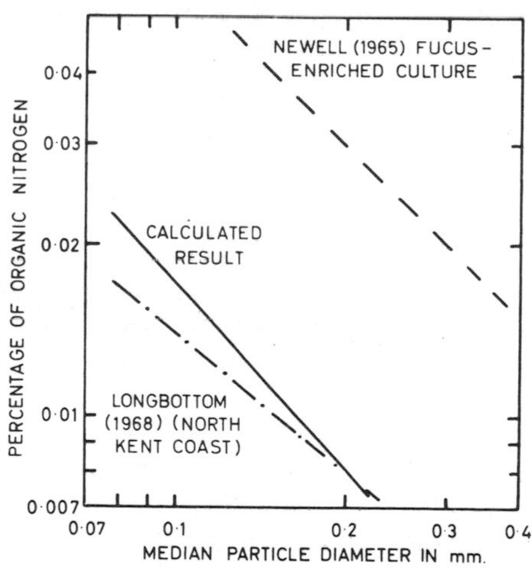

Fig. 6.11. Diagram showing the relationship between the percentage of organic nitrogen and particle size in intertidal deposits and in *Fucus*-enriched culture. Also shown is the line calculated assuming that a layer of bacteria surrounds each particle in the deposit, (Longbottom, personal communication.)

be calculated by assuming that it comprises approximately 50% of the organic matter present. These values are shown in Figs. 6.11 and 6.12 together with the shore data of Longbottom (1968) and the laboratory results of Newell (1965).

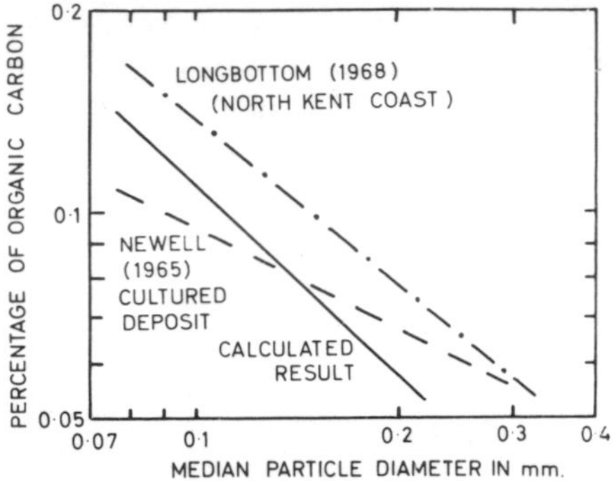

Fig. 6.12. Diagram showing the relationship between the percentage of organic carbon and particle size in intertidal deposits and in cultured deposits. Also shown is the line calculated assuming that a layer of bacteria surrounds each particle in the deposit. (Longbottom, personal communication.)

It is obvious that, despite the approximations used in this calculation, the total values for organic nitrogen and carbon obtained on the shore are similar to those which would be expected if a layer, one bacterium thick, surrounds each particle in the surface of the deposits. The laboratory results of Newell (1965) show rather high values for organic nitrogen since large amounts of powdered *Fucus* were used which might have formed an additional substratum for the bacteria. Also, as Longbottom (1968) points out, the absence of wave action might have allowed the development of a considerable interstitial bacterial population rather similar to those causing abnormally high values for organic carbon and nitrogen in semi-fluid muds on the north Kent coast (p. 249). Meadows and Anderson (1966, 1968) have, for instance, shown that the population of micro-organisms adsorbed onto the surface of sand grains is limited by abrasion, so that it would be reasonable to expect an increase in the population in the less turbulant conditions of estuaries. The predicted values for organic carbon are also strikingly similar to those obtained under culture conditions in the laboratory by Newell (1965), but rather lower than in the natural deposits sampled by Longbottom (1968). It seems likely that the increase in the level of organic carbon on the shore is due to the presence of non-bacterial organic matter such as the cellulose, lignin and chitin comprising the bulk of the organic debris in the deposits (also Newell, 1965; Seki and Taga, 1963).

3. SEASONAL VARIATIONS IN THE POTENTIAL FOOD OF DEPOSIT FEEDERS

From the results cited above, it is clear that the values for organic carbon and nitrogen in intertidal deposits represent a reliable estimate of the standing crop of bacteria and associated micro-organisms, although the value for organic carbon is partly an indication of the proportion of inert organic debris in the deposit. Longbottom (1968) has shown that the level of organic nitrogen in the deposits is subject to a seasonal variation: low in the winter and rising during the spring to a maximum during the summer. Such variation is relatively slight, however, and probably represents a reduction in the rate of multiplication of the bacteria. The rate of predation on the bacteria by deposit feeders (p. 261) may, under these conditions, reduce the standing crop causing observed fall in organic nitrogen values during the winter. Such seasonal variation in deposits of median particle diameter 0·19 mm and 0·13 mm are shown in Fig. 6.13.

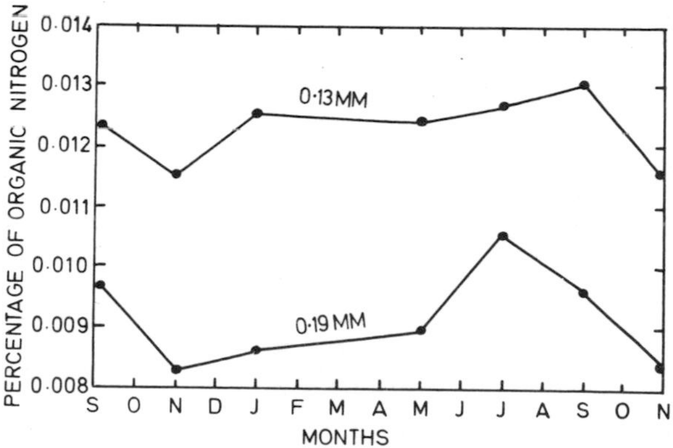

Fig. 6.13. Graph showing the seasonal variation in organic nitrogen in intertidal deposits from the north Kent coast of median particle diameter 0·13 mm. and 0·19 mm. (After Longbottom, 1968.)

The level of organic carbon is more variable than that of organic nitrogen, representing in part the organic debris falling onto the sea floor. Longbottom (1968) has also demonstrated a seasonal variation in the organic carbon content of intertidal deposits of the north Kent coast. Such variations are shown in Fig. 6.14 from which it will be noted that a pronounced peak occurs in May and in September, corresponding with the phytoplankton blooms. It seems likely, therefore, that the seasonal variation in organic carbon noted here reflects in part variations in the organic debris falling onto the sea floor rather than

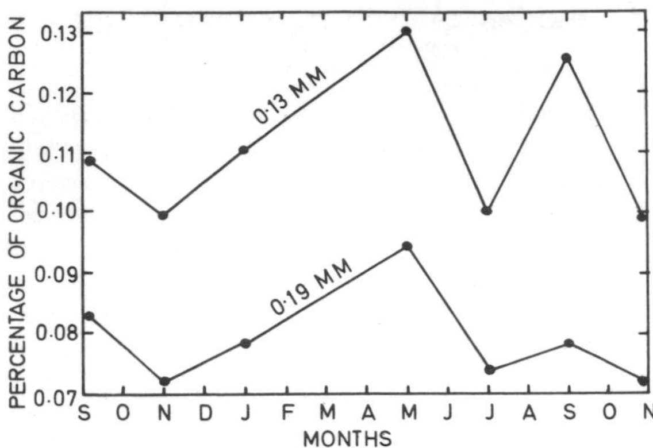

Fig. 6.14. Graph showing the seasonal variation in organic carbon in intertidal deposits from the north Kent coast of median particle diameter 0·13 mm. and 0·19 mm. (After Longbottom, 1968.)

variations in the standing crop of micro-organisms, since such peaks are not accompanied by similar increases in the level of organic nitrogen (cf. Fig. 6.13).

The data for seasonal variations in the level of organic carbon and nitrogen in the deposits of the north Kent coast are summarised in Fig. 6.15 which also indicates that the C/N ratio of the deposits is normally less than 10. Rullier (1959) suggested that C/N ratios of 10 or less indicate a preponderance of animal material, whilst values of 20 or more indicate plant material. The values of the C/N ratio for various seasons in deposits of median particle diameter 0·1 mm 0·15 mm and 0·2 mm collected from the north Kent coast are shown in Table XI.

Since the ratio of C/N in bacteria is of the order of 7 (approximately 50% organic carbon and 7·5% organic nitrogen; p. 255), it is clear that the

Table XI. Table showing the C/N ratios in deposits with median particle diameters of 0·10 mm., 0·15 mm and 0·20 mm from the north Kent coast at various seasons during 1964/1965. (Abstracted from Fig. 6.15. Data from Longbottom 1968.)

Season	*Median particle diameter*		
	0·10 mm	*0·15 mm*	*0·20 mm*
September 1964	9	8·9	8·7
January 1965	8	8·5	9
May 1965	11	10·6	10
July 1965	8·9	7·8	7
September 1965	10	9·5	9
November 1965	8·5	8·7	8·9

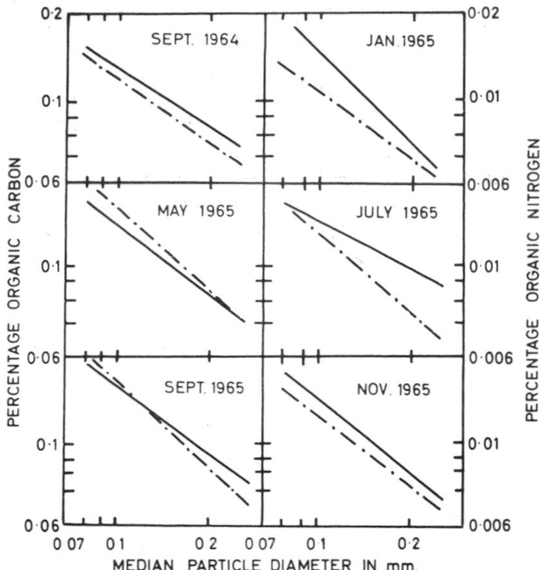

Fig. 6.15. Graphs showing the seasonal variation in the levels of organic carbon and organic nitrogen in intertidal deposits of the north Kent coast. Continuous line indicates organic nitrogen; broken line, organic carbon. (After Longbottom, 1968.)

ratios in Table XI support the conclusion that micro-organisms represent the dominant part of the organic carbon and organic nitrogen values. The higher values recorded during May and September may be due to the increase in the organic debris which probably occurs during these months (p. 258)

4. THE ROLE OF MICRO-ORGANISMS IN THE NUTRITION OF DEPOSIT FEEDERS

Although the standing crop of micro-organisms might appear to be rather small to support large populations of deposit feeding animals, it is of importance to take into account the rate of multiplication of bacteria. This factor has been emphasised by Mare (1942) in her pioneer studies on the role of micro-organisms in the nutrition of invertebrates (also Zobell, 1946). The importance of micro-organisms in the nutrition of deposit feeders is demonstrated by two main lines of evidence. Firstly there is a good correlation between the biomass of certain selected deposit feeders and the organic nitrogen and carbon content of the deposits (≡ the micro-organisms). Secondly, direct feeding of invertebrates with bacteria has indicated that a variety of

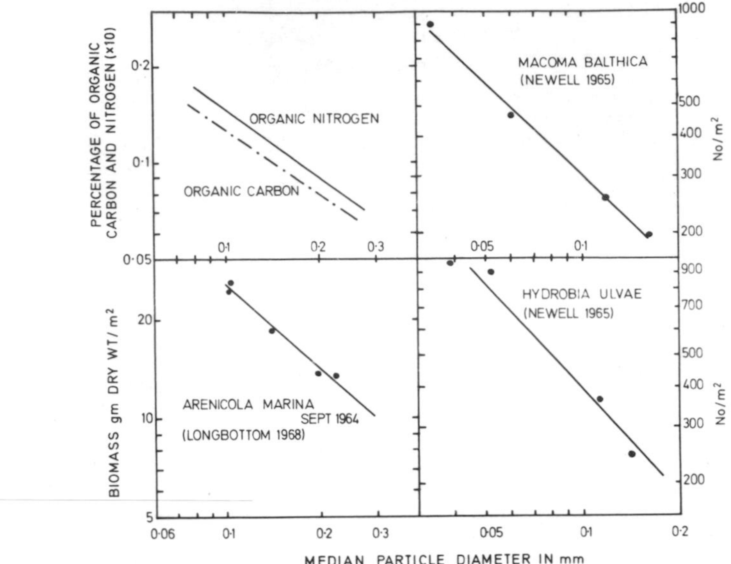

Fig. 6.16. Graphs showing the similarity of slopes of lines relating the percentage organic carbon and nitrogen to grade of deposit with those relating the abundance of certain deposit feeders to grade of deposit. (Redrawn after Newell, 1965 and Longbottom, 1968.)

animals are capable of surviving for long periods when fed only on bacteria.

Detailed correlation between the abundance of deposit feeders and the organic matter in the deposits has been attempted for a few species only. Newell (1965) showed that the numbers of the prosobranch *Hydrobia ulvae* and the deposit feeding bivalve *Macoma balthica* both varied with grade of deposit in a similar way to that shown for the organic content of the deposits. Fig. 6.16 shows the numbers of these two species plotted against the median particle diameter, and the data for organic carbon and organic nitrogen in the deposits. It is apparent that the lines for population density and those for organic matter vary in a similar way with grade of deposit. Longbottom (1968) has obtained similar data for the biomass of the lugworm *Arenicola marina* on the north Kent coast (Fig. 6.16), a result which tallies with those of Popham (1966) and a number of other workers (for review Longbottom 1968). It follows, therefore, that there is a direct correlation between the abundance of *Hydrobia ulvae, Macoma balthica* and *Arenicola marina* and the organic nitrogen (≡micro-organism component) of the deposits. Fig. 6.17 illustrates the relationship between the abundance of these three species and the organic nitrogen in the

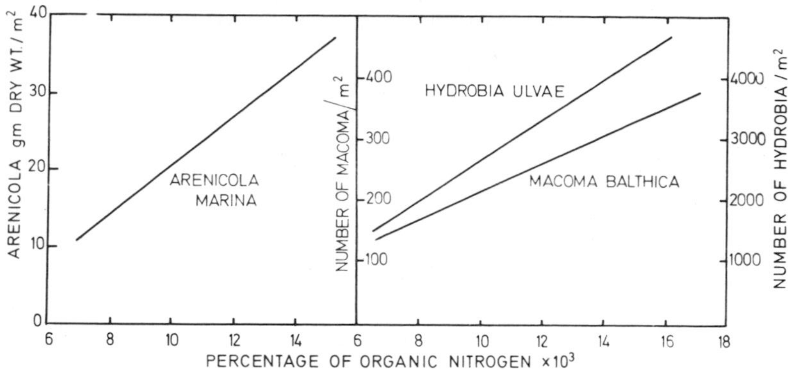

Fig. 6.17. Graphs showing the relationship between the percentage organic nitrogen in intertidal deposits of the north Kent coast and the abundance of the deposit feeders *Arenicola marina*, *Macoma balthica* and *Hydrobia ulvae*. (Compiled from Longbottom, 1968, and Newell, 1965; also Fig. 6.16)

deposits and it seems likely, as suggested by Newell (1965), that the abundance of other deposit feeders, such as *Corophium volutator* (p. 233), which is also extremely numerous in fine deposits, would show a similar relationship to the organic component of the soil.

The animals cited are, however, not found where the salinity is low. This may in part be due to the limited salinity tolerances of the deposit feeders themselves, but it is of interest to note that the number of marine bacteria attached to the particles comprising the deposits also declines with reduced salinity (Meadows, 1965). The muds may therefore be less nutritive in regions of low salinity. A second factor may limit the abundance of deposit feeders even where the organic content of the deposits is high, as for example in semi-fluid muds. It has been pointed out by Yonge (1952) that the increase in organic content of muds results in extensive deoxygenation which may preclude colonisation of this otherwise suitable habitat. To avoid deoxygenation, many of the inhabitants of muds retain contact with the oxygenated water above by means of tubes or burrows, but this is rarely possible in semi-fluid muds. *Arenicola marina* therefore does not occur in such areas even though the concentration of suitable food is high. Small deposit feeders, such as *Hydrobia ulvae, Macoma balthica* and *Corophium volutator* are able to survive because they are able to colonise the oxygenated surface layers of the deposits.

Direct experimental evidence for the nutritive importance of microorganisms is contradictory and has been extensively reviewed by Jørgensen (1966) and by Darnell (1967a, b). A number of marine invertebrates such as *Mytilus* and *Urechis* for example, have been shown to grow on a diet of bacteria (Zobell and Landon, 1937; MacGinitie,

1932) whereas certain sponges are apparently adversely affected by bacteria (Kilian, 1952) although high concentrations of dead bacteria (*Escherichia coli* and *Staphyloccoccus aureus*) induce growth in gemmula sponges (Rasmont, 1961). Bacteria have also been found unsuitable for the nutrition of veligers of the oysters *Crassostrea virginica* and *Ostrea edulis* (Davis, 1953; Walne, 1958), although this adverse effect appears to be confined only to certain species of bacteria (Guillard, 1959). Despite such conflicting experimental results, a wide variety of intertidal invertebrates are found to ingest large quantities of detritus (for example, Jensen and Petersen, 1911; Blegvad, 1914; Jensen 1915; Baier, 1935; Mare, 1942; Fox and Coe, 1943; Coe, 1948, Fox, 1950. For review, MacGinitie, 1932; Jørgensen, 1966; Darnell, 1964; 1967a, b). All of these workers have emphasised the importance of organic debris and associated micro-organisms in the diet of intertidal animals, and it seems clear that a considerable variety of intertidal animals can thrive on micro-organisms. Newell (1965) attempted to demonstrate experimentally that the prosobranch *Hydrobia ulvae* and the bivalve *Macoma balthica* abstract proteins in the form of micro-organisms from the deposits in which they live. Since the abundance of these animals was shown to be correlated with the organic content of the deposits (p. 260), it is of interest to consider such nutrition experiments in more detail.

A large number of *Hydrobia ulvae* were placed in filtered seawater and their faecal material collected, a sub-sample being analysed for organic carbon and nitrogen. The organic carbon was found to be high (10% of the dried weight of the faeces) whilst the organic nitrogen comprised less than 0·025% of the dried faecal material. The remainder of the collected faeces was cultured at 18° C and sub-sampled over a period of 3 days. The results are shown in Fig. 6.18 from which it will be noted that the organic nitrogen component rose to as much as 1·75%, the organic carbon value meanwhile falling to approximately 8% of the dried weight of the cultured faeces. This was taken to indicate that a population of micro-organisms had built up on the indigestible material voided as faeces, and had broken down a proportion of the organic carbon in the process much as described on p. 252 for *Fucus*-enriched deposits. Starved specimens of *Hydrobia ulvae* were then allowed to feed on the cultured faeces, and after 1·5 days their freshly voided faecal pellets were collected and analysed. It was found that the organic nitrogen component was again low (c 0·075%) whilst the organic carbon component was similar to that in the material fed to the animals. The percentage recovery of faecal material approached 95% and subsequent culturing of the faeces again resulted in an increase in the organic nitrogen component (Fig. 6.18).

From these results it was suggested that the faecal material, con-

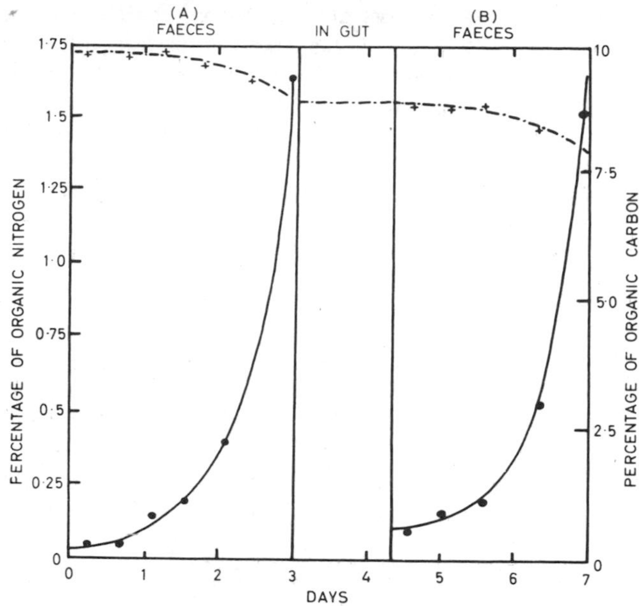

Fig. 6.18. Graphs showing the percentage organic carbon and nitrogen in faeces of *Hydrobia ulvae* cultured at 18°C in seawater under a neon light. (A) before and (B) after feeding to *Hydrobia ulvae.* Broken line indicates organic carbon; continuous line, organic nitrogen. (After Newell, 1965.)

taining indigestible chitin, cellulose and lignin which together probably make up the bulk of organic debris in estuaries, is subjected to a cycle in which bacteria develop on its surface and are subsequently digested by deposit feeding organisms. The voided faecal material is then recolonised by bacteria and thus again becomes suitable for food. The whole cycle may be repeated many times until the organic carbon comprising the debris is used up (also Seki and Taga, 1963).

Similar analyses have been attempted with the bivalve *Macoma balthica* (Newell, 1965) and with the lugworm *Arenicola marina* (Longbottom, 1968). In both of these animals, re-feeding of the cultured faeces has not proved possible. Nevertheless, it has been demonstrated that the organic nitrogen content of the freshly voided faeces is low but that this value increases on culturing. It therefore seems likely that in these animals too, micro-organisms play an important part in the diet. This being so, it follows that any inorganic particulate, or sub-particulate, material which is encrusted with bacteria will prove of equal nutritional value to organic debris itself, since the latter appears to act mainly as a substrate for bacteria (Baier, 1935;

Newell, 1965; Darnell, 1967a, b; Longbottom, 1968). It is not surprising to find, therefore, that not only many deposit feeders but also some suspension feeders ingest large quantities of silt as well as organic debris, and have not evolved a means of distinguishing between the two components. It is not surprising either, that in deposits where the particles are fine, and in which adsorbed micro-organisms are abundant, deposit feeding animals dominate the fauna.

CHAPTER 7

Respiratory mechanisms

A. INTRODUCTION

We have seen in Chapter 6 that not only the nature of feeding mechanisms, but also in many instances the rate of feeding, is appropriate to the particular type of substratum and tidal level at which the animals are to be found. In much the same way, the ebb and flow of the tide and consequent variation in the submersion period profoundly affects the mechanism of respiration adopted by intertidal animals. Those in the lower region of the intertidal zone, for example, are subjected to almost continuous submersion and gas exchange is consequently *via* the surrounding seawater, whilst increasing periods of emersion are experienced by animals living high on the shore. Under such conditions desiccation becomes an important factor, and the animals might be expected to reduce water loss by the closure of shell valves as in mussels and oysters (Dodgson, 1928; also Schlieper, 1957), *Mya arenaria* (van Dam, 1935; Ricketts and Calvin, 1948), *Syndosmya* (= *Abra*) *alba* (Moore, 1931), *Venus mercenaria* (Dugal, 1939), and in a number of other bivalves (for review, von Brand, 1946). Prosobranchs such as *Littorina neritoides* and *L. punctata* (Patané, 1946a,b, 1955) are able to close the operculum and survive for several weeks in the absence of oxygen, whilst intertidal barnacles such as *Balanus balanoides, B. crenatus* and *Chthamalus stellatus,* are all able to survive for several days in moist nitrogen (Barnes, Finlayson and Piatigorsky, 1963). Alternatively the animals may continue to respire aerobically whilst closed, utilising a store of oxygen combined with a respiratory pigment. Finally, in moist air the animals may retain a connection with the outside environment and use the atmosphere as a source of oxygen.

Indeed, intertidal animals such as the barnacles mentioned above have been shown to be able to air-breathe or to accumulate lactic acid as an end product of anaerobic glycolysis according to the conditions of desiccation which prevail (Monterosso, 1928a,b,c, 1930; Barnes, Finlayson and Piatigorsky, 1963; Barnes and Barnes, 1964; Grainger and Newell, 1965). Several gastropods including *Monodonta lineata* (Micallef, 1966) and the Mediterranean trochid *Monodonta turbinata* (Micallef and Bannister, 1967), *Thais* (= *Nucella*) *lapillus, Littorina littorea, L. littoralis* and *L. saxatilis* (Sandison 1966, 1967) have been shown to be capable of aerial respiration, which also occurs in the isopod *Ligia oceanica* (for review, Edney, 1960) the crab *Carcinus maenas* (M. Ahsanullah, personal communication), the intertidal fish *Blennius pholis* (M. J. Daniel, personal communication), the eel *Anguilla vulgaris,* (Berg and Steen, 1965, 1966) and the goby *Gillichthys mirabilis* (Todd and Ebeling 1966; for review, Saxena, 1963). It seems clear, therefore, that on rocky shores variations in the pattern of respiration are correlated primarily with the degree of desiccation which prevails at any particular shore level. Where the animals are submersed, as in rock pools or on the lower shore, gas exchange is with the surrounding water *via* the gills and general body surface, whereas in moist air, such as occurs in shaded crevices or under seaweed, aerial gas exchange may occur. Finally, under conditions of desiccation, closure of the shell valves in bivalves and barnacles, or of the operculum in prosobranchs, may result in anaerobic respiration occurring with the production of lactic acid. It should be noted, however, that the body fluids have limited buffering powers so that lactic acid cannot be retained indefinitely without a marked fall in *pH*.

On particulate shores the availability of oxygen also varies at any one tidal level according to the grade of deposit, being low in regions where the deposits are poorly drained (p. 39). Hence animals living within such substrata tend to show both morphological and behavioural specialisations for drawing oxygenated water through the burrow or tube (p. 293), and may also possess a respiratory pigment capable of functioning at relatively low oxygen concentrations. Added to such physiological and behavioural variations which can be interpreted in ecological terms, the level of oxygen uptake in intertidal animals is influenced by physical factors such as temperature fluctuation and the oxygen concentration of the surrounding medium, as well as by a variety of physiological factors including the size of the animal, its level of activity and whether or not an "oxygen debt" has been incurred. The following account is therefore concerned firstly with the major steps involved in anaerobic and aerobic respiration and with the biochemical implications of such respiratory systems, and secondly with ecological variations in the mechanisms of respiration.

B. THE PROCESS OF RESPIRATION

Respiration commonly involves the oxidation of glucose or an equivalent 6-carbon sugar in such a way that the energy content can be liberated in a large number of steps and stored in the form of adenosine triphosphate (ATP). If a mole of glucose were completely oxidised in air, the end product would be carbon dioxide and water plus 686,000 calories of heat representing the energy contained in the molecule of glucose. We may represent this basic reaction thus:—

$$C_6H_{12}O_6 + 6O_2 \longrightarrow 6CO_2 + 6H_2O + 686{,}000 \text{ calories/mole}$$

Now it is obvious that the energy cannot be stored in the form of heat, nor can the body convert the total 686,000 calories in one step into the energy contained by the ATP. An enzymatic sequence involving the gradual degradation of the glucose molecule into carbon dioxide and water under aerobic conditions is therefore used; the energy is liberated in relatively small amounts which can be stored as ATP. When energy is required for a particular reaction sequence, the terminal phosphate group of the ATP molecule is transferred to a second molecule called a "phosphate acceptor", the energy originally stored in the ATP molecule being released, leaving adenosine diphosphate (ADP). This can in turn be rephosphorylated back to ATP by the transfer of phosphate from a high energy compound called a "phosphate donor" which has gained its energy from one of the energy-yielding steps in the degradation of the original glucose molecule.

The structure of ATP is shown in Fig. 7.1. from which it is seen that the molecule is built up of a heterocyclic ring derivative of purine called adenine, attached to which is the 5-carbon sugar D-ribose. Attached to the D-ribose sugar are three phosphate groups, the last containing the high energy which is released when the terminal phosphate is removed by enzymatic hydrolysis. Thus ATP acts as a store of the energy which is liberated by the stepwise oxidation of glucose to carbon dioxide and water, and it has been calculated that each mole of ATP represents the capture of approximately 7,000 calories from the original 686,000 calories contained per mole of glucose. For convenience, we may consider the capture of such energy in the form of ATP as occurring in a first anaerobic phase and a later aerobic phase.

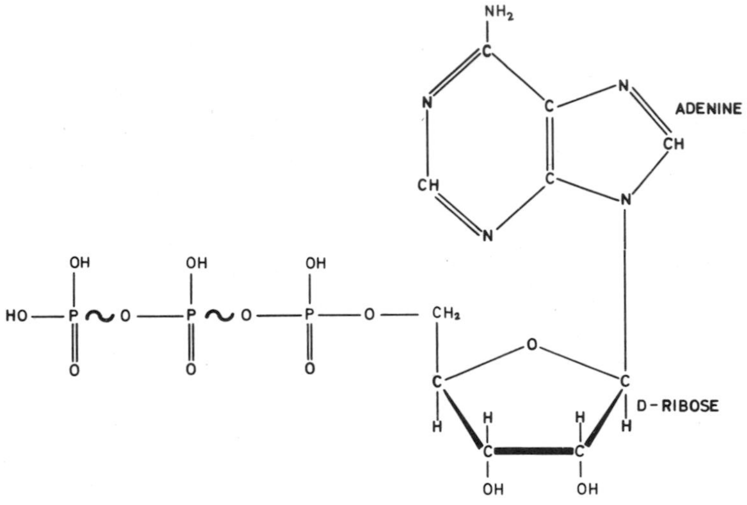

Fig. 7.1. Diagram illustrating the structure of adenosine triphosphate (ATP). High energy bonds are denoted ~.

1. ANAEROBIC RESPIRATION

The first stage in the oxidation of glucose is the anaerobic phase called glycolysis, or the Embden-Meyerhoff sequence, in which each molecule of the 6-carbon sugar glucose is broken down to two molecules of pyruvic acid, each of which contains three carbon atoms. Under anaerobic conditions, the pyruvic acid is then reduced asymmetrically to L–(+)–lactic acid. This reaction is reversible and is catalysed by the enzyme lactate dehydrogenase (LDH). Nicotinamide adenine dinucleotide (NAD), which is sometimes known as diphosphopyridine nucleotide (DPN), acts as an electron carrier in this process and donates its electrons to the pyruvate to form lactate, being itself therefore oxidised in the process. Under aerobic conditions, the lactic acid is oxidised back to pyruvic acid and the NAD becomes reduced again.

$$\begin{matrix} CH_3 \\ | \\ C=O \\ | \\ COOH \end{matrix} + NAD_{red} \underset{\text{dehydrogenase}}{\overset{\text{Lactate}}{\rightleftharpoons}} \begin{matrix} CH_3 \\ | \\ CHOH \\ | \\ COOH \end{matrix} + NAD_{ox}$$

Pyruvic acid — Lactate dehydrogenase — Lactic acid

In fact the conversion of glucose into pyruvate under aerobic conditions, or lactate under anaerobic conditions, is far from simple and is known to involve the sequential action of at least eleven different enzymes the steps of which are shown in Fig. 7.2. The first important point is that one molecule of ATP is used to convert glucose to glucose–6–phosphate, and another is used to convert fructose–6–phosphate to fructose–1–6–diphosphate. On the other hand two steps in the sequence each yield two molecules of ATP. In the first of such steps glyceraldehyde–3–phosphate is oxidised enzymatically by NAD (Fig. 7.2) and includes the incorporation of phosphate to form a high energy compound 1, 3–diphosphoglycerate:–

$$R-\underset{\underset{O}{\|}}{C}-H + HPO_4^{--} + NAD_{ox} \rightarrow \boxed{R-\underset{\underset{O}{\|}}{C}-O-\underset{\underset{O}{\|}}{\overset{\overset{O^-}{|}}{P}}-O^-} + NAD_{red}$$

glyceraldehyde–3–phosphate 1, 3–diphosphoglycerate

This high energy derivative of glyceraldehyde–3–phosphate then donates a phosphate group to ADP:–

$$\boxed{R-\underset{\underset{O}{\|}}{C}-O-\underset{\underset{O}{\|}}{\overset{\overset{O^-}{|}}{P}}-O^-} + ADP \rightarrow R-COO^- + ATP$$

1, 3–diphosphoglycerate adenosine triphosphate

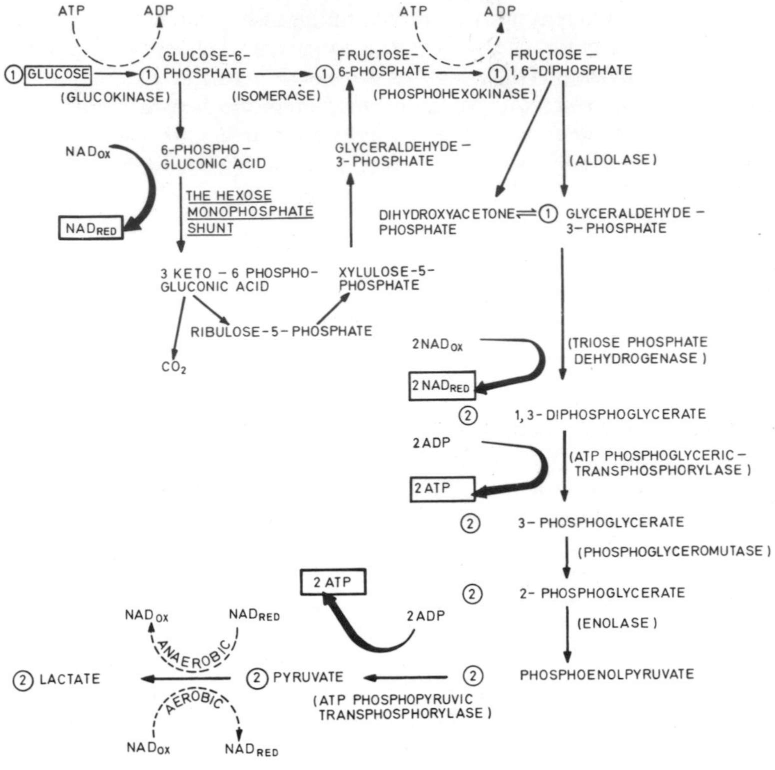

Fig. 7.2. Diagram illustrating the more important steps in the glycolytic sequence. Enzymes responsible for catalysing the steps are given in brackets whilst the number of molecules formed from the original glucose is indicated within circles. The steps where high energy products are formed are indicated by heavy tapering arrows.

The second step in glycolysis at which energy is conserved again involves a two-stage reaction: (1) A shift of electrons within the molecule of 2–phosphoglycerate (Figure 7.2) to form phosphoenolpyruvate:–

$$\begin{array}{c} CH_2OH \\ | \\ H-C-O-PO_3^{2-} \\ | \\ COO^- \end{array} \rightleftharpoons \begin{array}{c} CH_2 \\ \| \\ C-O-PO_3^{2-} \\ | \\ COO^- \end{array} + H_2O$$

2-phosphoglycerate phosphoenolpyruvate

(2) The phosphoenolpyruvate donates a phosphate group to ADP, this reaction being catalysed by the enzyme pyruvate phosphokinase (ATP phosphopyruvic transphosphorylase).

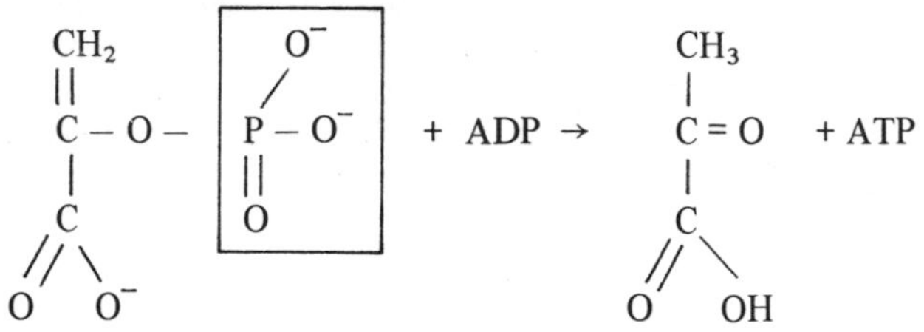

$$\begin{array}{c} CH_2 \\ \| \\ C-O-\boxed{P(=O)(O^-)_2} \\ | \\ COO^- \end{array} + ADP \rightarrow \begin{array}{c} CH_3 \\ | \\ C=O \\ | \\ COOH \end{array} + ATP$$

phosphoenolpyruvate → pyruvic acid

Thus two molecules of ATP are lost in the process whilst four are gained, giving a net gain of two molecules of ATP which is equivalent to a capture of 2 × 7,000 = 14,000 calories per mole of glucose. In animal

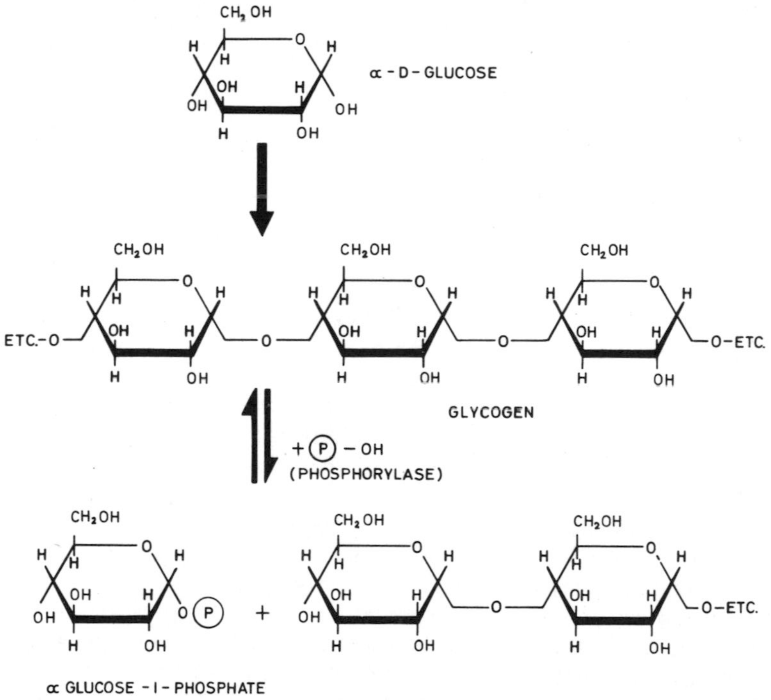

Fig. 7.3. Diagram showing the structure of α – D – glucose and glycogen. The first step in the phosphorylysis of glycogen is shown with the enzyme catalysing the reaction shown in brackets.

tissues, glycogen rather than glucose is the starting point of glycolysis. Glycogen is composed of a sequence of approximately 40,000 molecules of α–D–glucose linked together as shown in Fig. 7.3. This factor is of some importance since glycogen is converted *via* a two stage process to glucose–6–phosphate, not by the action of ATP but by the elements of phosphoric acid. The first reaction, which is reversible and is known as phosphorylysis, is catalysed by phosphorylase and yields α–glucose–1–phosphate. (Fig. 7.3) The second reaction converts the α–glucose–1–phosphate into glucose –6–phosphate and is catalysed by the enzyme phosphoglucomutase.

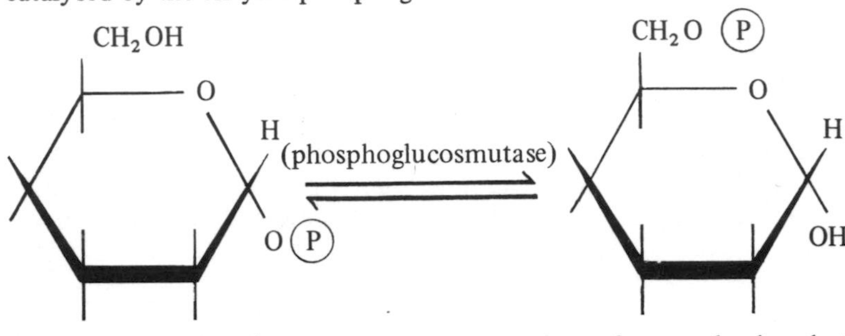

From then on, the sequence is as outlined in Fig. 7.2 so that in animal tissues, as distinct from fermentation processes, there is a net gain of three ATP molecules for each 6–carbon unit of glycogen metabolised in contrast to the net gain of two molecules of ATP in the fermentation process. The whole of the process of glycolysis described above can be carried out in the absence of oxygen, and, since there is a net gain of three molecules of ATP, this sequence results in the capture of approximately 21 Kcals per mole from the potential 686 Kcals contained in the 6–carbon compound. The process so far is thus only $\frac{21}{686}$ 100% = 3·1% efficient, although it should be noted that the total free energy loss from the carbon compound to lactic acid is equivalent to approximately 52 Kcals per mole, so that approximately 40% of the total free energy exchange up to the formation of lactic acid is retained in the form of ATP. Nevertheless, $\frac{686-52}{686} \times 100 = 92{\cdot}4\%$ is still rejected in the form of lactic acid; it is not surprising to find, therefore, that anaerobic cells use relatively vast amounts of metabolic fuel to achieve a comparable amount of work to that carried out by more efficient energy extraction processes. The price being paid by the anaerobic cell for being able to extract energy from glycogen in the absence of oxygen is a wastage of approximately 92% of the energy available in the original 6–carbon compound.

In many animals there is an additional anaerobic oxidative sequence called the hexose monophosphate shunt which is an alternative pathway from glucose–6–phosphate to fructose–6–phosphate (Fig. 7.2) and which varies in importance from species to species and within different tissues from any one species. The sequence becomes complicated in its later stages since a number of derivatives, including fructose–6–phosphate and glyceraldehyde–3–phosphate, may be formed. The important point, however, is that by the use of the hexose monophosphate shunt an additional pair of electrons is released and carried by reduced pyridine nucleotide (NAD_{red}) to the site of oxidative phosphorylation where, under aerobic conditions, ATP is formed. The sequence of reactions involved in the shunt is shown in a simplified form in Figure 7.2.

2. AEROBIC RESPIRATION

In aerobic forms the lactic acid is not excreted and lost to the system as in anaerobes, but is oxidised in a stepwise fashion to carbon dioxide and water with the efficient extraction of much of the remaining 92% of the energy from the original 6–carbon sugar. The enzymes responsible for this are not distributed freely within the cytoplasm, as are those responsible for glycolysis, but are located within special respiratory particles, mitochondria, which occur in especially large numbers in cells which perform a great deal of work. Essentially, the aerobic phase of respiration consists of the Krebs tricarboxylic acid cycle (or citric acid cycle) followed by a process of oxidative phosphorylation. Enzymes associated with the Krebs cycle are found in the inner matrix of the mitochondria whilst those concerned with oxidative phosphorylation are located in an orderly array on the inner mitochondrial membranes or cristae.

(a) The Krebs tricarboxylic acid cycle

One of the important features of the Krebs cycle is that it is itself not directly concerned with the formation of ATP from ADP and phosphate. Instead, pairs of electrons are carried by special enzymes from each energy-yielding step in the Krebs cycle and, together with those produced in glycolysis, are transferred to the site of oxidative phosphorylation where reaction with ADP occurs and ATP is formed.

Before the products of glycolysis or any other food materials such as fatty acids and proteins can be oxidised in the Krebs cycle, they must be converted either to acetyl co-enzyme A directly or *via* pyruvate, or to intermediate compounds of the Krebs cycle such as α – ketoglutaric acid or oxaloacetic acid (Fig. 7.5). It is convenient to consider here the fate of pyruvic acid which, as we have seen, is formed as an end product of glycolysis under aerobic conditions. If lactic acid has accumulated under anaerobic conditions and aerobic conditions subsequently prevail, then it is oxidised back to pyruvate, the enzyme NAD_{ox} being restored back to NAD_{red} in the process (p. 269). The three-carbon compound pyruvate, like other components entering the Krebs cycle, is then oxidised enzymatically into a 2–carbon compound, acetic acid, in this case by the action of pyruvate dehydrogenase, liberating a molecule of carbon dioxide in the process. Here, then, there is an oxidation process, namely that of the 3–carbon compound pyruvate to the 2–carbon compound acetic acid. In this process, the electron pairs removed from each of the two molecules of pyruvate formed from the original 6–carbon sugar are accepted by the electron carrier NAD, which is thus converted from its oxidised form NAD_{ox} to its reduced form NAD_{red} . Each pair of electrons is carried by a molecule of NAD and is transferred to the site of oxidative phosphorylation where ATP is formed (p. 278). The acetic acid formed in the reaction is, in fact, not in a free form but is attached to coenzyme A (CoA) forming the compound acetyl coenzyme A. The acetylation of CoA occurs inside the mitochondria. The structure of CoA is shown below (Fig. 7.4) from which it will be noticed that the compound resembles ATP and also NAD insofar as it contains adenine, D–ribose and phosphate groups. Attached to the terminal phosphate group is the chain compound pantothenic acid whose terminal –SH (thiol) group attaches to the acetate to form a thio ester. The reaction may be depicted as follows:–

$$\underset{\text{Acetic acid}}{CH_3COO^- H^+} + \underset{\text{Co-enzyme A}}{\text{CoA-SH}} \longrightarrow \underset{\text{Acetyl co-enzyme A}}{\text{CoA-S-}CH_3CO} + H_2O$$

So far, apart from the molecules of ATP formed directly by glycolysis, we have obtained two pairs of electrons in the conversion of glyceraldehyde–3–phosphate to 1,3–diphosphoglycerate during glycolysis, and a further pair in the oxidation of each of the two molecules of pyruvate to the combined acetate in the form of acetyl coenzyme A (i.e. a total of 4 pairs from the original 6–carbon sugar), all of these electron pairs being used at a later stage in the process of oxidative phosphorylation (p. 278). Having obtained acetyl coenzyme A, the "raw material" of the Krebs cycle is present and we may now consider the fate of this compound of acetic acid.

The main object of the Krebs cycle is to oxidise acetyl coenzyme A in a large number of steps to carbon dioxide and water, but, as mentioned on p. 273, ATP is not produced directly at the energy-yielding steps of oxidation. Instead, the energy is carried in the form of electron pairs by the electron carrier NAD (p. 269) and by the active group of an oxidising enzyme called flavoprotein whose hydrogen acceptor is flavin adenine dinucleotide (FAD). In each case, the electron pairs are transported as before to the site of oxidative phosphorylation. The sequence

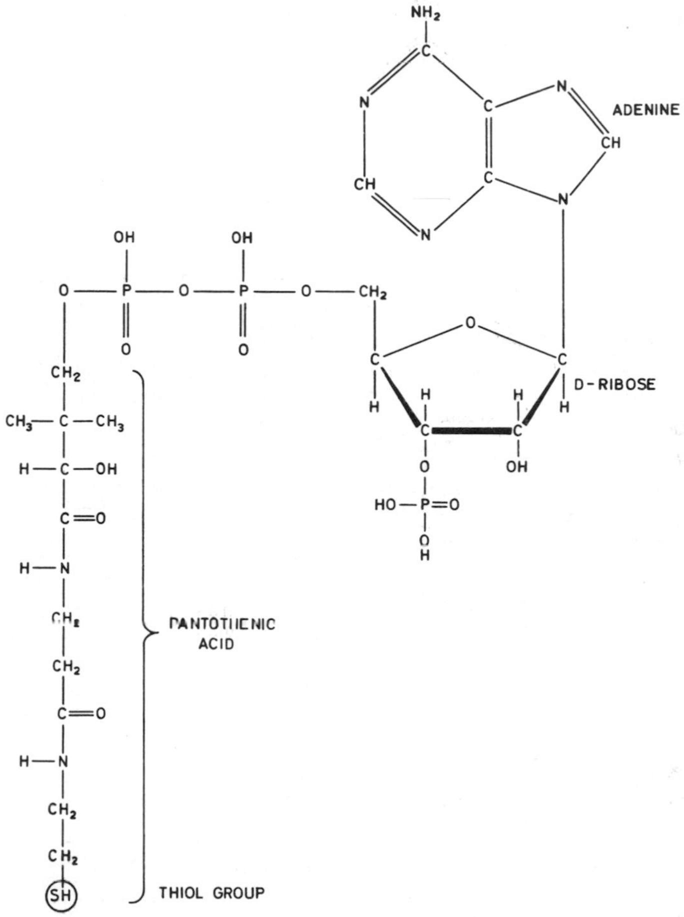

Fig. 7.4. Diagram illustrating the structure of Coenzyme A. Note the general similarity in structure to ATP (Fig. 7.1.)

of processes in the Krebs citric acid cycle is shown in Fig. 7.5, the first stage being the enzymatic transfer of the acetyl group of acetyl coenzyme A to the 4–carbon dicarboxylic acid oxaloacetic acid to form the 6–carbon tricarboxylic acid citric acid from which the cycle takes its name. The coenzyme A is regenerated in the process and thus becomes available to accept further acetyl groups.

$$\underset{\text{Acetyl coenzyme A}}{CH_3 - \underset{\overset{||}{O}}{C} - S - CoA} + \underset{\text{oxaloacetic acid}}{COOH - \underset{\overset{||}{O}}{C} - CH_2 - COOH} \longrightarrow \underset{\text{coenzyme A}}{CoA - S - H}$$

$$+ \underset{\text{citric acid}}{COOH - CH_2 - \underset{\underset{COOH}{|}}{\overset{\overset{OH}{|}}{C}} - CH_2 - COOH}$$

The next reaction is reversible and converts citric acid first of all into the tricarboxylic acid *cis*-aconitic acid by the removal of the elements of water, and then into iso-citric acid by the addition of the elements of water into other parts of the molecule, both of these reactions being catalysed by the enzyme acontinase. Now, an important energy-yielding step is reached in which the 6–carbon iso-citric acid is enzymatically oxidised by the enzyme isocitrate dehydrogenase to the 5–carbon compound α –ketoglutaric acid plus a molecule of carbon dioxide. In the process, two electrons are removed from the isocitric acid by the electron carrier NAD which is thus converted from its oxidised form to its reduced form. This electron pair, the first removed in the Krebs cycle, is transported to the site of oxidative phosphorylation (p. 278).

A second energy yielding step occurs in the next reaction which converts the 5–carbon α –ketoglutaric acid into the 4–carbon compound succinic acid by the action of the enzyme α –ketoglutarate dehydrogenase. Again, carbon dioxide is produced, and a pair of electrons accepted by NAD_{ox} which is converted in the process to the reduced state (NAD_{red}) and carries the pair of electrons to the site of oxidative phosphoylation. A third energy-yielding step converts the 4–carbon compound succinic acid by the removal of two hydrogen atoms (dehydrogenation) to form the 4–carbon compound fumaric acid. The enzyme catalysing this process is flavoprotein (FAD_{ox}) (p. 279) which itself acts as the electron carrier and in its resultant reduced form (FAD_{red}) carries an electron pair to the site of oxidative phosphorylation.

The next reaction, which is not an energy-yielding step, involves the hydration of fumaric acid at the site of the double bond to form malic acid, this reaction being catalysed by the enzyme fumarase. Lastly, dehydrogenation of the malic acid to form oxaloacetic acid is catalysed by malic dehydrogenase and constitutes the final energy-yielding step in the Krebs cycle. The pair of electrons are accepted by NAD_{ox} which becomes reduced in the process and transports them to the site of oxidative phosphorylation as before.

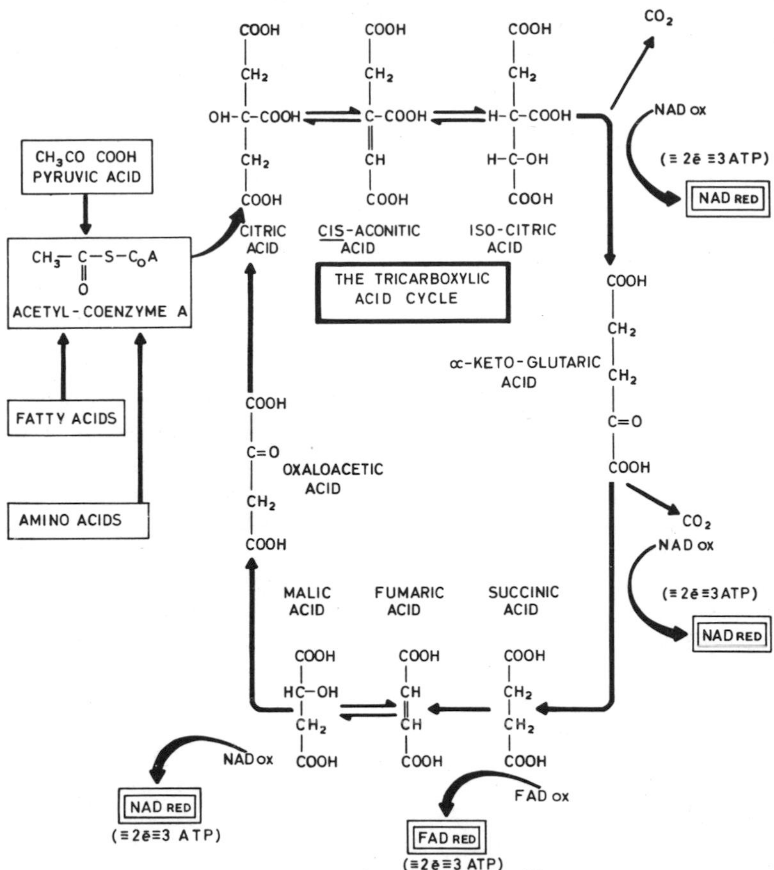

Fig. 7.5. Diagram illustrating the Krebs tricarboxylic acid cycle. The energy yielding steps are indicated together with their approximate value in terms of ATP produced by oxidative phosphorylation.

Both coenzyme A and oxaloacetic acid which are required in the initial reaction (p. 276) are thus regenerated with each cycle, whilst the 2–carbon acetyl group is oxidised to two molecules of carbon dioxide, the stored energy being carried in the form of either reduced NAD or reduced FAD to the site of oxidative phosphoryation. In all, so far, neglecting electrons donated *via* the hexose monophosphate shunt (p. 273), there have been two molecules of reduced NAD produced in glycolysis under aerobic conditions (Fig. 7.2), and one molecule of reduced NAD for each of the two acetyl-coenzyme A molecules formed from the original 6–carbon sugar (i.e. two molecules of reduced NAD are derived from this step); then for each of the two molecules of acetyl coenzyme A formed from the original sugar there is one revolution of the Krebs cycle, each revolution yielding four pairs of electrons (i.e. a total of 8 electrons per molecule of the original 6–carbon sugar). The final pathway in respiration is therefore a common pathway along which the energy, transported by the electron carriers NAD_{red} and FAD_{red} to the site of oxidative phosphorylation from a variety of energy yielding steps in the respiratory sequence, is converted into ATP. Therefore 12 electron pairs are derived from each 6–carbon sugar (one pair being carried by each molecule of NAD_{red} and FAD_{red}). The energy contained in these electron pairs must be converted into ATP before it can be used as an energy source and, together with the net gain of three molecules of ATP during glycolysis from glycogen (p. 272), represents the total energy extracted from the original 6–carbon units of glycogen.

Whilst we have considered glycogen as the prime metabolic fuel, it is by no means true that this is the only possible source of energy; many animals use lipids and amino-acids as an energy source. The terminal steps of oxidative phosphorylation are also known to be modified from the scheme shown in figure 7.8 in some animals. Fats in particular yield large quantities of ATP and may well be of great importance in the metabolism of intertidal invertebrates. The main steps involved in the metabolism of fatty acids are reviewed by Baldwin (1963), whilst those involved in amino-acid metabolism are described by Cohen and Brown (1960; also Hoar, 1966).

(b) Oxidative phosphorylation

We come now to the mechanism by which the energy contained in the electron pairs carried to the enzyme sites on the mitochondrial membranes, or cristae, by NAD and FAD is converted into the energy rich phosphate compound ATP. Essentially the process consists of the donation of the pairs of electrons contained in each of the three NAD_{red} molecules and the FAD_{red} molecule derived from each revolution of the Krebs cycle, plus those obtained from reactions preceding the Krebs cycle (see above), to a series of enzymes located in an orderly array on the mitochondrial membranes and called the respiratory chain enzymes. These enzymes are compounds of porphyrin and iron called cytochromes, and eventually facilitate the combination of each electron pair with one atom of oxygen. This reduction of oxygen is a stepwise process, the passage of each pair of electrons along the respiratory chain having been found to yield a total of three molecules of ATP from ADP and phosphate, one molecule of ATP being formed at each of three distinct sites which correspond with regions of relatively large energy drop in the respiratory chain. This production of ATP during electron transport along the respiratory chain is known as oxidative phosphorylation; it represents a mechanism by which the energy contained in NAD_{red}, which would yield 52 Kcal/mole when oxidised, is removed in three stages in the conversion of ADP to ATP. The energy required to convert ADP plus phosphate into ATP is approximately 7 Kcals/mole so that the total energy stored up by the conversion of three moles of ADP to three moles of ATP is 21 Kcals. Thus the efficiency of energy conservation during oxidative phosphorylation is $\frac{21}{52} \times 100 = 40{\cdot}4\%$; it is obvious that this last sequence in the respiratory process is of prime importance in the conservation of the energy contained in the original 6–carbon sugar compound.

We may now look at the process of oxidative phosphorylation in more detail. Electrons pass into the respiratory chain either from NAD_{red} to a flavoprotein, or directly from the Krebs cycle where the step from succinic acid to fumaric acid is associated with the formation of reduced flavoprotein (FAD_{red}) (p. 276). Both of these flavoproteins are thus hydrogen acceptors, but are followed by cytochrome *b* which has a stronger affinity for electrons than the reduced flavoproteins which therefore donate their electrons to cytochrome *b* and themselves are restored to the oxidised state in the process. The same process occurs between cytochrome *b* and cytochrome *c*, and also between cytochrome *c* and cytochrome *a* as shown in Fig. 7.6. Finally, the electrons pass to the terminal link, cytochrome oxidase, which donates its electron to oxygen to form water. An important point is that whilst NAD and the flavoproteins carry a pair of electrons each, the cytochromes carry only one electron at a time.

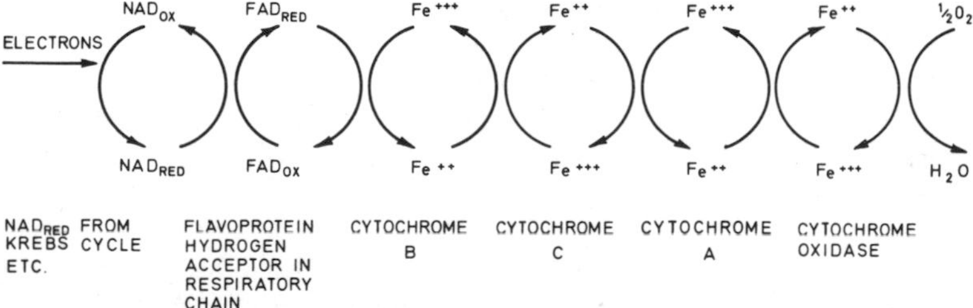

Fig. 7.6. Diagram illustrating the major coupled reactions involved in electron transport along the respiratory chain. Oxidised cytochrome is indicated by Fe^{+++} and reduced cytochrome by Fe^{++}.

Now the three points at which there is a relatively large fall in energy are firstly, in the transfer from NAD_{red} to flavoprotein, secondly from cytochrome *b* to cytochrome *c* and finally from cytochrome *a* to cytochrome oxidase. It is at these points that the liberated energy is stored in the phosphate groups of high energy intermediates which in turn donate a phosphate group to ADP to form ATP. We may now modify the scheme shown in Fig. 7.6 and put in the sites at which ATP is produced (Fig. 7.7). For the passage of the two electrons down the respiratory chain, three ATP molecules are formed, the end product of this process of oxidative phosphorylation being water.

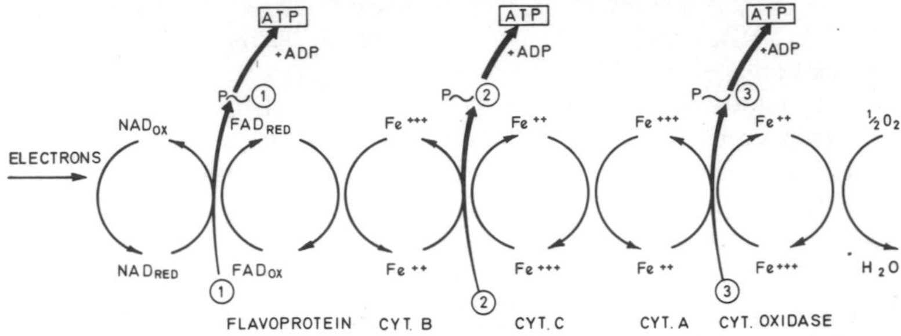

Fig. 7.7. Diagram showing the sites of ATP formation during the passage of one electron pair along the respiratory chain. The compounds 1, 2 and 3 combine with NAD, Cytochrome *b*, and Cytochrome *a*, respectively, to form the high energy intermediates P~ 1, P~ 2 and P~3 during electron transfer. The high energy intermediates then donate their high energy phosphate groups to ADP to form ATP.

(c) The energetics of glycogen oxidation

The object of our digression into the biochemical processes in respiration has been twofold. Firstly, it is of interest to understand the stage at which acid end products of anaerobiosis may accumulate, and the stages at which carbon dioxide is evolved as well as that at which oxygen is used. Secondly, it is only by a consideration of such processes that the relative efficiency of anaerobic and aerobic respiration may be compared. The energy-yielding steps in the whole process of respiration, starting from the original process of glycolysis are indicated in Fig. 7.8.

Glycolysis, as shown on p. 272 (Fig. 7.2), yields a net gain of three molecules of ATP when the starting compound is glycogen (p. 271), which may be regarded as roughly equivalent to 21 Kcals/mole out of the total of 686 Kcals per mole of the 6–carbon hexose sugar unit. Under anaerobic conditions, lactic acid is accumulated and often excreted to avoid excessive changes in the *pH* of the tissues. The efficiency of this conversion, as pointed out on p. 272, is thus $\frac{21}{686} \times 100 =$ 3·1%, although, as also mentioned on p. 272, the glycolytic sequence is extremely efficient at trapping that small proportion of the total energy which is liberated. The energy actually liberated during glycolysis amounts to only 52 Kcals/mole, so that the 21 Kcals represents the very high efficiency of transfer of $\frac{21}{52} \times 100 = 40\%$. Nevertheless, insofar as the vast majority of the energy available in the original hexose sugar unit is not utilised, anaerobic respiration uses large amounts of glycogen for the quantity of ATP produced.

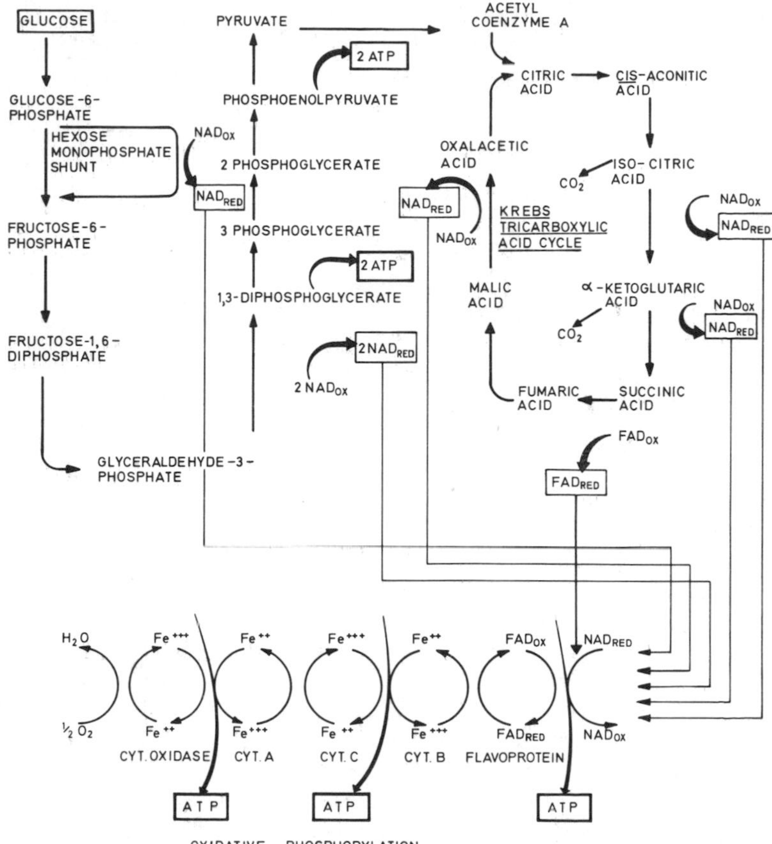

Fig. 7.8. Diagram summarising the three main stages involved in aerobic respiration – glycolysis, the Krebs tricarboxylic acid cycle and oxidative phosphorylation. The main energy yielding steps are indicated by heavy curved arrows and the path of the electron pairs carried by reduced NAD and FAD are indicated by slender arrows. For details of each of the stages, see text and Figs. 7.2, 5 and 7.

Under aerobic conditions, however, pyruvate is formed either by enzymatic oxidation from lactate or directly from the original hexose sugar unit. During glycolysis, as we have seen on p. 270 (Fig. 7.2), not only are there three molecules of ATP formed but also two molecules of reduced NAD corresponding with two electron pairs. Added to these may be other electron pairs formed in the hexose monophosphate shunt (p. 273; Fig. 7.2), but since the importance of this varies from species to species and from tissue to tissue we shall neglect it from the calculation here. The molecules of reduced NAD donate their electrons to the hydrogen acceptors of the respiratory chain and, as was shown above (p. 279), the passage of each electron pair corresponded with the liberation of three molecules of ATP. We may therefore write the equation for this reaction which represents the overall reaction of oxidative phosphorylation in the electron transfer chain and describes the fate of all electron pairs donated to the cytochrome system:–

$$2NAD_{red} + 6ADP + 6P + O_2 \longrightarrow 2NAD_{ox} + 8H_2O + \boxed{6ATP} \qquad (1)$$

Now the formation of two molecules of acetyl coenzyme A from two molecules of pyruvate is accompanied by the production of two molecules of reduced NAD, one NAD_{red} being formed for each molecule of pyruvate (p. 274). We may write this as shown below:–

$$2NAD_{ox} + \underset{\text{pyruvic acid}}{2CH_3.CO.COOH} \xrightarrow{\text{Coenzyme A}} \underset{\text{acetyl coenzyme A}}{2CoA\text{-}S\text{-}CH_3CO} + 2CO_2 + 2NAD_{red} \qquad (2)$$

These electron pairs are converted by oxidative phosphorylation in the electron transfer (respiratory) chain to a further six molecules of ATP according to Equation 1 above. Finally, as we have seen on p. 278, the Krebs cycle contributes four pairs of electrons for each molecule of acetyl coenzyme A, and there are two molecules of acetyl coenzyme A formed from each molecule of the six carbon sugar. So the Krebs cycle contributes eight pairs of electrons which, after oxidative phosphorylation, may be regarded as equivalent to a total of twenty-four molecules of ATP according to Equation 1 above.

By summing the molecules of ATP formed, we obtain three directly from glycolysis, plus a further six by oxidation of the two molecules of reduced NAD produced during glycolysis (neglecting the hexose monophosphate shunt). Added to these are six molecules formed by the phosphorylation of the NAD_{red} produced in the conversion of pyruvate to acetyl coenzyme A (Equation 2 above), and a further twenty-four derived from the Krebs cycle. This gives a total of thirty-nine molecules of ATP, which is approximately equivalent to 273 Kcals/mole since each mole of ATP represents the capture of approximately 7 Kcals of energy. Thus the total efficiency of conversion of the 686 Kcals contained per mole of the original hexose sugar unit is $\frac{273}{686} \times 100 = 39{\cdot}8\%$ of which approximately 3·1% can be obtained under anaerobic conditions. This is, of course, only a very approximate calculation, for the production of NAD_{red} at other sites may greatly enhance the efficiency of the process. It is clear, however, that whilst closure with subsequent anaerobiosis may allow intertidal animals to survive adverse conditions such as desiccation or reduced salinity, the process is accomplished at the expense of relatively large amounts of metabolic reserves, since lactic acid is often excreted to prevent a lethal reduction of the *pH* of the body fluids. Aerobic respiration, on the other hand, conserves a considerable proportion of the energy locked up in the glycogen molecule. That is, there are sound economic reasons for aerobic respiration and it would be expected that wherever suitable conditions prevail, intertidal animals would abstract oxygen for this process either from the surrounding seawater or from air.

C. ECOLOGICAL VARIATION IN THE MECHANISM OF RESPIRATION

We have seen that anaerobic respiration yields acid end products. Lactic acid is commonly formed, but in many invertebrates other products such as acetic acid may be produced as a result of anaerobiosis. This situation in itself is likely to result in a serious reduction in the *pH* of the tissue fluids unless the accumulated acid end product is excreted, but this represents a vast wastage of the energy locked up in the original 6-carbon sugar compound (p. 272). Anaerobiosis depends, therefore, not only upon the ability to excrete acid end products but also on the presence of adequate carbohydrate supplies (Beadle, 1961). On the other hand, in some animals the period of anaerobiosis may be of sufficiently short duration to allow the accumulation of, for example, lactic acid without an excessive reduction in the *pH* of the tissue fluids. The accumulated lactic acid can then be converted back to pyruvate when aerobic conditions return, approximately 20% of the lactic acid being oxidised in the process to carbon dioxide and water. Such oxidation results in an increase in the oxygen demand of the tissues over normal metabolic requirements, and constitutes the repayment of an "oxygen debt" whose magnitude is commonly related to the duration of the period of anaerobiosis as well as to a number of other factors which include the temperature. In this way the metabolic reserves are conserved and one of the serious drawbacks of anaerobiosis averted.

It would be expected, therefore, that where intertidal organisms are exposed to anaerobic conditions of relatively short duration, as on the lower shore, no special mechanisms for the excretion of acid end-products would occur. Such animals would also be expected to repay an "oxygen debt" when resubmersed in aerated seawater. On the other hand, conditions of desiccation in the upper parts of the intertidal zone may be such that anaerobiosis must be adopted for prolonged periods and here lactic acid, or an equivalent end-product of anaerobiosis, is excreted, so that when the animals are resubmerged in aerated seawater the oxygen consumption is comparable with that which occurred before anaerobiosis. There are, of course, many animals such as barnacles and littorinids which extend over a wide range in the intertidal zone and are able to air-breathe, accumulate acid end-products of anaerobiosis, or excrete such products, according to the conditions which prevail. Again, some animals, although not subjected to desiccation, nevertheless experience an oxygen lack due to the restricted diffusion of oxygen into the interstitial water. Burrowing animals such as the lugworm *Arenicola marina,* for example, must be adapted to periods of oxygen depletion which may last for as much as 9 – 10 hr for animals living near the top of the shore. It would be expected that this animal would excrete the end-products of anaerobiosis and in fact this has been shown to occur, acid end products other than lactate being discharged into the surrounding water (Dales, 1958), although here again, *Arenicola* is endowed with considerable versatility and is also able to air-breathe during the intertidal period (van Dam, 1938; Wells, 1945, 1949a).

A further factor which tends to minimise the depletion of the metabolic energy reserves, including proteins and fats as well as carbohydrates, and also to minimise the accumulation of the end-products of anaerobiosis in intertidal animals, is a reduction in the level of metabolism during certain phases of the tidal cycle. A great many animals become quiescent during the period of emersion, a suppression of the heart rate having been shown to be correlated with closure of the valves in the mussel, *Mytilus edulis,* under natural conditions and in the laboratory (Helm and Trueman, 1967; also Schlieper, 1955). A similar situation exists in the cockle *Cardium edule,* (Trueman, 1967), whilst the lugworm *Arenicola marina* also greatly reduces its activity when emersed (Wells, 1949b). Dales (1958; also Von Brand, 1946) found that in the polychaete *Owenia fusiformis* there was no significant reduction in the glycogen content of the coelomic cells over a period of 10 days, and suggested that here, too, the ability of this animal to survive long periods in the absence of oxygen was at least partly due to the reduction in activity which occurs under such conditions.

Despite such adaptations to accommodate the drawbacks of anaerobiosis, it remains true that aerobic respiration is vastly more economical of metabolic fuel and, where possible, oxygen is utilised for respiratory processes. It is of interest, therefore, to consider the ways in which intertidal animals obtain the oxygen they require for aerobic respiration before going on to review the adaptations shown to oxygen deficiency.

1. MECHANISMS OF OXYGEN UPTAKE FROM SEAWATER

Seawater surrounding intertidal animals, under normal circumstances, is saturated with atmospheric air owing to the turbulance which characterises inshore waters. The composition of the atmosphere with which the water is in equilibrium is constant, containing 20·948% oxygen and 0·030% carbon dioxide, the rest being represented by approximately 78% nitrogen and 0·94% argon (Krogh, 1941). It follows that the oxygen content of the seawater covering the intertidal zone at full tide is controlled predominantly by the temperature and salinity. The oxygen content of seawater in equilibrium with the atmosphere varies inversely with both temperature and salinity (Fox, 1907). This relationship is shown in Fig. 7.9. from which it will be noticed that at

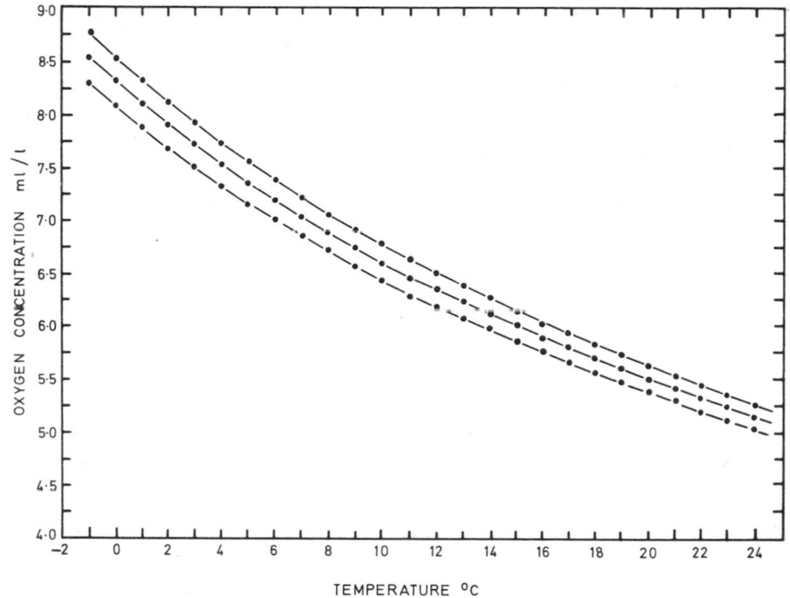

Fig. 7.9. Graphs showing the relationship between temperature and the oxygen concentration of seawater in equilibrium with the atmosphere. The upper curve is for seawater of 27·11‰ salinity (chlorinity 15‰), the middle curve is for seawater of 30·72‰ salinity (chlorinity 17‰) and the lower curve for seawater of 34·33‰ salinity (19‰ chlorinity). Based on data from Fox, 1907.)

the normal seawater salinity of 34·4‰, the oxygen available to intertidal animals falls within the range 8·5 ml O_2/l to 4·95 ml O_2/l. In fact, values of approximately 6·5 ml 0_2/l are those commonly available to intertidal animals and it is at once apparent that the availability of oxygen in seawater is much less than the 21% which occurs in air.

This reduction in the availability of oxygen to aquatic organisms necessitates the development of a relatively large respiratory surface through which gas exchange can occur. This factor has been clearly illustrated by Harvey (1928) who pointed out that in animals without a circulatory system diffusion of oxygen is sufficient to meet the metabolic requirements provided that the animal is small. The relationship for a spherical organism is given by:–

$$C_O = \frac{Ar^2}{6D}$$

Where C_O is the tension of oxygen at the surface in atmospheres, A is the rate of oxygen consumption (ml O_2/g/min), r the radius of the spherical animal in centimetres, and D the diffusion coefficient of oxygen in atmospheres /cm/cm^2. The diffusion coefficient for oxygen in air may be defined as the volume of oxygen (ml) which diffuse per minute through 1 cm^2 area and 1 cm length at a pressure difference of 1 atmosphere (760 mm mercury). This value is 11·0 for oxygen in air at 20° C and has been determined for a number of substances by Krogh (1919). Some of the values are shown in Table XII from which it is apparent that the rate of diffusion through common biological substances e.g. connective tissue, is of the order of one million times less than in air.

Table XII. Table showing the diffusion coefficient and diffusion constant of some common biological media. (Based on Krogh, 1919.)

Substance	*Diffusion coefficient at 20 ° C*	*Diffusion constant at 20 ° C*
Water	0·000034	0·34
Gelatine	0·000028	0·28
Muscle	0·000014	0·14
Chitin	0·000013	0·13
Connective tissue	0·000011	0·11

Assuming a spherical organism to be composed of connective tissue, it is possible to calculate the maximum size which could be attained before the rate of diffusion of oxygen failed to supply the metabolic requirements of the animal. Krogh (1941) took as an example an animal of 1 cm radius having a metabolic rate which required 100 ml O_2/kg tissue/hr. This is equivalent to 1/600 ml O_2/g tissue/min. The diffusion coefficient of connective tissue is 0·00001, so the oxygen tension required at the surface to supply the metabolic demands is given by:–

$$C_O = \frac{1}{600 \times 6 \times 0{\cdot}000011} = 24{\cdot}25 \text{ atmospheres}$$

That is, oxygen at normal atmospheric pressures (*c.* 0·21 atmospheres) would be unable to supply the metabolic needs of such an animal. Transformation of Harvey's (1928) expression given above shows that the maximum radius of an organism whose metabolic demands can be met by diffusion from the atmosphere is:–

$$r = \sqrt{\frac{C_O \times 6D}{A}}$$

where r is the radius in cm, D is the diffusion coefficient in atmospheres, A is the oxygen consumption in ml O_2/g/min, and C_O is the tension of oxygen at atmospheric pressure (0·21 atmospheres). Assuming the same metabolic rate and diffusion constant as that used above, the maximum radius which could be attained would be approximately 1 mm.

In fact, as will be seen later, small animals respire considerably faster per gm weight than large ones so that the maximum size which can be attained with a simple diffusion system is likely to be considerably less than 2 mm diameter. Clearly, one way in which more oxygen could reach the tissues is by the use of a gas transport system. In this case

$$C_O = \frac{A\,r\,T}{3D}$$

where C_O is the tension of oxygen at the surface in atmospheres, A is the oxygen consumption in ml O_2/g/min, r is the radius of the sphere in cm, T is the thickness of the diffusion barrier or cuticle surrounding the body, and D is the diffusion coefficient for oxygen passing through the substance of which the diffusion barrier is composed. Krogh (1941) calculated that for an animal in which the oxygen consumption, A, was 100 ml/kg/hr (= 1/600 ml/g/min) the radius, r, was 1 cm, the thickness of the cuticle, T was 0·005 cm and the diffusion coefficient, D, was 0·000011, the value of C_O was 0·25 atmospheres. That is, by the use of an oxygen transport system which arrived at the surface completely oxygen free and left it fully saturated, a spherical animal of 1 cm radius (= 2 cm diameter) could survive by simple diffusion through the body wall alone. In fact, as we shall see later, and as has been mentioned above, the oxygen requirement per gm of tissue (A) is less in large animals than in small ones; this means that provided an efficient circulatory system is established in contact with the surface through which diffusion occurs, the metabolic requirements of even quite large animals could be met by simple diffusion, provided that the thickness and permeability of the cuticle remained the same as in small animals.

Now in most animals the diffusion of oxygen cannot take place through the whole of the general body surface owing to the development of thickened structural and protective skeletons. In such organisms, special thin-walled outgrowths of the body surface are developed for the function of gas exchange; in many animals such as serpulid and sabellid polychaetes, bivalves and tunicates, such gills also play an important part in food collection (Chapter 5). As would be expected, the structure of such respiratory organs is subject to enormous variations from group to group, being particularly well-developed in active animals such as the shore fishes and in animals such as crabs and bivalves in which oxygen diffusion through the general body surface can take place only to a limited extent. The importance of diffusion through unspecialised parts of the body should, however, not be underestimated as many organisms such as anemones, prosobranchs belonging to the families Omalogyridae, Rissoellidae and Pyramidellidae and some nudibranchs (for review Newell, 1966) are without gills, whilst others such as *Sabella pavonina* can survive for extensive periods after the branchial crown has been removed (Wells, 1952). The structure of such organs of gas exchange has been reviewed by Krogh (1941), Nicol (1960) and Barrington (1967) and will be discussed only briefly here except where some special adaptation to intertidal life is demonstrated.

Amongst the polychaetes, the gills show a wide range of structure from the branchial crown of the sabellids (Fig. 5.10) and serpulids through which a stream of water is drawn, to the bunches of purely respiratory gills in the terebellids (Fig. 5.35). The respiratory significance of the crown of *Sabella pavonina* and *Myxicola infundibulum* has been studied by Wells (1951, 1952). In *Sabella pavonina,* the worm passes a series of peristaltic waves down the body which serve to irrigate the tube. Decapitation does not prevent such irrigation and under such circumstances a new crown can be regenerated. The crown seems, therefore, to serve primarily as an organ of food collection as described on p. 191 although Wells (1952) was able to show that when worms were placed in a tube the lower end of which was blocked, animals with crowns were able to survive whilst those without crowns died or left the tube. Thus under certain circumstances diffusion through the crown is able to meet the oxygen demands of the tissues, although such gas exchange normally occurs through the general body surface.

In *Myxicola infundibulum* the crown has undergone considerable modification from the condition in *Sabella,* the membrane which unites the bases of the branchial filaments extending nearly up to the tips.

Water is therefore drawn into the feeding crown not from a ventro-lateral direction as in *Sabella*, but down between the membrane and the pinnules before passing into the central part of the funnel. The difference in the feeding currents of *Sabella* and *Myxicola* is shown in Fig. 7.10. Unlike *Sabella, Myxicola infundibulum* lives in a gelatinous tube which is closed at its posterior end, no current of water passes through this tube and hence the crown in this genus is responsible for gas

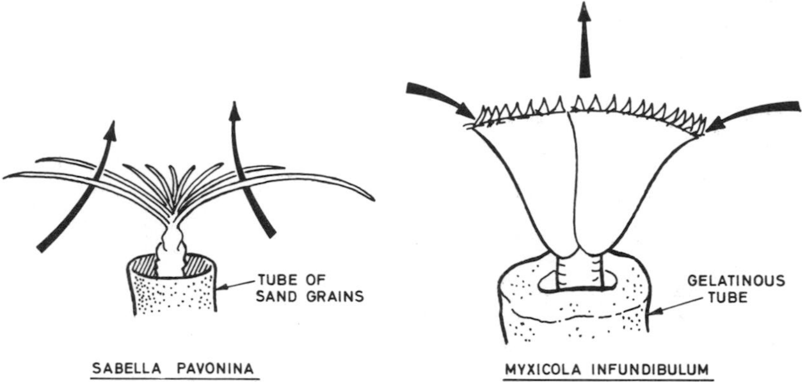

Fig. 7.10. Diagrams showing the direction of the feeding currents in *Sabella pavonina* and *Myxicola infundibulum.* (Based partly on Wells, 1952.)

exchange. As would be expected, the animals live for long periods in tubes closed at the lower end but can survive only a short period if the whole crown is removed, although some animals can survive and regenerate the branchial filaments if the stump of the crown is left intact. Under these circumstances it seems that the ciliary activity of the small stump of the crown may be sufficient to cause a stream which meets the oxygen requirements of the worm, particularly as it is commonly quiescent until a new crown has been formed.

The respiration of terebellid polychaetes has been studied by Dales (1955, 1961). In these animals the gills are separate from the tentacles which are used in food collection as described on p. 221. The gills are small dorso-lateral filamentous or branching structures and in genera such as *Amphitrite johnstoni* irrigation of the burrow is from tail to head. In other polychaetes, such as the nereids, nephtyids and eunicids, the respiratory structures represent foleaceous or filamentous modifications of the parapodia; in the case of *Nereis* water is drawn over them by undulations of the body (Lindroth, 1938) or by cilia in *Nephtys* (Coonfield, 1931; also Jones, 1954). In *Aphrodite,* gas exchange is through the dorsal surface of the body which is covered by a series of flattened plates called elytra. The animal lives buried in sand

and retains a connection with the surface at the posterior end. Water is drawn through the burrow along the under surface of the animal from the tail end by the raising of the body away from the burrow. The ventral surface is then depressed and water is forced dorsally between the parapodia and underneath the raised elytra where gas exchange takes place. The elytra are then depressed and water is forced out over the dorsal surface towards the tail (van Dam, 1940) (Fig. 7.11).

The sites of gas exchange in crustaceans are varied. In the barnacles there are gills or branchiae (Fig. 5.28) through which gas exchange may occur but since circulatory movements within the branchial sacs are

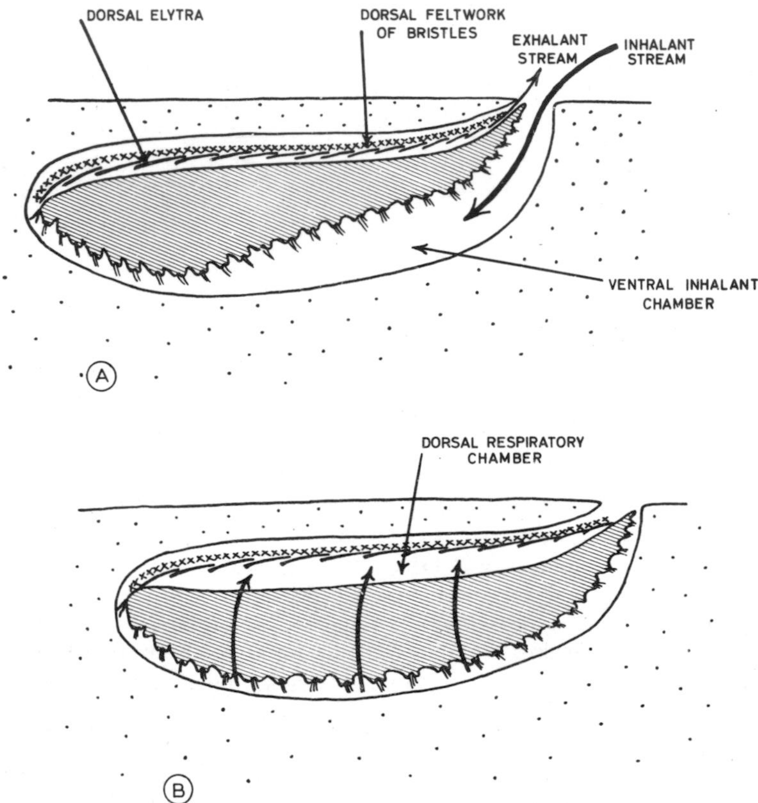

Fig. 7.11. Diagram illustrating the respiratory movements in the sea mouse *Aphrodite aculeata*. (Based on van Dam, 1940.) (A) Depression of elytra forcing fluid out of the dorsal respiratory chamber. Elevation of ventral surface drawing water in. (B) Elevation of elytra and depression of ventral body surface forcing water up between the parapodia into the respiratory chamber. The exhalant aperture is closed.

undescribed, it seems likely that other parts of the body, such as the cirri, are also important in gas exchange (see Newman, 1967). As pointed out on p. 213, extrusion of the cirri is accomplished by the action of the coelomic fluid, and thus the cirri contain a suitable circulatory medium into which oxygen can diffuse from the surrounding seawater. Among the malacostracan crustaceans, the gills are typically found on the thoracic appendages and may be enclosed in a special respiratory chamber as in crabs. In isopods, however, the gills are abdominal, although as much as 50% of gas exchange may take place through the general body surface (Edney and Spencer, 1955). Amongst the Mollusca, gas exchange is typically through the ctenidia which are paired in some archaeogastropod gastropods, and in the bivalves and cephalopods. In the mesogastropods and neogastropods, the ctenidium has undergone modification and has lost the filaments of one side. There is also only one ctenidium, that of the right having been lost. In many opisthobranch gastropods, as well as in some limpets, the ctenidium has been lost entirely, to be replaced by a series of secondary gills which are no longer located within the mantle cavity. The general form of the ctenidium has been described in Chapter 5 together with the ways in which this organ is used for food collection in the prosobranchs and bivalves. The respiratory structures in the Mollusca are reviewed by Ghiretti (1966) and those of the Crustacea by Wolvckamp and Waterman (1960).

In the echinoderms, respiration is commonly through sac-like branchial pouches and through the tube-foot system. In some holothurians, however, there is a cloacal pumping system in which aerated seawater is taken into the cloaca and then driven into branching thin-walled diverticula which extend through much of the coelom. In *Holothuria grisea* it has been found that approximately 10 ml of seawater are taken into the respiratory trees before the fluid is expelled and more aerated water taken in, each contraction of the cloacal muscles

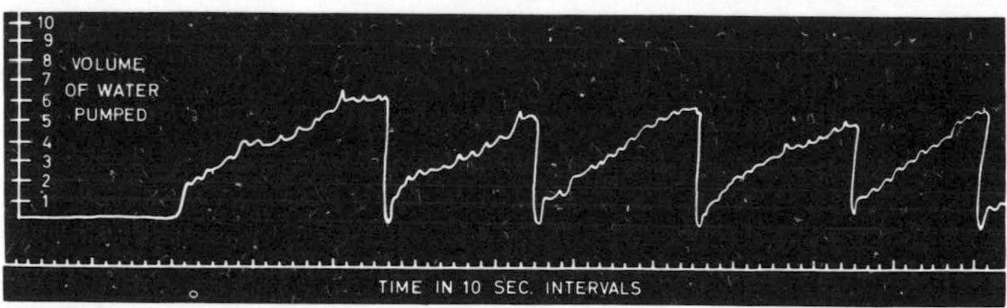

Fig. 7.12. Part of a trace showing the cloacal respiratory movements of *Holothuria forskali* in oxygenated seawater. Upward movements of the lever show an increase of the volume of the holothurian, each minor fluctuation corresponding with a cloacal contraction. After approximately 8 ml had been taken into the respiratory trees in this animal, the water was rapidly ejected before the cycle was repeated. (After Newell and Courtney, 1965.)

transferring approximately 1 ml of aerated seawater into the respiratory trees (Fig. 7.12). In general, it has been found that cloacal gas exchange accounts for approximately 60% of the total oxygen consumption of a variety of holothurians (Winterstein, 1909; Budington, 1937; von Buddenbrock, 1938; Pantin and Sawaya, 1953; Newell and Courtney, 1965). This mechanism may represent an adaptation to reduced oxygen in the surrounding seawater and will be discussed in more detail on p. 253. Finally, amongst the chordates the gill of the ascidians is primarily a highly specialised filter feeding system and as such the flow of water maintained is far in excess of the respiratory needs. Much the same applies to the lancelet, *Branchiostoma lanceolatum* (Courtney and Newell, 1965) as well as to the filter feeding bivalves described in Chapter 5. The structure and functioning of the gills of fishes have been described in detail by Hughes and Shelton (1962; for review, Fry, 1957; Hughes, 1963; Hoar 1966). The blood flow through the gills has been shown to be opposite in direction to that of the water flow; this results in an extremely efficient arrangement known as a counter-current system which may result in 85% of the oxygen being extracted from the water passing over the gills (Saunders, 1962).

In all such gills, the respiratory fluid, which may be either coelomic fluid or blood, is separated from the oxygenated seawater by only a thin layer of tissue, and gas exchange is by diffusion across the gill membrane. In many animals, particularly inactive forms with a high respiratory rate, the surface area of the gills is greatly enlarged by folding, whilst in other animals, such as bivalves, ascidians and some prosobranchs (Chapter 5) the suspended material in the water flowing through the gills is utilised as a source of food. Such gills are used mainly to abstract oxygen from seawater which is in equilibrium with air, but at some stages in the tidal cycle the oxygen concentration is likely to become reduced; special adaptations are then required to enable gas exchange to meet the metabolic demands of the tissues; some of these adaptations are discussed below.

2. MECHANISMS OF GAS EXCHANGE UNDER CONDITIONS OF REDUCED OXYGEN CONCENTRATION

It has been shown on p. 286 that at the partial pressure of oxygen commonly occurring in seawater, diffusion would be adequate to supply the metabolic needs of a spherical animal with no gills, provided that an internal circulatory system was present. Many intertidal animals, however, experience conditions of oxygen lack during the intertidal period. Rock pools at night, for example, often contain low concentrations of oxygen owing to the respiration of the animals and plants and the limited circulation when the tide has ebbed (p. 65). Much the same applies to the interstitial water of fine sands and muds where microbial oxidation reduces oxygen to a negligible level only a few centimetres below the surface (p. 39). Animals living within such habitats are thus exposed to prolonged periods, possibly as much as 10 hr, during which the availability of oxygen is reduced. Under these circumstances it is not surprising to find that such animals show special adaptations which enable them to survive. The most important of such adaptations are firstly the ability to pump oxygenated water into the immediate vicinity of the animal, as is shown by many polychaetes; secondly the possession of an efficient means of extraction of the oxygen from the surrounding seawater, a process which is often aided by the use of a respiratory pigment both at high and low partial presence of oxygen; thirdly, the use of an oxygen store either combined with a pigment or in solution in the blood or coelomic fluid; and finally, the ability to reduce the level of metabolism so that the oxygen demand of the tissues on any available store of oxygen is minimised. Any one species may show one or more of these adaptations which are followed, in the absence of oxygen, by anaerobiosis in many animals. This is discussed in more detail on p. 363.

(a) Irrigation rhythms

Intertidal polychaetes which live in muddy sands are subjected to reduced oxygen concentrations for much of the time even when the deposits in which they live are covered by the tide. This is due to the oxidative activity of the micro-organisms living within the deposits and necessitates pumping movements in order to draw oxygenated water over the respiratory organs. Such pumping rhythms are widespread amongst polychaetes but one of the best-known examples is that of the lugworm *Arenicola marina*. (for review, Wells, 1959).

The pumping activity of *Arenicola marina* under experimental conditions has been extensively studied by Wells (1949a,b, 1950; Wells and Albrecht, 1951a,b). As described on p. 236, the lugworm lives in a U-shaped burrow (Fig. 5.48) through which oxygenated seawater is drawn by bursts of activity which alternate with periods of rest. A system for recording such movements is shown in Fig. 7.13, the animal being allowed to burrow in sand which is "sandwiched" between a pair of glass plates separated by a rubber strip. The lugworm soon constructs a U-shaped burrow and begins its normal defaecation and irrigation movements. Such movements result in a change in the level of the water above the tail shaft, the fluctuations being recorded by means of a float attached to a lever trailing against a smoked drum. The head shaft compartment has free access to water from an outside tank whilst that of the tail shaft communicates with the tank by a capillary tube. In this way, persistent fluctuations in the level of the water above the tail shaft

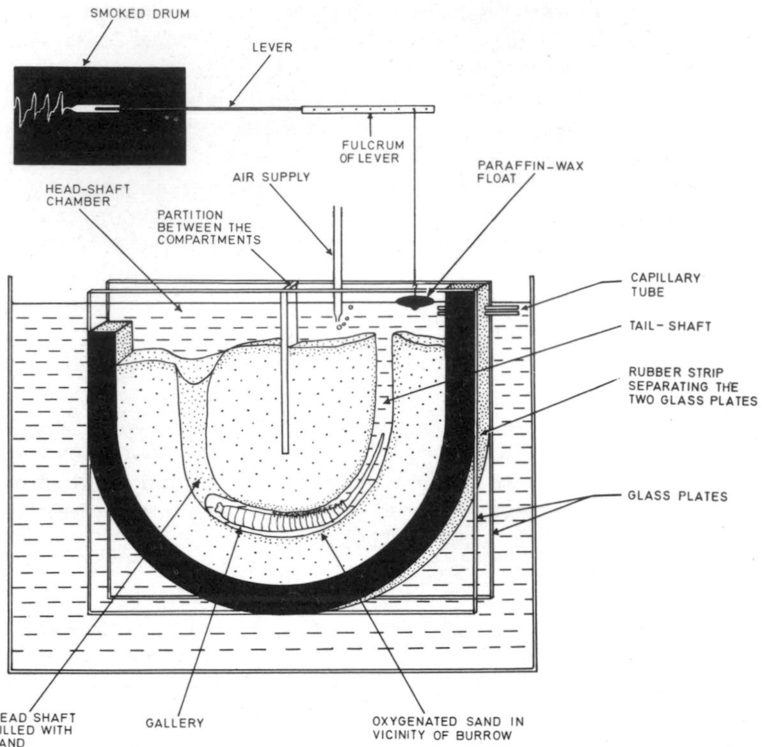

Fig. 7.13. Diagram showing the apparatus used for recording spontaneous activity cycles in *Arenicola marina*. For simplicity, details of the recording drum and water bath containing the seawater have been omitted. (Based on Wells, 1949a, 1959.)

are avoided, the head of water above the head shaft and tail shaft being maintained approximately equal.

A typical trace is shown in Fig. 7.14, from which it can be seen that there is a regularly repeating rhythm which, in fact, occurs approximately every 40 mins. The trace shown in the figure was recorded from the tail shaft compartment and thus, as will be seen from Fig. 7.13, tailward movement of water results in a downward movement of the lever. This trace may be interpreted in terms of the behaviour of the animal in the following way. The lugworm first moves slowly backwards until its tail is at the exit of the tail shaft. This movement causes a slow rise in the water level of the tail shaft compartment and the trace moves irregularly downward. At the end of such movements, the faeces are extruded onto the surface of the sand resulting in a rapid rise in the level of the water above the tail shaft and a consequent sharp fall in the

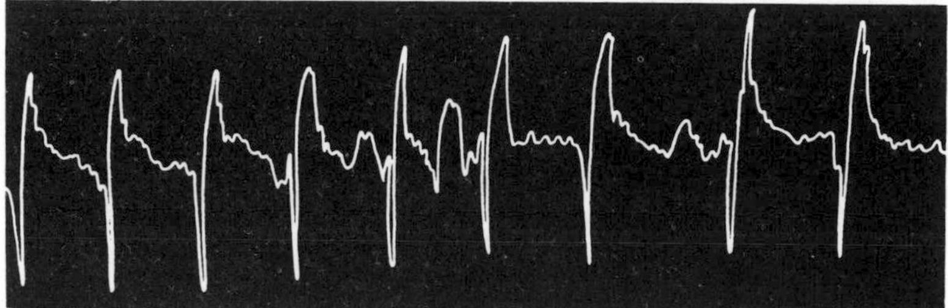

Fig. 7.14. Trace made by an actively-feeding lugworm over a period of approximately 6 hrs. (After Wells, 1949b). The irregular downward movement of the trace records the backward movement of the worm which defaecated at approximately 40 min. intervals, causing the downward spikes. The sharp upward spikes record the headward irrigation which follows defaecation. Notice also the minor fluctuations of the lever with a periodicity of 7–8 min; these record a feeding cycle whose frequency is controlled by an oesophageal pacemaker (Wells, 1949b; Wells and Albrecht, 1951a,b.)

trace. The lugworm then moves anteriorly down to the bottom of the burrow, and pumps aerated seawater vigorously from the tail shaft towards the head. This irrigation results in a sharp rise of the lever (Fig. 7.14), after which the cycle is repeated. Such a sequence occurs even in a fasting worm although, as Wells (1949b) pointed out, only about half the records obtained may be interpreted in terms of such defaecation and irrigation activities. At other times the lugworm may be quiescent for as much as 24 hr before the rhythmic patterns reappear. On other occasions, amounting to as much as 35% of the total recording time made on six worms, there were periods of unexplained activity which could not be readily interpreted. Where rhythmical activity occurs, however, it appears that the irrigation-defaecation cycle is controlled by a pacemaker located in the nerve cord (Wells and Albrecht, 1951a).

Inspection of Fig. 7.14 shows that there are minor fluctuations of the lever superimposed upon the general defaecation-irrigation cycle described above. Such fluctuations have a periodicity of approximately 7–8 minutes and have been attributed to a feeding cycle whose frequency is controlled by an oesophageal pacemaker (Wells, 1949b; also Wells and Albrecht, 1951a,b). Wells (1949b) was able to record from the branchiate segments of a lugworm and from the proboscis synchronously, and found that the proboscis showed regular outbursts of activity which reduce the amplitude of the slight spontaneous movements of the body wall. Part of a trace showing the synchronous recordings made on the proboscis and body wall is shown in Fig. 7.15, from which the inhibition of the spontaneous contractions of the body wall during each burst of oesophageal activity is clearly visible.

Arenicola, therefore, is able to pump oxygenated seawater through its U-shaped burrow when immersed by the tide and is thus able to live

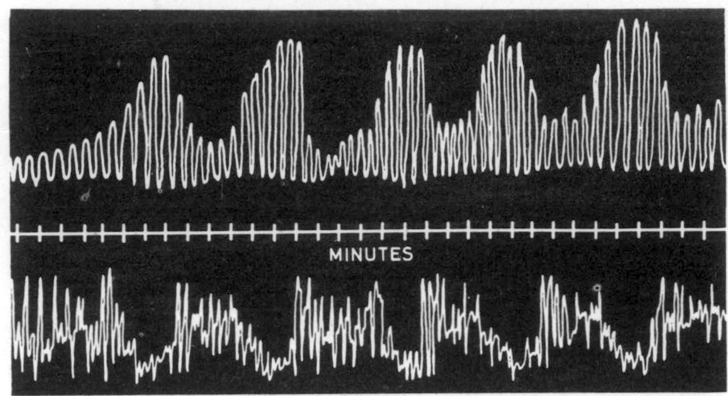

Fig. 7.15. Diagrammatic traces of the contraction rhythms shown by the body wall (above) and the oesophageal extrovert (below) of *Arenicola marina.* Note the reduction in the amplitude of the body wall contractions with each oesophageal outburst. Time scale in minutes. (Redrawn after Wells, 1949b.)

within deoxygenated deposits. However, as we shall see later, special adaptations are required to enable survival during the intertidal period when water can no longer be drawn through the burrow. Many other polychaetes, such as *Chaetopterus variopedatus,* also survive in deposits by means of their ability to draw oxygenated water through a tube which, in this case, is constructed of a parchment-like material. The form of the tube and of the body has been described in Chapter 5 and is illustrated in Fig. 5.9. It will be remembered that on the dorsal surface of the body there are three pairs of fans, one pair on each of body segments 14, 15 and 16, which are responsible for drawing water in through the anterior end of the tube and pumping it out of the rear. It was pointed out on p. 189 that this stream of water carried suspended material with it which was filtered through a mucous bag and subsequently engulfed (Enders, 1909; MacGinitie, 1939a). Not surprisingly, the stream of water is also respiratory in function. Wells and Dales (1951) have shown that the irrigation movements in *Chaetopterus* may usually be grouped into one of four main patterns, although sometimes, as in *Arenicola,* it is not possible to interpret the records.

The first type of behaviour often occurred when the tube of the animal was attached at each end to a water-filled chamber; it involved bursts of activity of the fans, each stroke transporting as much as 3 ml into the posterior float chamber. Between such active periods the animal lay motionless; Wells and Dales (1951) interpreted such "expulsion behaviour" as a response to mechanical or chemical disturbance. The second type of trace involved the "periodic reversal" of the worm. Here the irrigation record was reversed at regular intervals of approximately 40 min. A third trace was obtained when the animal was

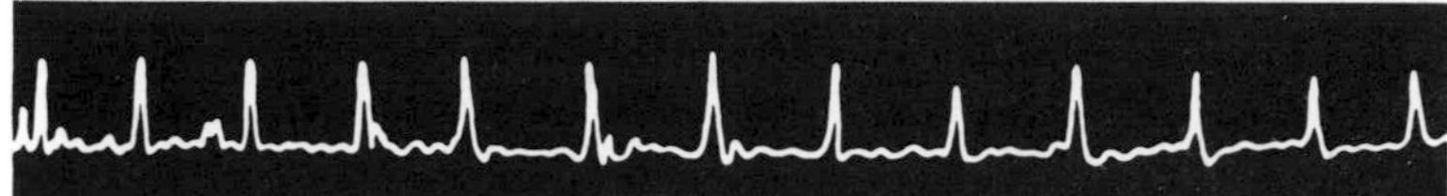

Fig. 7.16. Diagram showing the irrigation record of *Chaetopterus variopedatus* during feeding. The upward spikes occur at intervals of approximately 18 min and represent pauses in pumping when the mucous feeding bag was engulfed. (Redrawn after Wells and Dales, 1951.)

feeding by means of its mucous bag in the way described by MacGinitie (1939a) (p. 189). In this pattern the worm irrigated more or less continuously with fan strokes of approximately 1 per sec, resulting in a flow rate of approximately 16 ml/min. At intervals of approximately 18 min there was a pause in pumping so that the level of seawater in the posterior chamber fell and the lever consequently moved upwards. This pattern is illustrated in Fig. 7.16, from which it will be noticed that the pause in pumping is of short duration. This represents the time taken for the mucous bag to be engulfed prior to a new bag being formed. Since a new bag is formed each 18 minutes, and 16 ml/min are pumped, it follows that each bag has 288 ml of water pumped through it before being engulfed. Finally, there is a phase of fairly continuous activity upon which a 5 min cycle is imposed. This cycle consists of a series of expulsion movements associated with the ejection of faeces.

Another example of a tubicolous polychaete whose activity cycle has been studied is *Sabella pavonina,* which is a common sabellid occurring in the lower parts of the intertidal zone in muddy sand; the tube in this instance consists of sand grains cemented together with a mucous secretion. Wells (1951) has compared the irrigation pattern of this species with that of the Mediterranean species *Sabella spallanzanii* and has shown that differences in the behaviour can be correlated with the form of the tube. In *S. pavonina,* the tube extends a few centimetres above the surface but its lower end terminates some 30 cm below the surface; in *S. spallanzanii* the tube is U-shaped and communicates at each end with the oxygenated seawater above the deposits. Wells (1951) removed the lower end of the tube in each species and connected it onto a recording system as shown in Fig. 7.17, the anterior end remaining free in an aquarium of aerated seawater whilst the posterior end connected to a float chamber and lever system.

In *S. spallanzanii* water is pumped almost continuously in one direction or another, probably mainly by means of peristaltic waves which frequently change direction (Fox, 1938a). Periods of quiescence never exceeded 30 min, and the patterns of irrigation were found to fall into three main types. Firstly there were brief irrigation bursts which could take place in either direction and which were rarely sustained for more than a short length of time. Secondly, sustained headward ir-

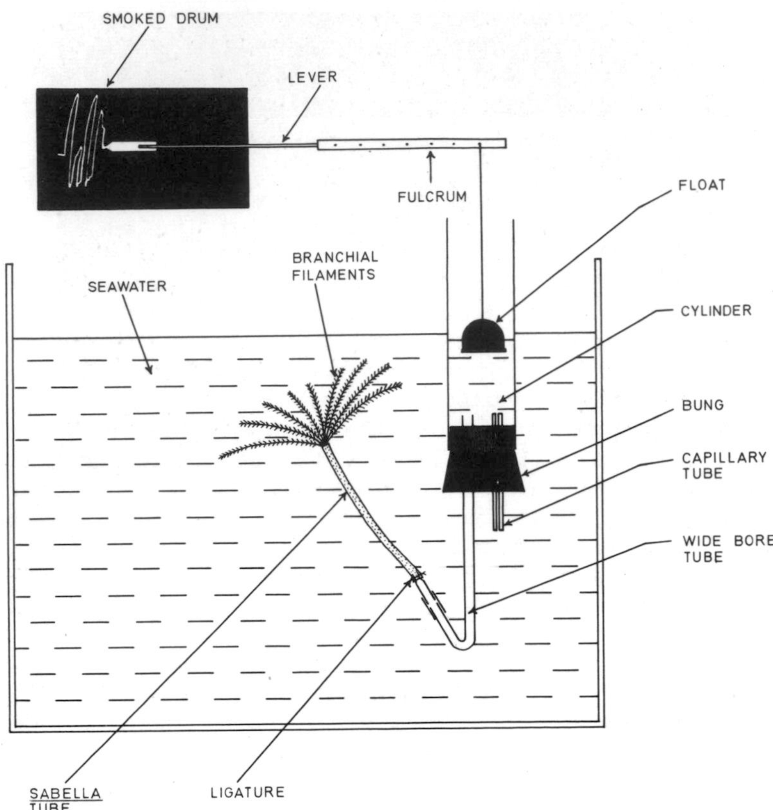

Fig. 7.17. Diagram showing the system used for recording pumping rhythms in *Sabella*. (Based on Wells, 1951, 1959.)

rigation occurred for periods between 2 and 3 hr and sometimes as long as 12 hr during which the velocity of irrigation varied considerably, resulting in a series of characteristic excursions of the kymograph trace which occurred at intervals of 30 – 90 min (Fig. 7.18a). Finally, periods of sustained tailward pumping occurred for 15 – 20 min. Figure 7.18b shows a recording of an animal which had withdrawn into the tube and was pumping in a tailward direction, but at approximately 30 min intervals it extruded the crown from the tube and pumped in a headward direction, this being indicated by the upward spikes on the trace. The irrigation movements of *Sabella pavonina* differ sharply from those of *S. spallanzanii.* In both species irrigation continues even when the crown and anterior segments are removed, but the most important difference is that although *S. pavonina* is able to pump in a headward direction, it pumps only in a tailward direction under normal circum-

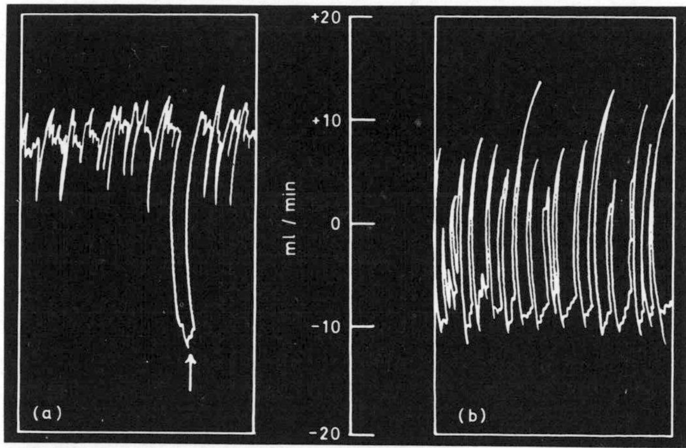

Fig. 7.18. Diagrammatic traces of the irrigation behaviour of *Sabella spallanzanii.* (a) Sustained headward irrigation with regular pauses giving the downward spikes which reach the zero level. At one point (indicated by an arrow) tailward irrigation occurred which resulted in a substantial fall in the trace. (b) Sustained tailward irrigation with brief pauses of headward pumping at intervals of approximately 30 minutes which resulted in the upward spikes. (Redrawn after Wells, 1951, 1959). Notice that during headward irrigation (a) the general level of the trace was approximately +10 ml/mm, whereas during tailward irrigation (b) the level was approximately −10 ml/mm.

stances. This is at least partly because the tube is not U-shaped so that headward pumping would merely clog the posterior end of the tube with debris from the substratum in which it is buried.

In contrast to *Sabella pavonina,* irrigation by the terebellids *Eupolymnia, Thelepus* and *Neoamphitrite* of comparable dry weight and *Neoamphitrite robusta* is almost always from tail to head by means of piston-like swellings which pass along the body (Dales, 1955, 1961). In both *Eupolymnia* and *Thelepus* the waves start nearer the posterior end than in *Amphitrite* and *Neoamphitrite,* and in all of these genera reversal of the direction of irrigation occurs occasionally. Dales has shown that the mean irrigation rate per gram dry weight of *Eupolymnia, Thelepus* and *Neoamphitrite* of comparable dry weight could be arranged in a similar series to their rates of oxygen consumption as is shown in Table XIII.

The percentage utilisation was found to be 50–60% for a 3·0 gm wet weight worm when the oxygen content of the water was high; this compares well with other polychaetes which irrigate their burrows for respiratory purposes. Hazelhoff (1938), for example, estimated a 50% utilisation from fully aerated water in *Eunice gigantea,* whilst van Dam (1937) obtained a value of 30 – 50% in *Arenicola marina* and Lindroth (1938) a value as high as 70% in *Nereis virens.* This is a very high

Table XIII. Table showing the mean irrigation rate (ml/gm dry wt/hr) and mean oxygen consumption (ml O_2/gm dry wt/hr) of the terebellid polychaetes *Eupolymnia heterobranchia, Thelepus crispus* and *Neoamphitrite robusta* (data from Dales, 1961).

Species	*Mean irrigation rate (ml/gm dry wt/hr)*	*Mean oxygen uptake (ml O_2/gm dry wt/hr)*
Eupolymnia	117·4	0·54
Thelepus	59·7	0·28
Neoamphitrite	31·4	0·20

utilisation rate compared with that obtained in animals such as bivalves and ascidians in which the stream is used also for feeding purposes; in such cases the utilisation is only some 5 – 10% (Hazelhoff, 1938; Jørgensen, 1955, 1966; Dales, 1961) which reflects the increased pumping rate which is necessary to supply the nutritive requirements of such animals.

Finally, we come to other examples of polychaetes which live in deoxygenated sands and muds and which, as in the examples cited above, show rhythmic irrigation movements. *Nereis diversicolor,* for example, has been studied by Wells and Dales (1951) and shown to exhibit several different behaviour patterns which replace one another intermittently as in *Chaetopterus.* Differences in the rhythmic activities of the polychaetes *Clymenella torquata, C. zonalis, C. mucosa, Petaloproctus socialis* and *Branchioasychis americana* have been correlated with their ecological distribution by Mangum (1964). She found that the *Clymenella* species, which live in black anoxic sediments, show a rhythmic activity; the largest deflections of the pen are caused by the irrigation of the tube, during which the worm moves tailwards and pushes deoxygenated water from the tube. The animal then descends slowly into the tube, drawing oxygenated water with it; this water is circulated within the tube by the peristaltic contractions which accompany feeding. Intermediate deflections of the lever record the defaecation cycle of the animal.

Mangum (1964) then showed that such irrigation was sufficient to meet the metabolic needs of *Clymenella torquata.* She pointed out that a 50 mg animal consumes 0·3528 ml O_2/day, and that the worm pumps 88·82 ml seawater per day. Since the oxygen content of the seawater above the surface of the substratum was 4·4 ml/l at 27 – 29°C and 33·6 – 34·2‰ salinity, a total of 0·3997 ml O_2/day was made available to the animal by its irrigatory activities; that is, the irrigation was adequate to supply the metabolic needs of the animal. On the other hand, *Branchioasychis* has gills but shows little evidence of irrigatory currents. In this case the animal appears to have developed a large surface area for gas exchange, and to utilise the small movements of the surrounding medium to replace the deoxygenated water rather than pump large volumes of oxygenated water over a relatively small gill area as in *Clymenella* spp. Finally, *Petaloproctus socialis,* which lives in coarse deposits, lacks gills and shows no irrigatory activity; it seems likely that during much of the tidal cycle its needs are satisfied by the oxygen content of the interstitial water.

Wells (1949a, 1951) and Wells and Dales (1951) have shown that in *Arenicola marina, Nereis diversicolor* and *Sabella spallanzanii* closure of the circulation system, the pumped water being then no longer aerated, has important effects on the nature of the irrigation sequence. It might be expected, for example, that as the oxygen concentration falls and the carbon dioxide concentration rises the irrigation might become more and more vigorous to compensate for the reduction in available oxygen. In fact this response does occur in *Chaetopterus,* which is only rarely intertidal and more commonly lives below the low water mark; a similar increase in pumping rate also occurs over a certain range of oxygen concentrations in the holothurian *Holothuria forskali* (Newell and Courtney, 1965). However, such a response to deoxygenation would be clearly disadvantageous in the case of *Arenicola* which lives intertidally; when the tide ebbs and the oxygen concentration falls, increased irrigation would draw water of low salinity or high temperature into the burrow (Linke, 1939, Wells, 1949a, b). It is not surprising, therefore, that the response of *Arenicola* to closure of circulation is the rapid reduction of irrigation movements, followed by hyper-irrigation when the oxygen concentration of the medium is restored. During the low tide period, intermittent 'testing movements' are made so that normal activity can be resumed as soon as the tide returns. Paradoxically, an abrupt cessation of irrigation movements during the intertidal period does not occur in *Nereis diversicolor* (Wells and Dales, 1951); instead, the irrigation movements become gradually reduced until finally the animal is quiescent. Under natural conditions, therefore, surface water from the deposits would be drawn into the burrow, but the animal is much more tolerant of reduced salinities and high temperatures than *Arenicola* and appears to be undamaged by such factors. There may even be the advantage that the water drawn in from the surface is oxygenated, and provided that the animal can tolerate any thermal or osmotic stresses which might occur, gas exchange can continue even when the tide has ebbed.

(b) Respiratory pigments

We have seen above and on p. 288 that, in order to meet their metabolic requirements, many intertidal animals have developed gills through which a gas-transport system is circulated, and that where the oxygen content of the surrounding seawater is reduced, pumping rhythms may be established to bring oxygenated seawater into the vicinity of the sites of gas exchange. Now the maximum oxygen which can be contained in simple solution in the blood is defined by the absorption coefficient of the blood and the partial pressure of oxygen with which it is in equilibrium. As mentioned on p. 285, the maximum partial pressure of oxygen in the atmosphere corresponds to 160 mm of mercury (approximately 21% oxygen) and, since this is normally a fixed value, it follows that an increased supply of oxygen to the tissues of active animals whose metabolic demands are high is dependent either upon an increased rate of circulation of the blood or upon an increase in its absorption coefficient. The absorption coefficient is increased in some animals by the use of respiratory pigments which combine reversibly with oxygen which is taken up from the surrounding medium at the gills and liberated at the tissues (carbon dioxide is transported in the reverse direction). Such respiratory pigments include haemoglobin, chlorocruorin, haemocyanin and haemerythrin, the structure and properties of which have recently been reviewed by Fox and Vevers (1960), Manwell (1960a, 1964), Wolvekamp (1961), Lehmann and Huntsman (1961) and Jones (1963).

(i) Nature of respiratory pigments

As mentioned above, there are four pigments which are known to either carry oxygen in the blood or coelomic fluid, or to store it in the tissues. These pigments resemble one another in that they consist of a conjugated protein which is linked to a prosthetic group containing a metal. Here the general similarities end, however, for haemoglobin and chlorocruorin are the only ones having a similar type of prosthetic group called haem which is a porphyrin linked to one atom of ferrous iron. In haemocyanin, despite the prefix, the prosthetic group is not a porphyrin but a polypeptide and is attached to copper and sulphur rather than iron; in haemerythrin, too, the prefix is misleading since the prosthetic group is not a porphyrin although it does contain iron.

(ii) Haemoglobin and the general properties of some other respiratory pigments

Haemoglobins are conjugated proteins in which an organic prosthetic group, haem, is attached by a covalent linkage to an unbranched chain of amino-acids, the polypeptide chain or globin. The structure of the haem unit is similar in all haemoglobins and also occurs in many other enzymes such as cytochromes (being almost identical with the haem unit of cytochrome *b* which occurs in all aerobic cells), catalases and peroxidases, but the length and sequence of amino-acids of the polypeptide chain varies enormously and partly accounts for the differing properties of the pigment which sometimes varies from species to species, in different animals of the same species and even in different tissues of the same animal (p. 339). In general, however, invertebrate haemoglobins differ from vertebrate ones in the possession of less lysine and histidine, but more arginine and cystine in the side chain. X-ray analysis of mammalian haemoglobins has shown that the globin is coiled round the haem group in a special fashion to form a spherical mono-haem unit and despite the variation in the nature of the side chain, it is assumed that a similar structure occurs in all haemoglobins.

The structural formula of haem is shown in Fig. 7.19 from which it is seen that there are four pyrrole rings united to one another by methene (–CH=) groups to form a disc-like super ring in the centre of which is a ferrous iron atom attached to the nitrogen atoms of the pyrrole rings by four of its six co-ordination bonds. The fifth bond from the ferrous iron atom attaches it internally to a nitrogen of the imidazole group of histidine in the polypeptide side chain (globin). The final bond is attached to a molecule of oxygen in oxyhaemoglobin, this oxygen being replaced by a molecule of water in deoxyhaemoglobin or by a molecule of carbon monoxide in carboxyhaemoglobin, and can also be readily replaced by nitric oxide and hydrogen cyanide.

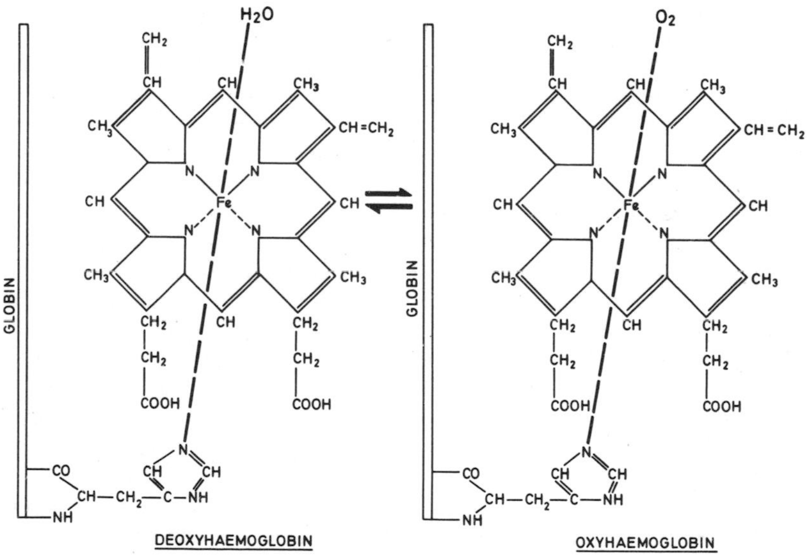

Fig. 7.19. Diagrams showing the structural formula of deoxyhaemoglobin and oxyhaemoglobin. (Based on Wolvekamp, 1961.)

The basic structure of this haem unit is constant in all haemoglobins and together with its side chain comprises a unit of molecular weight of approximately 16,500 – 17,000. However, in addition to variation in the number and sequence of amino-acids in the globin side chain, two or more units may polymerise. Thus in the polychaetes *Glycera xouxii* and *Notomastus latericeus* two units are combined to give a molecular weight of 34,000 – 36,000 whilst that of all vertebrates except cyclostomes corresponds to the aggregation of four units, giving a molecular weight of approximately 66,000 – 68,000. In most invertebrates, however, the molecular weights correspond to approximately 20 or more units although there is an enormous variation ranging from 23,600 in the holothurian *Thyone briareus* and 33,600 in the bivalve *Arca pexata* to composite molecules with a molecular weight as high as 3,000,000, corresponding with an aggregation of 180 monohaem units in polychaetes such as *Nereis virens, Arenicola marina* and *Pectinaria belgica* (for review, Prosser and Brown, 1961).

As Florkin (1949) has pointed out, the haemoglobin molecules of low molecular weight tend to be confined to corpuscles as in vertebrates (molecular weight 66,000 – 68,000) and polychaetes such as *Glycera* and *Notomastus* (molecular weight 34,000 – 36,000) whilst extracellular pigments have a high molecular weight as in *Arenicola*. This may in part serve to prevent the loss of the extracellular respiratory pigments through the filtration system of the kidney, a molecular weight of at least 50,000 being necessary to avoid the loss of protein. However, one other reason for an increase in the molecular weight of extracellular haemoglobin is that the oxygen capacity of the blood is a function of the number of conjugated protein units, since only one molecule of oxygen can be bound to each atom of ferrous iron. If the oxygen capacity of the blood were increased simply by increasing the number of monohaem units, then there would be a great rise in the osmotic pressure of the blood. This problem can be overcome by polymerising the simple monohaem units into larger aggregates which contribute less to the osmotic pressure for a given oxygen-carrying capacity as in certain polychaetes. Nevertheless this tends to raise the viscosity of the blood and sets a limit to the total oxygen capacity of the blood of such invertebrates. Whilst the metabolic requirements can be met in the invertebrates by a compromise of this kind between the colloid osmotic pressure and the viscosity of the blood, increased oxygen capacity in more active animals, such as vertebrates, necessitates the use of blood corpuscles which allow the high concentrations of haemoglobin to be accumulated without a corresponding increase in the colloid osmotic pressure of the blood. Florkin (1949) illustrated the importance of the use of blood corpuscles in man by pointing out that the corpuscular haemoglobin gives an oxygen capacity of 20 volumes % whilst the presence of plasma proteins result in a colloid osmotic pressure of 30 mm of mercury. Now, if such haemoglobin (molecular weight 66,000 – 68,000) were freely dispersed in the plasma instead of being contained in corpuscles, the colloid osmotic pressure would be as high as 175 mm Hg which would seriously interfere with the water balance of the tissues. On the other hand, aggregation of the molecules into units of molecular weight of 400,000, which would reduce the osmotic pressure to normal levels, would result in an enormous increase in the viscosity of the blood.

As has been mentioned earlier, haemoglobin occurs in a wide variety of invertebrate phyla and is especially widespread amongst the annelids and entomostracan crustaceans, but its distribution is sporadic in any one group of invertebrates. It is found in a variety of tissues, including coelomic corpuscles, muscles and ganglia, and is commonly found freely dispersed in the plasma. In this respect invertebrate haemoglobins differ from those of vertebrates which, when occurring in the blood, are confined to corpuscles and occur elsewhere only in muscle (myoglobin). Amongst the polychaetes, haemoglobin is found freely dispersed in the plasma of *Nereis virens, Arenicola marina* and *Pectinaria belgica,* whilst it is confined to the coelomic corpuscles of forms such as *Notomastus latericeus,* glycerids (e.g. *Glycera xouxii*) and capitellids. Haemoglobin is also found in the muscles of some polychaetes, it is located in the body wall muscles as well as in the blood of *Arenicola marina,* and in the ganglia and pharyngeal muscles of *Aphrodite.* The pigment also occurs in the eggs of *Scoloplos armiger* and is found in the coelomic corpuscles and freely dispersed in the plasma of *Terebella lapidaria. Nephtys hombergi* also has haemoglobin both in the coelom and in the plasma but contrasts with other polychaetes in the fact that the haemoglobin is freely dispersed in the coelomic fluid in addition to the plasma. In most polychaetes the fact that the coelomic haemoglobin is contained within corpuscles prevents its excretion through the coelomducts or nephridia. However, *Nephtys* has no coelomducts, excretion being through protonephridia, so that loss of coelomic haemoglobin is avoided even though it is not confined to coelomic corpuscles (Jones, 1963).

Amongst the entomostracan crustaceans, haemoglobin occurs freely dispersed in the plasma of the Notostraca, Anostraca, Conchostraca, in many of the Cladocera, and in many of the Ostracoda and some parasitic copepods such as *Mytilicola* (which occurs in the mussel *Mytilus edulis*) and rhizocephalan cirripedes. It also occurs in the trematode parasite *Proctoeces subtenuis* which lives in the tellinid *Scrobicularia plana* (Freeman, 1963; for review of the occurrence of haemoglobin in parasites, see Lee and Smith, 1965). Haemoglobin is also widely distributed amongst the molluscs (for review Read, 1966) but is found freely dispersed in the blood in only certain genera of freshwater snails belonging to the family Planorbidae and in the bivalves *Astarte alaskensis* (Astartidae) and *Cardita floridana* (Carditidae). It occurs in the corpuscles of the haemocoel of a large number of bivalves including *Anadora* sp (Arcidae), *Arca* sp., *Glycymeris* sp., tellinids such as *Gastrana* sp., and *Tellina planata, Cultellus* sp. and *Solen legumen* (Solenidae) and *Poromya granulata* (Poromyidae). Elsewhere, the haemoglobin is confined mainly to the muscles and nervous system; it is found in the radular muscles of many chitons and of *Dentalium,* pharyngeal muscles of *Patella, Littorina* and *Buccinum,* muscles associated with the buccal mass of *Bulla* and *Aplysia* sp. and in the heart and adductor muscles of the bivalve *Mercenaria* sp (Veneridae). In *Teredo,* and in certain other bivalves such as *Saxidomus* sp., the haemoglobin occurs solely in the adductor muscles.

Haemoglobin is also found in the coelomic corpuscles of the echiuroid *Urechis,* and in the coelomic corpuscles and coelomic epithelium, body wall musculature, gut wall, nerve cord, anal vesicles and eggs of another echiuroid *Thalassema. Phoronis,* too, has haemoglobin but in this case it is confined to the plasma corpuscles whilst in the holothurians *Cucumaria, Caudina* and *Thyone briareus* the haemoglobin is found in the coelomic corpuscles. Thus haemoglobin is widely distributed amongst intertidal animals, and in invertebrates in general, but is absent in the malacostracan crustaceans and phyla such as the Porifera, Coelenterata, Sipunculoidea and Rotifera and several others. The occurrence of haemoglobin is discussed by Fox and Vevers (1960), Prosser and Brown (1961) and by Read (1966) whilst its role in the life of invertebrates is reviewed by Jones (1963) and will be discussed in detail for intertidal animals on pp. 338–352.

Despite such variations in the location of the respiratory pigment, as well as in the degree of aggregation of the haemoglobin units, and of the number and sequence of amino-acids in the side chains of different species, the general spectroscopic features of haemoglobins are similar. If blood is viewed through a micro-spectroscope, a series of dark bands are apparent which correspond with the absorption of light of that particular wavelength by the pigment. Early investigators were able to match the absorption bands against a spectrum and to read off with surprising accuracy the wavelengths at which absorption of light occurred. However, analysis of such absorption bands are now measured in terms of the percentage absorbance (or transmission) of the light of various wavelengths, so that a graph or absorption spectrum is obtained showing the absorption of light plotted against the wavelength, rather than a spectrum with a series of dark bands. Four main peaks, termed α , β, γ, and δ, may be recognised in the absorption spectrum of a typical vertebrate oxyhaemoglobin (Fig. 7.20). The

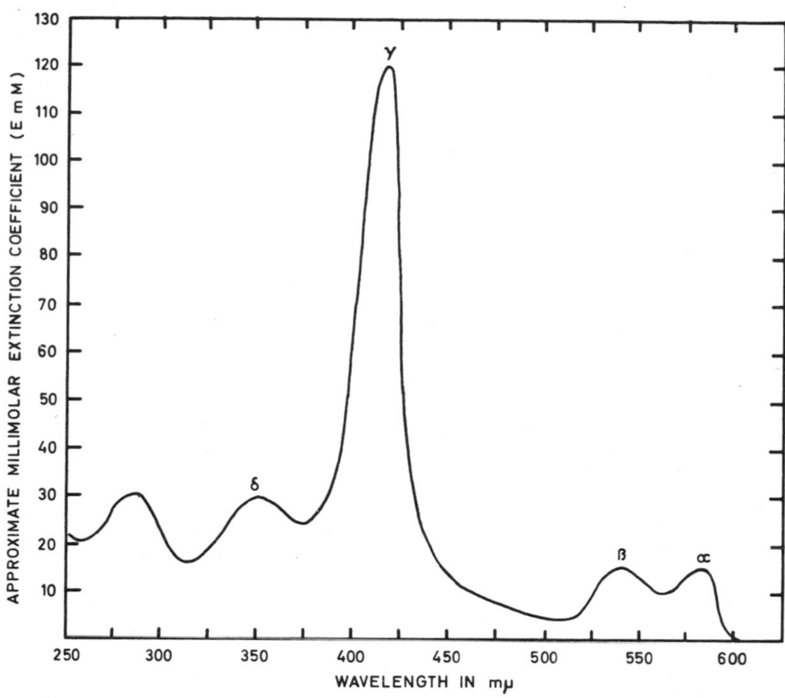

Fig. 7.20. Absorption spectrum of a typical vertebrate oxyhaemoglobin. (Based on Lee and Smith, 1965.)

visible part of the spectrum extends from approximately 400 – 700 mμ; higher wavelengths are in the infra-red range and lower ones in the ultra-violet. The α peak commonly occurs at 580 mμ and the β peak at 540 mμ as is shown in Fig. 7.20. and sometimes forms one peak which is then referred to as the visible peak. The γ peak (or Soret region) occurs between 440 and 390 mμ, and the δ band below 400 mμ in the near ultra-violet region; this band is due partly to the absorbance of haem, but also to that of aromatic amino-acids such as tyrosine, phenylalanine and tryptophan whose maximum absorption occurs at approximately 280 mμ. Since different haemoglobins may have differing numbers of haem units, and this will affect the density of the absorption band (and thus the height of the peaks), it is customary to express the percentage absorbance per millimole of single haem units. This value is then the absorbance of a millimolar solution of the haemoglobin in a standard cell (normally 1 cm) and is termed the millimolar extinction coefficient. Such values are indicated on the y axis of Fig. 7.20.

Now the absorption spectrum of oxyhaemoglobin is different from

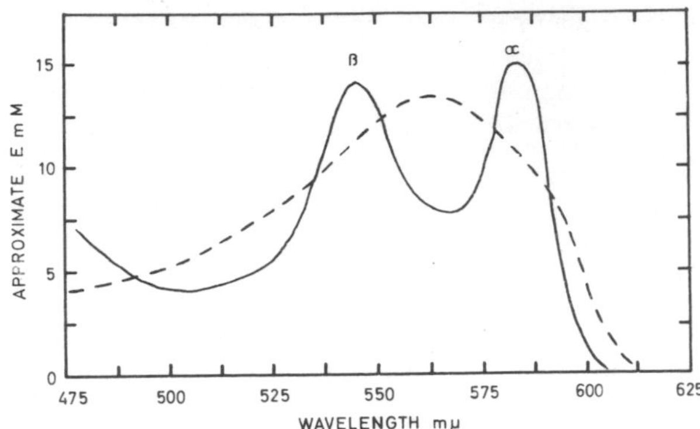

Fig. 7.21 Absorption spectrum of a vertebrate oxyhaemoglobin (continuous line) and deoxygenated haemoglobin (broken line). (Based on Lee and Smith, 1965.)

its carboxyhaemoglobin and deoxygenated derivatives, and this factor enables important measurements to be made on the proportion of oxyhaemoglobin present under a variety of partial pressures of oxygen. Fig. 7.21 shows the absorption spectrum of the oxyhaemoglobin obtained when haemoglobin is fully oxygenated (continuous line), the characteristic double peak of the curve between 540 and 580 mμ is obviously different from the curve obtained for deoxygenated haemoglobin (broken line), which shows only one peak between the two peaks of oxyhaemoglobin. Since at any one wavelength (e.g. 560 mμ) the millimolar extinction coefficient of the curve for oxyhaemoglobin differs from that of deoxygenated haemoglobin, and these points are known to correspond with 100% oxyhaemoglobin and 0% oxyhaemoglobin respectively, it is possible to calculate the percentage saturation of the pigment at various partial pressures of oxygen from the following expression:–

$$\frac{y}{100} = \frac{K_d - K}{K_d - K_o}$$

Where y = the percentage saturation of the pigment
K = the measured optical density at a particular wavelength
K_o = the optical density of oxygenated haemoglobin at that wavelength
K_d = the optical density of deoxygenated haemoglobin

In practice, the absorption spectrum of reduced haemoglobin is determined at a fixed wavelength (e.g. 560 mμ) in a gas tight tonometer which consists of an upper glass equilibration chamber and a lower cuvette which fits into the spectrophotometer. Known amounts of air are then injected into the tonometer with a calibrated syringe and, knowing the barometric pressure, temperature, and humidity, the partial pressure of oxygen within the vessel can be calculated from:–

$$pO_2 = \frac{T}{V}\left(\frac{P_O V_O}{T_O} - nR\right)$$

Where pO_2 = the oxygen pressure
T = the temperature in the tonometer
T_O = the room temperature
V = the volume of the tonometer
V_O = the volume of air injected
n = the number of moles of oxygen combined with the haemoglobin
R = the gas constant
and P_O = $0{\cdot}21\,(P - Hp)$
Where P = the barometric pressure
H = the relative humidity
and p = the vapour pressure of water (Riggs, 1951)

Fig. 7.22 shows a family of such absorption spectra obtained between 520 and 590 mμ on a haemoglobin which had been

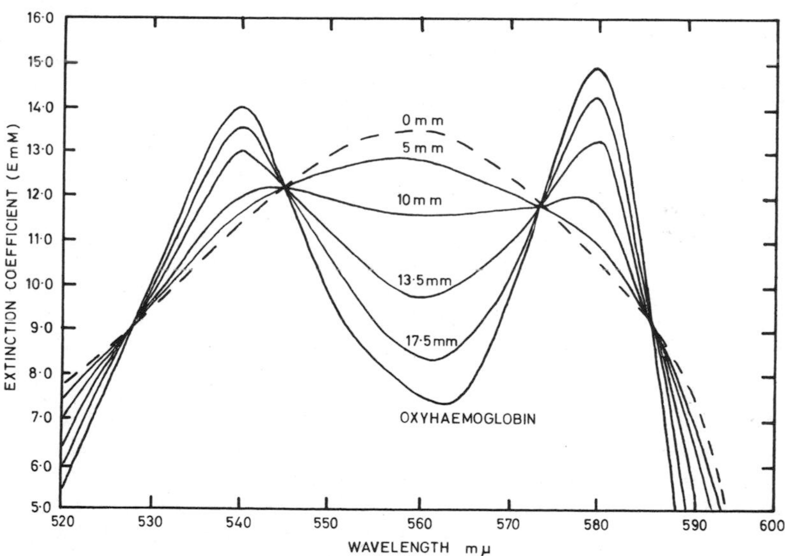

Fig. 7.22. Absorption spectra obtained when a vertebrate haemoglobin is equilibrated with partial pressures of oxygen corresponding to 0, 5, 10, 13·5 and 17·5 mm Hg respectively.

equilibrated to a variety of partial pressures of oxygen; it is obvious that the spectra vary between the extremes set by the spectra for oxyhaemoglobin and deoxygenated haemoglobin respectively. The levels of these absorption spectra can then be interpreted in terms of the percentage saturation (= percentage of oxyhaemoglobin) by means of the expressions referred to above.

In this way we obtain data for a graph called an absorption or, more usually, dissociation curve, which relates the percentage saturation of the pigment with oxygen to the partial pressure of oxygen with which it is in equilibrium. The form of this curve is of fundamental importance in an understanding not only of the mode of action of haemoglobin but also of other respiratory pigments. The curve obtained from similar data to those shown in Fig. 7.22 is illustrated in Fig. 7.23

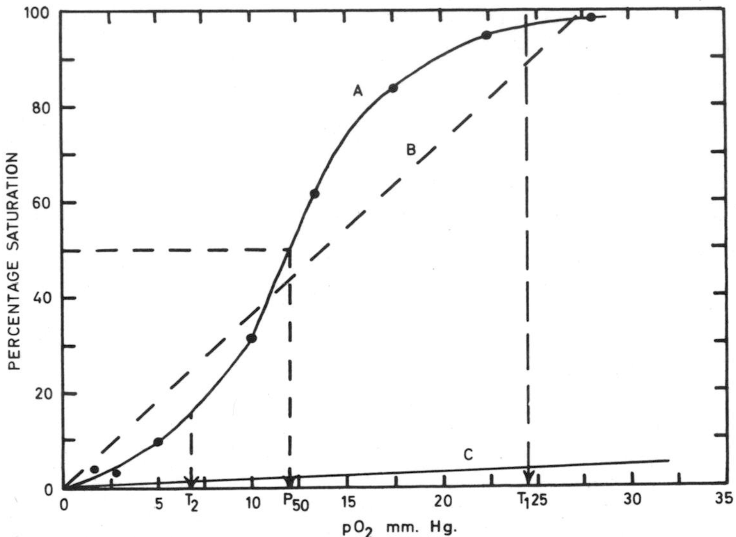

Fig. 7.23. Dissociation curve obtained from data similar to that shown in Fig. 7.22 p_{50}= 12·0 mm. Hg; p_{95}= 23·5 mm Hg For explanation see text.

The first point which is obvious from an inspection of the oxygen dissociation curve of the haemoglobin illustrated in Fig. 7.23. (continuous line, curve A), is that it is sigmoid in form, becoming 50% saturated at a definite partial pressure of oxygen. The exact combined oxygen content of the blood at 100% saturation will, of course, vary from species to species according to the concentration of the pigment since a fixed number only of oxygen molecules can combine with each haemoglobin molecule depending upon the number of monohaem units from which it is aggregated (p. 304). This value is referred to as the oxygen capacity of the blood and is commonly expressed as volumes of oxygen % or ml 0_2/100 ml blood. Oxygen in solution in the blood accounts for only some 0·5 vols %, compared with a combined oxygen capacity of 4 – 20 vols % in fishes, 1 – 10 vols % in the plasma of annelids and 1 – 6 vols % in that of molluscs (for review, Nicol, 1960; Prosser and Brown, 1961; Hoar, 1966); the presence of haemoglobin thus vastly enhances the total oxygen content of the respiratory fluid. As Lehmann and Huntsman (1961) have pointed out, if each haemoglobin molecule consists of one monohaem unit only, then the oxygen dissociation curve is a hyperbola (Curve A, Fig. 7.24). But when

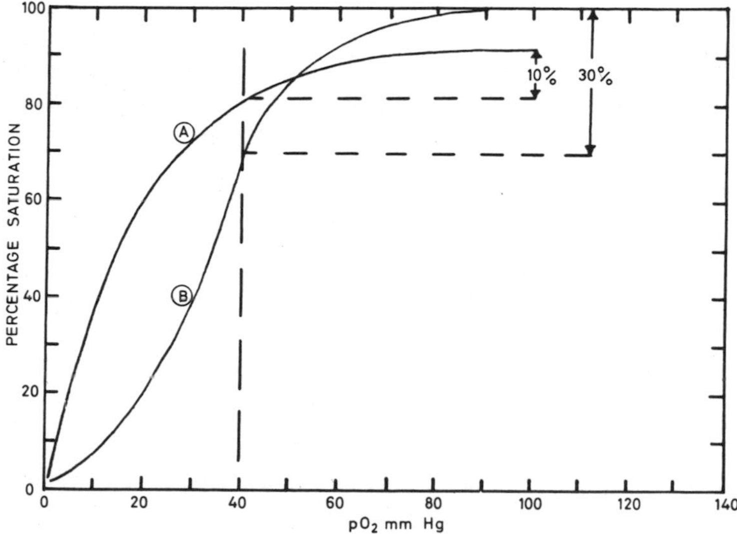

Fig. 7.24. Diagram illustrating the increase in oxygen turnover which is made possible by the possession of a pigment with a sigmoid oxygen dissociation curve. In a pigment with the hyperbolic curve (A) the turnover between 100 mm and 40 mm Hg is only some 10% of the oxygen capacity of the pigment whereas in a pigment with the sigmoid curve (B) the turnover is 30% between these partial pressures of oxygen. (After Lehmann and Huntsman, 1961.)

two or more monohaem units are aggregated then the curve becomes sigmoid (curve B), partly owing to interaction between the active oxygen-combining sites on the haemoglobin as is discussed on p. 316. The advantage of a sigmoid dissociation curve over a hyperbolic one is especially marked at high partial pressures of oxygen. From the two hypothetical curves shown in Fig. 7.24A and B, it is seen that the pigment with a hyperbolic dissociation curve can unload only some 10% of its capacity between a partial pressure of oxygen from 100 mm Hg to 40 mm Hg. On the other hand, the pigment with the sigmoid oxygen dissociation curve can unload more than 30% of its total capacity between these two partial pressures of oxygen. Thus one further advantage of the polymerisation of the monohaem units into complex haemoglobin molecules, apart from the increase in the total oxygen capacity of the blood and the prevention of the excretion of freely-dispersed plasma haemoglobin (p. 304), is that the possibility of interaction between the oxygen-combining sites on the haem units of any one molecule alters the dissociation curve from a hyperbola to a sigmoid, with a consequent increase in the efficiency of function at high partial pressures of oxygen.

Other important features of the sigmoid form of the oxygen dissociation curve of many haemoglobins, as well as of other respiratory pigments, are best illustrated by comparison with the straight line (Line B) in Fig. 7.23; they have been discussed in some detail for a variety of invertebrates by Jones (1963). As he pointed out, for effective functioning the pigment must be fully oxygenated at the respiratory surface but donate most of its oxygen by the time the partial pressure has fallen to that in the veins. If the partial pressure in contact with the blood in the respiratory organs is indicated by T_1 and that of the tissues by T_2, then the amount of oxygen released by the haemoglobin during the passage of the blood from the respiratory organ to the tissues is T_1 (= 98% of its saturation value) minus T_2 (= 15% of its saturation value) which is equivalent to 83% of its total oxygen capacity. By comparison, if the dissociation curve was the straight line B, then only 90 – 24 = 66% of its total oxygen capacity would be donated to the tissues. Of course, even the oxygen in simple solution (Curve C in Fig. 7.23) will be released to some extent during the passage of the blood from the respiratory organ to the tissues, but will amount to only 4·2% (T_1) – 0·75% (T_2) = 3·45% of the total oxygen. Thus by an increase in the steepness of the curve over the effective range of partial pressures of oxygen (i.e. between that occurring at the respiratory surface and that at the tissues) the amount of oxygen released is vastly augmented by means of a respiratory pigment.

There is one further factor which enhances the release of oxygen from haemoglobin during its passage from the respiratory surface to the tissues; this is the effect of *pH* on the oxygen affinity of haemoglobin. In many haemoglobins, as well as in other respiratory pigments, low *pH* values such as are caused by high partial pressures of carbon dioxide normally tend to shift the oxygen dissociation curve to the right and to make the curve less steep (Fig. 7.25). This is called the Bohr effect,* and results in the respiratory pigment having a lower affinity for oxygen

* The Bohr effect, in which the dissociation curve is shifted to the right at low *pH* (= high CO_2) values, is to be distinguished from the Root effect in which there is a *lowering* of the saturation with acidification.

than it would have at higher *pH* values. The importance of this factor is that low *pH* values tend to occur at the tissues where carbon dioxide is released, and this automatically enhances the oxygen released from the respiratory pigment. In the hypothetical oxygen dissociation curve illustrated in Fig. 7.25, for example, the percentage oxygen which

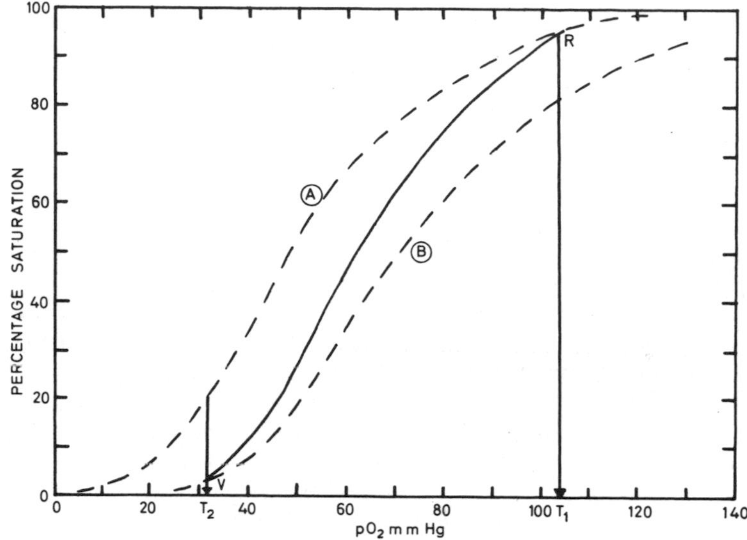

Fig. 7.25. Oxygen dissociation curves of a pigment in the presence of (A) 2 mm CO_2 and (B) 6 mm CO_2 showing a Bohr effect. The point *R* represents the condition of the blood at the respiratory surface and *V* represents the condition at the tissues. The curve joining the two points represents the shift to the right of the dissociation curve as the pCO_2 increases towards the tissues. In the absence of such a Bohr effect the total turnover between the pO_2 at the respiratory surface (T_1) and that at the tissues (T_2) would be 97–20 = 77% whilst the actual turnover due to the Bohr effect is 97–5 = 92% of the total oxygen capacity of the pigment.

would be released at the tissues, assuming a constant *pH,* is 97% (T_1) minus 20% (T_2) = 77% of the total oxygen capacity of the pigment. But when carbon dioxide at a partial pressure of 6 mm Hg is present, the curve is displaced to the right as shown in the figure. If we assume that such a partial pressure of carbon dioxide occurs in the tissues but not at the respiratory surface, then point *R* represents the conditions in the blood leaving the respiratory surface and point *V* the conditions in the blood of the veins, and the sigmoid curve between the two shows the shift of the dissociation curve to the right as the tissues are approached. Under these circumstances, the amount of oxygen released is equivalent to 97 – 5 = 92% of the total oxygen capacity of the pigment rather than the 77% which is released in the absence of a normal Bohr effect. A possible mechanism underlying the Bohr effect has been summarised by Wolvekamp (1961). He pointed out that the link between the ferrous iron and oxygen may resonate between a left-handed and a right-handed configuration. When carbon dioxide is released from the tissues it combines partly with water to give H_2CO_3 and partly with NH_2 groups on the globin chain to form carbamic acid. Both of these substances lower the *pH* and would result in less resonance of the $Fe - O_2$ link. The stability of the oxygen bond would thus decrease and oxygen be released (the Bohr effect).

Since pigments with a high affinity for oxygen are fully saturated at low partial pressures of oxygen, it follows that their dissociation curve is displaced to the left of those with a low oxygen affinity. The partial pressure of oxygen in mm Hg at which the pigment is 50% saturated is referred to as the p_{50} and is easier to read precisely from the dissociation curve than the pressure at which the pigment is 100% saturated; the p_{50} is commonly used to define the oxygen affinity of the pigment. This value is sometimes called the "unloading tension" as distinct from the "loading tension" which is the partial pressure of oxygen in mm Hg at which the pigment is 95% saturated (designated p_{95}).

Although we have discussed the importance of the sigmoid form of the oxygen dissociation curve, it is often convenient to replot the graph in such a way that the data approximate to a straight line. Fig. 7.26A shows an arbitrary dissociation curve of the characteristic sigmoid form which is displaced to the right with low *pH* values. On the *y* axis on the left hand side of the figure is shown the percentage of the oxygenated form of the pigment (% saturation) and on the right hand side is shown the ratio of oxygenated to deoxygenated haemoglobin ($y/100-y$). When the data for bloods which show a normal Bohr effect are read off in terms of this ratio, and plotted on a logarithmic scale against the log of the partial pressure of oxygen, a straight line is obtained which is displaced to the right with lowered *pH* values. Fig. 7.26B shows the hypothetical dissociation curves illustrated in Fig. 7.26A but plotted on such a double logarithmic scale; the value for p_{50} is, of course, easy to read off with accuracy at the point where $y/100-y = 1{\cdot}0$.

It is obvious that high values of p_{50} indicate a low oxygen affinity, and that when lowered *pH* values cause an increase in the value of p_{50}, then the pigment shows a positive Bohr effect. Some pigments show a reversed Bohr effect in which the p_{50} is reduced with low *pH* values although such reversed Bohr effects are never so great as a positive Bohr effect. Their significance is discussed on p. 335. It should be noted, however, that the value for p_{50} varies not only with *pH* but with temperature (p_{50} increasing as the temperature rises) as well as with the concentration of the pigment in some mammalian haemoglobins, but not in that of *Arenicola* or in haemoglobins from a variety of other

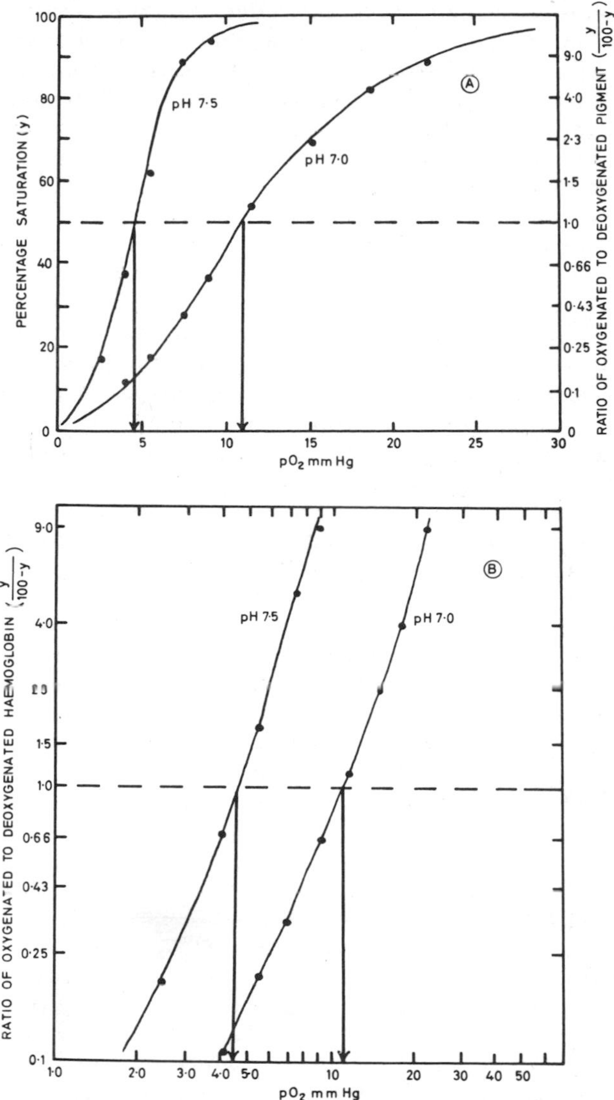

Fig. 7.26. A. Oxygen dissociation curves of a haemoglobin showing a normal Bohr effect. The values of the ratio of oxygenated to deoxygenated haemoglobin $(\frac{y}{100-y})$ are shown on the right hand side of the figure. B. Log-log plot of $\frac{y}{100-y}$ against pO_2 showing that the slope of the lines remains constant despite displacement to the right which occurs with the change from *pH* 7·5 to *pH* 7·0.

animals (Barcroft and Barcroft, 1924; Wolvekamp and Vreede, 1941). In those haemoglobins where this effect has been noted, the p_{50} becomes lower with increasing dilution (i.e. the oxygen affinity increases with dilution of the pigment). Now, although one molecule of oxygen combines reversibly with each atom of ferrous iron in each monohaem unit, and this reaction occurs in accordance with the law of Mass Action, it is known that in many of the multihaem units of which most invertebrate haemoglobins are comprised, combination with oxygen does not proceed according to the law of Mass Action. One haem unit when combined with oxygen may effect the ability of other haem units on the same molecule to combine with oxygen, converting the dissociation curve from a hyperbolic to a sigmoid form (also p. 322 and Fig. 7.27). Such "haem-haem interactions" may be either negative, in which the presence of oxygen combined with one haem unit inhibits the interaction between another molecule of oxygen and another haem unit on the same haemoglobin molecule, or they may be positive, in which case oxygen combined with a haem unit facilitates the combination of oxygen at other sites on the same molecule.

The extent of such interaction between the active sites on the pigment molecule profoundly alters the shape of the oxygen dissociation curve (= oxygen equilibrium curve). The shape of such a curve is approximated by the Hill equation:–

$$\frac{y}{100} = \frac{Kp^n}{1 + Kp^n} \qquad (1)$$

Where y = the percentage oxyhaemoglobin (percentage saturation)
K = the overall equilibrium constant
p = the partial pressure of oxygen (mm Hg)
and n = the Hill constant

The Hill constant n is found to give a numerical value for the extent of haem-haem interaction or for similar interaction between the active sites of other respiratory pigments and may therefore be termed the "interaction coefficient". Where $n = 1$ there is no interaction and the curve is a hyperbola, where n is greater than 1·0 there is a positive haem-haem interaction (i.e. combination with oxygen at one site facilitates further combination at other sites), whereas a value of less than 1·0 indicates a negative haem-haem interaction.

Positive values for n are common in haemoglobins; they vary between 1·0, when there is no interaction, and 6·0 which is the highest recorded value, n normally being between 2 and 3 in vertebrate haemoglobins. Negative values, on the other hand, have been reported in only a few instances, e.g. $n = 0{\cdot}82$ in the haemoglobin of the holothurian

Caudina sp, n = 0·1 in certain fish, and n = 0·8 in the haemoglobin of the clam *Cardita* (for review Manwell, 1964). In either case, the value of n indicates the degree of sigmoid nature of the curve, values of n = 1·0 indicating a hyperbola and $n > 1·0$ indicating sigmoid curves. Thus the important sigmoid properties of the dissociation curve which have been referred to on p. 310 may be caused by interactions between the active sites of oxygen attachment, such interaction clearly being only possible

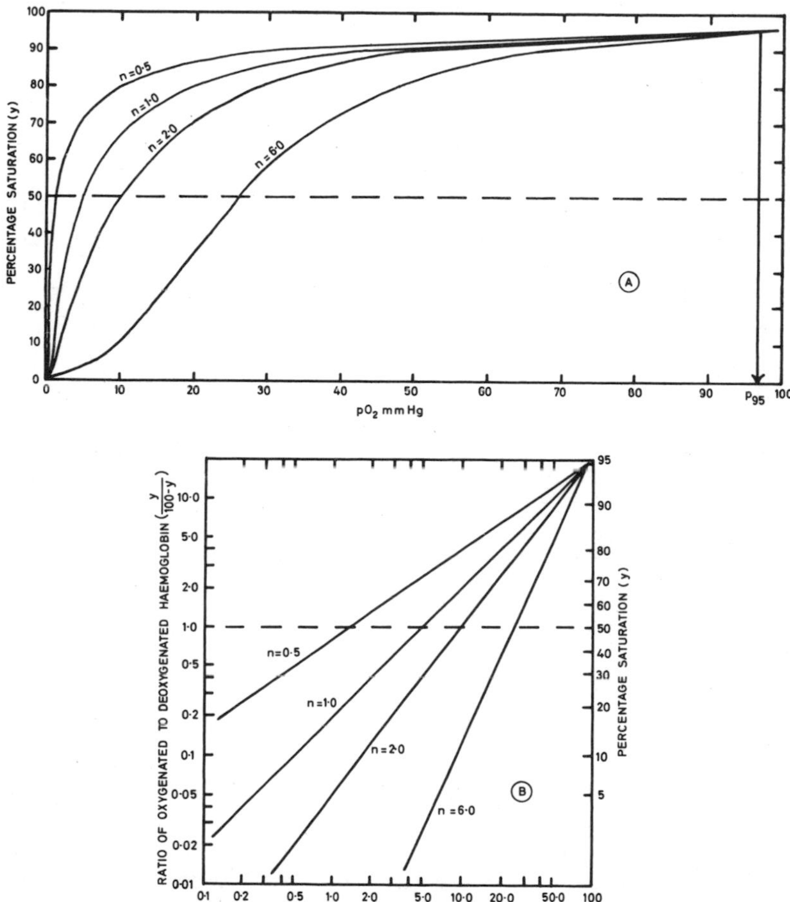

Fig. 7.27. Diagram showing the oxygen dissociation curves of four different haemoglobins each with the same p_{95} but with different values of the interaction coefficient n. A. Linear plot showing the increase in sigmoid nature of the curve with increase in n. B. Log-log plot of the ratio of oxygenated to deoxygenated haemoglobin ($\frac{y}{100-y}$) against pO_2 of the same curves as in A.

in haemoglobin molecules consisting of two or more monohaem units. Fig. 7.27A plots four dissociation curves on a linear scale when the interaction coefficient n is 0·5, 1·0, 2·0, and 6·0 respectively, whilst in Fig. 7.27B they are plotted as a log-log plot of $y/_{100-y}$ against pO_2, as discussed on p. 314. It can be seen from this last set of lines that the value of n determines the slope of the curve, so that if a dissociation curve is plotted in this way, the slope gives a direct value for the interaction coefficient n and can be read off from the graph at the point where $y/_{100-y}$ = 1·0 (i.e. where the pigment is 50% saturated) and is equivalent to the tan of the angle of slope.

It is also possible to calculate the value of n using the Hill equation and the data for the oxygen dissociation curve. By taking logarithms, the Hill equation becomes:–

$$\log \frac{y}{100} = \log \frac{Kp^n}{1 + Kp^n} \tag{2}$$

whence $\log y - \log 100 = \log K + \log p^n - \log (1 + Kp^n)$

and $\log y = \log K' + n \log p - \log (1 + Kp^n)$

Thus $\log y = A + n \log p - \log (1 + Kp^n)$ (3)

(Where $K' = 100K$; also $\log K' = A$)

Now this gives linear values for the Hill equation at low partial pressures of oxygen i.e. at the bottom of the dissociation curve. It is clearly more convenient if the value for the interaction coefficient n can be calculated from the whole dissociation curve. In order to do this, values for both y (the percentage oxygenated pigment or % saturation) and $100 - y$ must be substituted into the Hill equation. Then:–

$$\frac{y}{100 - y} = Kp^n \tag{4}$$

and

$$\log \frac{y}{100 - y} = \log K' + n \log p \tag{5}$$

$$= A + n \log p \tag{6}$$

Thus, plotting the logarithm of the ratio of oxygenated to de-oxygenated respiratory pigment against the logarithm of the partial pressure gives a straight line for the full Hill equation.

Some examples of positive and negative haem-haem interactions in the haemoglobin of two species of holothurians from Friday harbour, Washington are illustrated in Fig. 7.28. A line steeper than that obtained when n = 1·0 indicates a positive haem-haem interaction, e.g. in *Cucumaria minuta*, and one shallower than the line obtained when n = 1·0, as in *Molpadia intermedia*, indicates a slight negative haem-haem interaction. Of special interest is the fact that in some haemoglobins, including that of man and the holothurian *Thyone* sp. (for review,

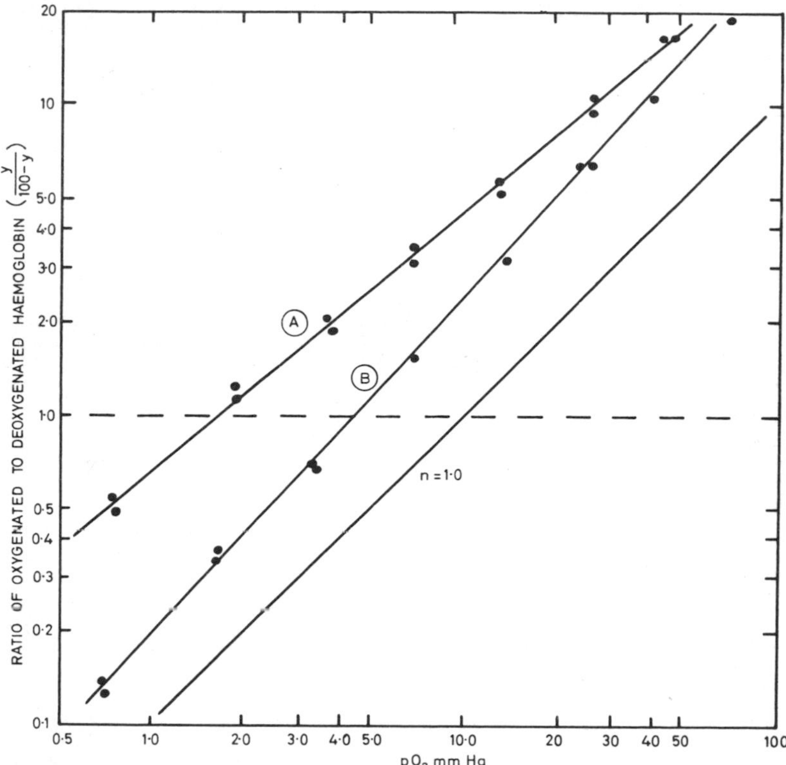

Fig. 7.28. Diagram showing negative haem-haem interaction in the haemoglobin of the holothurian *Molpadia intermedia* (A) and positive haem-haem interaction in that of *Cucumaria minuta* (B). There was no Bohr effect at *pH* 7·4 – 7·0. The temperature was 20–22° C. A line with a slope of n = 1·0 is also shown for comparison. (After Manwell, 1964.)

Manwell, 1964), the oxygen dissociation curve may be diphasic, having an initial shallow slope with a value of n = 1·0 or less, whilst at higher partial pressures of oxygen positive haem-haem interactions occur and the curve becomes much steeper with values for n greater than 1·0. Fig. 7.29 shows an oxygen dissociation curve of *Thyone* sp. from which it can be seen that when the ratio of the percentage oxyhaemoglobin to deoxyhaemoglobin is 3·0 (i.e. the percentage saturation is 75%) an increase in the steepness of the slope occurs and the value of the interaction coefficient becomes greater. Manwell (1964) has also found that

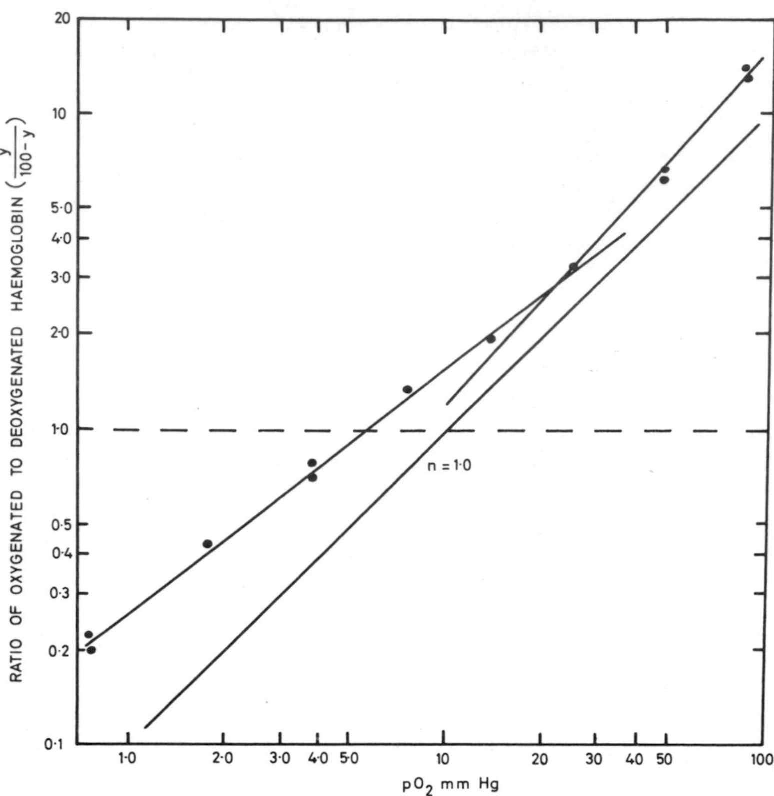

Fig. 7.29. Diagram showing the discontinuous oxygen dissociation curve of the haemoglobin of the holothurian *Thyone* sp. at *pH* 7·0 – 7·5 and a temperature of 20–22° C. Note the change in the interaction coefficient *n* which is $<$ 1·0 when the pigment is less than 75% saturated ($\frac{y}{100-y}$ = 3·0) and $>$ 1·0 at higher saturation. (After Manwell, 1964.)

similar interaction between oxygen binding sites (chlorocruorohaem interactions) occur in the chlorocruorin of the sabellid polychaete *Eudistylia vancouveri,* the first phase having a value of n = 1·0 and the second 4·8. Unlike the haemoglobins of the holothurians mentioned above, the chlorocruorin of *Eudistylia* has a strong Bohr effect so that the exact position of the diphasic curve depends upon the *pH.* The curves obtained by Manwell (1964) after the blood had been dialysed against potassium phosphate buffers of *pH* 7·93, 7·51 and 7·00 are shown in Fig. 7.30.

Manwell (1964) has recently reviewed the evidence which suggests that not only in mammalian haemoglobins but also in the haemoglobins

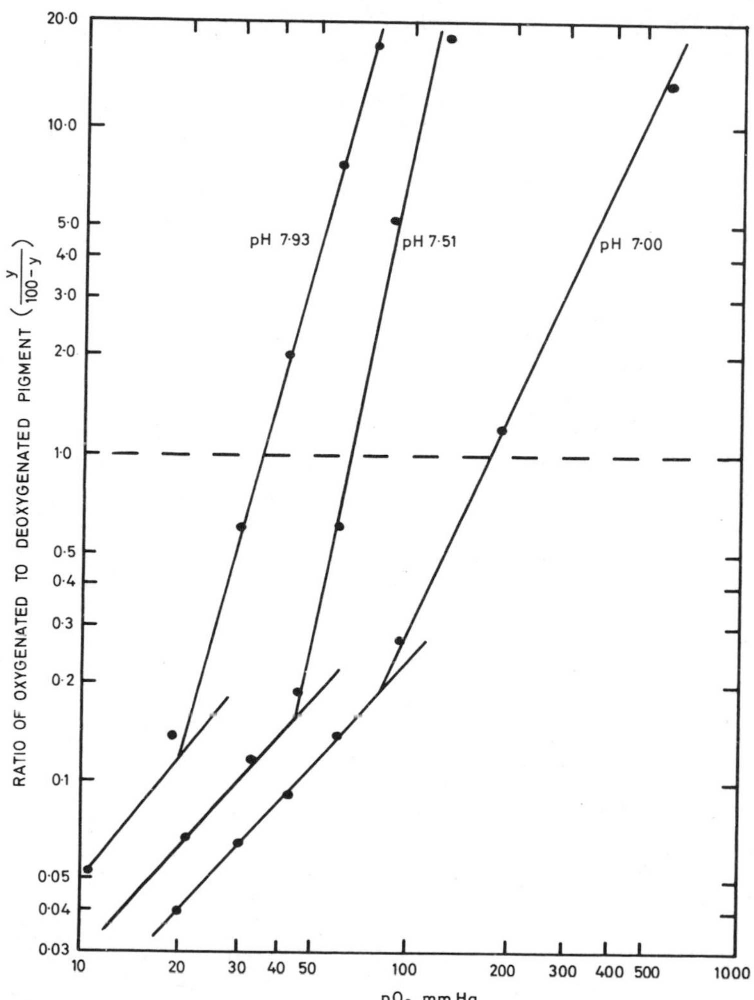

Fig. 7.30. Oxygen dissociation curves of chlorocruorin from the sabellid polychaete *Eudistylia vancouveri* at *pH* of 7·93, 7·51 and 7·0 at 22°C. Note the diphasic nature of the dissociation curve, the first phase having a value of $n = 1{\cdot}0$ and the second phase as high as 5·0. The marked Bohr effect results in the lines being displaced to the right at low *pH* values. (After Manwell, 1964.)

of the holothurians *Molpadia intermedia, Cucumaria minuta* and *Thyone* sp, as well as in the haemocyanin of the gastropod *Busycon* and in the cephalopod *Octopus dofleini* and in the chlorocruorin of the sabellids *Eudistylia vancouveri, Schizobranchia insignis* and *Sabella* sp., the combination of the respiratory pigment with oxygen is associated with configurational changes of the protein molecule. Such changes involve alterations in the secondary, tertiary and sometimes in the quaternary structure of the protein, and appear to be fundamental to the oxygen equilibrium not only of mammalian haemoglobin but also of haemocyanin and chlorocruorin.

In *Octopus* haemocyanin, for example, positive interactions between oxygen combining sites (in which the addition of an oxygen molecule facilitates the combination of another oxygen molecule at another site on the same protein molecule) and the large normal Bohr effect are eliminated, as in mammalian haemoglobin, by treatment with –SH blocking reagents such as p-chloromercuribenzoate (PCMB), p-hydroxymercuribenzoate, iodoacetamide and N-ethylmaleimide, or with 6M urea (for review, Manwell, 1964). Again, as with studies on mammalian haemoglobin, the actions of such –SH blocking reagents on *Octopus* haemocyanin are not identical, the p-hydroxymercuribenzoate reducing the Bohr effect but having less effect than iodoacetamide and N–ethylmaleimide on the oxygen combining centre interactions, particularly when calcium is bound to the haemocyanin. The oxygen dissociation curves of *Octopus* haemocyanin treated with –SH blocking reagents are shown in Fig. 7.31. Curve B indicates the dissociation curve of the unpurified pigment in which some calcium ions remained bound to the haemocyanin and in which the interaction coefficient n, was 2·0. The 50% saturation value (i.e. the ratio of oxygenated to deoxygenated haemocyanin $y/_{100-y}$ = 1·0) is indicated by a horizontal broken line. Now it is clear that the action of p-hydroxymercuribenzoate (Curve C) is to displace the dissociation curve to the right; that is, to eliminate the Bohr effect. But it is important to notice that the slope of the line decreases only slightly from a value of n = 2·0 to n = 1·6. This indicates that interactions between oxygen combining sites are only slightly affected by p-hydroxymercuribenzoate. On the other hand the line obtained after treatment with iodoacetamide and N-ethylmaleimide (Curve A) gives a value for the interaction coefficient, n, of 1·0 so that interactions between oxygen combining sites are completely abolished. But the line is shifted (in this case to the left) only very slightly, the value for p_{50} being 17 mm Hg in the control and 15 mm Hg after treatment with the reagents so that the Bohr effect is almost unaffected. Thus the action of these non-mercurial agents is upon the interactions between the oxygen combining sites rather than upon the Bohr effect. It seems possible, however, that the Bohr effect and the haem-haem interactions may both be the result of configurational changes in the haemoglobin molecule. The effects can be reversed, as in mammalian haemoglobin, by the addition of glutathione in the case of the positive interactions between the oxygen combining sites, and

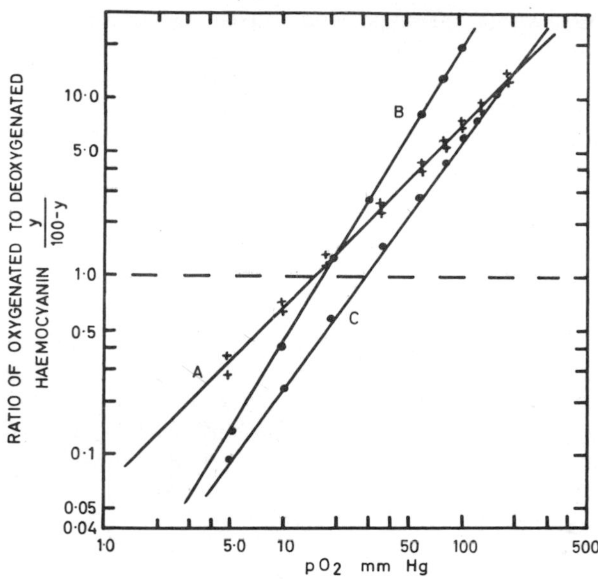

Fig. 7.31. Oxygen dissociation curves of the haemoglobin of *Octopus dofleini* treated with –SH blocking agents. (A) In the presence of iodoacetamide and of N-ethylmaleimide. (B) Control with some calcium ions remaining bound to the protein. (C) In the presence of p-hydroxymercuribenzoate. Note that the value for n in the control (B) is 2·0 whilst the action of the non-mercurial – SH Blocking drugs iodoacetamide and N-ethylmaleimide (A) is to block interaction between oxygen-combining centres so that n=1·0. But they produce little effect on the oxygen affinity, p_{50} being 15 as compared with 17 mm in the control. On the other hand p-hydroxymercuribenzoate (C) has little effect on the interaction between oxygen-combining sites (n=1·6), but the Bohr effect is partially eliminated by the shift to the right which occurs, p_{50} being 27 mm Hg compared with 17 mm in the control. (After Manwell, 1964.)

in the case of the Bohr effect by dialysis. Essentially similar results have been obtained with the chlorocruorin of *Sabella* sp., *Eudistylia vancouveri* and *Schizobranchia insignis* (for review, Manwell, 1964) in which treatment with 6 M urea again appears to induce configurational changes in the respiratory pigment resulting in a reduction of the Bohr effect. Unlike the experiments on mammalian haemoglobin and invertebrate haemocyanin, however, the Bohr effect is not restored by subsequent dialysis.

Manwell (1964) has pointed out that the "active sites" of haemocyanin and haemerythrin are different from those of chlorocruorin and haemoglobin. There are also differences in the size of the polypeptide chains which may suggest that haemocyanin, haemerythrin and haemoglobin/chlorocruorin have originated independently. It also seems possible that haemoglobin and haemocyanin may have a polyphyletic origin (Manwell, 1964). Nevertheless, despite such differences in the possible origin of respiratory pigments, there appears to be a fundamental similarity between the configurational alterations associated with changes in the properties (e.g. the chemical elimination of oxygen combining centre interactions and the Bohr effect) of a wide variety of respiratory pigments. We may therefore now look in more detail at the properties of respiratory pigments other than haemoglobin.

(iii) Chlorocruorin

Chlorocruorin is a green respiratory pigment of restricted occurrence which is found dissolved in the blood of polychaetes belonging to the families Chlorhaemidae, Ampharetidae, Sabellidae and Serpulidae. It is of limited occurrence amongst invertebrates in general, but in those polychaete families where chlorocruorin does occur it is found in most species and is sufficiently concentrated to colour the body green in the Chlorhaemidae and Ampharetidae. Chlorocruorin is found as a blood pigment in all the members of the Chlorhaemidae and in four genera of the Ampharetidae. All members of the Sabellidae have clorocruorin dissolved in the blood except for *Fabricia sabella* which has haemoglobin. Finally, all the Serpulidae have chlorocruorin in the blood except for *Spirorbis militaris* which has no blood pigment, and *S. corrugatus* which has haemoglobin. The genus *Serpula* is most unusual in the possession of both haemoglobin and chlorocruorin in the blood, the former being more common in young animals and being replaced by more chlorocruorin in adults (for review Fox and Vevers, 1960). *Potamilla* is the only example with a tissue pigment, chlorocruorin occurring in the blood and haemoglobin in the muscles.

Chlorocruorin closely resembles haemoglobin in chemical structure, the iron porphyrin which comprises the prosthetic group differing from that of haemoglobin only in the presence of a formyl (–CHO) group instead of the vinyl ($-CH{=}CH_2$) group on one of the four pyrrole rings. As with haemoglobin, one molecule of oxygen combines with one atom of ferrous iron. The prosthetic group has been thought to be identical with that of cytochrome *a* which, like cytochrome *b*, occurs in all aerobic cells, but this is now thought to be untrue. The haem of cytochrome *a* differs from both chlorocruorohaem and haem itself, although it does contain a formyl residue like chlorocruorohaem (for summary Manwell, 1964). The structural formula of the haem unit of chlorocruorin, or chlorocruorohaem as it is called, is shown in Fig. 7.32. The amino-acid composition and molecular weight of *Sabella pavonina* chlorocruorin is very similar to that of the haemoglobins of some other annelids (*c* 3,000,000), but the protein is not a globin. Thus, although chlorocruorin is clearly built on the same structural plan

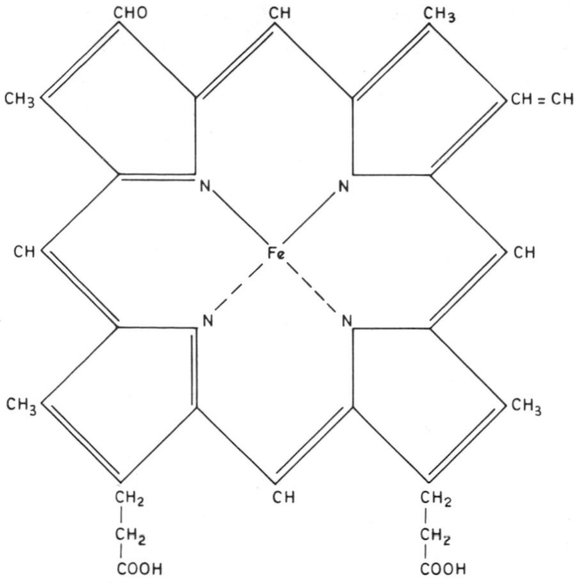

Fig. 7.32. Diagram showing the structural formula of the prosthetic group of chlorocruorin, chlorocruorohaem. Note the replacement of the vinyl ($-CH=CH_2$) group of one of the pyrrole rings of protohaem with a formyl ($-CHO$) group.

as that of haemoglobin, neither the haem nor the protein are the same as in haemoglobin.

As in haemoglobin, chlorocruorin exists in the oxygenated and deoxygenated states, the absorption spectra of the two states being different from each other. In the oxychlorocruorin of *Sabella spallanzanii* there is a strong α absorption band at 604 mμ and a much weaker β band at 558 mμ these bands being displaced towards the red end of the spectrum as compared with those of haemoglobin. Different chlorocruorins differ from one another in the wavelength of the α band. There is also a weak band in the green part of the spectrum at 515 mμ and a strong γ band in the violet range at 430 mμ. On the other hand deoxygenated chlorocruorin, as in haemoglobin, has a broad weak absorption spectrum which does, however, show some indications of the α and β peaks (for review Fox and Vevers, 1960). The oxygen capacity of the blood depends, as in haemoglobin, on the concentration of respiratory pigment; in *Sabella spallanzanii* it is 9·1 ml $0_2/100$ ml blood (vols %) when in equilibrium with air and 10·2 vols % in equilibrium with oxygen at 760 mm Hg. This is the highest value for any invertebrate although it is approached by the 7·3 vols % of *Arenicola* haemoglobin. The oxygen affinity is, however, rather low in the chlorocruorins of *Sabella pavonina, S. spallanzanii* and *Branchiomma* so that the p_{50} is high and this may be a feature of chlorocruorins in general. The implications of the form of such dissociation curves have been discussed on p. 310 and their role in the gas exchange of intertidal animals is discussed in detail on pp. 338–352 (also reviews by Fox and Vevers, 1960; Wolvekamp, 1961; Jones, 1963; Manwell, 1964).

(iv) Haemocyanin

Haemocyanin is a copper-containing respiratory pigment which, when present, is always dissolved in the blood plasma and, like chlorocruorin, is never confined to corpuscles. It is characteristic of cephalopods, gastropods, decapod crustaceans and some spiders but is absent from insects, from entomostracan crustaceans, and probably also from bivalves. Although it is a protein belonging to the globulins, there are no porphyrin structures. The pigment is colourless in the reduced state and blue when combined with molecular oxygen; it has been found that one molecule of oxygen is bound between two atoms of copper in oxyhaemocyanin, in contrast to one atom of iron as in haemoglobin. One further difference is that whereas in haem the iron atom is in the ferrous state in both reduced haemoglobin and in oxyhaemoglobin, in haemocyanin the formation of oxyhaemocyanin is accompanied by the oxidation of some of the copper. In fact, by using 2,2′–bichinoline, which gives a pink colour with cuprous ions, it has been found that one half of the copper is in the cupric state in oxyhaemocyanin. Thus the combination with oxygen involves the oxidation of one of the copper atoms from the cuprous to the cupric state and the oxygenation of the other (for review Wolvekamp, 1961; cf Fox and Vevers, 1960). Although one molecule of oxygen is bound by haemocyanin containing two copper atoms, a large number of copper atoms are present in one molecule of haemocyanin. Indeed, the molecular weight of haemocyanin is very high indeed, being greatest in some molluscan haemocyanins which have sedimentation constants equivalent to molecular weights of approximately 6,500,000; examples are *Buccinum undatum, Littorina littorea, Busycon canaliculatum* and the land snail *Helix pomatia.* In *Octopus vulgaris* and other cephalopods the molecular weight is 2,785,000. In the Crustacea, however, the molecular weights are of the order of several hundred thousand e.g. 397,000 in *Pandalus* sp., 447,000 in the rock lobster *Palinurus vulgaris* and 803,000 in the common lobster *Homarus vulgaris* (Prosser and Brown, 1961).

The axis of the absorption spectrum of oxyhaemocyanin lies in the orange/green region between 600 and 550 mμ according to the species of animal. There is, in addition, a Soret band at 278 mμ due to the presence of cyclic amino-acids, whilst a second band in the ultra-violet region occurs at 350–346 mμ. This second band is due to that part of the oxyhaemocyanin where attachment to oxygen takes place and indicates that the copper is bound to sulphydryl groups (for review Wolvekamp, 1961). In reduced haemocyanin it is replaced by one of increased absorption between 380 and 300 mμ whilst the main band in the visible part of the spectrum at approximately 550 mμ disappears completely.

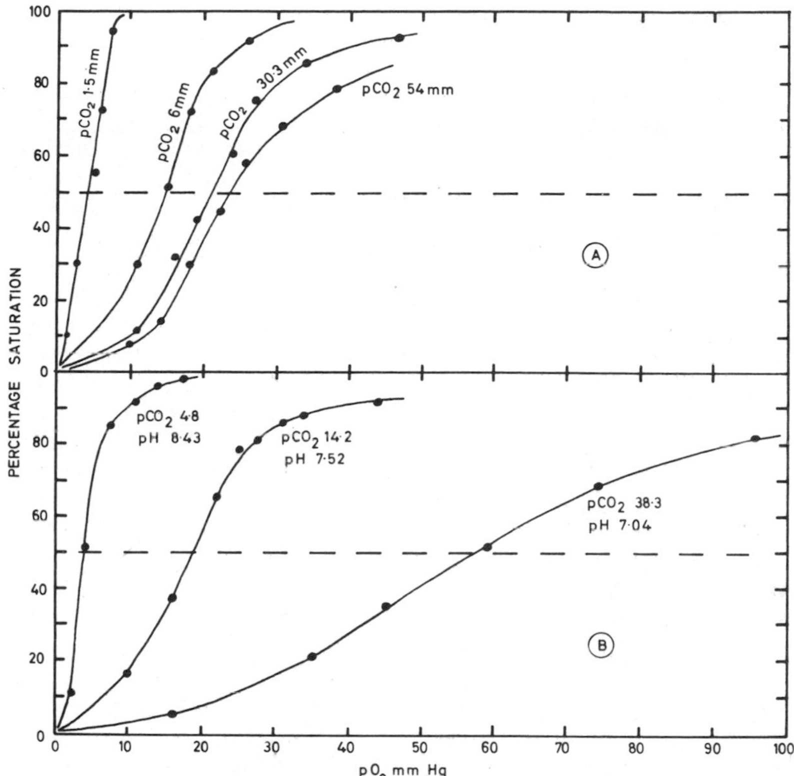

Fig. 7.33. Oxygen dissociation curves for the haemocyanin of (A) *Cancer pagurus* and (B) *Homarus gammarus* at 15°C.
Note the large normal Bohr effect. T=15°C, Cu content 0·060 mg/ml. (After Spoek, 1958 and Baggerman, 1950, both unpublished quoted by Wolvekamp and Waterman, 1960.)

The oxygen capacity of haemocyanin per atom of metal is half that of haemoglobin, and it has been found that in the haemocyanin of the rock lobster *Palinurus vulgaris,* 100 g of haemocyanin is able to bind 25

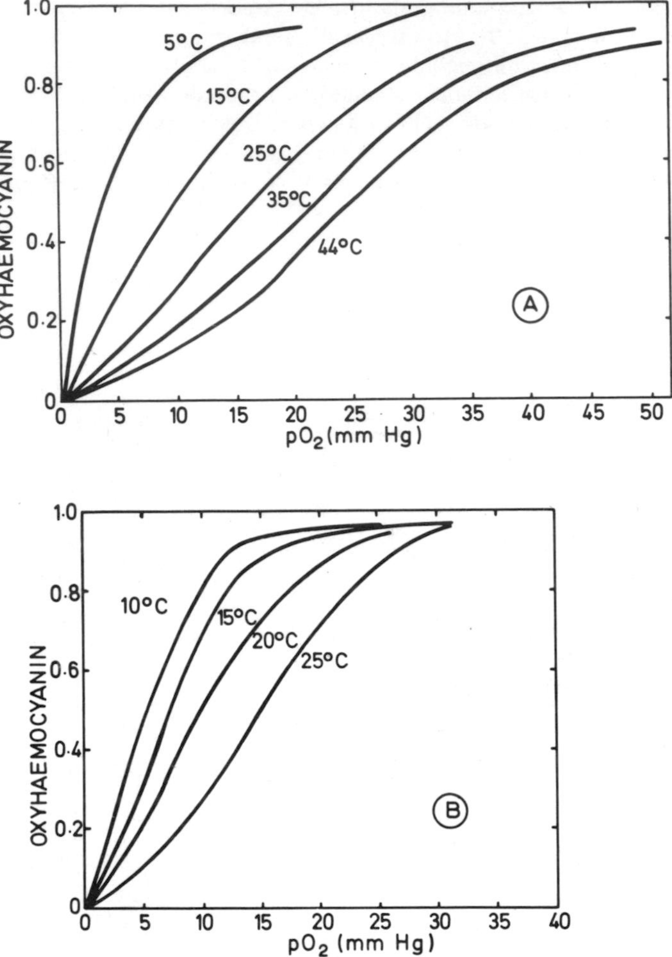

Fig. 7.34. Diagram showing the effect of temperature on the oxygen dissociation curve of the haemocyanin of *Limulus* (A) and of *Panulirus interruptus* (B) at *pH* 7·53. (After Redfield and Ingalls, 1933; Redmond, 1955.)

ml of oxygen compared with the 134 ml of oxygen bound by 100 g of haemoglobin. Of course, as in other blood pigments, the total quantity of oxygen carried by the blood depends upon the concentration of respiratory pigment but, here again, bloods containing haemocyanin are found to be less effective than invertebrate bloods containing haemoglobin or chlorocruorin. The blood of the lobster *Homarus*,, for example, has been found to contain 3·1 ml $0_2/100$ ml blood (vols %)

when in equilibrium with air, and that of the octopus is 4·5 vols %; this is less than half that of *Sabella* blood. By comparison, seawater contains approximately 0·5 vols %.

Despite its low oxygen capacity, haemocyanin shows the normal sigmoid oxygen dissociation curve and in many species such as *Limulus* there is a positive Bohr effect, although the partial pressure of oxygen at which this operates varies considerably from species to species. The edible crab, *Cancer pagurus,* for example, shows a positive Bohr effect at low partial pressures of oxygen whilst in the lobster *Homarus vulgaris* the Bohr effect is marked even at high partial pressures of oxygen (Wolvekamp, 1961). Fig. 7.33 shows the oxygen dissociation curves of the haemocyanin of these two decapods at a variety of partial pressures of carbon dioxide whilst Fig. 7.34 shows a typical temperature effect on the oxygen dissociation curve of the haemocyanin of *Panulirus interruptus* and of *Limulus.* The significance of the differing oxygen dissociation curves of haemocyanin in the gas exchange of intertidal animals is discussed on pp. 347–349.

(v) Haemerythrin

Haemerythrin is a respiratory pigment containing iron but, as in haemocyanin, the prosthetic group is not a porphyrin. The pigment is never found in solution and occurs in the sipunculids *Phascolosoma, Sipunculus* and *Dendrostomum*, in *Priapulus,* and in the brachiopod *Lingula.* In these last two organisms the pigment is confined to coelomic corpuscles, whilst in *Sipunculus* it occurs not only in coelomic corpuscles but also in the epithelium of the gut. Finally, haemerythrin occurs in one polychaete, *Magelona,* and in this case is confined to anucleate blood corpuscles; *Magelona* is thus the only annelid with a respiratory pigment carried in corpuscles in the blood although, as we have seen, some other polychaetes have haemoglobin in corpuscles in the coelom.

In oxyhaemerythrin, one molecule of oxygen is united to three atoms of iron all of which are in the trivalent state, but by using reagents such as 2,2′–bichinoline (or phenanthroline) which stain ions in the bivalent state, it has been found that in reduced haemerythrin two out of every three of the iron atoms are ferrous. Thus one iron atom appears not to be directly involved in the reaction with oxygen. The two active iron atoms are thought to be bound to the sulphur atom of a sulphydryl group whilst the third iron atom may join the two side chains of the protein molecule which contains the iron (for review, Fox and Vevers, 1960; Prosser and Brown, 1961; Wolvekamp, 1961). The bonding of the iron may, however, be unlike that in simple ferrous/ferric complexes and a resonance system appears to explain the properties of the pigment most satisfactorily.

The molecular weight of haemerythrin ranges from 66,000 in *Sipunculus* to 119,400 in *Phascolosoma.* The visible absorption spectrum of oxyhaemerythrin shows none of the sharp absorption bands which characterise porphyrin derivatives, the extinction coefficient of the oxyhaemerythrins of sipunculoids, *Priapulus* and *Lingula* increasing regularly from the red to the blue end of the spectrum with small maxima at 500 mμ, 330 mμ and 280 mμ. Deoxygenated haemerythrin, on the other hand, has no absorption peak in the visible range. The oxygen dissociation curve of sipunculoid and brachiopod haemerythrins is hyperbolic. There is no obvious Bohr effect in *Phascolosoma, Dendrostomum* and *Sipunculus,* but in the brachiopod *Lingula,* in contrast to the sipunculoid haemerythrins, a positive Bohr effect occurs. Examples of such dissociation curves for haemerythrin are shown in figures 7.42 and 43 and their significance in the life of the animals is discussed on pp. 342–343.

(c) The transport of carbon dioxide

We have seen that the blood of a wide variety of intertidal organisms carries a great deal more oxygen than the amount in simple solution. This increase in oxygen-carrying capacity is necessary in organisms whose respiratory demands are high. Much the same argument applies to the transport of carbon dioxide since this will be produced in greater quantities by tissues of active organisms than by those of inactive ones. Now the solubility coefficient of carbon dioxide in seawater at, for example, 24°C is 0·71 vols % and when in equilibrium with air which has a partial pressure of carbon dioxide of 0·23 mm Hg, the amount of carbon dioxide dissolved in seawater would be only 0·0215 vols % even though carbon dioxide is approximately thirty times more soluble in water than oxygen. In fact seawater contains as much as 4·8 ml CO_2/100 ml, this great increase in the carbon dioxide content being due to the fact that the CO_2 combines with the excess cations from the various buffers in seawater which displaces reaction (1) to the right causing the further formation of carbonic acid. We may represent these reactions as follows and their nature is similar to those which occur between carbonic acid and the inorganic buffering agents of the plasma.

$$CO_2 + H_2O \rightleftharpoons \underset{\text{carbonic acid}}{H_2CO_3} \quad (1)$$

$$H_2CO_3 + Na_2HPO_4 \rightleftharpoons NaH_2PO_4 + NaHCO_3 \quad (2)$$

Because of this increase in the capacity of buffered solutions such as seawater for carbon dioxide, it is not surprising to find that in some sedentary and relatively inactive animals, such as the sea hare *Aplysia* and certain ascidians, the carbon dioxide capacity of the blood does not exceed that of seawater, this capacity being apparently sufficient to meet the metabolic output of carbon dioxide. Indeed, in some ascidians the supply of cations in the plasma is so low that the total carbon dioxide capacity is lower than that of seawater (Florkin, 1934). This may suggest that much of the carbon dioxide produced by the tissues diffuses directly out into the surrounding seawater so that a plasma with a high capacity for this gas is unnecessary. The carbon dioxide dissociation curves for *Aplysia* blood and seawater are shown in Fig. 7.35. In more active animals in which a greater carbon dioxide capacity

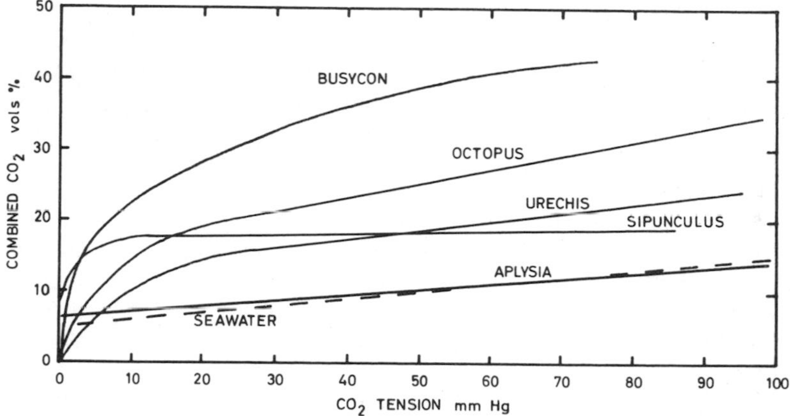

Fig. 7.35. Carbon dioxide dissociation curves of the bloods of various marine invertebrates compared with that of seawater (broken line). (Based on Florkin, 1934.)

is achieved, the blood must have two main properties. Firstly there must be no great reduction in the *pH* with the increased carbon dioxide content, and secondly the dissociation curve must be of such a shape that the system will load and unload at the partial pressures of carbon dioxide occurring at the tissues and respiratory surface respectively.

The most important agents which prevent excessive changes in the *pH* of the blood with varying carbon dioxide content are the blood proteins (Duval and Portier, 1927). All blood proteins, whether they are respiratory pigments or not, have a buffering capacity, but since the respiratory pigments form the dominant blood protein in the plasma of most invertebrates, they are of more importance as buffers than the others. The respiratory pigment such as haemoglobin or haemocyanin is in the form of its potassium salt, potassium haemoglobinate, and reacts with carbonic acid to form potassium bicarbonate according to equation (3)

$$\begin{array}{lcl} CO_2 + H_2O & \rightleftharpoons & H_2CO_3 \\ K\,Hb + H_2CO_3 & \rightleftharpoons & KHCO_3 + HHb \\ \text{Potassium haemoglobinate} & & \text{Potassium bicarbonate} \end{array} \quad (3)$$

Respiratory pigments are the only blood proteins in some marine invertebrates such as *Urechis,* in which the plasma has little buffering capacity, but the corpuscles are comparable with those of vertebrates for buffering power. In echinoids, whose coelomic fluid has little protein, there is correspondingly little buffering power (Pantin, 1932). The buffering ability of the blood proteins is due to the *pH* of the blood being on the alkaline side of the isoelectric point of the respiratory pigment. Thus the respiratory pigments combine with alkali which can be taken up by weak acids (such as the carbonic acid formed by solution of carbon dioxide into the blood plasma) in accordance with the reactions shown above. The haemocyanin in the blood of *Limulus* is efficient in this respect and allows a large carbon dioxide capacity in this animal with relatively little change in the *pH*; the total capacity is related to the total nitrogen content of the blood which in turn represents the haemocyanin concentration. Such buffering capacity is best expressed in terms of a titration curve in which the total amount of acid or alkali added to the blood is plotted against *pH*. Thus in *Limulus* the curve is steep, indicating a small *pH* change for a large flux in alkali or acid ($\equiv CO_2$), whilst in *Echinus* the curve is shallow. The importance of such buffering power in increasing the total CO_2 capacity of the blood is illustrated in Fig. 7.35, in which the total carbon dioxide content of the blood of *Urechis* is compared with that of seawater and of *Aplysia* blood. It is clear that the presence of blood proteins in *Urechis* allows a CO_2 content of approximately double that of *Aplysia* blood and seawater at partial pressures in excess of 10 mm Hg. Curves for *Sipunculus, Octopus* and *Busycon,* are also shown in the graph.

As well as the limited buffering capacity of the inorganic salts of the plasma (Equation (2), p. 330), and the considerable buffering powers of the blood proteins (Equation (3)), both of which increase the carbon dioxide capacity of the blood, carbon dioxide can be transported by direct combination with respiratory pigment. Small amounts of carbonic acid may be bound to haemoglobin to form haemoglobin bicarbonate, whilst CO_2 itself combines directly with amino groups on the protein molecule to form a carbamino compound or carbamate:–

$$\text{Pr–N}\begin{array}{l}\diagup H\\ \diagdown H\end{array} + CO_2 \longrightarrow \text{Pr–N}\begin{array}{l}\diagup H\\ \diagdown COO^-H^+\end{array} \quad (4)$$

Carbamino compound

Although such mechanisms of carbon dioxide transport may account for as much as one fifth of the CO_2 content of the blood of some vertebrates, carbon dioxide transport in invertebrates is normally mainly in the form of bicarbonate accumulation made possible by the buffering capacity of inorganic buffering agents in the plasma and, above all, by the blood proteins.

Apart from increasing the total carbon dioxide content of the respiratory fluid without causing a great reduction of the *pH* value, the respiratory pigments must have a CO_2 dissociation curve which loads and unloads at the appropriate tension of the tissues and the respiratory surface. It will be noticed in, for example, the curve for *Sipunculus* haemerythrin that over a wide range of partial pressures (50 mm Hg to 10 mm Hg) the total CO_2 content is almost constant. This would be a most unsuitable form of curve if the partial pressure of carbon dioxide of the arteries and veins was high, as it is in lung-breathing animals, since at the respiratory surface the pigment would not automatically unload its carbon dioxide. In fact the pCO_2 of the bloods of marine organisms is always low, that of the arteries being in equilibrium with the low partial pressures of carbon dioxide occurring in seawater (approximately 0·23 mm Hg, p. 330). It is this region which is the steepest part of the CO_2 dissociation curve of the pigments of most marine organisms and allows the pigment to unload much of its carbon dioxide despite the small differences in the pCO_2 between the arteries and veins. In lung-breathing animals the alveolar pCO_2 is high, so too is that of the arteries, being as much as 40 mm Hg or more. Thus the dissociation curves for the respiratory pigments of these organisms tend to be steep even towards high partial pressures of carbon dioxide.

Now just as the oxygen transporting function of many haemoglobins is facilitated by the Bohr effect which enhances the liberation of oxygen at the tissues and its uptake at the respiratory surface, so this phenomenon has its counterpart in carbon dioxide transport, for oxygenated blood has a lower affinity for carbon dioxide than the deoxygenated pigment. This is known as the Haldane effect, and although it was described first for mammalian blood, it is now known to occur in some invertebrate haemoglobins and haemocyanins as well as possibly in other respiratory pigments. The principal effect is that the CO_2 dissociation curve for fully oxygenated pigment lies to the right of the deoxygenated pigment. The importance of this factor may be illustrated by reference to the idealised CO_2 dissociation curve illustrated in Fig. 7.36. Let us suppose that the pCO_2 of the tissues is that indicated by T and that of the respiratory surface by R. Then with oxygenated blood the amount of CO_2 which would be released at the respiratory surface is T (= 54 vols %) minus R (= 49 vols %) = 5 vols %. But the blood at the tissues is in the deoxygenated state, and owing to

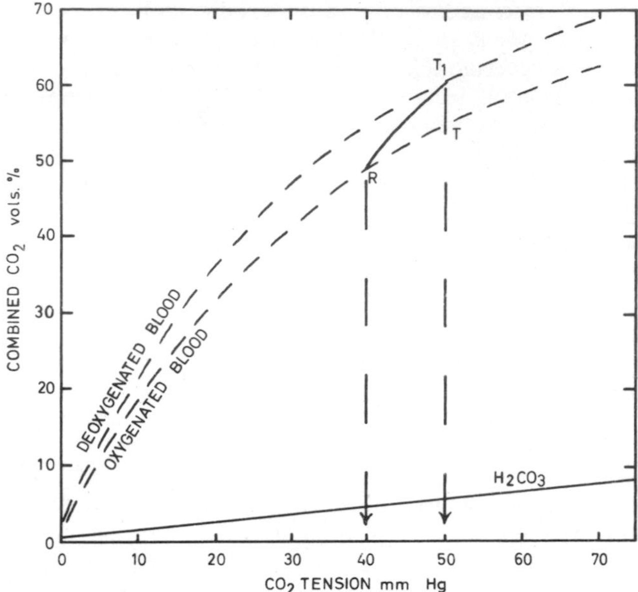

Fig. 7.36. Diagram illustrating the role of the Haldane effect in increasing the turnover of CO_2 from the tissues to the respiratory surface. The dissociation curve for H_2CO_3 at 20°C is shown by the line at the bottom of the figure. (Based on Jones, 1963.)

the Haldane effect its CO_2 affinity is raised and the carbon dioxide dissociation curve is displaced to the left. Thus the CO_2 content of the deoxygenated pigment at the tissues is given by T_1 (= 60 vols %), and by the time it reaches the oxygenated state at the respiratory surface the curve has become displaced to the right. Thus the carbon dioxide given up at the respiratory surface is $T_1 - R = 11$ vols % rather than the 5 vols % which would have been liberated in the absence of the Haldane effect.

This Haldane effect, like the Bohr effect, is related to the acidity of oxyhaemoglobin compared with the alkaline deoxygenated pigment (Wolvekamp and Waterman, 1960; Bard, 1961). The same is probably true of other respiratory pigments such as haemocyanin. When the titration curves of oxyhaemoglobin and deoxygenated haemoglobin are compared, it is found that they are parallel with the more acid oxyhaemoglobin binding 0·7 equivalents more base than the deoxygenated haemoglobin. The two titration curves are shown in Fig. 7.37., the horizontal distance between them indicating the difference in acidity between oxyhaemoglobin and deoxygenated haemoglobin. The vertical difference at any one *pH* indicates the buffering capacity or

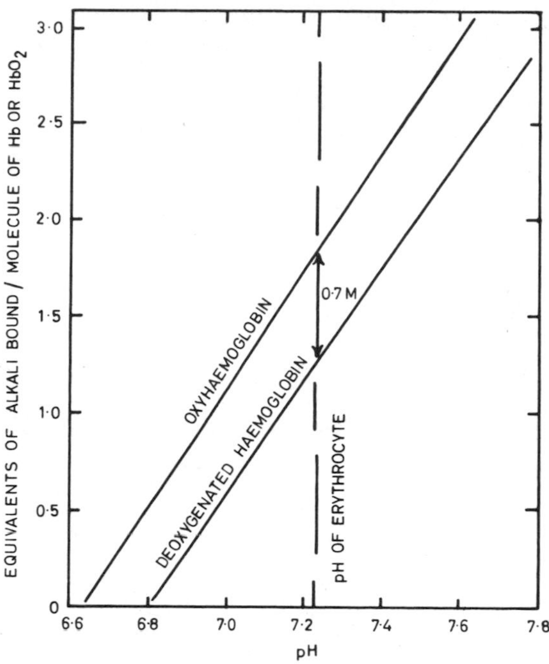

Fig. 7.37. Titration curves of oxyhaemoglobin and deoxygenated haemoglobin (Based on Bard, 1961 and Hoar, 1966.) Note that the horizontal distance between the two lines is the *pH* difference between oxyhaemoglobin and deoxygenated haemoglobin. Oxyhaemoglobin binds 0·7 equivalents more base than does deoxyhaemoglobin, thus when one mole of oxygen is released from oxyhaemoglobin at the tissues 0·7 M of H^+ ions formed from the carbonic acid produced by the tissues can be neutralised by the deoxygenated haemoglobin without altering the *pH*.

number of equivalents of alkali which can be added or subtracted without a change in *pH*. The importance of these curves is that they show that when one mole of oxygen is released from oxyhaemoglobin, 0·7 M of H^+ ions can be neutralised by the deoxygenated haemoglobin without altering the *pH* from that of the blood saturated with the more acid oxyhaemoglobin. Thus the discharge of oxygen in the tissues makes the haemoglobin more alkaline, and allows the combination of more carbonic acid with the respiratory pigment.

Carbon dioxide transport is achieved in a number of ways in vertebrates, the process being complicated by the presence of blood corpuscles which contain the respiratory pigment. Firstly, carbon dioxide diffuses into the plasma and a little carbonic acid is formed, although the amount is small because the reaction is slow:—

$$CO_2 + H_2O \underset{}{\overset{\text{slow}}{\rightleftharpoons}} H_2CO_3 \rightleftharpoons H^+ + HCO_3^-$$

More important, the carbon dioxide diffuses into the erythrocytes where there is a zinc-containing enzyme called carbonic anhydrase. This enzyme facilitates the first part of the reaction outlined above and the carbonic acid so formed dissociates into H^+ and HCO_3 ions. Since the respiratory pigment may be regarded as being present in the form of its potassium salt, potassium haemoglobinate (p. 332), excess H^+ ions formed by the dissociation of the carbonic acid do not result in a fall of the *pH* of the contents of the erythrocyte.

$$KHb + H^+ + HCO_3^- \rightleftharpoons HHb + K^+ + HCO_3^-$$

This absorption of the excess H^+ ions displaces the reaction to the right so that further CO_2 is absorbed. We may thus write the overall reaction as follows:–

$$CO_2 + H_2O \underset{\text{FAST}}{\overset{\text{carbonic anhydrase}}{\rightleftharpoons}} H_2CO_3 \rightleftharpoons (H^+) + HCO_3^-$$

Clearly, an accumulation of the negatively charged HCO_3^- ions inside the erythrocyte would upset the chemical and electrical equilibrium (Donnan equilibrium) across the cell membrane. There is thus an outward movement of bicarbonate ions into the plasma and a corresponding inflow of negatively charged chloride ions into the erythrocyte to maintain the electrical equilibrium. The plasma thus becomes enriched in sodium bicarbonate whilst the erythrocyte contains chloride together with its original potassium ions. This compensating inflow of chloride ions is called the chloride or Hamburger shift, the removal of bicarbonate ions from the corpuscle further increasing the displacement of the above reactions to the right and enhancing the further combination of carbon dioxide with water under the influence of carbonic anhydrase (Fig. 7.38.).

Carbon dioxide also reacts directly with amino groups on the protein chains to form carbamino-compounds (p. 332) which normally account for less than 10% of the CO_2 transported, though in some vertebrates as much as one fifth of the total CO_2 content may be represented by carbamino compounds. The reaction of the carbon dioxide with the protein is reversible, so that the carbamino compound dissociates liberating CO_2 at the respiratory surface. The transport of carbon dioxide both as the carbamino compound and as the bicarbonate is

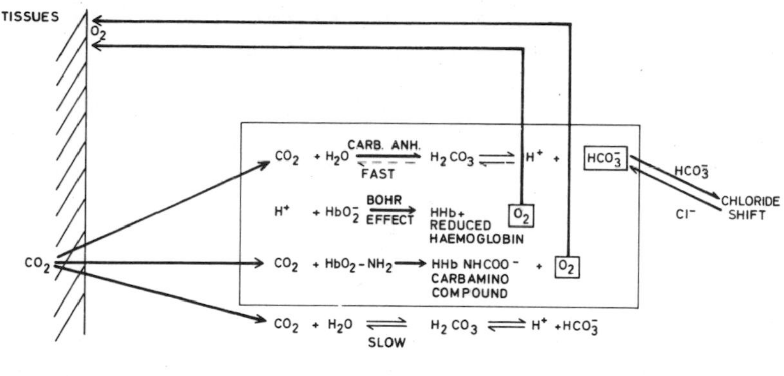

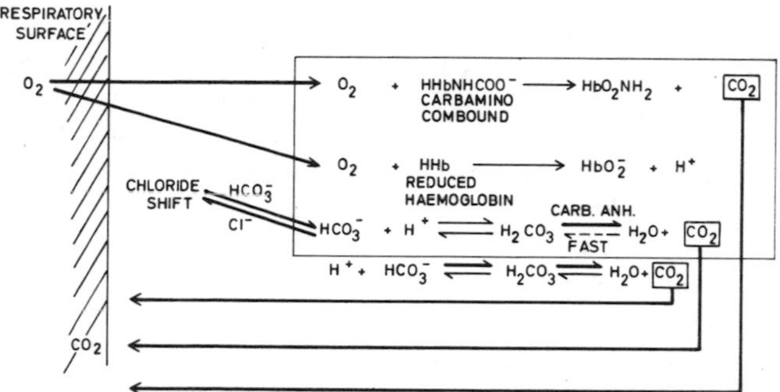

Fig. 7.38. Diagram illustrating the reactions involved in gas transport in vertebrates. The upper part shows the reactions occurring at the tissues and the lower part those occurring at the respiratory surface. (Based on Hughes, 1963.)

greatly enhanced by the Haldane effect in which the CO_2 capacity of deoxygenated haemoglobin is higher than that of oxyhaemoglobin.

We may now review the situation when the venous blood of a vertebrate reaches the respiratory surface. Firstly the oxygen combines with the reduced haemoglobin to form oxyhaemoglobin which is more acid than in the reduced state (Fig. 7.37).

$$HHb + O_2 \longrightarrow HbO_2^- + H^+$$

The H^+ ions combine with the HCO_3^- ions and the reaction being displaced to the right by the removal of carbon dioxide and catalysed by carbonic anhydrase.

$$H^+ + HCO_3^- \rightleftharpoons H_2CO_3 \xrightarrow[\leftarrow\text{-----}]{\text{carbonic anhydrase}} H_2O + CO_2$$

Further HCO_3^- ions diffuse into the erythrocyte from the plasma as CO_2 is evolved and there is a corresponding outflow of Cl^- ions to maintain the Donnan equilibrium.

The presence of carbonic anhydrase in the coelomic fluid of *Sipunculus* and *Arenicola* and in the gills of many invertebrates is significant, although it is normally absent from the blood. It seems likely that particularly in the gills its role is, as in vertebrate erythrocytes, to facilitate the formation and breakdown of carbonic acid in much the same way as has been described above for vertebrates. The sequences of reactions occurring at the tissues and at the respiratory surface are illustrated in Fig. 7.38.

(d) The role of respiratory pigments in gas exchange

Respiratory pigments, as we have seen, show a wide variety of properties; these may often be related to the environmental conditions in which the particular organisms live, although in many instances the role of the pigments is uncertain. Nevertheless, there is sufficient information to suggest the probable part respiratory pigments play in the gas exchange of a number of intertidal organisms. Examples of such pigments are given below whilst those of other invertebrates are extensively reviewed by Prosser and Brown (1961) and Jones (1963).

The first and most obvious function is that of an oxygen transport system operating at both high and low partial pressures of oxygen. In this case the function of the pigment is to increase the total oxygen capacity of the blood, and no special adaptation is shown to reduced partial pressures of oxygen. As would be expected, animals with such respiratory pigments are only rarely exposed by the tide or, if they are exposed, maintain contact with aerated seawater or with the air itself. Into this category we may include the chlorocruorin of the sabellid *Sabella pavonina* which has been studied by Ewer and Fox (1940), and that of two other sabellids, *Sabella spallanzanii* and *Branchiomma vesiculosum* (Fox, 1932). The chlorocruorins from all of these animals show a characteristically low oxygen affinity, although the oxygen capacity of the blood of *S. spallanzanii* is as high as 9·1 vols % in equilibrium with air and 10·2 vols % at a pO_2 of 760 mm. The oxygen dissociation curve of *S. spallanzanii* chlorocruorin is shown in Fig. 7.39. All the chlorocruorins mentioned show a positive Bohr effect and would appear to be capable of loading and unloading under normal physiological conditions, and thus functioning as respiratory pigments although the pigment does not become fully saturated even at high partial pressures of oxygen. There is no information on the partial pressures of oxygen in the blood at the gills and at the tissues so that the extent to which the pigment unloads can only be inferred by indirect methods. One of such methods has been used extensively by Fox and

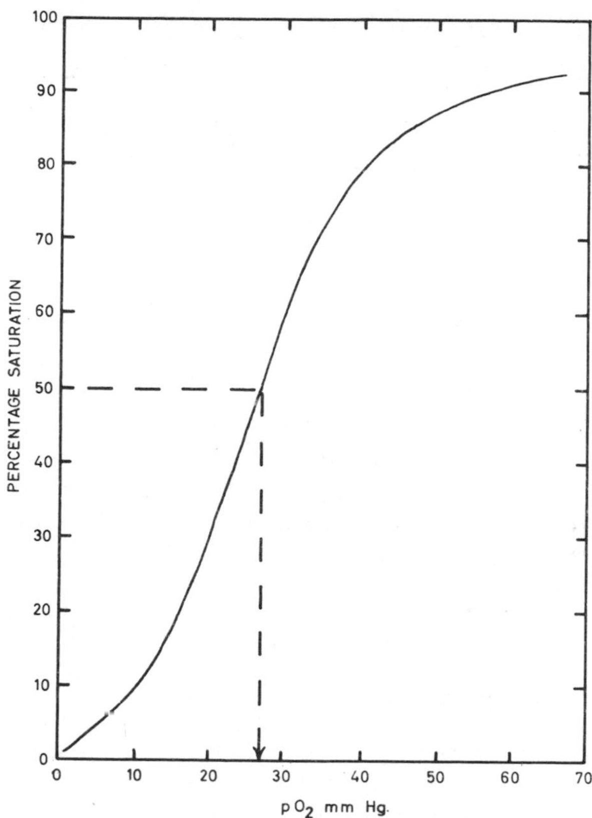

Fig. 7.39. Oxygen dissociation curve of the chlorocruorin of *Sabella spallanzanii* at 20°C and *pH* 7·7. (After Fox, 1932.)

others and involves the exposure of the animal to carbon monoxide of sufficient concentration to saturate the respiratory pigment with gas and so prevent its participation in gas transport, but not so as to disrupt cellular respiratory enzymes. The difference between the respiratory rate of animals with and without functional respiratory pigment over a wide range of partial pressures of oxygen then gives an indication of the role of the respiratory pigment in gas transport. Carbon monoxide is mixed with the oxygen to keep the pigments inactive and its continued saturation with carbon monoxide is checked with a spectroscope at the end of the experiment. The results of such an experiment on *Sabella pavonina* are shown in Fig. 7.40; untreated animals were used as controls.

It will be noticed that the oxygen consumption fell with decreasing

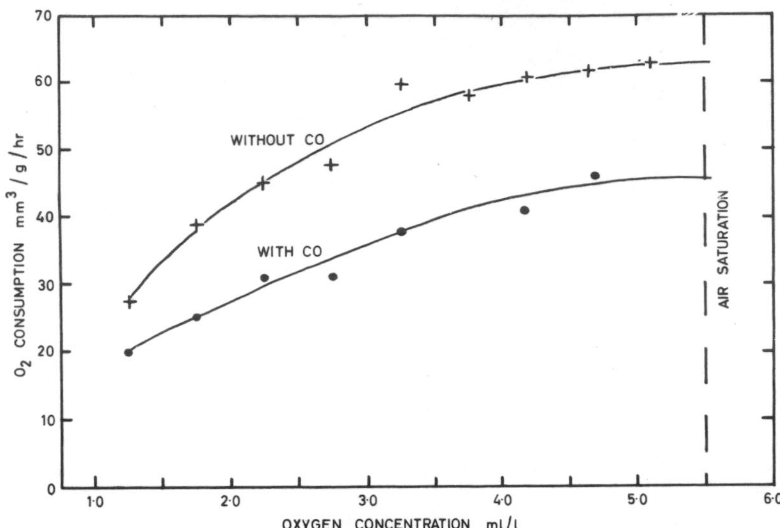

Fig. 7.40. The oxygen consumption of *Sabella pavonina* in seawater without carbon monoxide (above) and when the chlorocruorin had been converted into carboxychlorocruorin by treatment with carbon monoxide (After Ewer and Fox, 1940.)

partial pressures of oxygen, and that approximately one third of the oxygen uptake of untreated animals is carried by the respiratory pigment. As Jones (1963) has pointed out, this amount is rather low if we assume that the total capacity of the blood is comparable to that of *S. spallanzanii,* for instead of an increase of some 30–40% above the treated animals, oxygen transport in intact animals would be expected to be as much as 18 times that of animals treated with carbon monoxide. There is thus a serious discrepancy between such indirect measurements and the contribution calculated on a basis of the oxygen capacity and oxygen affinity of the blood. Such problems may be overcome by the use of micro oxygen electrodes to determine the pO_2 of the tissues and respiratory surface; this would enable an estimate to be made of the total amount of oxygen turned over to the tissues. It seems safe to assume that the chlorocruorin of these sabellids is adapted to function in well-aerated seawater under conditions of almost continuous irrigation, and these are known to occur (p. 298).

Another polychaete which has a respiratory pigment which may function as an oxygen-transport system operating at high partial pressures of oxygen is *Nephtys hombergi.* This animal lives in impermanent burrows in muddy sand through which oxygenated water is drawn by ciliary activity when the sand is submersed by the tide. Under these

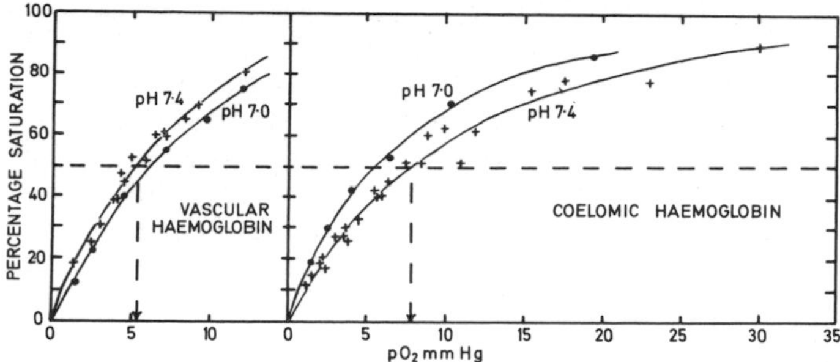

Fig. 7.41. Oxygen dissociation curves of the vascular and coelomic haemoglobins of *Nephtys hombergi* at *pH* of 7·4 (crosses) and 7·0 (dots) at 15°C. Note that although the curves are comparable at a *pH* of 7·0, because of the reversed Bohr effect in the coelomic haemoglobin and the normal Bohr effect in the vascular haemoglobin, at the normal *pH* of 7·4 oxygen will be transferred from the coelomic haemoglobin (p_{50} 7·5 mm) to the vascular haemoglobin (p_{50} 5·5 mm) (After Jones, 1954.)

conditions the pigment would be expected to serve as an oxygen transport system capable of loading and unloading at high partial pressures of oxygen. In fact there is haemoglobin in the vascular system and in the coelom, both of these pigments having a relatively low oxygen affinity. But the coelomic haemoglobin differs from the vascular pigment in having (a) a reversed Bohr effect and (b) a lower oxygen affinity at the normal *in vivo pH* of 7·4. The dissociation curves of the vascular and coelomic haemoglobin of *Nephtys hombergi* are shown in Fig. 7.41. It seems possible, therefore, that oxygen is transferred from the coelom to the blood by means of the different oxygen affinities of the haemoglobins from these two regions rather as in the vascular and tissue pigments of a number of other animals (see p. 342).

Although a pigment with a low oxygen affinity could function at high partial pressures of oxygen, as in the sabellids mentioned above, when the tide ebbs the burrow of *Nephtys* collapses and the oxygen tension of the interstitial water falls rapidly to a partial pressure of approximately 6 mm Hg. Under such conditions it might be supposed that the pigment could be used as an oxygen store, but Jones (1954) has shown that the total oxygen-combining potential amounts to only some 2·5 – 4·0 μl O_2/gm of worm whilst the lowest level of respiration was 20 μl/gm wet wt/hr. This means that the combined oxygen could last only some 7 – 12 minutes of intertidal exposure. It seems likely, as Jones (1954, 1963) points out, that the abrupt change from high partial pressures of oxygen to low ones offers no possibility for transport at reduced partial pressures of oxygen, and that the animal therefore resorts to anaerobiosis during the intertidal period. Although no accumulation of lactic acid was detected, it is possible that other acid end products of glycolysis may have been excreted as is known to occur in *Arenicola* (p. 284). Other polychaetes such as the terebellid *Eupolymnia,* which possesses haemoglobin, have pigments adapted to transport at high partial pressures of oxygen (Manwell, 1959). In *Eupolymnia* the p_{50} at 10°C was 36 mm and there was no Bohr effect between *pH* 7·2 and 7·7. As in the haemoglobin of *Nephtys hombergi,* the dissociation curve is hyperbolic, not sigmoid, the value for *n* being close to 1·0, whilst the pigment is only 85% saturated at the atmospheric partial pressure of oxygen.

We have seen that in *Nephtys hombergi* there is some indication from the form of the oxygen dissociation curves that oxygen transfer from the coelomic to the vascular haemoglobin may take place, since the vascular haemoglobin has a higher affinity for oxygen than the coelomic haemoglobin. This difference is enhanced by the reversed Bohr effect of the coelomic haemoglobin. This phenomenon is similar to the relationship between the vascular haemoglobin and the myoglobin of the muscles of vertebrates and has been established for a number of other invertebrate respiratory pigments. Manwell (1960b), for example, has shown that an oxygen transfer system of this kind operates in the haemerythrins of the sipunculids *Dendrostomum zostericolum* and *Siphonosoma ingens,* although the details of the mechanism in these two genera are different.

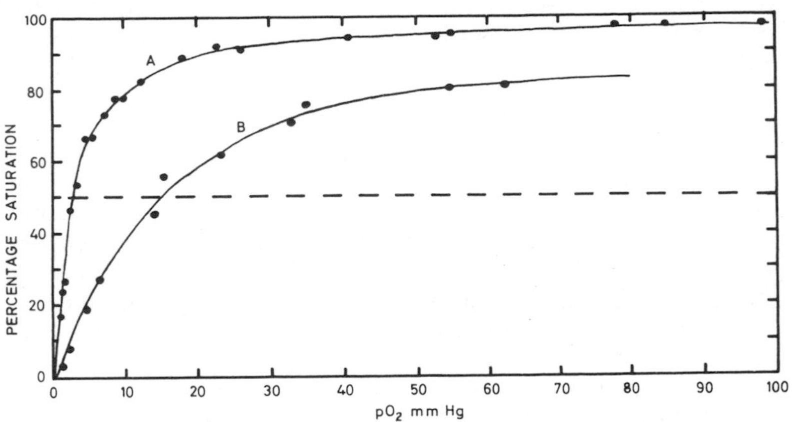

Fig. 7.42. Oxygen dissociation curves of the coelomic (A) and vascular (B) haemerythrins of the sipunculoid *Dendrostomum zostericolum.* The points on each curve were obtained at 20°C at a variety of *pH* values, there being no Bohr effect. Note that transfer of oxygen would take place from the vascular (curve B) to the coelomic (curve A) pigment. Gas exchange in *Dendrostomum* is through the tentacles hence oxygen combines first with the vascular haemerythrin and then passes to the coelomic haemerythrin. (After Manwell, 1960b.)

The oxygen dissociation curves for the vascular and coelomic haemerythrins of *Dendrostomum zostericolum* are shown in Fig. 7.42. In this animal the tentacles are the respiratory organs, oxygen diffusing through and combining with the vascular haemerythrin. The vascular haemerythrin is conveyed into thin walled blood-filled diverticula which intrude into the coelom and through which the oxygen passes to the coelomic haemerythrin. This transfer is facilitated by the high affinity of the coelomic haemerythrin compared with the vascular pigment. The relationship between the coelomic and vascular haemerythrin is quite the reverse in *Siphonosoma ingens.* Gas exchange in this animal is through the general body surface so that oxygen passes through the body wall into the coelom where it combines with the coelomic haemerythrin. The vascular pigment has a higher affinity for oxygen than the coelomic pigment (as in *Nephtys hombergi*), and the oxygen thus passes from the coelom into the vascular system (Fig. 7.43).

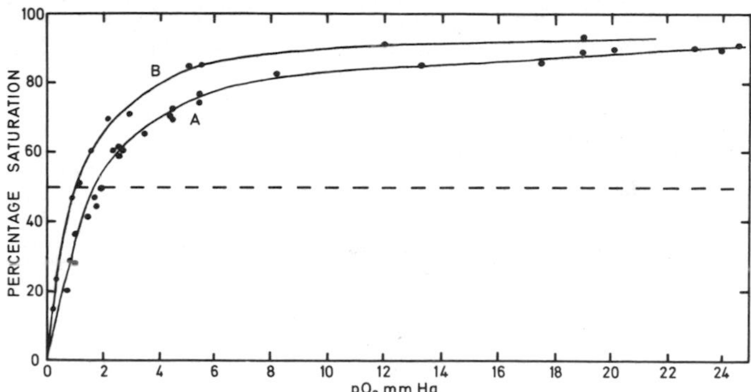

Fig. 7.43. Oxygen dissociation curves of the coelomic (A) and vascular (B) haemerythrins of *Siphonosoma ingens* at 18°C. The points on each curve were obtained at a variety of *pH* values and there was no Bohr effect.
Note that the transfer of oxygen in this animal would take place from the coelomic (curve A) to the vascular (curve B) haemerythrin (cf Fig. 7.42). Gas exchange in *Siphonosoma* is through the general body surface and oxygen combines first with the coelomic haemerythrin, being subsequently transferred to the vascular haemerythrin. (After Manwell, 1960b.)

A closer resemblance to the vertebrate oxygen transfer system between vascular haemoglobin and myoglobin in the muscles is seen in some chitons (Giese, 1952; Manwell, 1958, 1960b) and in the gastropod *Busycon.* In *Cryptochiton stelleri, Ischnochiton* and *Busycon*, the blood pigment is haemocyanin (p. 326) which shows a low oxygen affinity, the p_{50} being approximately 10 mm Hg, and rather weak Bohr effect. The oxygen dissociation curve of the haemocyanin of *Ischnochiton* is shown as both a linear plot and as a double logarithmic

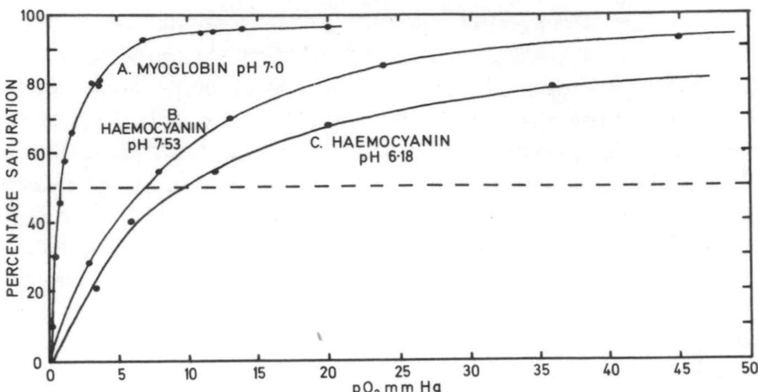

Fig. 7.44. Oxygen dissociation curves of the radular muscle myoglobin (curve A) and of the vascular haemocyanin at a *pH* of 7·53 (curve B) and 6·18 (curve C) at 20°C from the chiton *Ischnochiton*. Note the rather weak Bohr effect in the haemocyanin and the fact that transfer of oxygen from the blood to the radular muscles would occur (based on Prosser and Brown, 1961 after Manwell, unpublished).

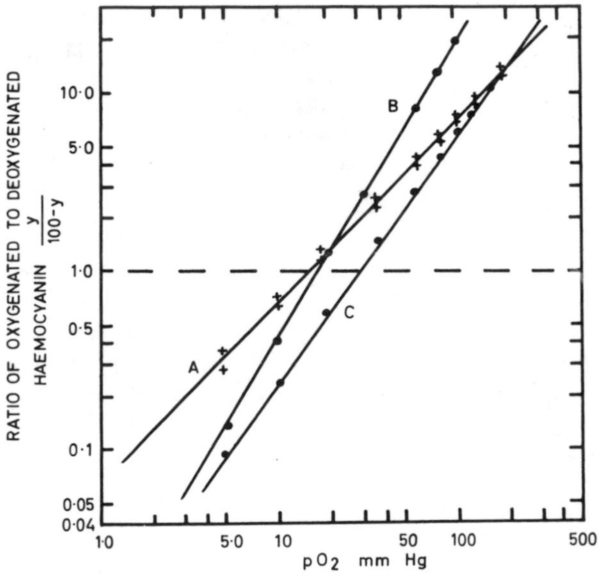

Fig. 7.45. Oxygen dissociation curves for the myoglobin and haemocyanin of *Ischnochiton*. Note that the value of the interaction co-efficient n of the haemocyanin decreases with low *pH* values. (Data as in Fig. 7.44.)

plot in Fig. 7.44 and 45 from which it will be noticed that the value for the interaction coefficient *n* becomes lower with reduced *pH* values. The radular muscles, on the other hand, contain haemoglobin (myoglobin) with a high oxygen affinity (p_{50} approximately 1·0 mm Hg) so that the transfer of oxygen from the haemocyanin of the blood to the radular muscles occurs (Manwell, 1960b). Again, in *Cryptochiton stelleri* an extract of the tissue pigment had a p_{50} of 2·5 mm Hg compared with 18 mm for the haemocyanin of the undiluted blood (Manwell, 1958). Hemmingsen (1962) has recently demonstrated that the presence of haemoglobin greatly enhances the passage of oxygen across a Millepore filter filled with haemoglobin. This enhancement of oxygen diffusion is known as specific oxygen transport and may play an important part in increasing the transport of oxygen through massive tissues such as the radular muscles of the chitons referred to above, as well as to tissues such as nervous systems which are extremely sensitive to oxygen lack and which also often contain respiratory pigments p. 305).

Under some circumstances, as we have seen on p. 284, intertidal animals are subjected to prolonged periods of reduced oxygen availability. Animals such as the lugworm *Arenicola marina,* for example, no longer pump oxygenated water through the burrow when the mud flats are exposed by the tide, and although aerial respiration may take place under some circumstances (Wells, 1949a), the oxygen available to the animal is normally that dissolved in the water of the burrow plus that diffusing into the burrow from the more superficial layers of the deposit. A respiratory pigment with a high affinity for oxygen, and one which is relatively unaffected by the low *pH* values which commonly occur in deoxygenated deposits, would aid oxygen uptake during this period of the tidal cycle; in fact the haemoglobin of *Arenicola* has been shown to possess such properties (Barcroft and Barcroft, 1924; Borden, 1931; Wolvekamp and Vreede, 1941; Jones, 1954; Manwell in Prosser and Brown, 1961). Wolvekamp and Vreede (1941) found that in undiluted blood the p_{50} was 2 mm in the absence of CO_2 and was only 8·3 mm in the presence of the very high pCO_2 of 120 mm. In addition, there appeared to be a small difference in the oxygen affinity of the blood of animals sampled in the winter (p_{50} 1·6 mm) and in the summer (p_{50} 2 mm). The oxygen dissociation curve of *Arenicola* haemoglobin at a variety of *pH* values between 6·8 and 5·4 is shown in Fig. 7.46.

A storage function has sometimes been attributed to the haemoglobin of *Arenicola* (for example, Barcroft and Barcroft, 1924; Borden, 1931; Barcroft, 1934; also Wolvekamp and Vreede, 1941; Jones, 1963) and for this reason the total oxygen capacity of the blood is of interest. Values of 2·0 – 6·2 volumes % have been obtained by

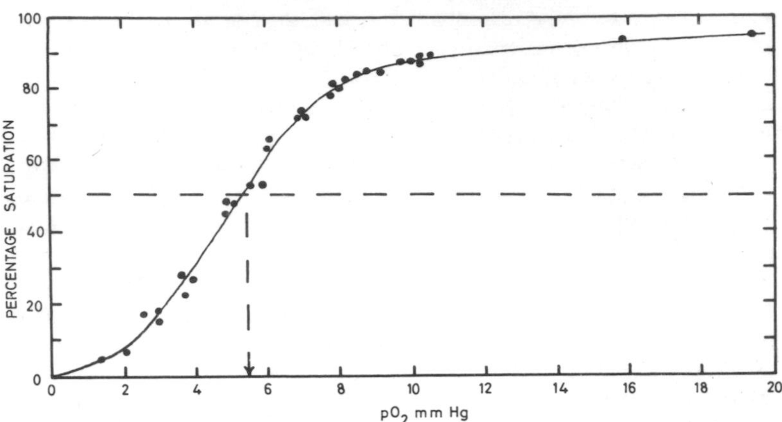

Fig. 7.46. Oxygen dissociation curve of the vascular haemoglobin of *Arenicola marina* at 20°C. Points represent data at *pH* values from 5·4–6·8. (After Prosser and Brown, 1961; from Manwell unpublished data.) Note the high oxygen affinity of the blood (p_{50} 5·5) and the absence of a Bohr effect. The interaction coefficient n = 3·5.

Barcroft and Barcroft (1924) and 8·4 – 9·9 volumes % by Borden (1931) whilst Wolvekamp and Vreede (1941) found intermediate values. This variation is possibly related to the fact that the water content of *Arenicola* varies with the osmotic concentration of the medium from a salinity of 35 ‰ down to as low as 8 ‰ (Wolvekamp and Vreede 1941). Despite the fact that the highest values recorded for oxygen capacity are nearly as high as in *Sabella spallanzanii* (Fox, 1932), it is generally agreed that the supply of oxygen would last only from 7 – 70 minutes. However, as Jones (1963) points out, a storage function may be of use during the intermittent pauses in irrigation (p. 294) which may be of 30 minutes duration. Again, when the animal is exposed by the tide the oxygen content of the water in the burrow itself may greatly enhance the oxygen available to the animal, and Jones (1963) has calculated that a reserve of oxygen to last as much as 2 hr may be available, especially if the partial pressure of oxygen in the veins is reduced to a very low level. Nevertheless, lugworms do live at high levels on the shore, and it must be assumed that aerial respiration (Wells, 1949a) or use of oxygen diffusing into the burrow must play a significant part in prolonging the period during which the lugworm can survive without resort to anaerobiosis. Inspection of Fig. 7.46 indicates that the haemoglobin could also be used for gas transport at the high partial pressures of oxygen which occur in the seawater pumped through the burrow when submersed by the tide. Although, as will be seen, the percentage turnover to the tissues is very small, the high oxygen capacity of the blood would mean that a considerable quantity of bound oxygen could still be donated to the tissues.

A pigment with a high affinity for oxygen, as in *Arenicola,* also occurs in some decapod Crustacea in which the respiratory pigment is haemocyanin. This is perhaps at first surprising since animals such as the American rock lobster *Panulirus interruptus,* the American lobster *Homarus americanus* and the sheep crab *Loxorhynchus grandis,* are not normally exposed to very low partial pressures of oxygen as in *Arenicola.* The oxygen capacities of decapod bloods are also very low, often being less than that of the aerated seawater in which the animals live, and rarely exceed 2·5 volumes %. These factors have in the past suggested that the haemocyanin of decapods plays little part in the transport of respiratory gases in these animals. Redmond (1955) has, however, shown that in the decapods mentioned above, and probably in others, the gills represent a major diffusion barrier such as that in *Panulirus.* Here the partial pressure of oxygen in the arteries is only some 7 – 12 mm Hg even when the gills are bathed in aerated seawater (pO_2 = 150 mm Hg), whereas that of the capillaries is 3 mm Hg. At these partial pressures only some 0·03 ml of oxygen would be dissolved in 100 ml of arterial blood and approximately 0·01 ml O_2/100 ml of venous blood. That is, only some 0·02 ml of dissolved oxygen would be donated to the tissues. Despite its low oxygen capacity, the haemocyanin of decapods may thus markedly improve the oxygen capacity of the blood compared with that carried in simple solution.

The oxygen dissociation curve in volumes per cent of combined oxygen as a function of partial pressure of oxygen in mm Hg of the haemocyanin of *Panulirus* is shown in Fig. 7.47. The levels of dissolved oxygen which occur in the blood over the range of partial pressures of oxygen are also indicated. At the partial pressure of oxygen occurring in the arteries (7 mm Hg) the blood was only 54% saturated at 15° C and contained 0·82 vols % of oxygen, whereas at the venous pO_2 (3 mm Hg) the blood was 22% saturated and contained 0·35 vols O_2%. Thus the total oxygen donated by the pigment over the range 7 mm Hg to 3 mm Hg was 0·47 vols % compared with only 0·02 vols % unloaded from simple solution. This means that despite the fact that the haemocyanin has only a low oxygen capacity, it enables a very much greater quantity of oxygen to be transported than by simple solution at the relatively low partial pressures of oxygen occurring in the arteries. The Bohr effect is positive, that is, at a low *pH* value the oxygen dissociation curve is displaced to the right, but is small in extent, and since the *pH* of the veins is only 0·02 of a *pH* unit lower than that of the arteries, the Bohr effect probably plays no significant part in oxygen turnover (Redmond, 1955; Jones, 1963).

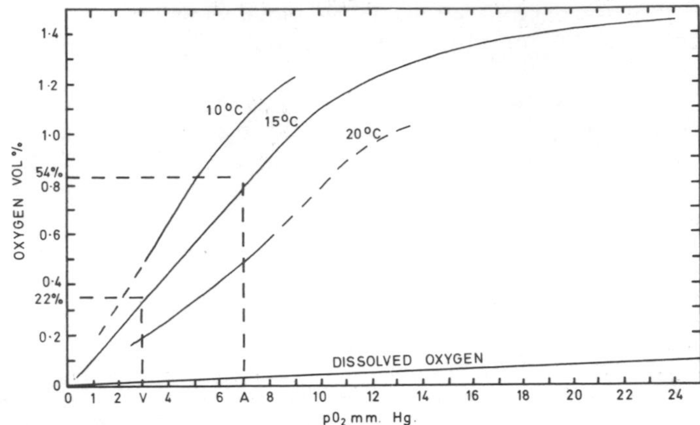

Fig. 7.47. Oxygen dissociation curves of the blood of *Panulirus* at 10°C, 15°C and 20°C. Note that at the oxygen tension of the arteries (A = 7 mm Hg) the pigment is only 54% saturated whilst at that of the veins (V = 3 mm Hg) the blood is 22% saturated. Thus the turnover of combined oxygen is 32% = 0·47 vols % compared with only 0·02 vols % unloaded from simple solution.
Note also the typical temperature effect on the oxygen capacity of the blood. (After Redmond, 1955 and Jones, 1963.)

Redmond (1955) has shown that a similar diffusion barrier exists at the respiratory surface of the gastropod *Busycon canaliculatum*, which also has haemocyanin as a blood pigment (also, Redfield, Coolidge and Hurd, 1926; Henderson, 1928; Jones, 1963). In this animal the pO_2 in the arteries is 36 mm Hg (A) and that of the veins is 6 mm Hg(V) when the partial pressure of oxygen in the seawater is 150 mm Hg. At a pO_2 of 36 mm Hg in the absence of CO_2 the haemocyanin is 90% saturated, whilst at 6 mm Hg it is 20% saturated (Fig. 7.48), so that the pigment is well adapted to function at the low partial pressures of oxygen which occur in the blood, a turnover of 70% of the oxygen capacity (1·4 – 2·0 vols %) occurring in the absence of CO_2. Of particular interest is the reversed Bohr effect which characterises the haemocyanin of this animal. Redmond (1955; also Jones 1963) has emphasised the possible functional importance of this effect by a consideration of the situation which occurs when the animal is exposed to low partial pressures of oxygen. If, for example, the partial pressure in the surrounding seawater falls so that the partial pressure of oxygen in the arteries is reduced to 12 mm Hg(A_1), (Fig. 7.48) then the amount of oxygen which could be donated to the tissues would be only 28% of the total capacity. But if at the same time there was a fall in the partial pressure of oxygen in the veins from 6 mm Hg to 3 mm Hg (V_1), then the turnover would be 40% of the total oxygen capacity. If the fall in partial pressure of the external oxygen is then associated with increased

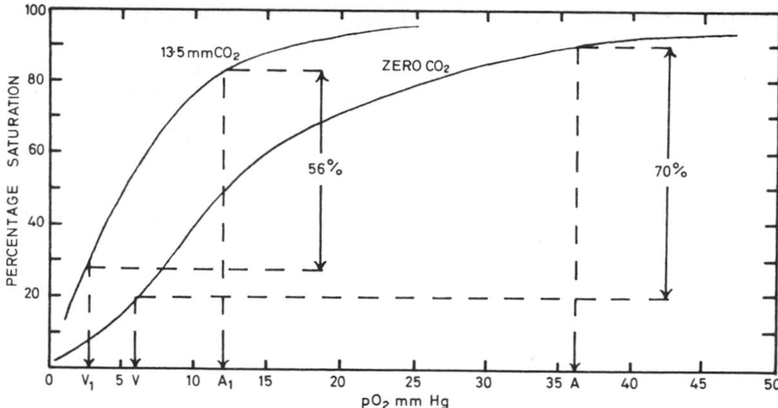

Fig. 7.48. Oxygen dissocation curves of the haemocyanin of *Busycon canaliculatum* in the absence of CO_2 and in the presence of 13·5 mm CO_2. (After Redfield, Coolidge and Hurd, 1926 and Jones, 1963). Note the reversed Bohr effect. The turnover between the arterial tension of 36 mm Hg (*A*) and 6 mm Hg (*V*) is indicated as well as the turnover between 12 mm (A_1) and 3 mm (V_1) in the presence of 13·5 mm. CO_2.

partial pressures of CO_2, such as would occur in association with fermentation, increasing the pCO_2 of the blood to 13·5 mm Hg for example, then because of the reversed Bohr effect the turnover between a pO_2 of 12 mm Hg and 3 mm Hg would be 82 – 26 = 56%, representing an increase of 16% on the turnover occurring in the absence of CO_2. It represents an increase of 56 – 28 = 28% over that which would occur in the absence of a reversed Bohr effect coupled with a reduction in venous pO_2. The extent to which *Busycon* would be exposed to high partial pressures of CO_2 is unknown, nevertheless the animal lives on mud-flats where fermentative processes are known to occur, and it seems significant that the king crab, *Limulus,* which is another inhabitant of mud-flats, also possesses haemocyanin which shows a reversed Bohr effect (Redfield, Coolidge and Hurd, 1926).

Although it might be supposed from the foregoing discussion that aquatic organisms would benefit by access to the maximum possible concentrations of dissolved air, Fox and Taylor (1954, 1955) have demonstrated that a large number of aquatic invertebrates, including several marine forms, survive for long periods in poorly oxygenated water. Indeed, some survive not only longer but grow faster in poorly oxygenated water than in fully aerated water. As might be expected *Sabella pavonina* was found to survive for at least 14 days at 10 – 12°C in 100%, 21% and 10% dissolved oxygen provided that the animals remained in their tubes, but survived less than 4 days in seawater containing only 4% dissolved oxygen. *Arenicola marina,* on the other hand, survived only two weeks in water saturated with 100% oxygen, but in seawater saturated with 21%, 10% and 4% oxygen survival was much longer, so that the animals appear to be unaffected by a reduction of 4/5 of the maximum oxygen available in air saturated seawater. *Scoloplos armiger* lived longer in 1/5 (4% oxygen) aerated seawater than under fully aerated conditions. After 35 days the animals in the 4% oxygenated water were larger and had 5 or 6 pairs of gills compared with only 2 or 3 pairs in the animals in 21% oxygenated seawater. After 75 days no animals survived in the 21% oxygenated water but in the 4% oxygenated water the animals were large and possessed up to 28 pairs of gills. Fox (1955) has also shown that in *Arenicola marina* and *Scoloplos armiger,* in contrast to some other invertebrates such as members of the Cladocera and conchostracan crustaceans, exposure to low oxygen for about 1 month fails to cause an increase in the haemoglobin concentration.

High concentrations of dissolved oxygen thus appear to have injurious effects on some organisms, especially those which normally live in mud-flats where the pO_2 of the environment is low. Manwell (1959) has suggested that in such animals which are susceptible to "oxygen poisoning" the respiratory pigment may serve to protect the organisms against high internal oxygen concentrations. This function of respiratory pigments is made possible by their ability to buffer changes of pO_2 in the blood. Jones (1963) has discussed the influence of the oxygen affinity of respiratory pigments, and a number of other factors, on the range over which the pO_2 of the blood may be buffered. Assuming that the pigment comes into equilibrium at the respiratory surface and at the tissues (an assumption which probably holds for most vertebrates but which may not apply to a number of invertebrates), a series of graphs can be drawn showing the change in oxygen concentration of blood passing through the capillaries of the respiratory surface and of the tissues.

Fig. 7.49A shows the change in the pO_2 of a hypothetical blood passing through the capillaries of a respiratory surface. If there is no respiratory pigment (as is shown in Curve a), then initially there is a large difference in pO_2 between the blood and external medium so that the pO_2 of the blood increases rapidly. This process reduces the difference in concentration of oxygen between the blood and outside medium so that the rate of increase in the pO_2 of the blood becomes gradually slower. The curve is thus asymptotic in form. Now if a respiratory pigment is present in the blood and the pigment has a relatively low oxygen affinity, then it will not load until the pO_2 in the blood has risen after contact with the respiratory surface. Thus the pO_2 of the blood rises initially (Curve b Fig. 7.49A) and then remains almost constant whilst the pigment combines with oxygen diffusing into the

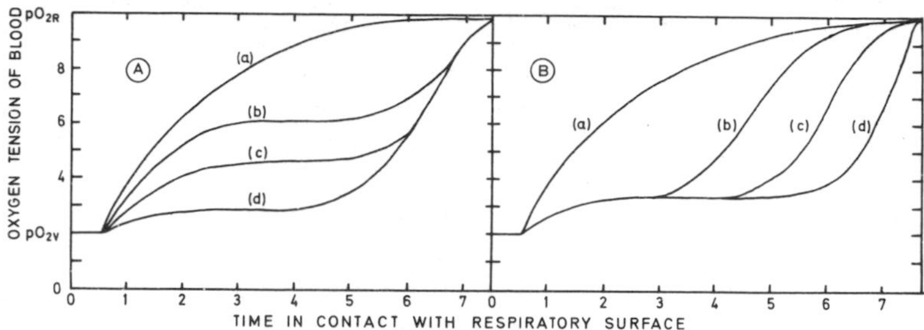

Fig. 7.49. Diagram showing the change in oxygen tension (arbitrary units) of venous blood passing through the capillaries of the respiratory surface. The pO_2 of the veins (pO_{2V}) and of the respiratory surface (pO_{2R}) are indicated. (A) Showing the change if there was no respiratory pigment (a) and in bloods of increasing oxygen affinity ($d > c > b$) but with the same oxygen capacity. (B) Showing the change if the bloods had the same oxygen affinity but differing oxygen capacity ($d > c > b$); (a) represents blood with no respiratory pigment. Note that in (A) the oxygen affinity of the pigment determines the level of oxygen tension in the blood, a high level being maintained by a pigment with a low affinity. In (B) the length of time the pO_2 of the blood is held constant is determined by the oxygen capacity of the pigment, high capacities tending to maintain a constant pO_2 for longer than low capacity pigments. (Based on Jones, 1963.)

blood. When the pigment is saturated, oxygen is no longer absorbed and the pO_2 of the blood rises to that contained in blood without pigment. If the pigment has a high affinity for oxygen then it loads at correspondingly lower pO_2 values and the pO_2 of the blood is maintained at a low value until the pigment becomes saturated (Curves c & d, Fig. 7.49A). Thus the oxygen affinity of the pigment determines the level at which the pO_2 of the blood is maintained. Clearly, in a pigment with a low oxygen capacity, the pO_2 of the blood would be maintained at a constant value for only a short time before the pigment was saturated and the pO_2 of the blood rose (Fig. 7.49B). On the other hand a pigment which has a very large oxygen capacity, as in *Arenicola* haemoglobin, might absorb oxygen during the whole time the blood is in contact with the respiratory surface, and thus maintain the pO_2 of the blood at a constant level (in this case a low level since the oxygen affinity of *Arenicola* haemoglobin is high).

Jones (1963) has constructed a similar series of curves depicting the effect of respiratory pigments of differing oxygen capacity on the pO_2 of the blood in the tissue capillaries rather than in the capillaries of the respiratory surface which we have considered above. These curves are shown in Fig. 7.50A, from which it will be noticed that a blood without a respiratory pigment (Curve a) would unload its oxygen over a wide range of pO_2 values whereas a pigment with a low oxygen affinity would tend to discharge its oxygen as the blood left the arteries, so

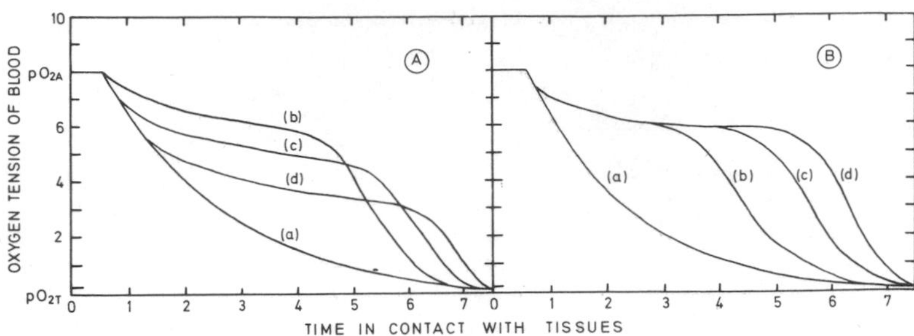

Fig. 7.50. Diagram showing the change in oxygen tension (arbitrary units) of the arterial blood passing through the tissue capillaries. The pO_2 of the arteries (pO_{2A}) and of the tissues (pO_{2T}) are indicated. (A) Showing the change if there was no respiratory pigment (a) and in bloods of increasing oxygen affinity ($d>c>b$) but of the same oxygen capacity. (B) Showing the change if the bloods had the same oxygen affinity but differing oxygen capacities ($d>c>b$); (a) represents blood with no respiratory pigment. (Based on Jones, 1963.)

maintaining the pO_2 of the venous blood relatively constant until all the combined oxygen had been unloaded. After this, the pO_2 of the veins would fall until it reached the value for blood without a pigment. If a pigment with a higher affinity for oxygen were present (Curve c) then the oxygen content of the blood would fall to a correspondingly lower level before the pigment unloaded its combined oxygen and maintained the pO_2 of the blood at a uniformly low level. As before, the duration of the period at which oxygen is unloaded and the pO_2 of the blood is kept constant, is determined by the oxygen capacity of the pigment, pigments with a low capacity maintaining a constant pO_2 for a shorter period than those with a high capacity (Fig. 7.50B).

Respiratory pigments with a high oxygen affinity may therefore act as oxygen buffers, and maintain the pO_2 of the blood at a uniformly low level despite great variations in the pO_2 of the medium. Whether this serves to mitigate the effects of "oxygen poisoning" as has been suggested by Manwell (1959) is not yet clear, but the possibility remains an attractive one.

(e) Oxygen stores

We have seen that irrigation rhythms are used in some intertidal animals to bring oxygenated seawater into the vicinity of the respiratory surface, whilst gas transport by the blood is often aided by the use of a respiratory pigment the properties of which are commonly appropriate to the partial pressures of oxygen occurring in the arteries. In some animals, however, there are pauses in the irrigation movements, whilst others may be subjected to oxygen lack at certain phases of the tidal cycle. Under such circumstances, oxygen may be stored either in simple solution in the body fluids or in combination with a respiratory pigment. Two animals in which the role of coelomic respiratory pigments as an oxygen store has been investigated are the echiuroid worm *Urechis caupo* (Hall, 1931; Redfield and Florkin, 1931) and the sipunculoid *Sipunculus nudus* (Florkin, 1933).

Urechis caupo normally lives in a U-shaped burrow through which water is pumped by means of peristaltic waves down the body. The animal feeds by means of a mucous funnel in which particles become entrapped much as in *Chaetopterus variopedatus* (p. 189) and also uses the water for respiratory purposes, gas exchange being *via* the thin walls of the hind-gut. During filter feeding the rate of pumping of animals in glass tubes was 50 ml/min but this rate occupied only 20% of the total time. At other times the rate fell to less than 25 ml/min during which water was pumped into the hind-gut by the activity of the muscular cloaca, up to 30 inhalations being followed by a single exhalation (Hall, 1931). Redfield and Florkin (1931) showed that the exhaled water contained 0·37 vols % of oxygen and 5·26 vols % of carbon dioxide compared with 0·56 vols O_2% and 4·77 vols CO_2% in the inhaled water. Thus the exhaled water was poorer in oxygen but richer in carbon dioxide than the inhaled water. The pO_2 of the water in the hind-gut was found to be approximately 100 mm Hg. Since samples of coelomic fluid drawn from animals kept in aerated water had a pO_2 of not less than 75 mm Hg, and the haemoglobin which occurs in the corpuscles of *Urechis caupo* has a p_{50} of 12·5 mm Hg and a p_{75} of 75 mm Hg at 19°C, it is clear that when the animal is pumping the pigment in the coelomic corpuscles is almost fully saturated. No Bohr effect was detected at CO_2 levels from 0·5 to 92 mm Hg. The oxygen requirement of the animal was found to be approximately 0·13 ml/min (Hall, 1931) which amounts to only $\frac{1}{60}$th of the total oxygen content of the coelomic fluid, so that the pigment is unlikely to unload its oxygen whilst pumping is taking place. However, pauses in irrigation occur for periods of up to 20 min, and under these circumstances the coelomic oxygen could function as a store (Jones, 1963). Assuming that the respiratory rate is maintained at the same level as in well aerated seawater, it has been calculated that the combined oxygen would last for 54 min, whilst the dissolved oxygen would allow a further 8 min and the water in the hind-gut a further 8 min giving a total oxygen supply sufficient to last 70 min (Redfield and Florkin, 1931). Thus the oxygen supply held within the animal would be more than sufficient to meet the requirements of the animal during its intermittent pauses in irrigation. Redfield and Florkin (1931) also showed that the dissolved oxygen in the burrow may be regarded as an oxygen store, much as has been described on p. 346, for the water in the burrow of *Arenicola marina.* They found that during the period of intertidal exposure the pO_2

of the water in the burrow fell to a minimum value of 14 mm Hg (0·06 vols %) after 4 hr exposure. At this pO_2 the haemoglobin is 60% saturated and thus could possibly aid oxygen transport during the exposure period giving a further supply of oxygen sufficient to last the animal 135 min.

Florkin (1933) has suggested that the haemerythrin of the coelomic corpuscles of *Sipunculus nudus* may also act as an oxygen store in a similar manner to that described above for *Urechis caupo.* In *Sipunculus* the coelomic oxygen tension was found to be approximately 20 mm Hg, at which the haemerythrin would be 85% saturated at 19°C. The oxygen affinity was rather higher than the haemoglobin of *Urechis,* p_{50} being 8 mm and p_{95} 45 mm at 19°C. A Bohr effect was lacking and the oxygen capacity of the coelom was found to be about 1·6 vols %. No function was attributed to the pigment as long as the animal was in aerated water. However, if the animal burrowed into the deposits and thus encountered low oxygen tensions, the presence of haemerythrin would be expected to increase the supply of oxygen by about four times compared with the oxygen contained in simple solution.

Oxygen may be stored not in combination with a respiratory pigment but in simple solution in the coelom, as in the sea cucumber *Holothuria forskali.* When this animal is placed in a sealed vessel of aerated seawater its rate of oxygen consumption is found to become reduced until at tensions corresponding to approximately 60 – 70% the rate of depletion is very slow indeed (Newell and Courtney, 1965). The overall rate of respiration of *Paracaudina* also declines as the oxygen in the surrounding water becomes depleted (Nomoura, 1926). In *Holothuria forskali,* as in *Urechis caupo,* water is driven by cloacal contractions into thin-walled respiratory structures, in this case termed the respiratory trees, which ramify through the coleom. One of the pumping rhythms is shown in Fig. 7.12 and it has been found that in *H. forskali* such cloacal pumping accounts for approximately 60% of the total oxygen uptake by the animal (Newell and Courtney, 1965), a value which agrees well with that obtained on other holothurians by Winterstein (1909). By separation of the water pumped from the cloacal region from that in contact with the general body surface by means of a rubber membrane it was found that the reduction in overall oxygen uptake by *H. forskali* was due to a fall in cloacal gas exchange. This fall was caused by the cessation of pumping when the oxygen in the pumped water fell below 60 – 70% of its air saturation value. Oxygen uptake through the general body surface meanwhile continued at a steady rate. After 1 hr during which oxygen uptake continued through the general body surface at a rate of 0·016 ml/g/hr but no oxygen uptake occurred through the cloaca, the oxygen concentration of the water in the cloacal chamber was restored to full air saturation and oxygen was found to be absorbed very rapidly at a rate of 0·057 ml/g/hr but fell to 0·013 ml/g/hr after 0·5 hr. This rate was comparable with the rate of absorption through the cloaca before the cessation of pumping. Thus a period of oxygen depletion and cessation of pumping was followed by a period of enhanced oxygen uptake when the oxygen was returned to full air saturation. These results are shown in Fig. 7.51

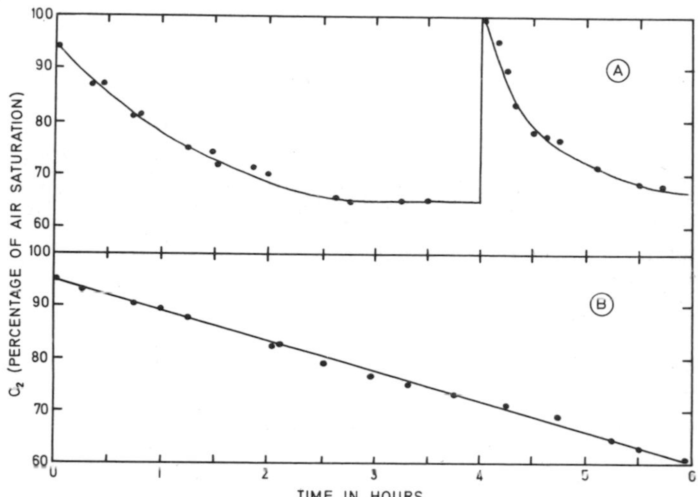

Fig. 7.51. Graphs showing the oxygen uptake through the cloaca (A) and through the general body surface (B) of *Holothuria forskali* at 17°C. (After Newell and Courtney, 1965.) Note that oxygen uptake through the general body surface (B) continued at a steady rate whilst uptake through the cloaca (A) ceased at 65% of air saturation due to the cessation of cloacal pumping. When the oxygen content of the water in the cloacal chamber was restored to 100% of air saturation, the oxygen again decreased due to the resumption of cloacal pumping.

and suggest that after cloacal pumping had ceased, the respiratory rate of the tissues was not reduced appreciably. Instead, either an oxygen store was used or an oxygen debt was incurred by the tissues, the oxygen being replaced in either case when the oxygen concentration of the pumped water was restored to normal.

Direct evidence of the use of dissolved oxygen in the coelom as a store was obtained by inoculation of a small oxygen electrode through a serum needle into the coelom and comparing the internal and external oxygen concentrations as an animal gradually used the oxygen in a sealed vessel of seawater. By correlating such changes with variations in the volume of the animal, it was found that the oxygen in the coelom was in equilibrium with that pumped by the cloaca until pumping stopped at approximately 60–70% of air saturation. Just before pumping ceased, however, the volume of the coelom was increased to as much as 75 ml in some animals by an alteration in the pumping rhythm, although the average volume of the coelomic fluid after such an increase was approximately 50 ml. This water was approximately 60% saturated and would supply rather less than 0·17 ml of oxygen to the tissues at 17° C. The rate of oxygen consumption of an average sized animal was 0·023 ml/hr, and, taking into account the uptake through the general body surface which occurs even when cloacal gas exchange has ceased, this gives an oxygen deficit of 0·019 ml/hr when pumping ceases. The approximately 0·17 ml of oxygen taken into the coelom prior to cessation of pumping might therefore be expected to augment the oxygen supply to the tissues for as much as 9 hr (Newell and Courtney, 1965). Unfortunately, as in *Sipunculus nudus,* there is little known of the oxygen availability in the natural environment; nevertheless, it seems clear that the coelomic oxygen supply is adequate to meet the metabolic oxygen requirements of the tissues when pumping has ceased.

f. Reduction in the level of metabolism

A final adaptation to reduced availability of oxygen, and one which leads to an economy of stored oxygen, is a reduction in the level of metabolism; this occurs in many intertidal animals when they are deprived of access to oxygenated seawater during the intertidal period. As has been mentioned on p. 284, the heart rate of the mussel *Mytilus edulis* has been shown to be suppressed when the animals are uncovered by the tide (Schlieper, 1955; Helm and Trueman, 1967), and this is also true of the cockle *Cardium edule* (Trueman, 1967). In *Arenicola marina* spontaneous activity rhythms are suppressed except for occasional 'testing movements' (Wells, 1949 a,b; also p. 301), whilst *Owenia fusiformis* appears to be able to survive anaerobic conditions for long periods partly by its ability to reduce the level of metabolism under such conditions (von Brand, 1946; Dales, 1958). Barnacles, too, become largely quiescent when emersed by the tide and the metabolic rate of several species has been shown to be lowered under such conditions (Barnes, Finlayson and Piatigorsky, 1963). Although such behaviour must obviously lead to an economy of metabolic energy reserves during periods of anaerobiosis, a survey of the literature makes it difficult to calculate the amount by which activity increases the oxygen demand compared with the consumption during quiescence. The main reason for this is that the oxygen consumption of aquatic organisms has commonly been recorded without reference to the activity level of the animals concerned. A number of fish physiologists (notably Spoor, 1946; Fry, 1947; Fry and Hart, 1948; Brett,

1962, 1963; Beamish, 1964; Beamish and Mookherjii, 1964), have recognised the importance of this factor and have established the quantitative relationship between swimming activity and oxygen consumption. By extrapolation, the respiration rate of a fish at zero activity can then be calculated (also Chapter 8). Similar measurements have also been made on a number of marine invertebrates including the shrimp *Leander adspersus* (Ivlev, 1963, cited by Halcrow and Boyd 1967) and the amphipod *Gammarus oceanicus* (Halcrow and Boyd, 1967). This method is of particular value in organisms which are continuously active, but in invertebrates which show periods of quiescence it is possible to record the oxygen consumption not only during the active phases but also during the quiescent ones. Adopting the nomenclature of Fry (1947, 1957), we may then recognise a rate of oxygen uptake during quiescence, the 'standard rate', and a rate corresponding with maximal activity, the 'active rate'. Between these two rates is a variety of 'routine rates' corresponding with spontaneous movements of the organism concerned. As Beamish (1964) has pointed out, the term 'standard rate of oxygen consumption' has often been misused in the literature, for even under conditions where quiescence is encouraged, a great variety of spontaneous movements may occur. The importance of these and other factors in controlling the rate of oxygen consumption of intertidal organisms is discussed more fully on p. 373.

Table XIV shows the standard and active rates of oxygen consumption of a number of common intertidal invertebrates, and allows an estimate to be made of the reduction in metabolism which is likely to occur simply by the onset of quiescence during the emersion period. The decrease in the rate of oxygen consumption which would occur from the active rate to the standard rate is also shown in the table. It is apparent that the proportionate decrease in the metabolism following the onset of quiescence varies a great deal, but even in *Gammarus oceanicus* and *Patella vulgata* which show the smallest reduction in metabolism of the animals listed, the rate of oxygen consumption is decreased by a factor of 2·5 and 1·4 respectively in the transition from the active to the standard rate. The differences between the active and the standard rates of metabolism shown in the table are all considerably less than the 170-fold increase noted in butterflies during flight (Krogh, 1941) (in other insects there may be a 50 – 100 fold increase) but in general the values correspond with those noted for other less active animals, including trout in which the active metabolism was approximately 4 times the standard metabolism (Fry, 1957; for review, Prosser and Brown, 1961).

Such a reduction in the metabolic demand of the tissues must have two important consequences in intertidal organisms subjected to a reduced oxygen availability. Firstly, the reduced oxygen demand will be

Table XIV. Table showing the standard and active rates of oxygen consumption of a variety of intertidal invertebrates. Most of the data are expressed in terms of an idealised animal of 100 mg or 500 mg dry tissue weight.

Species	*Temp °C*	*Dry Weight of animal mg*	*Standard rate µl/mg dry wt/hr*	*Active rate µl/mg dry wt/hr*	*Increase*	*Source*
Gammarus oceanicus	15°	150–350	0·61*	1·5*	x2·5	Halcrow and Boyd (1967)
Arenicola marina	17·5°	500	0·2	0·8	x4·0	Unpublished data**
Patella vulgata	15°	500	0·42†	0·60†	x1·4	Davies (1965)
Actinia equina	15·5°	100	0·5	18·0	x36·0	Newell and Northcroft (1967)
Nephtys hombergi	15°	100	0·12	0·8	x6·6	,,
Littorina littorea	15°	100	0·25	3·0	x12·0	,,
Cardium edule	15°	100	0·3	1·9	x6·3	,,
Balanus balanoides	14·0°	100	0·25	1·2	x4·8	Newell and Northcroft (1965)
Branchiostoma lanceolatum	15°	100	0·1	0·75	x7·5	Courtney and Newell (1965)
Diopatra cuprea	17·5°	*ca* 880	0·29*	1·0*	x3·45	Mangum and Sassaman (1969)

* recalculated from the original data assuming a wet weight/dry weight ratio of 10:1

** I should like to thank two of my students, Miss B. Newby and Miss M.S. Waite for permission to publish these data.

† recalculated from the published data, assuming a wet weight/dry weight ratio of 10:1 and refitting regression lines to maximal and minimal rates of high-level forms.

met by simple diffusion from correspondingly lower partial pressures of oxygen in the surrounding medium. The animal may thus be able to continue to respire aerobically until the pO_2 in the surrounding medium falls below that necessary to meet even the reduced oxygen demand of the tissues (Fig. 7.55 p. 370). Secondly, if the pO_2 falls below that required to supply the quiescent animal with oxygen and anaerobiosis occurs, then the rate of depletion of the metabolic reserves will be a small fraction of the rate which would occur in the absence of a reduction in the level of metabolism. Indeed there is some evidence that the metabolic demand may be still further reduced during anaerobiosis (for example, von Brand, 1946; Barnes, Finlayson and Piatigorsky, 1963). As has been mentioned on p. 266, many intertidal organisms are now known to air-breathe and this represents another possible mechanism by which anaerobiosis is avoided. We may therefore first examine the occurrence of air-breathing in intertidal animals and then review the conditions under which anaerobiosis might be expected to occur.

3. AERIAL RESPIRATION

The ability to utilise the air as a source of oxygen when uncovered by the tide is widespread amongst intertidal animals. Amongst the polychaetes, some hesionids swallow gas bubbles produced by algae, whilst others, such as *Arenicola marina,* may force a bubble of air anteriorly from the exit of the tail shaft over the gills by a modification of the normal headward irrigation movements (p. 294) when the level of water falls below that of the entrance and exit of the burrow (Wells, 1949a). *Nereis diversicolor* crawls to the entrance of the burrow until the head emerges from the water and then passes undulatory waves down the body; such movements comprise the normal irrigation pattern when the burrow is submersed by the tide, but at low water they serve to draw in a series of air bubbles which are trapped between the body and the walls of the tube. When the bubbles have reached the posterior end of the body, irrigation movements cease so that the bubbles are held against the body, being replaced after a few minutes by bubbles from a further set of irrigation movements. *Thoracophelia mucronata,* which is a polychaete found intertidally in California, is also able to air-breathe; this animal exposes the hind end of the body through a funnel-like depression in the sand, gas exchange being through the walls of the rectum (for review, Dales, 1963).

Amongst the Crustacea, air-breathing has been shown to occur in a wide variety of crabs of which some, such as members of the family Gecarcinidae, are adapted for a permanently terrestrial mode of life. Edney (1960) has pointed out that amongst the semi-terrestrial crabs there is a progressive reduction in the number of gills or gill volume relative to body volume in a series of species characteristic of aquatic through to more terrestrial habitats (also Pearse, 1929a,b; Ayers, 1938; Gray, 1953). Even the lobster *Homarus vulgaris,* which must be only rarely exposed to aerial conditions, is able to breathe both in air and in water (Thomas, 1954), the rate in air being approximately $\frac{1}{7}$ th of the rate in aerated seawater. Some intertidal crabs such as *Uca, Grapsus* and *Ocypode* can, as Edney (1960) points out, circulate air through water which is carried in the gill chambers, but in general the gills are less important as sites of gas exchange in the semi-terrestrial forms than in aquatic genera. Special vascular outgrowths or vascularisation of the walls of the branchial chambers may be developed in some crabs, but amongst the intertidal isopods such as *Ligia oceanica,* gas exchange is through the pleopods which are relatively unmodified from the condition in aquatic forms where the pleopods act as gills. Gas exchange can also occur through the general body surface in some isopods; in *Ligia* the oxygen uptake is only reduced to 50% of the normal rate when the pleopods are blocked (Edney and Spencer, 1955). The intertidal cirripedes *Balanus balanoides* and *Chthamalus stellatus* are both able to air-breathe through a micropylar opening or pneumostome between the opercular valves which allows the access of air into the mantle cavity, and the air may well be circulated by means of ventilation movements (Darwin, 1854; Monterosso, 1928a, b, c; 1930; Barnes, Finlayson and Piatigorsky, 1963; Grainger and Newell, 1965, also p. 364).

The ability to air-breathe has also been extensively studied in intertidal molluscs. Amongst the bivalves, the salt-marsh mussel *Modiolus demissus,* which lives amongst *Spartina* as well as in more exposed situations, has been shown to leave the valves ajar and to air-breathe when emersed by the tide (Kuenzler, 1961; Lent, 1968). Other mytilids such as *Brachyodontes demissus plicatulus* and *Mytilus edulis* are also able to breathe both in air and in water (Read, 1962). Since then, Sandison (1966, 1967) has made a comparative study of the ability of *Littorina saxatilis* (= *rudis*), *L. littoralis* (= *obtusata*), *Littorina littorea* and *Thais* (=*Nucella*) *lapillus* to breathe in air and in water. Micallef (1966), in a comparative survey of the intertidal adaptations of the trochid series *Monodonta lineata, Gibbula umbilicalis, G. cineraria* and *Calliostoma zizyphinum,* has shown that these animals can also respire both in air and in water. Similarly the mediterranean trochid *Monodonta turbinata* has recently been shown to be well-adapted for aerial respiration (Micallef and Bannister, 1967).

Many intertidal fishes also air-breathe. Perhaps the best-known example is the eel *Anguilla vulgaris* in which the oxygen uptake in air is approximately one half of that in water, about 33% of the oxygen uptake in air occurring through the gills compared with 85–90% in water. During the first hour of exposure to air, oxygen uptake through the gills is supplemented by oxygen removal from the swimbladder, but after 20 hr in air at 15°C a substantial oxygen debt has developed although this does not occur at 7°C (Berg and Steen, 1965a, b, 1966). Thus the eel appears to be well adapted for survivial in air at low temperatures, although it can survive for only a limited period at high temperatures without resort to anaerobiosis. Many other fishes are able to air-breathe (for review, Saxena, 1963), the families Gobiidae and Bleniidae containing several species which regularly resort to aerial respiration during the emersion period. The mudsucker *Gallichthys mirabilis* (Gobiidae), for example, gulps air at the surface of the water when the environmental dissolved oxygen falls below 2·0 mg/l, gas exchange occurring through the highly vascularised walls of the buccopharynx (Todd and Ebeling, 1966). Again, many intertidal fishes such as the Cornish sucker fish *Lepadogaster lepadogaster* (Gobiesocidae), as well as many of the blennies, and the butterfish *Pholis gunnellus*, are often found in damp situations but entirely uncovered by the tide. Under such circumstances the blenny *Blennius pholis* gulps air and may survive for several days under humid conditions at low temperatures in the laboratory (M. J. Daniel, personal communication).

In view of the widespread occurrence of air-breathing amongst such intertidal organisms, it seems likely that many other intertidal animals are capable of aerial respiration where suitable conditions prevail. Indeed, the prime requirement for aerial respiration is the presence of a suitable surface through which gas exchange can occur, and in many organisms such as the intertidal anemones *Actinia equina* and *Anemonia sulcata*, in nudibranchs and in many errant polychaetes a degree of aerial respiration must occur through the general body surface. However, although such animals may be able to air-breathe, the very presence of a large surface area through which gas exchange may occur renders them liable to desiccation, so that they are normally restricted to the damper situations within the intertidal environment. On the other hand animals such as bivalves, gastropods and barnacles, which are able to resist desiccation by virtue of the shell, have a correspondingly reduced surface area through which gas exchange can occur. In the gastropods of the lower shore, which are generally assumed to be less perfectly adapted to a semi-terrestrial existence than species characteristic of the upper shore, the oxygen demand of the tissues can still be met, as we shall see below, by passive diffusion through the restricted surface area because of the reduction in activity which occurs on emersion. Deshpande (1957) has, however, shown that in a series of trochids ranging from the lower shore *Calliostoma zizyphinum*, through *Gibbula cineraria* and *G. umbilicalis* to the high level *Monodonta lineata,* there is an increasing vascularisation of the mantle skirt which compensates for the reduced surface area through which gas-exchange can occur, and which has been shown to be correlated with the ability to maintain high levels of activity during the emersion period (Micallef, 1966). Comparison of the levels of respiration in air and in water of a series of intertidal animals from different shore levels may thus yield important information on the activity level which can be maintained under conditions of emersion.

One such study is that made by Sandison (1966) on *Thais* (=*Nucella*) *lapillus, Littorina littoralis* (=*obtusata*), *Littorina littorea* and *Littorina saxatilis* (=*rudis*). She found that the respiration rate of each of these animals was higher in air than in water, that of a 100 mg animal being approximately 8 μl/hr in water and 18 μl/hr in air in all cases except *L. littoralis.* This latter animal was found to have a rather lower respiratory rate than the other species in water but a higher rate in air, this difference being largely due to the browsing excursions which are made on its exposure to air. The other species all respired at a comparable rate to one another once the differences in their body weight had been taken into account.

Micallef (1966) was, however, able to demonstrate a more obvious ecological variation in the level of activity and respiration in air in the trochid series comprising the lower shore *Calliostoma zizyphinum,* the lower eulittoral *Gibbula cineraria* which overlaps and is replaced upshore by *Gibbula umbilicalis,* followed by the upper shore *Monodonta* (= *Osilinus*) *lineata.* The upper shore *Monodonta lineata,* despite its larger size, was found to respire faster per unit weight than the trochids from lower shore levels. Large animals respire considerably slower per unit weight than small ones (p. 378) so that if comparisons had been made between identical sized animals, the respiratory rate of *Monodonta* would have been very much higher than that of the other species. Fig. 7.52. shows the mean aerial and aquatic respiration of groups of each of the four species at temperatures between 3·5° C and 30° C and shows that the level of aquatic respiration was higher than aerial respiration at all temperatures in the low-tide forms *Calliostoma zizyphinum* and *Gibbula cineraria.* In *Gibbula umbilicalis* the level of aerial respiration was lower than the level of aquatic respiration between 3·5° C and 19° C but of a comparable level between 20° and 26° C. Finally, in *Monodonta lineata* the rates in air and in water were comparable from 3·5° to 14° C but higher in air than in water from 14° to 25° C. The relative importance of aquatic respiration compared with aerial respiration thus declines in animals adapted to life on the upper shore. There is some indication, too, that despite variations in the mean size of the four different species of trochids, those characteristic of the upper shore showed a higher level of respiration per unit weight of

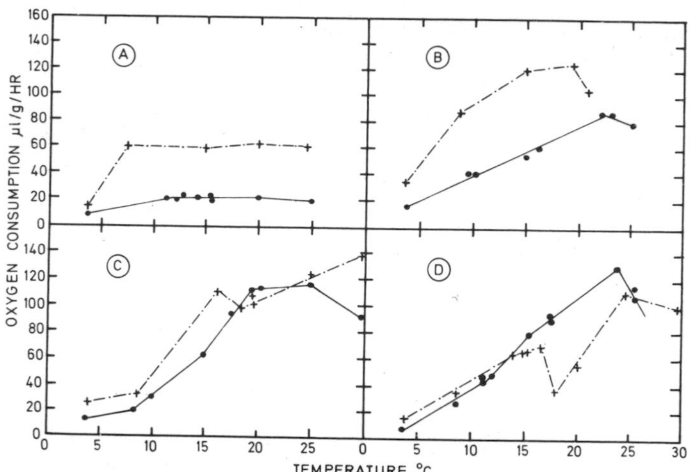

Fig. 7.52. Graphs showing the rates of oxygen consumption of a variety of Trochids in air (·—·) and in water +—+). A = *Calliostoma zizyphinum*, B = *Gibbula cineraria*, C = *Gibbula umbilicalis*, D = *Monodonta lineata*. (Data from Micallef, 1966.) Note that the relative importance of aerial respiration increases throughout the series.

tissue than lower shore forms, although this is not true of the metabolism of sessile animals such as barnacles (Barnes and Barnes, 1959) or of the limpet *Patella vulgata* (Davies 1966, 1967). Both the higher intrinsic respiration rate and the increase in level of aerial respiration in the trochids have been associated with the increase in activity during the emersion period shown by the upper shore trochids compared with the lower shore species (Micallef, 1966).

Aerial respiration is thus shown by a variety of intertidal organisms particularly under humid conditions; nevertheless the ability to respire air involves a compromise between the access of air to the respiratory surface which is necessary to maintain aerobic metabolism and the evaporative water loss which must occur at the higher shore levels. Where the humidity is high, as under boulders or amongst the intertidal algae, aerial respiration may continue throughout the emersion period; indeed, many animals actively seek damp sheltered situations when the tide has ebbed. On the other hand sessile organisms such as barnacles and mussels are unable to avoid environmental stresses, and may close the valves and resort to anaerobiosis during the unfavourable period of the tidal cycle.

4. ANAEROBIOSIS IN SOME INTERTIDAL ANIMALS

The occurrence and nature of anaerobic respiration in invertebrates has been reviewed by von Brand (1946). As mentioned on p. 266, anaerobiosis is known to occur in *Syndosmya* (=*Abra*) *alba* (Moore, 1931), *Mya arenaria* (Ricketts and Calvin, 1948), *Melaraphe* (=*Littorina*) *neritoides* and *Littorina punctata* (Patané, 1946a,b, 1955), *Venus mercenaria* (Dugal, 1939), mussels (Schlieper, 1957) and oysters (Dodgson, 1928) as well as in the lugworm *Arenicola marina* and in *Owenia fusiformis* (Dales, 1958), in barnacles (Monterosso, 1928a,b,c; Kreps, 1929; Barnes Finlayson and Piatigorsky, 1963) and in several seaweed-inhabiting invertebrates (Wieser and Kanwisher 1959). In some animals a period of anaerobiosis is followed by the repayment of an oxygen debt (p. 283) whilst in others the acid end products of anaerobiosis are excreted rather than reoxidised.

Although there are numerous instances in which the ability to respire anaerobically under conditions of extreme stress have been recorded, the precise conditions under which anaerobiosis is likely to occur on the shore have been demonstrated in only comparatively few animals. The intertidal barnacles *Balanus balanoides* and *Chthamalus stellatus,* for example, have been shown to air-breathe under humid conditions through a small micropyle or pneumostome between the opercular valves (Darwin, 1854; Monterosso, 1928a, b, c, 1930; Barnes and Barnes, 1957, 1958; Barnes Finlayson and Piatigorsky, 1963; Grainger and Newell, 1965). Grainger and Newell (1965) showed that in *Balanus balanoides* at Whitstable, Kent, animals at all tidal levels maintained a connection between the mantle cavity and the air, although the proportion of animals with an open micropyle was greater at low shore levels. This difference was found to be associated with the lower humidities which occur at high shore levels compared with the lower shore; thus when a series of humidity chambers was set up, each chamber containing 100 *Balanus balanoides,* more animals were found to show a micropyle in the high humidities than in the low ones (Fig. 7.53). This differing response of animals in high and low humidities appeared to be related primarily to the total weight loss, so that a similar proportion of animals showed a micropyle in low humidities after a short time as animals exposed to a higher humidity for a longer time. Fig. 7.54 shows that after a weight loss of approximately 0·5g/100 animals, the majority of the animals closed the micropyle, although a small proportion remained open even when as much as 1·75 g of water/100 animals had been lost. These results imply that under shore conditions *Balanus balanoides* maintains a connection *via* the micropyle, and aerial respiration must occur under such circumstances. Such aerial respiration has indeed been shown to occur in both *Balanus balanoides* and *Chthamalus stellatus* (Barnes, Finlayson and Piatigorsky, 1963; Grainger and Newell, 1965). The predominantly sublittoral *Balanus crenatus,* on the other hand, does not form a micropyle

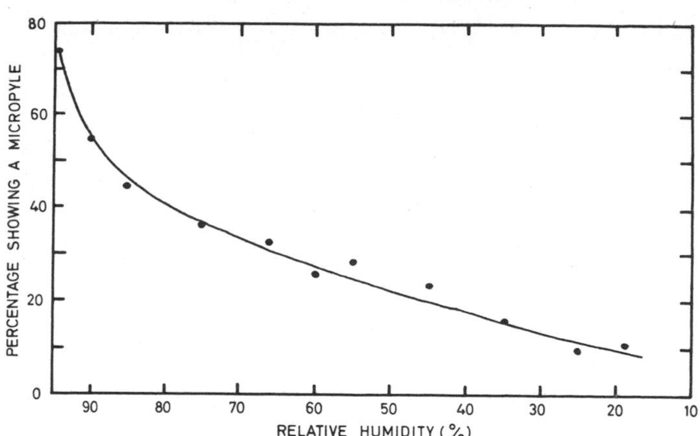

Fig. 7.53. The relation between percentage of *Balanus balanoides* having a micropyle and humidity after 3 hr at 18°C. Based on 100 animals in each chamber. (After Grainger and Newell, 1965.)

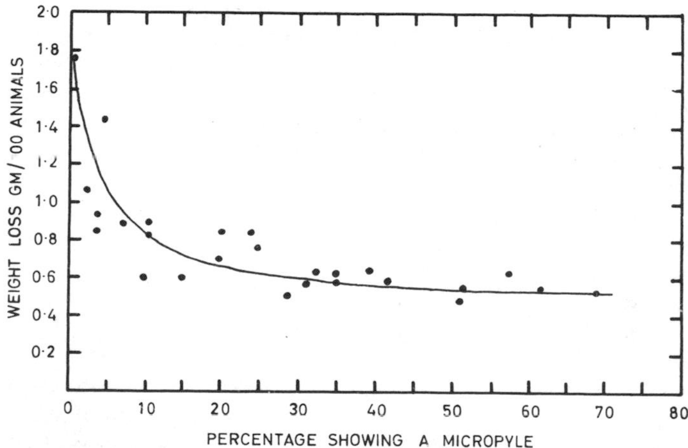

Fig. 7.54. Graph showing the relation between the total weight loss (g/100 animals) of *Balanus balanoides* and the percentage animals showing a micropyle.

Barnes *et al.* (1963) showed that the passage of air through the micropyle could be considered in the same way as the static diffusion of gases through stomata (Brown and Escombe, 1900). Assuming that all the gas entering the mantle cavity through the micropyle is used for respiration, the rate of diffusion (Q) through the micropyle is given by:–

$$Q = \frac{k\rho.\ A.\ 3600}{l + 2x} \quad \text{ml/hr.}$$

Where k = the diffusibility of the gas in C.G.S. units.
ρ = the density of the gas in the outer medium (atmospheres).
A = the area of pore opening (cm^2).
l = the length of the pore (cm).
x = a constant $\frac{1}{8\pi}$ x the pore diameter in cm.

Barnes *et al* (1963) found that for a specimen of *Balanus balanoides* of basal length 1 cm, in air $k = 0{\cdot}18\ cm^2/sec$, $\rho = 0{\cdot}2$, $A = 3{\cdot}14 \times 10^{-4} cm^2$ (radius = 0·01 cm) $l = 0{\cdot}1$ cm and $x = 0{\cdot}008$.

$$\text{Thus } Q = \frac{0{\cdot}18 \times 0{\cdot}2 \times 3{\cdot}14 \times 10^{-4} \times 3600}{0{\cdot}1 + 0{\cdot}016}$$

$$= 0.350 \text{ ml/hr}$$

$$= 350\ \mu\text{l/hr.}$$

Since the oxygen requirement of an animal of the size given above was only 2·7 μl/hr, it is obvious that even when the micropyle is relatively smaller than that given above, and even though all the oxygen entering the micropyle is not used, the rate of diffusion through the micropyle is more than adequate to meet the metabolic requirements of the barnacle.

It is clear that, as Barnes *et al.* (1963) stated, the ability to air-breathe involves water loss, and that when such desiccation becomes severe, or when other adverse factors such as reduced salinity occur, the valves are closed. Although such closure reduces the risk of desiccation or of osmotic stress, the products of anaerobiosis are likely to accumulate whilst excretion of such end products tends to deplete the metabolic energy reserves. Barnes *et al* (1963) have therefore studied the survival of the intertidal barnacles *Balanus balanoides* and *Chthamalus stellatus* compared with that of the sublittoral barnacle *Balanus crenatus* in dry nitrogen, wet nitrogen and dry air. In the dry air the dehydration stress was maximal and the anaerobic stress minimal, whereas in the nitrogen saturated with water vapour dehydration stress was absent but that due to oxygen demand was maximal. Finally, in dry nitrogen both desiccation and oxygen demand were maximal. The mean survival time (or days taken for 50% mortality) in the three species is shown in Table XV from which it is apparent that in dry air the two intertidal barnacles survived very much better than the sub-littoral *Balanus crenatus* which did not form a micropyle and which periodically extruded the collapsed cirri even when uncovered by the tide. In nitrogen saturated with water vapour the survival time of *Balanus crenatus* was greatly increased compared with that in dry air; the animal is evidently better able to survive anaerobic conditions than desiccation. Both *Balanus balanoides* and *Chthamalus stellatus* survived a similar time in moist nitrogen, which may suggest that the end products of anaerobiosis were responsible for the eventual death of the animals (Barnes *et al.,* 1963). In dry nitrogen the survival of all three species was similar and, since the micropyle of the intertidal species was held open, water loss was maximal.

Table XV. Table showing the mean survival in days of *Balanus balanoides, Chthamalus stellatus* and *Balanus crenatus* in dry air, wet nitrogen and dry nitrogen. (Data from Barnes, Finlayson and Piatigorsky, 1963.)

Species	*Balanus balanoides*	*Chthamalus stellatus*	*Balanus crenatus*
Habitat	Intertidal	Intertidal	Sublittoral
Mean survival in DRY AIR	4·0 days	5·2 days	0·6 days
Mean survival in WET NITROGEN	5·0 days	5·7 days	3·2 days
Mean survival in DRY NITROGEN	1·8 days	2·0 days	1·2 days

Thus the ability of the intertidal barnacles *Balanus balanoides* and *Chthamalus stellatus* to survive in dry air depends on their ability to reduce water loss, mainly by the formation of a micropyle and its periodic closure. This results in a rate of water loss in dry air of only 2 mg/hr in *Balanus balanoides* compared with 15 mg/hr in *B. crenatus.* By comparison, the water loss of a whole *B. balanoides* dissected free from the shell and suspended in dry air was as high as 24 mg/hr over the first 10 hr. This ability to avoid weight loss by closure of the micropyle in the two intertidal barnacles, *Balanus balanoides* and *Chthamalus stellatus,* but not in the sublittoral *Balanus crenatus,* implies that the intertidal forms resort to anaerobiosis under conditions of desiccation.

Barnes *et al.* (1963) found that when 10 specimens of *Balanus balanoides* were dissected from the shells and suspended under anaerobic conditions in seawater, lactic acid was accumulated in the tissues but that a similar amount was lost to the medium. When the animals were washed with oxygen-free water each 30 min and the lactic acid remaining in the bodies was estimated, it was found that lactic acid accumulated in the tissues at a rate of 0·3 mg/g tissue/hr, the total production including that lost to the medium probably approaching double this rate. Since starvation reduced the level of glycogen in the tissues, it would be expected that lactic acid production would also be lower following a period of starvation. This was shown to be the case in specimens of *Balanus balanoides* which were starved in filtered seawater for 60 days, after which the lactic acid production in oxygen-free seawater was only 0·14 mg/g tissue/hr. The rate of production of lactic acid by unstarved dissected *Chthamalus stellatus* was similar to that of unstarved *Balanus balanoides* (i.e. 0·3 mg/g tissue/hr) but in the sublittoral *Balanus crenatus,* the rate was only 0·09 mg/g tissue/hr over the first 10 hr after which there was no further increase.

Although the isolated tissues of the intertidal species were thus able to accumulate lactic acid under anaerobic conditions, when intact *Balanus balanoides* were stored in air there was little accumulation of lactic acid. This lack of anaerobiosis was due to aerial respiration, but as the stress of dehydration increased there was a sharp increase in the lactic acid content of the tissues and shell. It was found that a total of 12·7 mg lactic acid accumulated over a period of 16 days, which corresponds to a mean rate of accumulation of 0·033 mg/g tissue/hr. This is approximately $\frac{1}{10}$ th of the rate of accumulation by isolated tissues in oxygen-free seawater (0·3 mg/g tissue/hr), and reflects the use of oxygen for respiration during the initial phase of exposure to air. In contrast, when intact animals were maintained in gaseous nitrogen, there was no initial lag in lactic acid production since aerial respiration was excluded. In this instance the mean rate of lactic acid production was 0·1 mg/g tissue/hr and only some 5% of the animals were living after 70 hr. In oxygen-free seawater the rate of accumulation in the tissues was only 0·04 mg/g tissue/hr owing to the loss of lactic acid to the medium. As would be expected, the onset of lactic acid accumulation occurred sooner in low humidities than in high ones since the animals closed the micropyle in response to dehydration. Barnes *et al.* (1963) have estimated that accumulation begins when approximately 25% of the water content of the tissues has been lost. After this, the valves remain closed for much of the time allowing lactic acid to accumulate.

When animals are returned to aerobic conditions after a period of anaerobiosis, they may either excrete the acid end products or reoxidise them and thus repay an ‘oxygen debt’ (p. 283). In both *Balanus balanoides* and *Chthamalus stellatus* there is little evidence of the reoxidation of the accumulated lactic acid. In *Balanus balanoides,* for example, storage for 3 hr under anaerobic conditions resulted in a production of 358 μg of lactic acid which would require 267 μl of oxygen to oxidise it. In fact the rate of oxygen consumption was rather higher after a period of anaerobiosis than before it, but fell after only 1·5 hr to a slower rate, comparable with that before anaerobiosis, the excess oxygen used during the period of increased uptake amounting to only 36 μl compared with the 267 μl required for the complete oxidation of the accumulated lactic acid. Much the same results were obtained with *Chthamalus stellatus* in which the excess oxygen used was only 45 μl whereas the lactic acid produced after 3 hr anaerobiosis was 230 μg which would require 171 μl O_2 for its complete oxidation. In contrast, *Balanus crenatus*, which produces only small amounts of lactic acid (p. 368), showed no evidence of increased oxygen uptake following anaerobiosis. If after a period of anaerobiosis specimens of *Balanus balanoides* were placed in a large volume of aerated seawater, the lactic acid in the tissues diffused out, the rate of loss falling as the lactic acid content of the tissues was reduced; if after such a period of anaerobiosis the animals were placed in air, the lactic acid was neither excreted nor reoxidised. Thus when the rate of oxygen consumption in seawater of both groups of animals was compared, those which had been in seawater respired at a rate which is only 6% greater than normal whilst those animals which had been stored in air following a period of anaerobiosis had an uptake which was 15% greater than normal, reflecting the greater lactic acid content. Nevertheless, in neither case was the increase in oxygen consumption sufficient to reoxidise the accumulated lactic acid, which is therefore mainly excreted.

Thus intertidal barnacles are well adapted to air-breathe and this must suffice to meet the metabolic requirements of the animals during much of the period they are uncovered by the tide. When desiccation is severe, or when other environmental stresses such as reduced salinity occur, the shell valves are closed and anaerobiosis ensues. There is, however, little attempt to reoxidise the lactic acid which is allowed to diffuse into the surrounding seawater when the animals are submersed again by the tide. The possible reason for this lack of resynthesis of carbohydrate is that when the ability to air-breathe is taken into account, anaerobiosis must be for only short periods, so that excretion of the accumulated lactic acid may not unduly deplete the metabolic reserves. Added to this is the observation that the rate of glycogen consumption is relatively low under anaerobic conditions (Barnes *et al.*, 1963).

There are few other studies which closely relate the ability to respire anaerobically with environmental conditions. Nevertheless Lent (1968) has recently shown that aerial oxygen consumption by the mussel *Modiolus demissus* is also closely associated with evaporative water loss during the intertidal period. Some species of mussel are also known to be able to respire anaerobically (for review, von Brand, 1946), and it seems likely that, as in intertidal barnacles, the same sequence of air-breathing followed by anaerobiosis under extreme conditions may occur. We must thus envisage a general sequence of responses which are likely to occur when the animals are uncovered by the ebbing tide. Firstly the full active rate of oxygen consumption falls to the quiescent or standard rate in all but those upper shore species whose respiratory mechanisms are especially adapted to maintain a high level of activity (p. 362). Aerial respiration is adequate to meet the oxygen demands of the animal during quiescence, but with the onset of excessive desiccation, the shell shuts and anaerobiosis ensues. Such anaerobiosis may continue until the acid end products accumulate to lethal levels, or may be partially reoxidised by intermittent air-gaping. On the other hand, in mud-dwelling animals such as *Arenicola marina,* the intertidal period is characterised not so much by evaporative water loss as by a fall of the available oxygen in the surrounding water. Thus when submersed by the tide the oxygen in the water is high and the rate of oxygen consumption is that characteristic of the active animal (i.e. the active rate) and is unaffected by an increase in the pO_2 of the medium up to toxic levels. This range of tensions is known as the 'zone of respiratory independence'. When the oxygen in the surrounding water falls below an 'incipient limiting tension' (P_c) after the tide has ebbed, the respiratory rate falls to that characteristic of the quiescent animal (i.e. the standard rate). A further fall in the oxygen tension below that necessary to meet

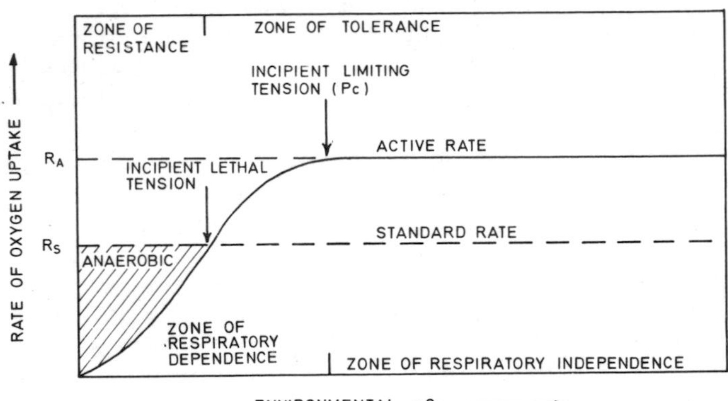

Fig. 7.55. Diagram showing the relation between the rate of oxygen consumption and environmental oxygen tension in an organism showing an 'independent' type of respiration. Note that when the environmental pO_2 is above the incipient limiting tension (P_c) oxygen uptake is independent of environmental pO_2. But below the incipient limiting tension the rate of uptake is dependent upon the pO_2 of the environment. This is due to a reduction in the level of activity until the standard rate is reached. At the incipient lethal tension the environmental pO_2 is no longer great enough to supply the respiratory needs even of the quiescent organism, and an increasing proportion of the metabolism is anaerobic. (Based on Shepard, 1955; Beadle, 1961 and Hoar, 1966.)

the metabolic demand even of the quiescent animal may then lead to anaerobiosis. The range of oxygen tensions below the incipient limiting tension (P_c), during which the rate of oxygen uptake declines with the external pO_2, is known as the 'zone of respiratory dependence' whilst the pO_2 which just fails to meet the metabolic demand of the quiescent animal is the 'incipient lethal tension' (Figure 7.55) (Shepard, 1955; also Fry. 1947; Beadle, 1961, Prosser and Brown, 1961; Hoar, 1966). The sequence of responses in mud-dwelling organisms subjected to a fall in the external oxygen tension is thus similar to that which occurs in animals which are exposed to conditions of desiccation during the emersion period except that the onset of anaerobiosis is controlled by a fall in the oxygen tension below the incipient lethal tension rather than by the degree of evaporative water loss. Such animals are, in addition, often able to tolerate longer periods of anaerobiosis since acid end-products of glycolysis can be excreted into the surrounding water rather than being allowed to accumulate to toxic levels in the tissues.

CHAPTER 8

Factors affecting the rate of oxygen consumption

A. INTRODUCTION

In the preceding chapter we have seen how the mode of oxygen uptake by intertidal organisms is profoundly influenced by the environmental factors prevailing at successive stages in the tidal cycle. We may now examine some of the parameters which influence the rate of oxygen consumption of intertidal animals, with particular reference to possible means by which such organisms adjust to the fluctuations in physico-chemical conditions which characterise the intertidal zone. Neglecting the increase in oxygen consumption which follows a period of anaerobiosis in many marine animals (for example van Dam, 1935; von Brand, 1946 and p. 283) such parameters may be divided into two main groups. Firstly there are endogenous factors, such as the size of animal, amount of activity, the nutritive state, and the degree of development of the gonads of the organisms concerned; the second group comprises environmental factors. These factors may, of course, interact to produce the wide variations in the respiratory rate which have been recorded in the literature for any one species.

B. ENDOGENOUS FACTORS

Assuming that environmental conditions are constant, the most important factors influencing the respiration rate of a particular organism are its level of activity and its body weight, although a large number of other factors such as the sex of the organism, its nutritive state, and the degree of development of the gonads all play a part in determining the level of oxygen consumption. In addition, many animals are known to show circadian rhythms in metabolic rate, the period length being independent of temperature (for review Harker, 1964).

1. THE INFLUENCE OF ACTIVITY ON OXYGEN CONSUMPTION

As has been shown on p. 356, the level of activity of a particular organism profoundly affects the level of oxygen consumption. In some fishes there is a linear relationship between oxygen uptake and swimming activity (Spoor, 1946; Fry and Hart, 1948; Beamish and Mookherjii, 1964; Muir, Nelson and Bridges, 1965) and this is also true of the intertidal amphipod *Gammarus oceanicus* (Halcrow and Boyd,

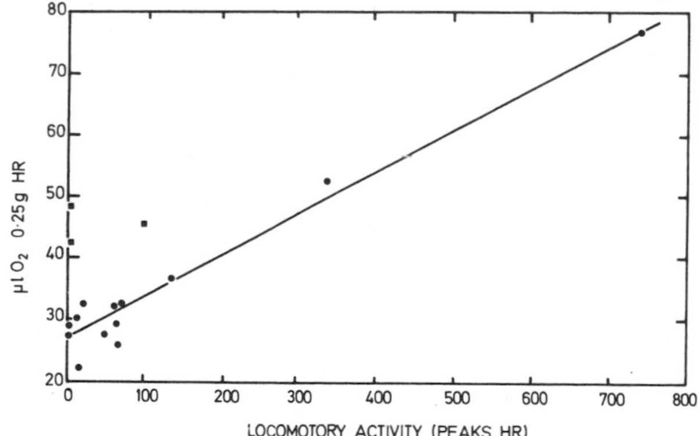

Fig. 8.1. Diagram showing the relationship between locomotory activity (peaks/hr) and oxygen consumption (μl 0_2/0·25g./hr.) of individual *Grammarus oceanicus* at 20°C. Solid circles indicate intermoult individuals and squares recently moulted individuals. (After Halcrow and Boyd, 1967.)

1967), but in *Leander adspersus* there is a semi-logarithmic relationship between oxygen uptake and swimming activity (Ivlev, 1963, cited in Halcrow and Boyd, 1967). The relationship between locomotory activity and oxygen consumption in individual *Gammarus oceanicus* at 20° C is shown in Fig. 8.1. The regression line was fitted by the method of least squares and the intercept *a* with the *y* axis was found to be 25·9 whilst the correlation coefficient was 0·97. The squares indicate the corresponding respiration rate of recently moulted animals and it is clear that the level of metabolism of such animals is very much higher than that of the intermoult individuals at comparable levels of activity. This may be related to the mobilisation of metabolic reserves which occurs at this time (Passano, 1960; also Halcrow and Boyd, 1967) and has also been noted in the crab *Pachygrapsus crassipes* (Roberts, 1957). Another factor which affected the level of oxygen consumption in *Gammarus oceanicus* was the duration of the experiment. When a re-

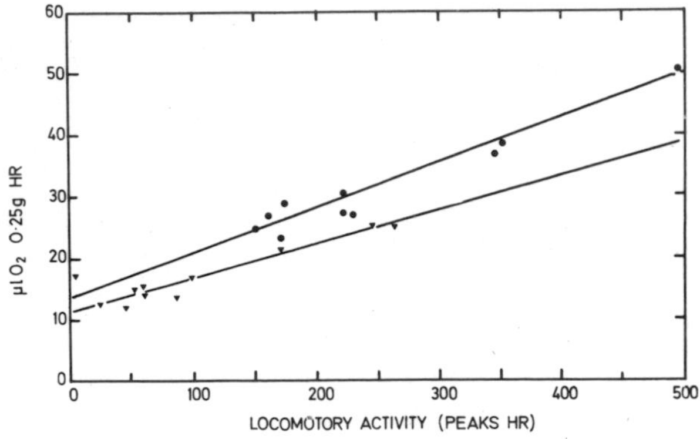

Fig. 8.2. Diagram showing the relationship between locomotory activity (peaks/hr) and oxygen consumption of individual *Gammarus oceanicus* at 10°C. The solid circles indicate the results obtained within 1 hr of insertion of the animal into the respirometer and the triangles are results from an experiment started 4–5 hr after insertion of the animals into the respirometer. (After Halcrow and Boyd, 1967.)

gression line relating activity to oxygen uptake was based on measurements made within one hour of putting the animals into the respiration chambers it was found to be higher than a similar line based on measurements started after 4 – 5 hr. These results are shown in Fig. 2, the value of the intercept *a* being 13·4 for the short-term experiment but only 11·3 for the long-term one. The correlation coefficients *r* were 0·93 and 0·79 respectively. Although the difference between the intercept values, which provide a convenient estimate of the respiration rate at zero activity (i.e. the standard rate, p. 357), was found to be insignificant in *Gammarus oceanicus*, a similar increase in the standard rate has also been noticed in certain fishes and has been attributed to the disturbance of the respiration chamber (for example Smit, 1965). The results on *Gammarus oceanicus* and fishes thus emphasise the importance not only of the correlation of oxygen uptake with activity level so that the standard rate under any particular set of environmental conditions can be estimated, but also of other endogenous factors such as stage in the moulting cycle and degree of "excitement".

Whilst the above method for the estimation of the standard rate of oxygen consumption and for the active rate under particular experimental conditions has been extensively used for fishes and also for some invertebrates, it is suitable only for organisms whose activity level can be easily recorded. It is correspondingly more difficult to record a wide variety of activity and respiration rates in the vast majority of

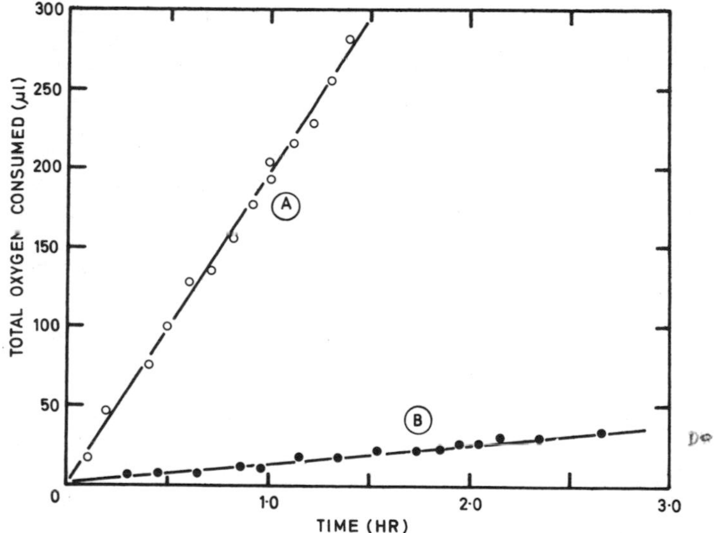

Fig. 8.3. Graph showing the oxygen consumption of a specimen of *Littorina littorea* at 10°C. Open circles show the oxygen consumption of the animal when it was actively crawling (A). Solid circles show results for the same animal when it was extruded but quiescent (B). (From unpublished data of Newell and Northcroft.)

intertidal animals such as anemones, polychaetes and gastropods. An alternative approach has therefore been used on a variety of intertidal and sublittoral organisms including the lancelet *Branchiostoma lanceolatum* (Courtney and Newell, 1965), the barnacle *Balanus balanoides* (Newell and Northcroft, 1965), the anemone *Actinia equina,* the polychaete *Nephtys hombergi,* the winkle *Littorina littorea* and the cockle *Cardium edule* (Newell, 1966; Newell and Northcroft, 1967). The animals were sealed into glass respiration vessels of approximately 100 ml capacity containing aerated seawater, and the fall in oxygen concentration was recorded at frequent intervals by means of an oxygen electrode over a period of several hours. Fig. 8.3 shows the results of such an experiment on the winkle *Littorina littorea* at 10° C. Curve A corresponds with the oxygen uptake of an animal which was actively crawling for much of the time during which measurements were made, whilst in curve B the same animal was quiescent (although it had not withdrawn into the shell) and its rate of oxygen consumption was correspondingly lower. The dried weight of the tissues was 48 mg, the rate of oxygen consumption during quiescence was 12·5 μl/mg dry weight/hr, and during crawling 190 μl/mg dry weight/hr. Now it is obvious that few experimental animals are likely to show such prolonged periods of quiescence and activity. Indeed, the specimen shown in Fig. 8.3 was one of only a few showing such uniform rates of oxygen consumption out of several hundred whose individual respiratory rates have been recorded (Newell and Northcroft, unpublished data). It is possible, however, to record fluctuations in the rate of oxygen consumption of an animal which is showing periods of activity alternating with quiescence, provided that sufficiently frequent recordings are

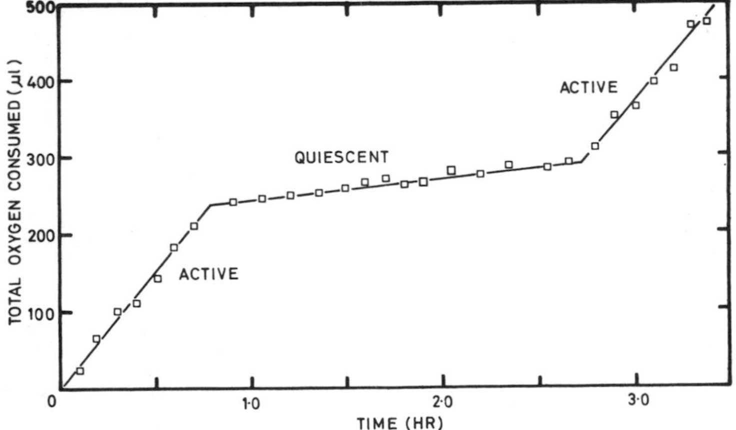

Fig. 8.4. Graph showing the oxygen consumption of a single specimen of *Littorina littorea* of dry weight 23 mg at 15°C. The periods of quiescence and crawling activity are indicated. (From unpublished data of Newell and Northcroft.)

made and that the periods of quiescence and activity are of long enough duration to allow a reliable estimation of the respiration rate during each phase. Figure 8.4 shows a typical record of the oxygen consumption of a specimen of *Littorina littorea* at 15° C which was showing an initial fast rate of oxygen consumption which corresponded with active crawling followed by a slower rate of respiration during which the animal was extruded but quiescent. The animal then began to crawl again and its oxygen consumption increased.

Mangum and Sassaman (1969) have likewise found that in the polychaete *Diopatra cuprea* variations in the rate of oxygen consumption could be correlated with the active and resting periods of the activity cycle of this animal. Measurements made on other intertidal organisms show similar variations in oxygen uptake and allow a maximal and minimal rate of oxygen consumption to be calculated. Such rates are related to the body weight in a predictable fashion (p. 386) and when observed to coincide with activity and quiesence may be regarded as corresponding with the "active rate" and the "standard rate" of oxygen consumption as estimated by direct measurements of the relationship between activity and oxygen consumption (p. 373). The values for the active and standard rates obtained by the method described above, as well as those obtained by Halcrow and Boyd (1967) on *Gammarus oceanicus* and by a number of other workers for other intertidal animals, are shown in Table XIV, p. 358).

It is possible that in an organism such as, for example, the limpet *Patella vulgata,* which may show extensive periods of quiescence, the rate of oxygen uptake may be approximately linear for long periods. In this case a line fitted to all the points relating oxygen uptake to time would give data similar to that shown in Fig. 8.3 (line B) for the occasional specimen of *Littorina littorea,* and would thus give a reliable estimate of the standard rate of oxygen consumption. Similarly, if the browsing excursions were maintained for long periods under experimental conditions, a rate approximating to the active rate would be anticipated. Data presented by Davies (1965) seem to show that the minimal and maximal rates of oxygen consumption in relation to body weight of high level specimens of the limpet *Patella vulgata* at 15° C are correlated to body weight in a similar way to the one common regression line relating all the points to body weight (p. 387); this does not apply to comparable measurements on low-level limpets. Accordingly, the estimated standard and active rates for high level *Patella vulgata* shown in Table XIV (p. 358) have been recalculated from the maximal and minimal lines. A very similar relationship between the maximal and minimal rates of oxygen consumption in relation to body size is also apparent in the data presented by Dehnel (1958) for the crab *Hemigrapsus oregonesis* (also p. 403).

Finally, the level of oxygen consumption during quiescence (the standard rate) has been estimated in barnacles by the careful excision of the animal from the shell (for review, Barnes, Barnes and Finlayson, 1963a,b). This method has proved of considerable value in an estimation of seasonal and thermal variations in the standard rate of oxygen consumption of barnacles, but does not yield information on the relation between the active and the standard rate. The results of work carried out on barnacles using the excision method is discussed in more detail on p. 384.

2. THE INFLUENCE OF BODY SIZE ON OXYGEN CONSUMPTION

The influence of body size on oxygen consumption has been discussed in detail for a large number of organisms by Kleiber (1932, 1947), Brody and Procter (1932), Brody (1945), Zeuthen (1947, 1953), Hemmingsen (1950, 1960) and Bertalanffy (1957). There is also a vast literature on the effects of size on the metabolism of particular poikilotherms, some of which is reviewed by Hemmingsen (1960). We shall deal here only with some of the general principles which have emerged from comparative studies on homoiotherms, poikilotherms and plants before going on to review examples of the effect of size on the metabolism of selected intertidal invertebrates.

According to Zeuthen (1947, 1953) and Hemmingsen (1950, 1960) and many other workers, the temperature corrected standard metabolism of protozoans, marine eggs, the larger metazoan poikilotherms and even homoiotherms is proportional to a constant power of the body weight. This factor has been shown to be 0·751 ± 0·015, although in some species there is some variation which may be partly attributable to the small weight range over which measurements have been made (Hemmingsen, 1960) (p. 390). The relation between the standard metabolism of whole animals and body weight is expressed as:—

$$\text{metabolism} = k \,.\, \text{body weight}^{\,b}$$

where k and b are constants.

Since the body weight is plotted on the X-axis and metabolism on the Y-axis it is convenient to rewrite the equation as:—

$$Y = k \,.\, X^b$$

where Y = the metabolism (in Calories/hr or in oxygen consumption) and X is the body weight (or body nitrogen). In its logarithmic form the equation becomes:–

$$\log Y = \log k + b \, . \log X.$$

When plotted on logarithmic axes, a straight line rather than an exponential curve is obtained relating standard metabolism to body weight; the slope of the line gives the constant b, and the intercept on the Y axis gives the value for the constant k. Since the intercept is normally designated a, it is convenient to write the overall equation as:–

$$Y = a \, . \, X^b$$

$$\text{or } \log Y = \log a + b \, . \log X$$

where Y = the metabolism, a = the intercept of the line on the Y axis, b = the slope of the line, and X = the body weight (or body nitrogen).

The straight lines relating the log metabolism (calories/hr) to log body weight in unicellular organisms and in the poikilotherms studied by Hemmingsen (1950) are shown in Fig. 8.5. It is clear that the line relating the log metabolism of unicellular organisms including marine eggs to the log body weight is of the same slope as that relating log metabolism of a variety of metazoan poikilotherms to log. body weight, but is at a lower level. Hemmingsen (1950, 1960) has shown that the slope, b, of the lines corresponds to 0·751 ± 0·015, whilst the difference in level between the lines for metazoan poikilotherms and for unicellular organisms is 0·913 ± 0·13 of a log decade and amounts to an 8·1 fold increase in the metabolism compared with unicellular forms. Between the upper end of the line for unicellular organisms and the lower end of the line for larger metazoans, there is a steeper line which represents the influence of body weight on the metabolism of minute metazoans.

Both Zeuthen (1947, 1953) and Hemmingsen (1950; 1960) have shown that there are thus three phases in the relation of the metabolism of poikilotherms to body size. The first phase ("phase 1" in the figure) includes unicellular organisms, the slope b (n of Hemmingsen, 1950) being 0·75 and the whole line being lower in level than the metazoan poikilotherm line. The constant log a, which represents the level of the line illustrated in Fig. 8.5 is 4·074 ± 0·110 (Hemmingsen, 1960) so that the overall equation relating the metabolism of unicellular organisms to body weight is:–

$$\log Y = 4{\cdot}074 + 0{\cdot}751 \, X.$$

All data are expressed in terms of the metabolism at 20°C.

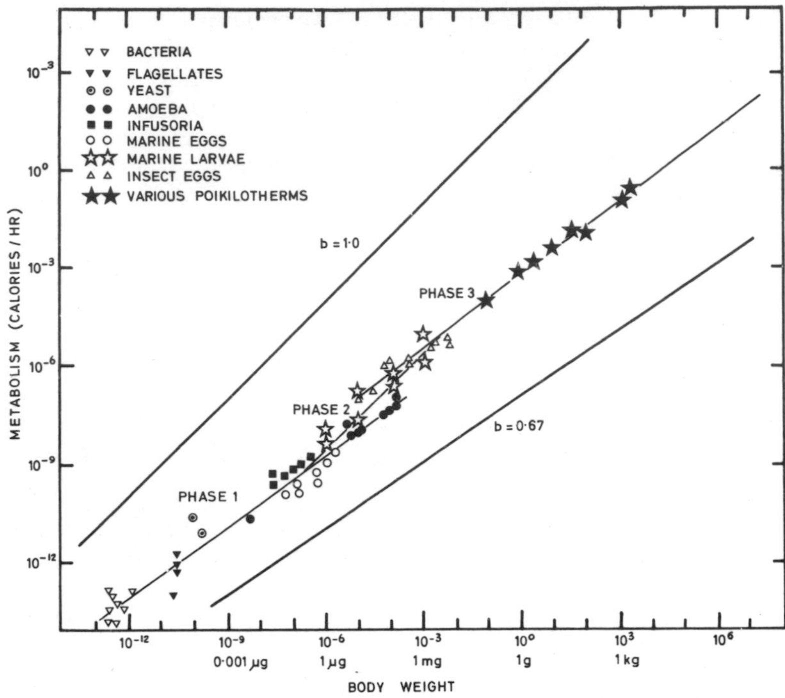

Fig. 8.5. Graph showing the relationship between metabolism (cals./hr.) and body size in a wide variety of poikilotherms. Data corrected to 20°C. (Based on Hemmingsen, 1960.)

The second phase ("phase 2" in the figure) applies to small metazoan poikilotherms weighing from 0·1 μg to 1·0 mg (Hemmingsen, 1950) or to 40 mg (Zeuthen, 1953). In these small organisms the slope of the line relating log metabolism to log body weight is very steep (0·9 – 1·0; Zeuthen, 1953; Hemmingsen, 1950, 1960). The third phase applies to all other metazoan poikilotherms and plants up to the largest sizes and starts at 1·0 – 40 mg, the slope for this phase again being 0·751. The constant log a for the metazoan poikilotherm line at 20° C is 3·161 ± 0·068 log decades, so that the equation relating metabolism at 20° C to body weight in such poikilotherms is:–

$$\log Y = 3{\cdot}161 + 0{\cdot}751\ X$$

In fact, as Hemmingsen (1960) points out, there is no sharp boundary between the weight of the largest unicellular organisms and that of the smallest metazoan poikilotherms of the "phase 2" line referred to above. Neither is there a sharp demarcation between the largest of such small metazoans and the smallest of the species which fall onto the main "phase 3" poikilotherm line. Phase 2 thus starts at body weights of approximately 0·1 μg before the upper end of the line for unicellular organisms (phase 1), giant amoebae having body weights of approximately 0·1 mg and approaching the end of the phase 1 line. Equally, the steep phase 2 line joins the metazoan poikilotherm line (phase 3) at approximately 40 mg even though the latter may actually begin with insect eggs at 0·01 mg. In some organisms such as mammals and birds and some insects the maximal or active rate of metabolism is also proportional to the 0·75 power of the body weight and exceeds the standard rate by about 20 times. (for review Hemmingsen, 1960).

In view of the apparently fundamental relationship between the log metabolism and log body weight, several workers have attempted to suggest a mechanism underlying the phenomenon. Zeuthen (1953) suggested that the metabolism of cells was related to their surface area, so that the total metabolism of metazoans would be related to their aggregate cell surfaces. In this case metabolism would be expected to increase with the 0·67 (= $\frac{2}{3}$) power of the body weight. In fact, as Hemmingsen (1950, 1960) has shown, the metabolism varies more nearly with the 0·751 (=$\frac{3}{4}$) power of the body weight; he has suggested that Zeuthen's hypothesis may be modified in such a way that the metabolism is proportional, not to the cell surface itself, but to additional factors such as internal convection, vascularisation and the development of complex respiratory surfaces. In unicellular organisms, (p. 379 and Fig. 8.5) the curve relating log metabolism to log body weight is linear with a slope of 0·751, so that the rate of increase above 0·67 may be related to internal membrane surfaces or to convection within the contents of the cell. By the time such spherical organisms had reached 0·5 – 1·0 mg they would eventually be limited in their metabolism by the diffusion of oxygen. This weight thus corresponds approximately with the maximum theoretical upper weight limit of the "unicellular line" (phase 1 in Fig. 8.5). Any increase in the entry of oxygen due to vascularisation, or alteration from a spherical shape to give a greater surface area, would allow a metazoan to metabolise faster than a spherical unicellular organism of the same weight. Such factors may account for the transition from phase 1 through phase 2 to the metazoan poikilotherm line phase 3. Now the slope of this last line is not 0·67 but 0·751, so that something other than a mere increase in the total cell surface must account for the slope of the line. It seems possible that the development of complex respiratory surfaces in the higher metazoans may account for the increase in the metabolism above the value of $b = 0{\cdot}67$ expected on a basis of the aggregate cell surface.

We have now seen that the metabolism of large animals exceeds that of small animals and is commonly related to the 0·751 power of the body weight in the unicellular line (phase 1) and in the metazoan poikilotherm line (phase 3). In phase 2, however, the slope b is nearly 1·0 and in this respect the metabolism is more nearly proportional to body weight rather than to other factors including the total surface area. It follows that as unicellular organisms (phase 1) and the larger metazoan poikilotherms (phase 3) grow larger, the metabolism does not increase at the same rate as the body weight. If this occurred, then the value of the slope would be 1·0. Thus the respiration or metabolism per unit weight which can be expressed as oxygen consumption in ml O_2 (at STP)/gm dry weight/hr (designated Q_{O_2}) of small animals is greater than that of large ones. This metabolism per unit weight is the weight specific metabolic rate of the organism. Since the metabolism $Y = aX^b$ (p. 379), then by dividing throughout by X, the weight specific metabolic rate is given by:–

$$\frac{Y}{X} = aX^{b-1}$$

where Y = the oxygen consumption, X = the body weight, and a and b are constants. As Davies (1966) pointed out, since b is less than 1·0 (see above), $b-1$ has a negative value and has often been referred to as either b or $-b$ in the literature. To avoid such confusion, he suggested that the equation relating weight specific metabolic rate to body size be rewritten as:–

$$Y' = aX^{b'}$$

Where Y' = the respiration rate $(\frac{Y}{X})$
X = the body weight
a = a constant.
and b' = the specific exponent of weight $(b-1)$.

Hence the equation becomes:–

$$\log Y' = \log a - b' \log X$$

(Where the constant $\log a$ is the intercept on the Y axis and b' = the slope $b - 1$) rather than:–

$$\log \frac{Y}{X} = \log a + (b-1) \log X.$$

As with the influence of size on the metabolism, there is abundant evidence in the literature of the negative exponential relationship between metabolic rate (or respiratory rate) and body weight (or body nitrogen). This relationship was established by Zeuthen (1947) for a wide variety of marine organisms, reptiles, birds and mammals and has since been confirmed in many intertidal organisms. Ellenby (1951, 1953; also Ellenby and Evans, 1956) for example, has shown that the oxygen consumption of the high shore level isopod *Ligia oceanica* is a

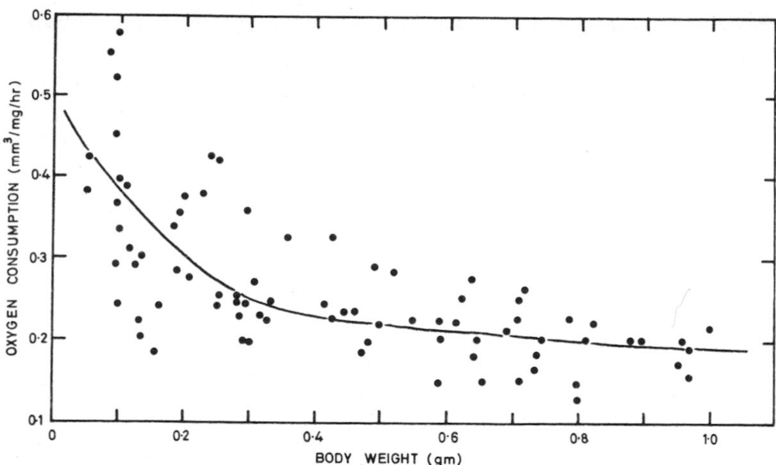

Fig. 8.6. Graph showing the rate of oxygen consumption (mm^3/mg/hr.) at 25° C of *Ligia oceanica* plotted as a function of body weight (grams). Oxygen consumption per mg was proportional to the $-0{\cdot}274$ power of the body weight. (After Ellenby, 1951.)

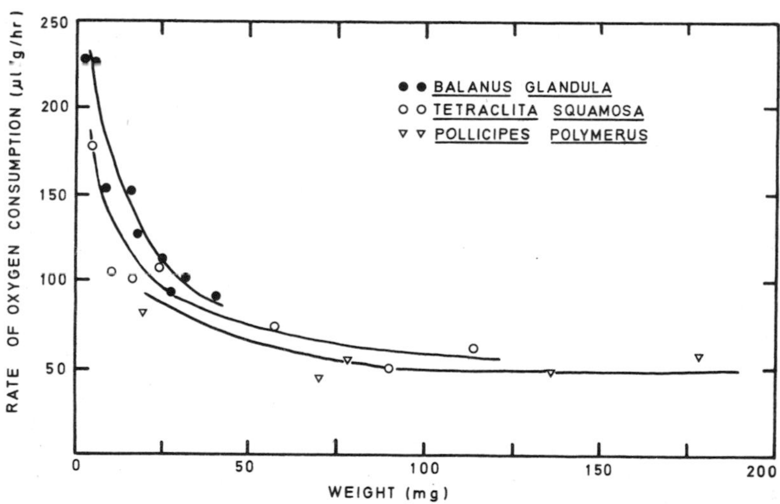

Fig. 8.7. Graph showing the relationship between the rate of oxygen consumption (µl/g/hr.) at 10°C and body weight (mg) of excised specimens of the barnacles *Balanus glandula, Tetraclita squamosa* and *Pollicipes polymerus.* (After Barnes and Barnes, 1959.) Note reduction of scatter compared with data illustrated in Fig. 8.6.

negative exponential function of the body weight. The total oxygen consumption was proportional to the 0·726 power of the body weight so that the value for $b-1$ (or b' of Davies, 1966) was −0·274. The results are shown in Fig. 8.6 from which it is evident that there is considerable scatter about the calculated curve, which may be partly attributable to the differing levels of activity of the animals concerned. Barnes and Barnes (1959) and Barnes, Barnes and Finlayson (1963a,b) have eliminated the scatter due to activity in many barnacles (with the exception of *Chthamalus* sp) by excision of the animals from the shell, and in this way have been able to compare the effect of size, as well as a number of other parameters (p. 392), on oxygen consumption. Barnes and Barnes (1959) studied the relation between oxygen uptake and body weight in several genera of intertidal barnacles from the Pacific coast of North America, and also compared variations in the metabolic rate with shore level. The relation between wet tissue weight and the respiration rate of *Balanus glandula, Tetraclita squamosa* and the stalked barnacle *Pollicipes polymerus* is shown in Fig. 8.7; these results are similar to those obtained on *Balanus cariosus, B. rostratus, B. crenatus, Chthamalus dalli* and *C. fissus*. Much the same type of curve has been obtained by Zeuthen (1947), who related the rate of oxygen consumption to body nitrogen in many animals including *Balanus* sp., *Nassa* (= *Nassarius*) sp., *Littorina* and *Mytilus.* Some of these data are shown in Fig. 8.8. Ganapati and Prasada Rao (1960) and, more recently, Prasada Rao and Ganapati (1969) have shown that the

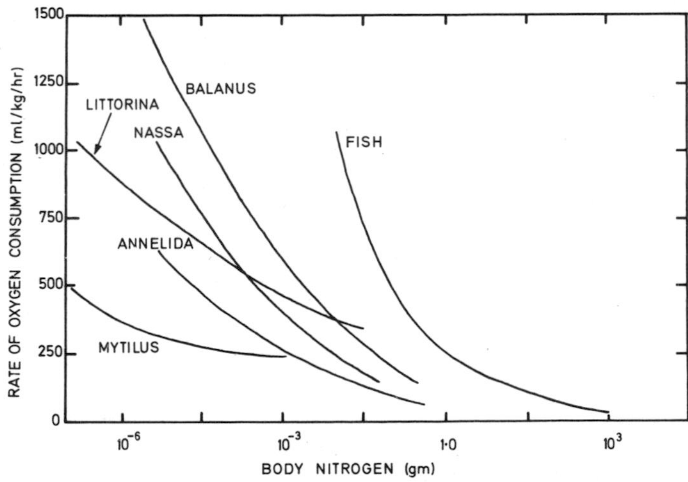

Fig; 8.8. Graphs showing the relation between the rate of oxygen consumption (ml O_2/kg/hr.) and body nitrogen. (After Zeuthen, 1947.)

relationship of the weight specific respiration rate to body size ($b-1$) varies significantly at 25°C in the barnacles *Balanus tintinnabulum tintinnabulum* and in *B. amphitrite amphitrite.* In the former, oxygen uptake increased with the 0·658 power of the body weight (i.e. $b-1 = -0{\cdot}342$) whilst in *B. amphitrite amphitrite* b was 0·872 (i.e. $b-1 = 0{\cdot}173$) so that even within one genus the metabolism may vary in a different way with body size. It may also vary with season and temperature as is discussed on p. 416.

It is often convenient to rectify the data so that a straight line is obtained relating the log oxygen consumption to the log body weight or log body nitrogen. Barnes and Barnes (1959), for example, replotted their data for the metabolic rate of excised barnacles and were able to show that the slopes of the lines for all the species were similar, the common value for the slope $b-1$ being 0·3333. On the other hand the level of the lines varied, being in general low for high shore barnacles and increasing in lower shore species (also Southward 1955b,c). Since the value of $b-1$ was 0·3333, the total metabolism varied with the 0·6667 power of the body weight, a factor which would occur if the metabolism was related to the total surface area of the barnacle (p. 381, and also Ellenby, 1951). The relationship between the logarithm of the respiration rate ($\mu l\ O_2$/g/hr) and the logarithm of the wet weight of the tissues for *Chthamalus* sp., *Pollicipes polymerus, Balanus glandula* and *B. rostratus* is shown in Fig. 8.9. The corresponding values for the intercept a which gives the level of the regression line, were 2·190 for

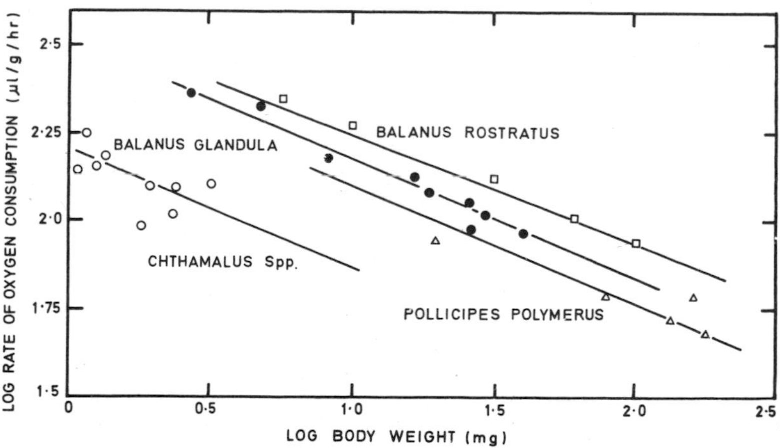

Fig. 8.9. Graphs showing the relation between the log oxygen consumption (µl/g/hr.) at 10°C and log body weight (mg) of excised barnacles. (After Barnes and Barnes, 1959.)

Chthamalus, 2·428 for *Pollicipes polymerus,* 2·504 for *Balanus glandula* and 2·575 for *Balanus rostratus.* Thus the sublittoral *B. rostratus* has a higher metabolic rate than the mid-tidal *Balanus glandula,* and this in turn has a higher metabolic rate than the upper shore *Chthamalus* sp. *Pollicipes polymerus* was the only pedunculate or stalked barnacle of the series, and its general level of metabolism was similar to that of the lower shore balanoids. This animal occurs on the lower shore and thus its level of metabolism fits in with the general reduction which occurs in species occupying the upper shore levels and which culminates in the high level chthamalids. As Barnes and Barnes (1959) pointed out, such a reduction in metabolism would be one way in which upper shore suspension feeding organisms might overcome the restriction in available feeding time which occurs in the upper intertidal zone (p. 15).

An alternative method of obtaining a linear relationship between the metabolic rate and body weight is to use a logarithmic scale on the ordinate and abcissa rather than to convert the X and Y values into logarithms. Davies (1965, 1966, 1967) for example studied the effect of size and a number of other parameters (p. 395) on the metabolic rate of the limpet *Patella* sp. He found that there was no significant difference in the effect of body weight on the metabolic rate in high and low shore *Patella vulgata* and in *P. aspera,* the value for b' (or $b-1$, p. 382) for high level *P. vulgata* was –0·2715, for low level *P. vulgata* –0·3003 and for *P. aspera* –0·3484. Since there was found to be no significant difference between these values, a common regression coefficient of –0·3042 ± 0·006 was adopted. This means that since $b' =$ –0·3042, the metabolism (as opposed to the metabolic rate) was proportional to the 0·6958 power of the body weight. The relationship between the respiration rate and the dry weight of the tissues of low shore level individuals of *Patella vulgata* at 15° C is shown in Fig. 8.10 B, whilst the relation between that of high level *P. vulgata* at the same temperature and wet weight of body tissues is shown in Fig. 8.10 A. Although the calculated regression lines clearly give a reliable index of the effect of size on the metabolic rate of *Patella* species and individuals from different shore levels (Davies, 1965, 1966, 1967), the fact that in Fig. 8.10 A the minimal respiration rates form a line which is parallel to that calculated for all the points, is of some significance. Similarly in Fig. 8.10 B the points forming the maximal rates recorded form a line which bears the same relationship to body weight as the line fitting the pooled data. This indicates that whilst the metabolic rate may vary considerably in any one animal according to a variety of factors including the level of activity, the active and standard rates set an upper and lower limit to the rate of metabolism.

As has been shown on p. 376, values approximating to the active and

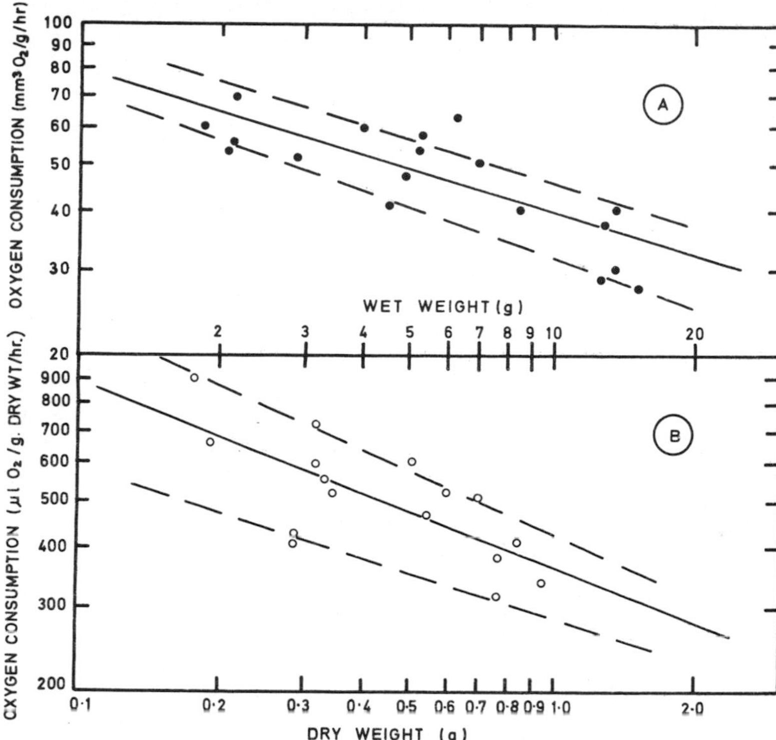

Fig. 8.10. Graphs showing the relationship between the rate of oxygen consumption and body weight in *Patella vulgata*. (A) High level animals at 15°C. (B) Low shore animals at 15°C. (Data from Davies, 1965, 1967.) The continuous lines indicate calculated regression line for the pooled data whilst the broken lines indicate the extremes of the data and have been added to indicate the levels of the active and standard rates of metabolism.

standard rates of oxygen consumption can be recorded in organisms which may vary the activity level from quiescence to the maximal level over a short period of time. Provided that sufficiently frequent recordings are made to provide a reliable estimate of the differing levels of metabolism associated with quiescence and with maximal activity, the data fall within two well defined limits when the log oxygen consumption is plotted as a function of log body weight. A line relating the maximal level of respiration may then be regarded as defining the active rate, whilst the regression line fitted to the minimal points defines the standard rate, much as has been indicated for the data on *Patella vulgata* (Fig. 8.10 A, B). Clearly, the simplest situation occurs in organisms which show a limited repertoire of activity levels. In the

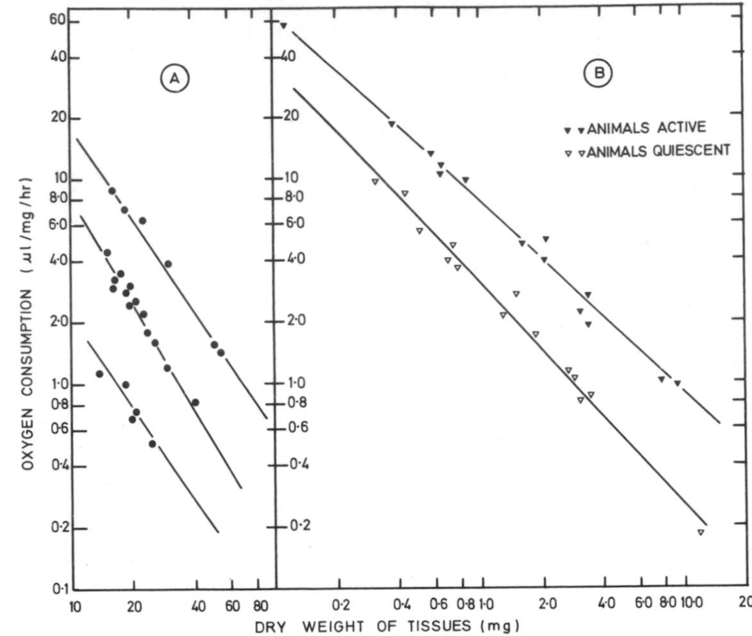

Fig. 8.11. Graphs showing the relationship between oxygen consumption (μ l mg/hr) and the dry weight of tissues (mg) in (A) *Branchiostoma lanceolatum* (15°C) and (B) *Balanus balanoides* (16·5°C). Regression lines have been fitted by the method of least squares. (Data from Courtney and Newell, 1965 and Newell and Northcroft, 1965.)

lancelet *Branchiostoma lanceolatum*, for example, the animal was found to show three different rates of oxygen consumption (Courtney and Newell, 1965). Consequently when the respiration rates of animals of a wide range of sizes were plotted against body weight, three different lines were obtained (Fig. 8.11A). Covariance analysis confirmed that these lines were significantly different in level but not in slope, so that each bore a similar relation to body weight. Such rates were shown in animals buried in sterilised gravel, and were unlikely to be an expression of variations in body movement since this would be expected to give rise to a continuous series of possible rates of oxygen consumption up to the active rate during full swimming activity. In any event the animals spent their time buried in gravel and extensive movement would be unlikely and rapid swimming impossible. In this animal, then, the maximal rate in gravel probably did not represent the maximal rate for the animal. However the three different rates of oxygen consumption were attributable to three different inhalant stream velocities which were noted in the lancelet. The fast inhalant stream was set up

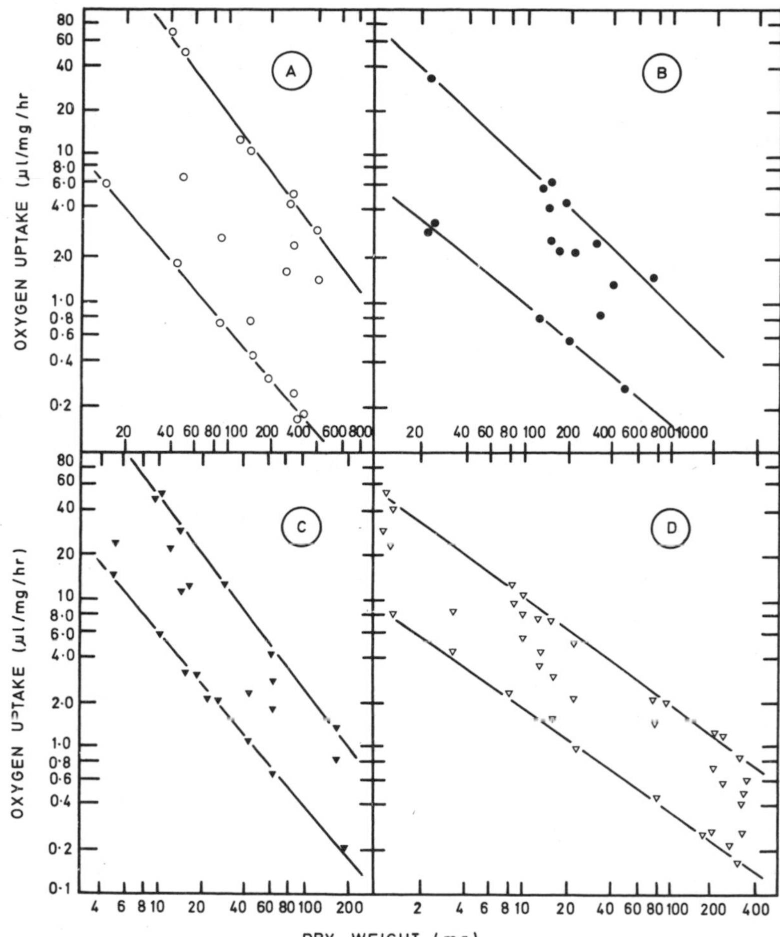

Fig. 8.12. Graphs showing the relationship between the rate of oxygen consumption (µl/mg/hr) and tissue weight (mg) in (A) the anemone *Actinia equina* at 15·5°C; (B) the polychaete *Nephtys hombergi* at 15°C; (C) the winkle *Littorina littorea* at 15°C and (D) the cockle *Cardium edule* at 10°C. (Data from Newell and Northcroft, 1967.) Note the wide scatter between the extremes set by the "active rate" and the "standard rate" both of which bear a similar relationship to body weight.

by all the cilia of the pharynx and comprised the feeding stream; this rate of flow was thought to correspond with the fast rate of oxygen consumption noted earlier. The second type of inhalant stream was slower and set up by lateral cilia which are under nervous control (Bone, 1958, 1961) and may have accounted for the intermediate rate of oxygen consumption. Finally, the slowest rate of oxygen uptake coincided with the cessation of ciliary activity in the pharynx and thus represented the standard rate of oxygen consumption.

A similar situation was found to exist in the barnacle *Balanus balanoides* (Newell and Northcroft, 1965). As shown on p. 384, measurement of the respiration rate of excised specimens yields interesting comparative data on variations in the standard rate between species occupying different zones on the shore, but does not give a measure of the active rate of oxygen consumption. Barnacles are sedentary and have a limited repertoire of movements (Crisp and Southward, 1961), the main one consisting of the well-known normal cirral beat which alternates with almost complete quiescence. Thus the respiratory rate at full and zero cirral activity closely corresponds with the active and standard rates of respiration. Instead of three different levels of oxygen consumption which vary in a similar way with body weight as in *Branchiostoma lanceolatum,* there are two main rates each of which has a similar slope when plotted on a logarithmic scale against the log dried weight of the tissues. These results are illustrated in Fig. 8.11. B.

Much the same relationship between the active and the standard rate of respiration and body weight has been established for a large number of other intertidal organisms including the anemone *Actinia equina,* the polychaete *Nephtys hombergi,* the winkle *Littorina littorea,* the cockle *Cardium edule* (Newell and Northcroft, 1967) and the mussel *Mytilus edulis* (Newell and Pye, in press). In these organisms, however, there was a continuous variation between the minimal recorded rates of oxygen consumption and the maximal recorded rates although, as in *Patella vulgata* (Fig. 8.10), *Branchiostoma lanceolatum* and *Balanus balanoides* (Fig. 8.11) the regression lines relating the maximal and minimal rates of respiration did not differ in slope from one another. A selection of such lines are shown in Fig. 8.12 from which the wide variation in respiratory rates between the extremes set by the active and standard rates will be noted. Such variation is to be expected in animals which show a greater variety of activities.

Thus studies on the effect of size on the respiration of intertidal organisms indicates that both the active and the standard rates of oxygen consumption vary in a similar way with tissue weight. In most instances the slopes of the lines indicate that the metabolism varies with approximately the 0·751 power of the body weight as described by Hemmingsen (1950, 1960), although exceptions are known in which the respiration is nearly independent of body weight (p. 420). Other exceptions may be due to the relatively small size range of organisms concerned. When all such lines are pooled, there is an approximately even distribution of those steeper and those shallower about the common regression line of $b = 0{\cdot}751$ (Hemmingsen, 1960). In many instances the respiration rates have not been measured at comparable levels of activity so that regression lines relating active and standard rates cannot be calculated; instead a single regression line for the pooled data with a correspondingly wide scatter about the line is obtained (for example, Rao, 1958; Dehnel, 1958; Davies, 1965, 1966, 1967). Such common regression lines are of value for comparative purposes, but do not allow an investigation of the effect of such parameters as tidal level, temperature and salinity to be made separately on the active and standard rates. This factor is of some importance since, as we shall see on p. 402, there is abundant evidence derived by a variety of methods that temperature at least, does not affect the active and standard rates of metabolism in the same way. Again, a fall in oxygen tension commonly leads to a reduction of the active rate but not of the standard rate until the incipient lethal tension is reached when anaerobiosis may ensue (p. 370).

3. OTHER ENDOGENOUS FACTORS AFFECTING THE RATE OF RESPIRATION

We have confined our attention mainly to the effects of activity and size on the metabolic rate of a variety of intertidal organisms, but the level of respiration is also affected by an enormous number of other factors. Some of these are endogenous insofar as they are the immediate factors which govern the respiratory rate, but nevertheless may be linked to other factors such as tidal and lunar cycles, temperature and abundance of food. The division into endogenous and environmental factors is thus somewhat arbitrary in this context, but forms a convenient basis for a discussion of some of the more complex factors which may influence the level of respiration.

One of the most significant of such factors is the degree of starvation to which the organism has been subjected during the experimental period. This factor has been clearly demonstrated by Saunders (1963) for the cod *Gadus morhua* which has been studied under carefully controlled conditions; this allows a comparison to be made with data obtained on intertidal invertebrates. Fig. 8.13 shows a double logarithmic plot of the oxygen consumption at 15°C of fishes of a wide variety of sizes against the weight of the body. The oxygen consumption of starved fishes was measured over a period of several days, the results being shown in the lower line of Fig. 8.13. After the routine level of respiration of the starved fishes had been determined, the fish were fed with herring and left for 1 hr after which any uneaten food was removed. The oxygen consumption was then measured and found to increase, reaching a maximum some 12 – 24 hr later (upper line of

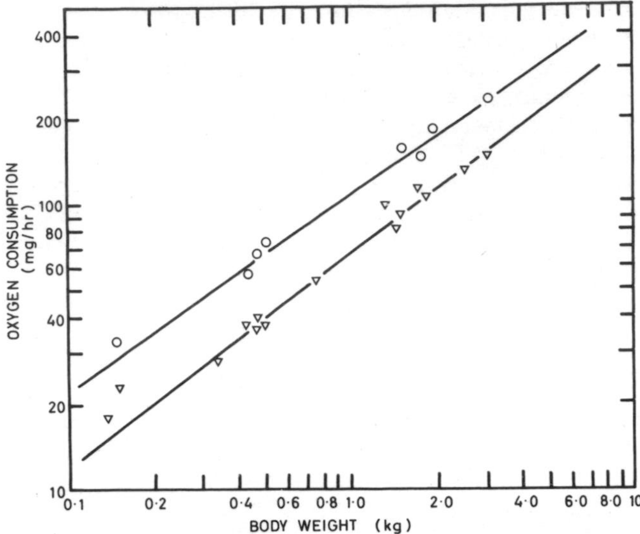

Fig. 8.13. Graphs showing the relation between the routine oxygen consumption of the cod *Gadus morhua* (mg O_2/hr) and body weight (kg). Open circles and upper regression line indicate fishes which had been recently fed whilst the open triangles and lower regression line indicate the results on starved specimens. (After Saunders, 1963.)

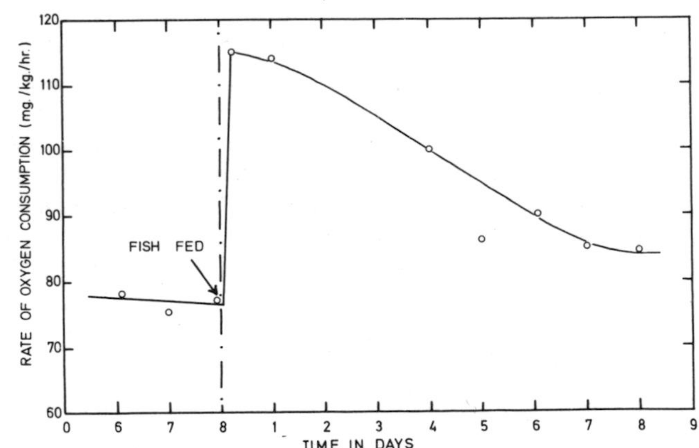

Fig. 8.14. Graph showing the duration of the period of enhanced oxygen consumption following feeding in the cod *Gadus morhua* at 10°C. The point at which the fish were fed following 8 days starvation, is indicated by an arrow and a broken line. (After Saunders, 1963.)

Fig. 8.13). The value for the slope *b* was 0·7572 for the fed animals and 0·8451 for the starved ones whilst the intercept *a*, indicating the level of the regression line, was 2·0857 for the fed animals and only 1·8901 for the starved fishes. The duration of the enhanced oxygen uptake at 10°C is shown in Fig. 8.14, from which it is evident that the cod, whose average weight was 1·07 kg, took approximately 7 days to return to the starvation level of oxygen consumption, although at higher temperatures the period of enhanced oxygen uptake was shorter. The respiratory rate was also shown to be influenced by a number of other factors such as crowding (Saunders, 1963).

Starvation has been shown to have a similar effect on the metabolism of the barnacles *Balanus balanoides* and *Balanus balanus.* Barnes, Barnes and Finlayson (1963a, b) studied the seasonal changes in the body weight and biochemical composition of the barnacles and showed that the body weight increases during the spring when reserves are laid down; the reserves are subsequently used in the development of the gonad. The oxygen consumption of whole dissected animals prepared as described on p. 384, was then shown to vary seasonally in both species in accordance with the amount of material laid down as reserves, and with the proportion of semen which has a very low rate of oxygen uptake. The results for *Balanus balanoides* and *B. balanus* are shown in Fig. 8.15. The respiration rate of *Balanus balanoides* becomes

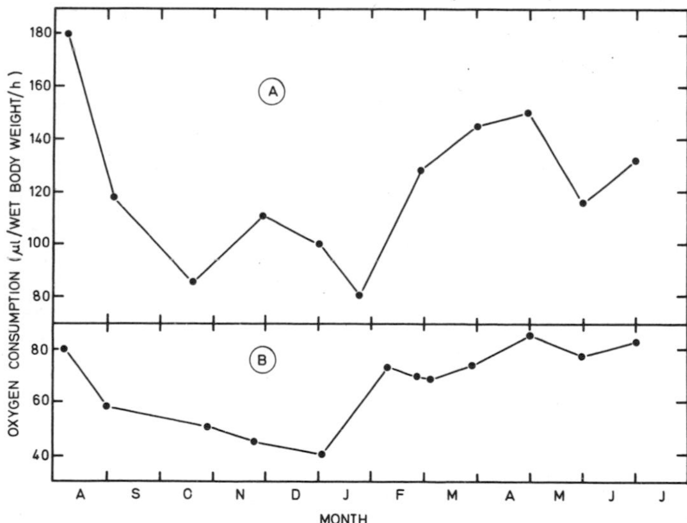

Fig. 8.15. Graphs showing seasonal changes in the rate of oxygen consumption of excised *Balanus balanoides* (A) and *Balanus balanus* (B). (After Barnes, Barnes and Finlayson, 1963a.)

very low by October by which time the relatively inert semen may comprise as much as 50% of the total body weight. Following copulation, in which the semen is discharged, the respiratory rate increases but nevertheless remains at a relatively low level throughout the winter. Barnes *et al.* (1963a) attributed this low level to the reduced food supply during the winter and to the uptake of water into the tissues which occurs at this time. During the late winter the metabolism increases to higher values, and during the spring and early summer there is an increase in the quantity of reserve material and non-active reserve material, the respiration rate of the excised animals showing little further rise during this period of the year. The seasonal changes in the oxygen uptake of *Balanus balanus* are similar to those of *Balanus balanoides* but differ in the details of the annual cycle (Fig. 8.15). In *Balanus balanus* the fall in the metabolism during the late summer and autumn is associated with the increase in relatively inert semen as in *Balanus balanoides,* but the post-copulatory rise in metabolic rate does not occur until early February. A further difference is that in *Balanus balanus* the respiratory rate falls from July onwards because of the early accumulation of semen, whereas in *B. balanoides* the respiratory rate does not begin to decline until early August when the semen begins to develop in this species.

Since starvation during winter appears to be related to a reduction in oxygen demand by the tissues in *Balanus balanoides,* and *B. balanus* (Barnes, Barnes and Finlayson, 1963a), Barnes *et al.* (1963b) studied the effects of starvation on the metabolism of *Balanus balanoides* during the winter when food reserves were scarce, and during the summer when food reserves were present in the tissues. They measured the respiration rate of excised animals and plotted the log respiration rate against the log dry weight and then compared the level of the regression line of starved and unstarved animals. The results are shown in Fig. 8.16. It will be noticed that the metabolism of the summer animals was reduced after a period of starvation but that of the winter animals was not. The reduction in the metabolic rate of the summer animals was associated with a shift in the type of substrate used from carbohydrate in the unstarved animals to protein and lipid following a period of starvation. The winter animals, however, were already using protein and lipid as a metabolic substrate and hence no further reduction could occur following a period of starvation. Thus in barnacles changes in the metabolic rate of starved animals, which must approximate to the standard rate of oxygen consumption, occur in response to the varying content of inert material such as the semen, as well as with the available carbohydrate; this substance is used first before protein and lipids are used as metabolic substrates.

A very similar sequence of seasonal changes in the respiratory rate

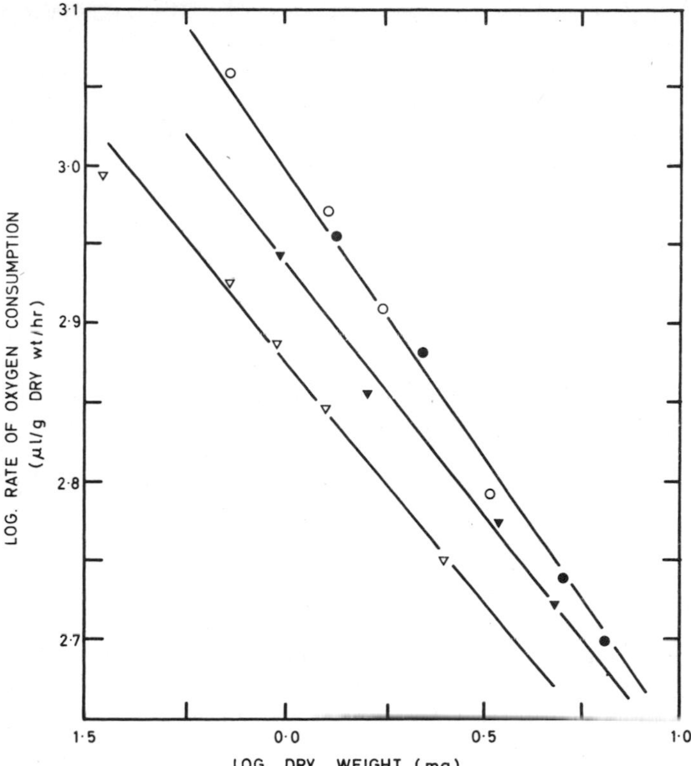

Fig. 8.16. Oxygen uptake of excised specimens of *Balanus balanoides* as a function of size (log mg dry wt). Solid triangles, specimens collected during summer/early autumn and measured directly; open triangles, the rate after starvation. Solid circles, specimens collected during autumn/winter and measured directly; open circles, the rate after starvation. (After Barnes, Barnes and Finlayson, 1963b.) Note that there was a significant reduction in the metabolism of summer/autumn specimens following starvation but no significant reduction in the metabolism of autumn/winter specimens.

was shown to occur in the limpets *Patella vulgata* and *Patella aspera* (Davies, 1967); superimposed upon such changes are the effects of tidal level which are comparable with those described on p. 384 for a variety of barnacles. Davies (1967) showed that in *Patella vulgata* the metabolic rate of lower shore animals was higher than that of upper shore animals during July, September and October, but that during November the respiration of the lower shore animals decreased at the same time as the post-spawning period (Orton, Southward and Dodd, 1956), the rate finally declining to lower levels than that of the upper shore animals. During January and March, however, the respiratory rate of the lower

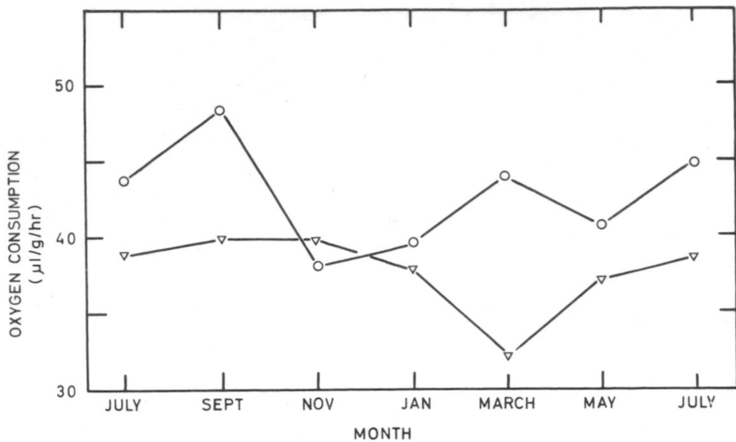

Fig. 8.17. Graphs showing the seasonal variation in the oxygen consumption of a 10g individual of *Patella vulgata* at 15°C. Circles indicate the rate for a lower shore animal, triangles indicate the rate for an upper shore animal. (After Davies, 1967.) Note the general lower level of oxygen consumption of the upper shore animal.

shore animals increased and was relatively unchanged from March onwards. The respiration rate of the upper shore animals was found to be relatively unchanged throughout the year. A similar trend was noted in *Patella aspera* in which the metabolic rate showed a decline during the winter and returned to a high summer level during January and March. The results for upper shore and lower shore *Patella vulgata* are shown in Fig.8.17. In addition to such changes which occur during the year, Davies (1967) showed that the metabolic rate of limpets living near algae was higher than that of animals living at a comparable tidal level but on bare rock. There was also a gradation in the level of metabolism from a high rate in limpets of the lower shore, to a lower rate in limpets from the upper shore. Such spatial and seasonal differences may all be related to the availability of food as demonstrated for barnacles by Barnes, Barnes and Finlayson (1963a, b). Indeed, there are many examples of interspecific and intra-specific differences in the growth rate, activity level, rate of beating of the heart, and metabolic rate which are correlated with level on the shore and which may in many instances be attributed to nutritional differences (Barnes and Barnes, 1959; Barnes, Barnes and Finlayson, 1963a,b; Davies, 1966, 1967), although in other instances the different temperature regimes of the upper and lower shore have been implicated (p. 417).

A final factor which may profoundly affect the level of respiration is the presence of circadian rhythms which may have a diurnal periodicity or a semi-diurnal periodicity in relation to the tidal-cycle. The nature and occurrence of such rhythms has been reviewed by Harker (1964). One of the organisms in which the details of the rhythms in metabolic rate are well-known is the fiddler crab *Uca.* Brown, Bennett and Webb (1954) described persistent rhythms in the oxygen consumption of two species of fiddler crab, *Uca pugnax* and *U. pugilator,* and the rhythms were found to include diurnal, semi-lunar and lunar ones. Webb and Brown (1958) subsequently investigated the persistence of the rhythm under conditions in which tidal effects were absent from the environment. The mean rate of oxygen consumption of *Uca pugnax* ranged from 28 – 69 ml/kg/hr between mid-July and mid-August on three successive years during which a diurnal rhythm of oxygen consumption of identical phase and form was noted. Fig. 8.18 shows such a curve in which there is a single maximum between 2 AM and 6 AM, which is followed by a broad minimum from noon to 7 PM, after which there is an increase in the rate which continues until midnight. Although the

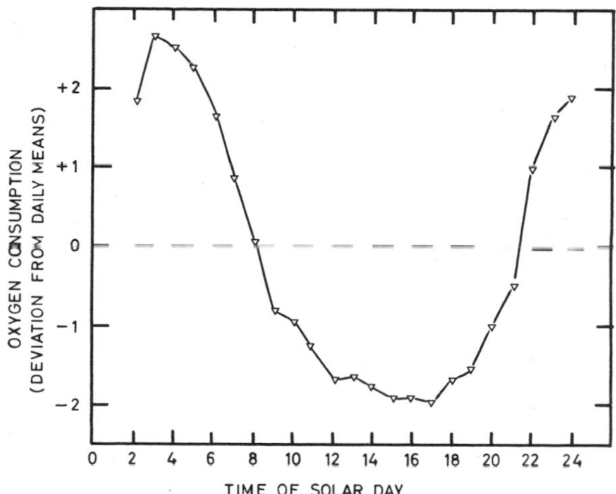

Fig. 8.18. Graph showing the mean diurnal variation in oxygen consumption of *Uca pugnax* over a 29 day period, July 13th – August 11th 1956. (Data from Webb and Brown, 1958.)

amplitude of the rhythm varied in the three years, the ratio of the maximum to the minimum value was similar (1·2 in 1955, 1·4 in 1956 and 1·2 in 1957).

In addition to such a diurnal rhythm in oxygen consumption, a lunar cycle was detected. This cycle is shown in Fig. 8.19, the lunar zenith being at 12 hr and the nadir at 24 hr. In this cycle, not only the form and phase relationships were constant at comparable times in three successive years, but also the amplitude of the rhythm in each year was very similar, the ratio of maximum to minimum values being 1·4. From Fig. 8.19 it will be noticed that there are two maxima and minima, the peak rates occurring at the lunar zenith and nadir. Both the maxima and the minima are of comparable level, so that there is in effect a rhythm with a period of 12 hr. As Webb and Brown (1958) pointed out, since the amplitude of the lunar rhythm is at least as great as the diurnal one, it would be expected that the lunar rhythm would appear in the respiration on single days. Further, there would be major maxima and minima in the daily cycle of metabolism repeating at 15-day intervals and this has been found to be so. Such rhythms in respiration were found to be maintained during the first seven days in the laboratory, but after this time there was a gradual decrease in the correlation between simultaneous hourly values of animals kept in the laboratory.

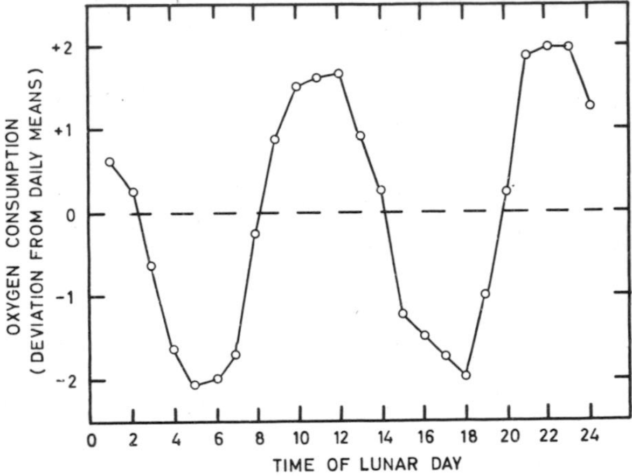

Fig. 8.19. Graph showing the primary lunar rhythm in the oxygen consumption of *Uca pugnax* for a 29-day period from July 13th – August 11th, 1956. (Data from Webb and Brown, 1958.)

Many other organisms show rhythmical fluctuations in activity, colour change, and general metabolism (for review Harker, 1964) although the precise controlling factors are in many cases obscure. In some examples, the lack of correlation with environmental variations has led to the conclusion that the rhythms are truly inherent, whilst in other organisms the cycles appear to be synchronised with complex changes in the environment, such as barometric pressure changes (Brown, Freeland and Ralph, 1955; Brown, Webb, Bennett and

Sandeen 1955; Brown, 1960) and cosmic ray cycles (Brown, Webb and Bennett, 1958). Terrestrial magnetism is known in turn to fluctuate rhythmically with solar and lunar periods and may influence circadian rhythms (Brown, Brett, Bennett and Barnwell, 1960; Brown, Webb and Brett, 1960; Brown, Bennett and Webb, 1960; for review, Harker 1964). Metabolic rhythms which show a clear correlation with some particular environmental parameter such as light or temperature are more properly included in the category of "environmental factors" which are discussed in more detail below.

C. ENVIRONMENTAL FACTORS

Before we discuss the environmental factors which influence metabolism, some consideration should be given to the nomenclature which Fry (1947) has adopted to describe the effects of the environment on animals. His examples apply principally to freshwater fishes in which the standard and active rates of metabolism have been measured; where similar data exist for marine organisms his terminology is of considerable value. He divided the environmental factors which govern metabolic rate into two classes (also Blackman, 1905). The first class may be termed "controlling factors" which operate in such a way as to govern both the maximal and the minimal metabolic rate, and in which several such factors may operate simultaneously. Controlling factors include, above all, temperature as well as photoperiod, salinity and in certain instances tidal level (see p. 396) and *pH*. The second category may be termed "limiting factors" and have been defined as "those factors which actually enter into the chain of metabolic processes of the organism". The rate of metabolism is thus limited by the pace of the slowest factor (Blackman, 1905). It follows that limiting factors might include oxygen availability and substrate supply.

1. CONTROLLING FACTORS

(a) Salinity and photoperiod

As has been shown on p. 373, in many instances measures of the respiratory rate of intertidal organisms have been made without reference to the level of activity of the organism concerned, so that many reported instances of variations in the rate of respiration may mainly reflect quantitative differences in the state of activity of the animal, rather than the effect of the particular environmental parameter on the standard rate of metabolism. Rao (1958), for example, has studied the respiration of marine and freshwater populations of the prawn *Metapenaeus monoceros* in relation to body size and salinity.

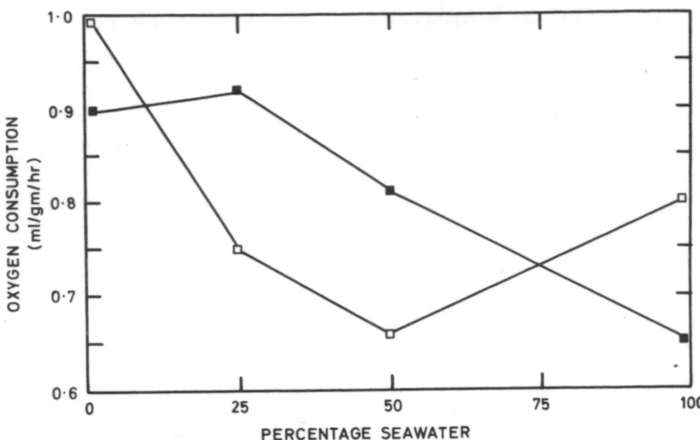

Fig. 8.20. Graphs showing the oxygen consumption of 3·5g specimens of the prawn *Metapenaeus monoceros* from a marine habitat (solid squares) and from a brackish-water habitat (open squares) in different dilutions of seawater. (After Rao, 1958.) Note that the respiration rate of the brackish-water prawns in their natural salinity of 50% seawater was comparable with that of the marine prawns in 100% seawater.

Regression lines were fitted to the data relating oxygen consumed in tap water, 25% seawater, 50% seawater and 100% seawater to the wet weight of the body. The metabolism of a medium-sized individual of 3·5 g fresh weight was then read off and plotted as a function of salinity. Such a graph is shown in Fig. 8.20, from which it is seen that the respiration rate of members of the brackish water population was low in 50% seawater but high in both tap water and 100% seawater. On the other hand, members of the marine population showed a maximum rate of respiration in 25% seawater after which the rate declined slowly towards tapwater and more steeply towards 100% seawater. Under natural conditions the brackish water prawns lived in water corresponding to approximately 50% seawater whilst the marine prawns lived in 100% seawater. Inspection of Fig. 8.20 thus shows that in their natural media the respiration of both groups of prawns was comparable in level. Rao (1958) was able to show that the level of respiration was related to the osmotic difference between the internal and external media, so that it seems likely that the increased level of respiration in such media is at least partly a reflection of the work needed to maintain an osmotic gradient. There was also a decline in the oxygen consumption after a prolonged period of sojourn in a hyper- or hypo-osmotic medium (i.e. brackish water prawns in 100% seawater or marine prawns in 50% seawater). Both of these observations could be explained equally satisfactorily in terms of an increase in activity

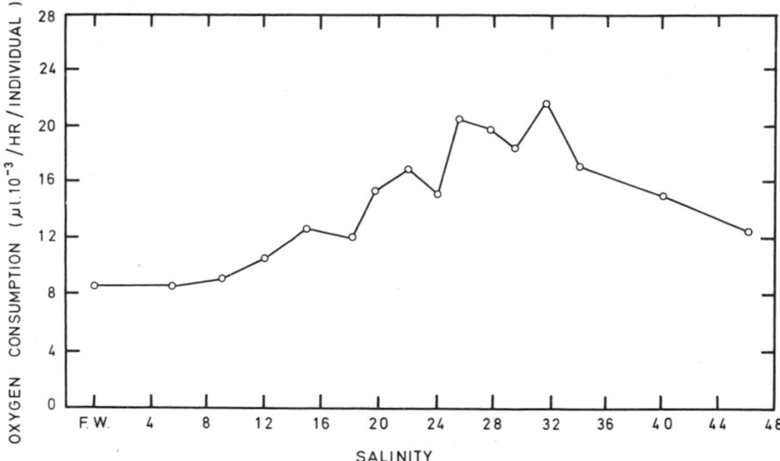

Fig. 8.21. Graph showing the effect of salinity of the medium (‰) on the oxygen consumption (μl.10^{-3}/hr/individual) of young freshwater *Potamopyrgus jenkinsi* at 21°C. Values represent the mean of 3–4 individuals except for the freshwater reading which represents the mean of a larger number of observations. (Data from Klekowski and Duncan, 1966.)

occurring under conditions of osmotic stress followed by a return to a normal level of activity after a period of acclimation to the osmotic stress.

Klekowski and Duncan (1966) and Duncan (1966), working on the small brackish water prosobranch *Potamopyrgus jenkinsi* have pointed out that in this animal much of the increased respiration which occurs in high salinities is attributable to the increased activity shown by the snails under such conditions. They observed an increase in the respiration of young freshwater *P. jenkinsi* of from 10–12 x 10^{-3} μl O_2/hr/individual at 10–15‰ salinity and 21°C to approximately 20 x 10^{-3} μl O_2/hr/individual at 25–30 ‰ salinity at 21°C (Fig. 8.21). Duncan (1966) then demonstrated that the speed of movement increased with salinity up to 18–22 ‰ depending upon the normal habitat of the animals. Thus animals from freshwater habitats increased the speed of crawling from freshwater up to 18‰, whilst animals from brackish water habitats of approximately 6‰ increased the speed of crawling up to 22 ‰ salinity (Fig. 8.22). She pointed out that the work involved in ion transport is only about 1% of the total metabolism in those animals where the work has been calculated (Potts and Parry, 1964), and that the large increase in the metabolism in *Potamopyrgus jenkinsi* thus mainly reflects the activity rather than the osmotic work involved in increased salinities. It would be of great interest to determine separately the effect of salinity change on the

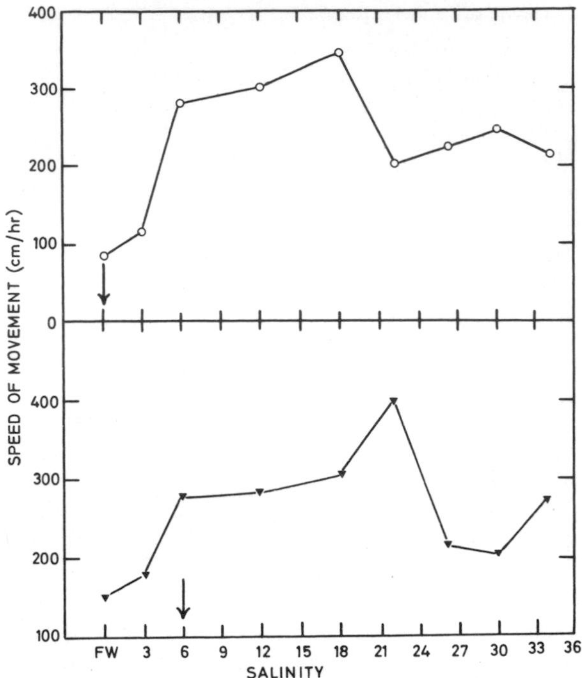

Fig. 8.22. Graphs showing the mean speed of movement at room temperature of 5 specimens of *Potamopyrgus jenkinsi* at each of nine different salinities (‰). The upper graph shows the results obtained with specimens from a freshwater population whilst the lower graph indicates the results for brackish-water individuals. (After Duncan, 1966.)

active rate of metabolism and on the standard rate as calculated from a line relating respiration rate to crawling activity (p. 373).

Photoperiod, as well as salinity, plays a part in controlling the level of metabolism in some marine organisms. Dehnel (1958) has investigated the effect of photoperiod on the oxygen consumption of two species of intertidal crab, *Hemigrapsus nudus* and *H. oregonensis.* He took the latter from the shore in summer and placed them under controlled conditions of salinity and temperature. One group of crabs was illuminated for 8 hr per day at 30 ft candles, alternating with 16 hr darkness. A second group was illuminated for 16 hr at 30 ft candles alternating with 8 hr darkness, whilst a control group was kept in total darkness. The respiration rate was measured at 15°C and the oxygen consumption plotted as a function of body weight on a log log scale. It is apparent from Fig. 8.23 that the respiratory rate of animals illuminated for 8 hr periods exceeded that of the animals illuminated for 16 hr and those kept in darkness. According to the regression lines which were fitted by eye, it is seen that a 2·0 g crab respired at approximately 54 mm^3/g/hr when in the dark and when illuminated for 16 hr periods but respired at 84 mm^3/g/hr when illuminated for 8 hr periods. Dehnel (1958) reported a similar effect in *Hemigrapsus nudus* in which the respiration rate of a 2·0 g animal was 64 mm^3/g/hr in the control and in those illuminated for 16 hr periods, but 87 mm^3/g/hr in the animals subjected to 8 hr illumination periods. Thus the increase in *Hemigrapsus oregonensis* was of the order of 55% whilst that of *H. nudus* was 36%. Such differences between the groups illuminated for 8 hr periods compared with those illuminated for 16 hr and the controls, tended to be enhanced at higher temperatures and lower salinities than those cited above. Dehnel (1958) pointed out that the summer crabs had raised their metabolism in response to illumination of a similar length to that occurring in the winter. This would tend to compensate for the reduction in the metabolism which would be expected to occur during the cold weather of the winter months.

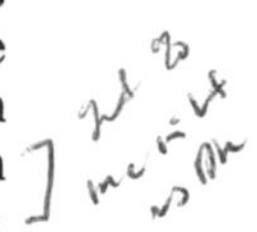

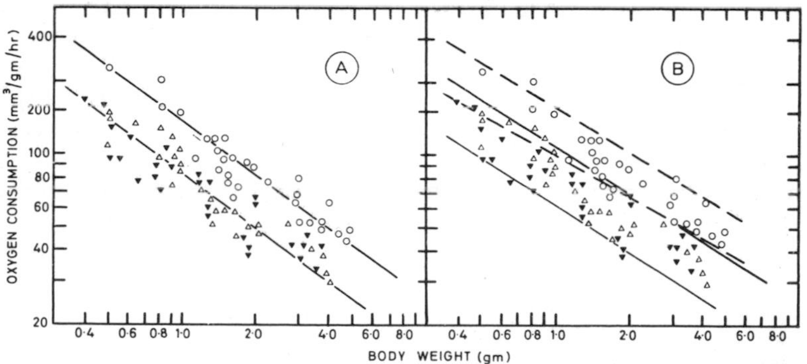

Fig. 8.23. Graphs showing the relationship between the oxygen consumption (mm^3/gm/hr) and body weight (gm) of the crab *Hemigrapsus oregonensis* at 15°C. The open circles indicate animals subjected to a daily illumination of 8 hr; the open triangles animals illuminated for 16-hr daily periods and the solid triangles indicate animals kept in darkness. (After Dehnel, 1958.) (A) Regression lines fitted by eye to each group of data, the animals kept in the dark and those illuminated for 16 hours having the same regression line. (B) Approximate regression lines fitted to the maxima and minima giving values for the active and standard rates for each group of animals. The lines for animals kept in the dark and those subjected to daily illumination of 16 hours (continuous lines) are lower than those for animals illuminated for 8 hours per day (broken lines).

Although Dehnel (1958) drew a regression line which apparently fitted the pooled data for the animals in the dark and for those illuminated for 16 hr (Fig. 8.23A), it is of some interest to note that for both of these groups of organisms the lowest points, representing the standard rate (p. 387), fall onto a common regression line, whilst the maximal rates for the two groups also fall onto one regression line. Such results confirm Dehnel's conclusions that the controls and the 16 hr illuminated group had the same level of metabolism, which he based on a single line fitting the mean of the data. The standard rate of 2·0 g individuals is, however, approximately 38 mm^3/gm/hr whilst the active rate is approximately 67 mm^3/gm/hr. The standard rate of the animals stored under conditions of 8 hr illumination is unfortunately less clear, mainly because of the lack of points in the smaller size range. However, the standard rate must approximate to 65 mm^3/gm/hr on a basis of the points illustrated in Fig. 8.23B, whilst the active rate of a 2·0 g animal approaches 100 mm^3/gm/hr. Thus the active rate exceeds the standard rate by the factor of 17·62 in the controls and in the 16 hr illuminated animals and by a factor of 15·38 in the 8 hr illuminated group. This quite remarkable correspondence suggests as Dehnel (1958) concluded on a basis of the common regression lines, that there is a rise in the metabolism of animals illuminated for 8 hr periods. It is possible to add to his conclusion that such an increase affects both the standard and active rates of metabolism to a similar extent, the increase in the standard rate being from 38 to 65 mm^3/gm/hr for a 2·0 g animal and that of the active rate from 67 to 100 mm^3/gm/hr, which indicates that the standard rate increases by a factor of 1·71 and the active rate by a factor of 1·49. A similar increase in the standard rate of oxygen consumption under the influence of a 9 hr photoperiod compared with a 15 hr photoperiod has been noted in the sunfish *Lepomis gibbosus* by Roberts (1964, 1967).

Other factors including the effect of tidal level and season have been discussed on p. 396. We may therefore now turn to the principal controlling factor, temperature, about which more is known.

(b) Temperature

(i) Effect on the active and standard rate of metabolism

Although the effect of photoperiod of 8 hr duration on the respiration rate of the crabs *Hemigrapsus nudus* and *H. oregonensis* is to raise the level of both the active and the standard rate to a similar extent, the measurement of both rates is of particular importance in the case of temperature change, since the two rates may be affected differently by this factor. The importance of this has been recognised in an extensive series of studies on the effect of temperature on the activity and metabolism of fishes (Fry, 1947; Beamish and Mookherjii, 1964; Beamish, 1964; for review Fry, 1957; Prosser and Brown, 1961), but has received little attention in studies on invertebrates, perhaps partly due to difficulties in quantifying the level of activity. Halcrow and Boyd (1967) have, however, recently applied identical techniques to those used on fishes in a study of the effects of temperature on the metabolism of the amphipod *Gammarus oceanicus.* As mentioned on p. 374, the respiration rate was measured at a variety of activity levels at any one temperature, and the standard rate calculated by extrapolation

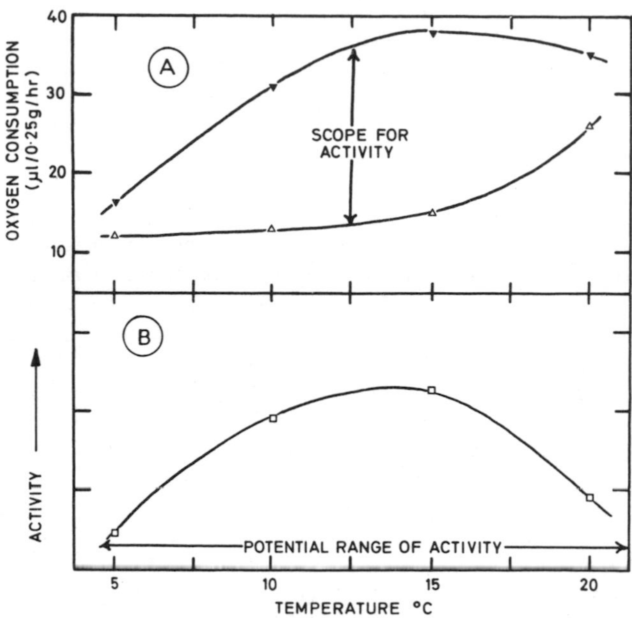

Fig. 8.24. Graphs showing (A) the effect of temperature on the active (solid triangles) and on the standard rate of oxygen consumption (open triangles) of the amphipod *Gammarus oceanicus* (After Halcrow and Boyd, 1967.). (B) The scope and potential range of activity. (Nomenclature after Fry, 1947.)

back to zero activity (Fig. 8.2, p. 374). Such measurements were made at 5°, 10°, 15° and 20°C, and the results for the active and the standard rate plotted as a function of temperature. The results are shown in Figure 8.24 from which the active and standard rates at 15°C were calculated for Table XIV (p. 358). The important feature of these graphs is that temperature evidently affects the active rate of oxygen consumption to a much greater extent than the standard rate. The slope of the line may conveniently be expressed in terms of the Q_{10} value or factor by which the rate of reaction is increased over a temperature interval of 10°C. Thus:–

$$Q_{10} = \left(\frac{R_1}{R_2}\right)^{\frac{10}{t_1 - t_2}}$$

where R_1 and R_2 are the rates of reaction at temperature t_1 and t_2 respectively.

The Q_{10} between 5°C and 15°C for an arbitrarily selected low activity level was approximately 1·4 as opposed to 2·4 for spontaneously active animals over this temperature range. Since the standard rate is the minimal rate of metabolism and the active rate is the level of oxygen consumption at maximal activity, the difference between these two rates defines the "scope for activity" in the organism (Fry, 1947). Where the two lines approach one another and finally meet at the upper and lower ends of the thermal range, there is no scope for activity and the metabolism represents that of the standard rate only. Such quiescence at high and low temperatures would represent the "heat coma" and "cold coma" which have been noted in a large number of marine organisms (Southward, 1958a; Sandison 1967). The scope for activity in *Gammarus oceanicus* is also shown in Fig. 8.24, whilst the range over which there is any scope for activity may be defined as the "potential range of activity".

The fact that the active rate of metabolism rises steeply with temperature is not surprising, since activity itself is markedly temperature-dependent. Many workers have recognised the importance of temperature in controlling the metabolism in animals (for example, Crozier, 1924; Janisch, 1925; Bělehrádek, 1930, 1935; Krogh, 1914, 1941; Kinne, 1963, 1964; Newell, 1969), whilst others have made direct studies of the effect of temperature on activity itself (Gray, 1923; Pantin, 1924; Clark, 1927; Fox, 1936, 1938b, 1939; Fox and Wingfield, 1937). Such studies have been summarised by Hoagland (1935), Barnes (1937), Sizer (1943), Schlieper (1952) and Prosser and Brown (1961). The main feature which has emerged is that activity, and in some instances the respiration of animals, increases logarithmically with temperature in approximate agreement with the principle of Arrhenius. This showed that if the velocities of reaction at any two temperatures T_1 and T_2 were denoted by k_1 and k_2 respectively, and if μ is a constant (the activation energy of the process) then:—

$$k_1 = k_2 e^{\left(\frac{\mu}{R} \times \frac{T_2 - T_1}{T_2 \cdot T_1}\right)}$$

where R is the gas constant of 1·98 calories.

This logarithmic increase in the rate of reaction approximates to the observed effects of temperature on ciliary activity (Gray, 1923; Schlieper, Kowalski and Erman, 1958; also Fig. 6.4), and on cirral activity in barnacles (Southward 1955b,c, 1957, 1962, 1964, 1965;

Ritz and Foster, 1968), and on activity in a large number of other invertebrates (Fox, 1936, 1938b, 1939; Fox and Wingfield, 1937). Unfortunately, as Fry (1947) pointed out, activity has often been equated with metabolism so that it has been assumed that the metabolism of quiescent organisms would be markedly dependent upon temperature. Nevertheless, Barcroft (1932) and Bullock (1955) have demonstrated that there are many instances where the metabolism does not increase regularly with temperature, and it seems likely that the confusion which exists depends at least in part upon the differing effects of temperature on the active and standard rates of metabolism (Newell, 1969). Other factors such as size and habitat have undoubtedly contributed to the discrepancies because, as is shown by Rao and Bullock (1954; also p. 415), the Q_{10} which gives a measure of the slope of the rate/temperature curve (or RT curve) may vary with size of organism, and also varies with the habitat temperature to which the organisms have been adapted. The significance of both of these factors is discussed later (p. 419).

We have seen that the RT curves for the active and standard metabolism of *Gammarus oceanicus* indicate that the standard rate is much less affected by temperature than the active rate. In this respect the metabolism resembles that of the goldfish *Carassius auratus* (Beamish and Mookherjii, 1964). The same is true also of the respiration of the larvae of certain Trichoptera which show wide ranges in temperature over which the metabolism is relatively unaffected by temperature fluctuation (Collardeau, 1961; Collardeau-Roux, 1966), whilst the range over which such temperature-independence is shown in freshwater amphipods appears to be related to the environmental temperatures in which they live (Roux and Roux, 1967). In *Orconectes immunis* and *O. nais,* too, the mean Q_{10} for oxygen consumption over the wide range of 16–31°C was only 1·42 (Wiens and Armitage, 1961). It is likely that such invertebrates were relatively quiescent in the respirometer, and thus their respiration approached that of the standard rather than the active rate (also Davies, 1966). Indeed, in the trochid *Calliostoma zizyphinum,* which is known to be quiescent during the intertidal period, the rate of respiration is almost independent of temperature (Micallef, 1966; also Fig. 7.52). Conversely in the tubicolous polychaete *Diopatra cuprea* the Q_{10} for the standard rate of respiration is just as affected by temperature as the active rate (Mangum and Sassaman, 1969). On the other hand, both the active and the standard rates of respiration of the shrimp *Palaemonetes vulgaris* have been shown to have Q_{10} values of less than 2·0 (McFarland and Pickens, 1965). Similarly Siefken and Armitage (1968) have shown that the rate of respiration of several species of the freshwater copepod *Diaptomus* show ranges of temperature-independence over as much as 10°C (Table

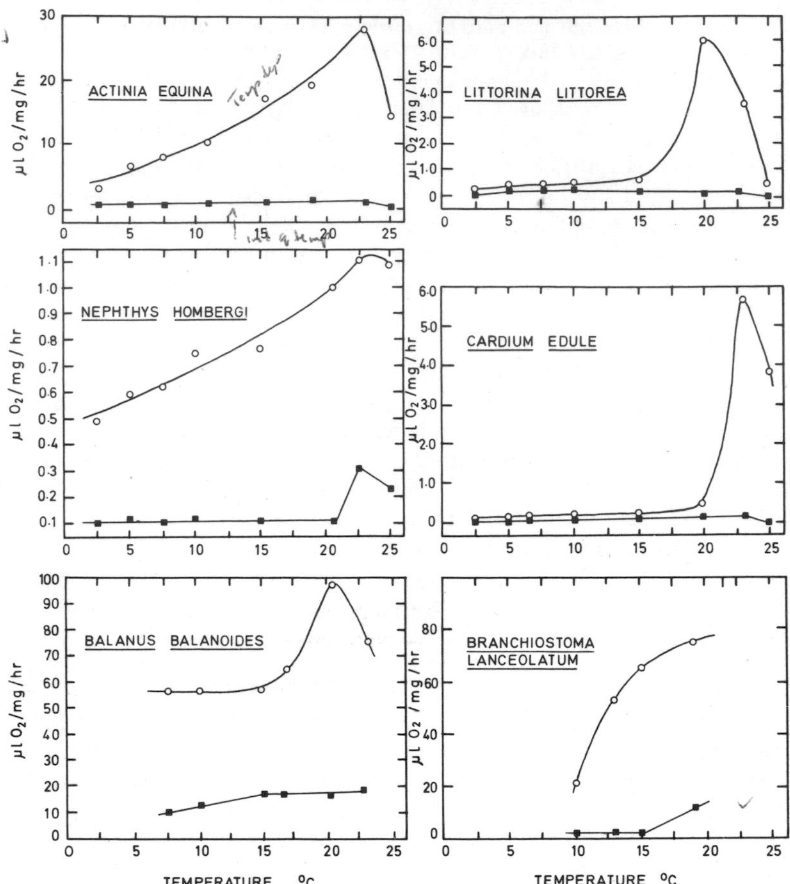

Fig. 8.25. Graphs showing the effect of temperature on the active and standard rates of oxygen consumption (µl/mg dry wt/hr) of 100 mg. dry weight specimens of the anemone *Actinia equina*, the winkle *Littorina littorea*, the polychaete *Nephtys hombergi*, the cockle *Cardium edule*, (data from Newell and Northcroft, 1967) and of a 1·5 mg dry weight specimen of the barnacle *Balanus balanoides* (Data from Newell and Northcroft, 1965.) The maximal and standard rates of a 20 mg. dry weight specimen of the lancelet *Branchiostoma lanceolatum* are also shown (data from Courtney and Newell, 1965) although the maximal rate here does not correspond with the active rate since the animals were in gravel. (See text.)

XVI). Daniels and Armitage (1969) have also reported the presence of low Q_{10} values in the metabolism of the freshwater snail *Physa hawnii.* The effect of temperature on the active and standard rates of respiration of a wide variety of intertidal invertebrates has been studied by Newell and Northcroft (1965, 1967) using the method described on

TABLE XVI. Table showing the temperature range over which the standard rate of oxygen uptake varies only slightly with temperature

Species	*Temperature range of shallow slope °C*	*Reference*
FRESHWATER INVERTEBRATES		
Trichoptera larvae		
Polycentropus flavomaculatus	4 – 22°C	Collardeau, 1961
Plectrocnemia conspersa	,,	,,
Limnophilus rhombicus	,,	,,
Micropterna testacea	3 – 7°C	Collardeau-Roux, 1966
Amphipoda		
Gammarus pulex	15 – 20°C	Roux and Roux, 1967
G. fossarum	5 – 10°C	,, ,,
Copepoda		
Diaptomus clavipes	15 – 25°C	Siefken and Armitage, 1968
Diaptomus pallidus	16 – 21°C	,, ,,
Diaptomus siciloides	4 – 11°C	,, ,,
Decapoda		
Orconectes immunis } *O. nais*	16 – 31°C	Wiens and Armitage, 1961
INTERTIDAL INVERTEBRATES		
Balanus balanoides	14 – 20°C	Newell and Northcroft, 1965
Patella vulgata	15 – 20°C	Davies, 1966
Actinia equina	7·5 – 25°C	Newell and Northcroft, 1967
Nephtys hombergi	3 – 20·5°C	,, ,,
Littorina littorea	10 – 23°C	,, ,,
Cardium edule	6·5 – 23°C	,, ,,
Gammarus oceanicus	5 – 15°C	Halcrow and Boyd, 1967
Calliostoma zizyphinum	8 – 24·5°C	Micallef, 1966
Eucidaris tribuloides	15 – 20°C	McPherson, 1968
FISHES		
*Lepomis gibbosus (sunfish)**	10 – 17·5°C	Roberts, 1964

* Acclimated to each experimental temperature.

p. 387 (Fig. 8.12), in which the lines relating the maximal and minimal rates of respiration to size were regarded as an index of the active and standard rates of metabolism. Similar regression lines were obtained over a range of temperatures, and the level of the regression lines read off for an idealised animal and plotted as a function of temperature. The results for the anemone *Actinia equina,* the polychaete *Nephtys hombergi,* the winkle *Littorina littorea* and the bivalve *Cardium edule* are shown in Fig. 8.25. together with those for the barnacle *Balanus balanoides* and the lancelet *Branchiostoma lanceolatum.*

The main feature which is common to all the graphs, with the exception of that for the predominantly sublittoral *Branchiostoma lanceolatum,* is that the level of the line relating the minimal rate of oxygen consumption to temperature (regarded here as the standard rate of oxygen consumption, p. 375) is relatively independent of temperature; the extent of the "plateau" varies from species to species (Table XVI). In this respect the results resemble those of Davies (1966) who reported the presence of a relatively low Q_{10} between 15° and 20° C in the metabolism of limpets. They also closely resemble the data already cited for the effect of temperature on the standard rate of oxygen consumption of the amphipod *Gammarus oceanicus* (Halcrow and Boyd, 1967; also p. 405), and also the results obtained on the aquatic Trichoptera and gammarids cited on p. 407 (Collardeau, 1961; Collardeau-Roux, 1966; Roux and Roux, 1967) as well as those for *Diaptomus* spp (Siefken and Armitage, 1968). In contrast, the minimal

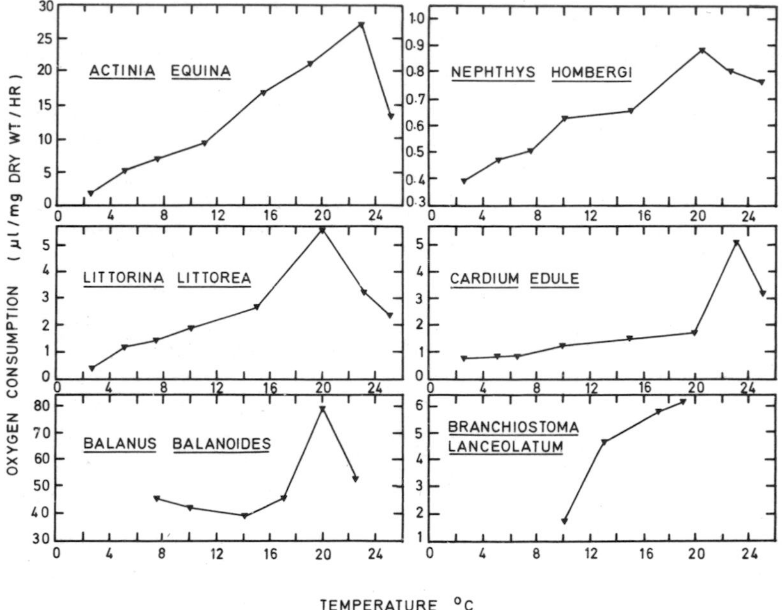

Fig. 8.26. Graphs showing the effect of temperature on the scope for activity in various marine invertebrates (obtained by subtraction of the standard from the active rate of oxygen consumption shown in Fig. 8.25.). It should be noted that in *Branchiostoma lanceolatum* the maximal rate was that obtained on animals in gravel and not on swimming animals. Thus the scope for activity would certainly be higher than that shown above and the temperatures at which activity could be maintained would be correspondingly lower.

rate of oxygen consumption in *Branchiostoma lanceolatum* was found to vary with temperature between 10° and 20°C; respiration rates above and below this range were not measured (Courtney and Newell, 1965). In the sunfish *Lepomis gibbosus* there is almost perfect compensation in the standard rate of respiration between 10·0° and 17·5°C when the fishes were acclimated to each experimental temperature, as opposed to the acutely-measured RT curves referred to above (Roberts, 1964, 1967).

These results for the effect of temperature on the standard rate of oxygen consumption are in marked contrast to the effect on the maximal or active rate, which in general increases rapidly with temperature before declining as lethal temperatures are approached. Although it would have been preferable to have obtained data at even higher temperatures than those shown in Fig. 8.25, it is possible to plot the effect of temperature on the scope for activity for such organisms which, in general agreement with that shown in Fig. 8.24 for *Gammarus oceanicus,* show an upper and lower temperature range at which the scope for activity is reduced (Fig. 8.26). It should perhaps be noted here that the curve for *Branchiostoma lanceolatum* represents the scope for ciliary activity, since the maximal rate was not obtained for swimming animals. Hence the point of thermal decline at low temperatures does not coincide with the point at which swimming activity has been noted to cease (Courtney and Webb, 1964).

In commenting upon a similar reduction in the scope for activity in fishes, Fry (1947) has made an interesting analogy between the active and standard rates of metabolism in animals and the fuel consumption and power output of an engine. In the analogy, the speed of rotation was equivalent to the level of the controlling factor (= temperature), whilst the power output was equivalent to the activity of an animal and the fuel consumption, the metabolic rate. It is obvious that at any given speed the engine may consume either a minimum amount of fuel when under no external load or else a maximum amount of fuel with a large external load, and it is the difference between the two rates which determines the power available to perform external work. These relationships are shown in Fig. 8.27 from which Fry (1947) pointed out that at 1000 rpm the difference between the maximal and minimal levels of fuel consumption was 12 lb/hr. At these revolutions the maximum power output was found to be 24 hp. Similar calculations over a range of engine speeds indicates that the maximum ability to carry out external work (or power output) is correlated, not with the maximum level of fuel consumption, nor with the minimal rate, but with the difference between the two (as shown in Fig. 8.27).

In much the same way, the activity level of a particular organism in relation to temperature would be expected to be correlated not with

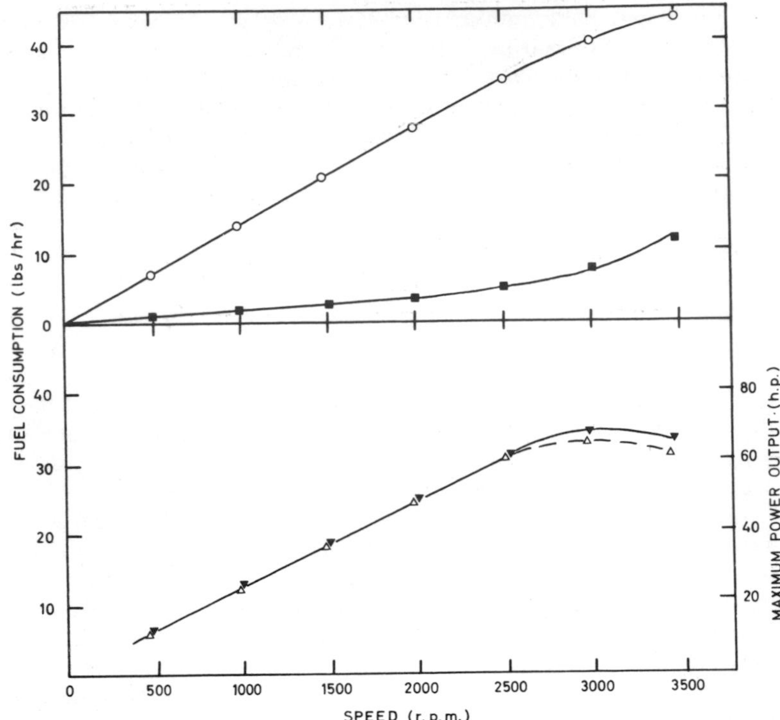

Fig. 8.27. Graphs showing a mechanical analogy of the effect of a controlling factor on the rate of metabolism and activity. (After Fry, 1947.) The graphs show the relation between the maximum power output (≡ maximum rate of activity) denoted by open triangles, and the maximum (circles) and minimum (squares) rates of fuel consumption (≡ the active and standard rates of oxygen consumption). It will be noticed that the maximum power output is closely correlated, not with the minimum nor with the maximum rate of fuel consumption, but with the difference between the two rates (≡ scope for activity) denoted by the solid triangles.

the active rate of metabolism, nor with the standard rate, but with the difference between the two (i.e. the scope for activity). Now it will be noticed from Fig. 8.24 and 26 that the scope for activity does not increase continuously with higher temperatures, and that with the exception of *Branchiostoma lanceolatum* the scope for activity is reduced at high temperatures when the standard rate and the active rate approach one another. A similar trend is apparent in the maximum and minimum fuel consumption of the engine (Fig. 8.27), and Fry (1947) has pointed out that apart from structural failure it might be imagined that the speed of the engine would go on increasing until all the fuel consumed was necessary for maintaining the speed of the engine, without any power being available for external work. Of course "limiting factors" (p. 447) intervene at this level and appear to do so also in many animals, since it is rare to find the maximum and minimum rates coinciding at their upper and lower ends. Nevertheless the decline in scope for activity at the extremes of the temperature range suggests why the optima for animal activity normally fall well within the range of tolerance of the organism and why cold and heat coma may occur at each end of the thermal range (Southward, 1958a).

Thus an estimate of the resting rate of oxygen consumption by the method employed by Halcrow and Boyd (1967) and by that used on a number of intertidal invertebrates by Newell and Northcroft (1965, 1967), as well as other measurements on the respiration of whole animals by Collardeau (1961), Collardeau-Roux (1966), Roux and Roux (1967), McFarland and Pickens (1965) Davies (1966) and Siefken and Armitage (1968), all indicate that the metabolism of a variety of invertebrates is relatively independent of temperature fluctuation, at least within the normal environmental temperature range. It is of interest to note here that Barnes, Barnes and Finlayson (1963b) noticed that the Q_{10} for the metabolism of excised *Balanus balanoides* was 1·0 when protein and lipid were being used as a substrate, but was steeper (2·0) when carbohydrate was used. Subsequently Barnes (personal communication), using excised barnacles (p. 423), has also found that in *Balanus balanoides* the rate of respiration is relatively unaffected by temperature during much of the year except for January (see below). His results certainly approximate to the standard rate and confirm those obtained on the whole organism by the methods described above. The rate of respiration of a variety of intertidal algae, too, is relatively independent of temperature over much of the normal environmental temperature range (Newell and Pye, 1968), so that the phenomenon appears to be of general occurrence amongst intertidal organisms; there is, however, some suggestion, as indicated by the data on *Branchiostoma lanceolatum,* that in sublittoral organisms the same situation may not exist (also Newell, 1966, 1969). A flattening of the curve over the temperature range to which the particular species is adapted has also been described for freshwater fishes by Fry (1947). The significance of this phenomenon has been discussed in some detail by Barcroft (1932), Bullock (1955) and more recently by Davies (1966), Newell (1966, 1969) and Newell and Northcroft (1967). It is generally agreed that a flattening of the curve relating the rate of metabolism to temperature would indicate an important homeostatic mechanism in a poikilotherm which would allow the rates of metabolic reactions to proceed at a relatively constant rate despite the large fluctuations in temperature which occur with each emersion/submersion sequence during the summer.

The tissue temperature of a series of intertidal animals at Plymouth, England for example, was found to be high during the emersion period in the summer (Southward, 1958a), and a similar situation has been found to occur in tropical intertidal animals (Lewis 1963); many motile organisms, however, seek the shaded areas and so evade thermal stress (Broekhuysen, 1940; Lewis, 1960). The data for selected British species has been summarised by Southward (1958a) who found that the body temperatures in many cases was higher than the local meteorological values because of warming by the sun. In limpets, for example, the tissue temperatures reached 24·9°C whilst that of barnacles reached 29·9°C in some cases, the corresponding sea temperature being 14·6°C. Thus, quite apart from fluctuations in the air temperature during the emersion period, a rapid reduction in temperature of over 10°C may occur on inundation by the tide. The low values for the Q_{10} of the standard metabolism of the many intertidal invertebrates described above have been obtained on animals taken from the shore during the summer when the range of such relatively temperature-independent metabolism would be expected to be great. The question therefore arises whether the temperature range over which such low Q_{10} values occur is a fixed phenomenon characteristic of a particular species or whether it is modifiable according to the temperatures prevailing in the habitat.

(ii) Modifications of the relationship between temperature and metabolism

As with the effect of the short-term fluctuations discussed above, long-term thermal effects on metabolism should ideally be considered with reference to the active rate and the standard rate separately. Again, activity itself gives little indication of changes in the metabolism of the organism concerned except insofar as it is correlated with the scope for activity (p. 411). The phenomenon of adjustment to long-term changes in temperature may be regarded as "acclimation" when in response to temperature itself, and "acclimatisation" when the adjustment is in response to some other factor such as photoperiod (p. 402). Nevertheless both responses result in a modification of the activity or metabolism to meet the thermal conditions prevailing within the environment (Prosser and Brown, 1961). Such adjustment has been known for a long time and is discussed in detail by Bullock (1951, 1955), Roberts (1952), Scholander *et al.* (1953), Rao (1953a,b), Rao and Bullock (1954), Prosser and Brown (1961) and Segal (1961, 1962), as well as by earlier workers such as Vernon (1897), Mayer (1914), Bělehrádek (1930), Fox (1936, 1938b, 1939), Fox and Wingfield (1937), Edwards and Irving (1943a,b) and Edwards (1946). We shall not deal in detail with the effects of acclimation on activity itself since this aspect is well summarised by a number of authors especially Bullock (1955), Fry (1958) and Prosser and Brown (1961). It is generally agreed that in many species cold-acclimation involves a compensatory rise in the level of activity in such a way that the rate remains comparable with that of warm-acclimated animals. Conversely, warm-acclimated animals show a suppression of activity compared with cold-acclimated forms. The result is that comparable sized individuals of any one species with a wide geographical range show similar levels of activity at the northern and southern limits of their distribution (Mayer, 1914). It follows from this that if the activity of a cold-acclimated animal and of a warm-acclimated animal is measured at the same intermediate temperature, that of the cold-acclimated animal will be higher than that of the warm-acclimated one. Unfortunately, as is pointed out by Rao and Bullock (1954) and by Bullock (1955), the effect of acclimation on a number of rate functions, such as the standard metabolism of the sand crab *Emerita* (Edwards and Irving, 1943a), and the beach amphipod *Talorchestia* (Edwards and Irving 1943b), and the rate of pumping of *Mytilus californianus* (Rao, 1953a; Rao and Bullock, 1954), all vary with body weight, so that unless

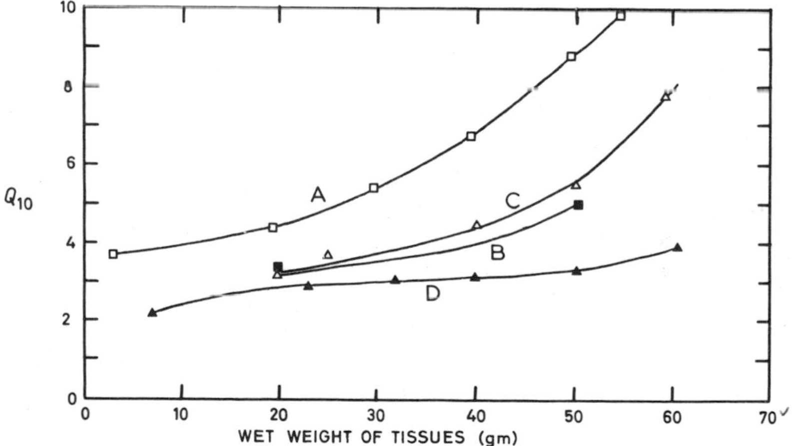

Fig. 8.28. Graphs showing the effect of size on the Q_{10} for water propulsion in the mussel *Mytilus californianus* (After Rao and Bullock, 1954.) (A) Open squares – sublittoral animals from Los Angeles, California measured at 8–16°C. (B) Solid squares – sublittoral animals from Los Angeles, measured at 10–16°C. (C) Open triangles – intertidal animals from Los Angeles, measured at 10–20°C. (D) Solid triangles – intertidal animals from Friday Harbour, Washington, measured at 5–15°C.

particular care has been taken to measure the rate of metabolism or pumping of animals of identical size, such effects will mask the acclimation pattern (Fig. 8.28). Acclimation of growth rates, too, is known in the gastropods *Thais emarginata, Crepidula nummaria* and *Lacuna carinata* whose larvae, whilst still confined within the egg capsule and living off the yolk, grew from 2 – 9 times faster when taken from Alaska and measured between 10 and 20°C than when taken from Southern California and measured at the same temperatures (Dehnel, 1955)

Perhaps the most reliable method of estimating the degree of acclimation is to measure the rate function at a number of temperatures (e.g. metabolic rate or pumping activity) of a wide range of sizes of animal from each acclimation temperature. A set of regression lines relating the rate function to body weight at each of the temperatures may then be drawn for animals from each acclimation temperature. An RT curve for an idealised animal (preferably near the middle of the weight range of the animals used) can then be plotted for animals from each acclimation temperature. The result which would be obtained is similar to that shown in Fig. 8.29 which shows the effect of latitude on

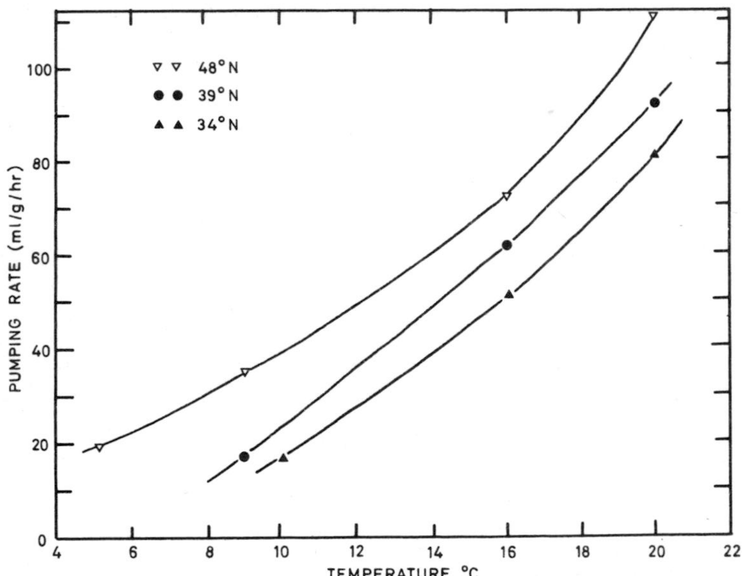

Fig. 8.29. Graphs showing the effect of temperature on the pumping rate of 50g specimens of the mussel *Mytilus californianus* from three different latitudes. (Data from Rao, 1953a.) Note the suppression of the RT curve in animals from lower latitudes compared with those from localities further north.

the pumping rate of 50gm specimens of *Mytilus californianus* (Rao, 1953a). It is seen that acclimation to high latitudes (= low temperatures) in this animal has involved a rise in the level of the RT curve (= an increase in the pumping rate) and that also in the highest latitude form the slope of the RT curve is somewhat shallower than in the other two groups. In this respect the results resemble those of Scholander, Flagg, Walters, and Irving (1953) who carried out a detailed survey on climatic adaptations in arctic compared with tropical poikilotherms, and found that adaptation involved a lateral displacement of the RT curve in many organisms. Their data also show that the Q_{10} in many cases tends to decrease in high latitudes (also Rao and Bullock, 1954). This effect is shown in Fig. 8.29 which illustrates the pumping rate of *Mytilus californianus.* Acclimation may thus involve both lateral translation and rotation of the RT curve. These and other effects of acclimation on RT curves are discussed by Precht, Christophersen and Hensel (1955), Prosser (1955, 1958) Precht, (1958) and Prosser and Brown (1961).

In the foregoing discussion we have considered organisms from widely differing latitudes, but it is also true that acclimation of activity occurs according to tidal level on any one shore (Segal, 1956). In summer, for example, the mean temperature of the upper shore may differ considerably from that of the lower shore and compensatory mechanisms might be expected to occur. Segal, Rao and James (1953) have shown that the heart rate of the limpet *Acmaea limatula* is faster in lower shore limpets than upper shore ones, indicating that the upper shore forms are acclimated to higher temperatures in the locality where this was observed. Similarly Rao (1953b) (also p. 416) has observed an acclimation effect in the pumping rate and shell/body ratio of the mussels *Mytilus californianus* and *M. edulis,* and points out that small vertical differences on the shore of 75 cm may be equivalent to latitudinal differences of 330 miles in a specimen of *M. californianus* with a tissue weight of 20 g and at 16°C.

A general survey of the effect of tidal level and temperature on activity in a variety of intertidal organisms has recently been made by Cornelius (1968). He found that, as had been reported by Southward (1955b), there was a considerable variation in the effect of tidal level upon the cirral activity of barnacles. In *Balanus balanoides* and *Chthamalus stellatus* this was found to be due to acclimation effects which became apparent even within a few hours. Since upper shore animals were exposed to higher temperatures during the summer and for longer periods than the lower shore animals, their activity on re-immersion was suppressed. This was shown to be true in the laboratory, and emphasised the importance of day to day variation in temperature on the shore. The feeding activity of the trochid *Gibbula umbilicalis,* as assessed by the number of radular movements per minute, was also found to be related to the temperature experienced during the low-tide period. In contrast, the activity rates of *Littorina littorea* and *Mytilus edulis* did not seem to be influenced by day to day variations in the environmental temperature. The results for *Balanus balanoides* and *Chthamalus stellatus* confirm the results of Southward (1964), who demonstrated a lateral shift of the curves relating cirral activity to temperature in barnacles exposed to different temperatures (also Crisp and Ritz, 1967a,b; Ritz and Foster, 1968), and show that the process is a rapid one. On the other hand the results for *Littorina littorea* and *Mytilus edulis* confirm the observations of Pickens (1965), who found that after transplantation of specimens of *Mytilus californianus* from one shore level to another, a period of two weeks was needed before the heart rate became that characteristic of the new shore level. Similarly Vernberg, Schlieper and Schneider (1963) found that acclimation of the upper limit of temperature tolerance of the gill cilia of *Modiolus demissus* took 17 days. In other animals such as barnacles and *Gibbula umbilicalis* where rapid acclimation of activity occurs, the results of Cornelius (1968) showing the importance of short-term temperature

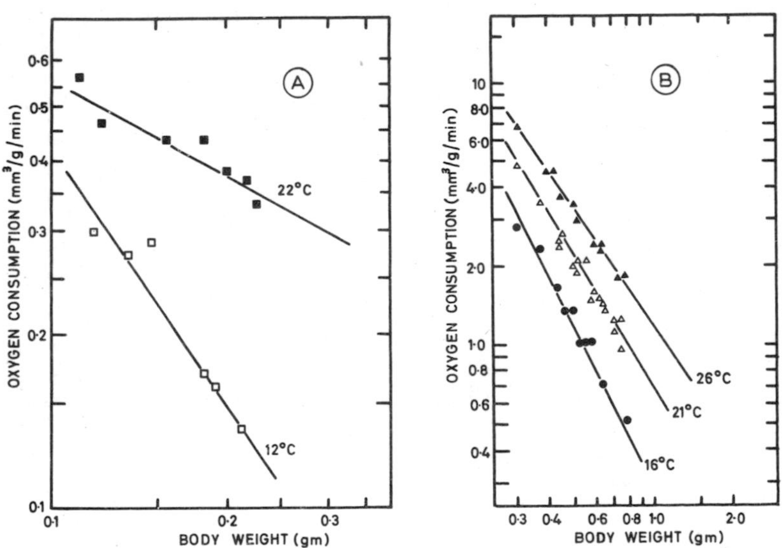

Fig. 8.30. Graphs showing the effect of body size on the rate of respiration of (A) *Talorchestia* and (B) *Emerita*. (Data of Edwards and Irving. 1943a,b, replotted after Rao and Bullock, 1954.) collected during the summer and measured at the temperatures indicated. Note that the rate of oxygen consumption of large animals is more affected by temperature change than small animals.

changes may suggest an explanation for the widely varying effects of tidal level which have been reported in the literature (for review Rao and Bullock, 1954; also p. 415).

Just as a variety of activities including growth rate, heart rate and pumping rate have been found to show acclimation to temperature, so too has the metabolism of a number of intertidal invertebrates. The lower metabolic rate of high-level limpets (Davies, 1965, 1966, 1967) and barnacles (Barnes and Barnes, 1959) has already been described on p. 386, whilst the acclimatisation of the metabolic rate of *Hemigrapsus nudus* and *H. californianus* (Dehnel, 1958) in response to photoperiod has also been discussed (p. 403). The data for excised barnacles and the recalculated "standard" rate of respiration of *Hemigrapsus* sp. indicate that in both instances the standard rate is elevated in lower shore forms or in response to winter conditions. As mentioned on p. 415, the conclusions of Edwards and Irving (1943a) for winter and summer acclimation of the metabolism of the sand crab *Emerita* and of the amphipod *Talorchestia* (Edwards and Irving, 1943b) have been criticised by Rao and Bullock (1954) and by Bullock (1955) on the grounds that the summer samples of *Emerita* were of a larger size than the winter ones, whereas in *Talorchestia* the winter samples were larger than the summer ones. Rao and Bullock (1954) replotted the data of Edwards and Irving (1943a) for specimens of *Emerita* collected during the summer (Fig. 8.30B), from which it is seen that the regression lines relating the respiration rate to body weight on a log log scale differ considerably in slope according to temperature. Similar data for *Talorchestia* collected during the summer are shown in Fig. 8.30A. The

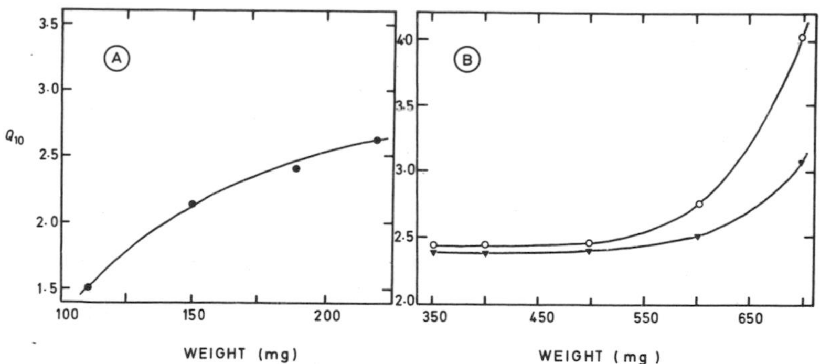

Fig. 8.31. Graphs showing the Q_{10} for oxygen consumption as a function of size in (A) summer specimens of *Talorchestia* between 12–22°C, (B) summer specimens of *Emerita* at 16–21°C (circles) and 16–26°C (triangles). (After Rao and Bullock, 1954.)

main feature which emerges is that there is a much greater increase in the respiration rate with increase in temperature in large animals than in small ones. Thus, as pointed out by Rao and Bullock (1954), the slope of the respiratory rate/weight regression line is not characteristic of a species but varies with temperature; it also varies with season (see replotted data of Edwards, 1946, in Rao and Bullock, 1954). The effect of size on the Q_{10} value for respiration in *Talorchestia* and *Emerita* is shown in Fig. 8.31. The result is that in some species the value for the Q_{10} increases with size and habitat temperature, although there are many other examples in which there is a decrease in Q_{10} with increase in size (Rao and Bullock, 1954).

Barnes (personal communication) has established a similar variation in the effect of temperature and season on the Q_{10} values for the respiration of excised barnacles. He has measured the effect of temperature on the log respiration rate/log body size regression line for animals taken from the shore in January, April, August and October. It was found that in specimens of *Balanus balanoides* collected in January the respiration rate when measured at 5° C, was almost independent of weight, ($b' = 0{\cdot}1038$), but that at 10° C the slope of the regression line was steeper ($b' = 0{\cdot}4076$) and was somewhat steeper still at 15° C ($b' = 0{\cdot}5739$). At 20° C however, the value was rather lower ($b' = 0{\cdot}3274$) but this was found to be within the error of the regression coefficient. These results are shown in Fig. 8.32A. Much the same was found to be true of *Balanus balanus* in which the respiration rate of January

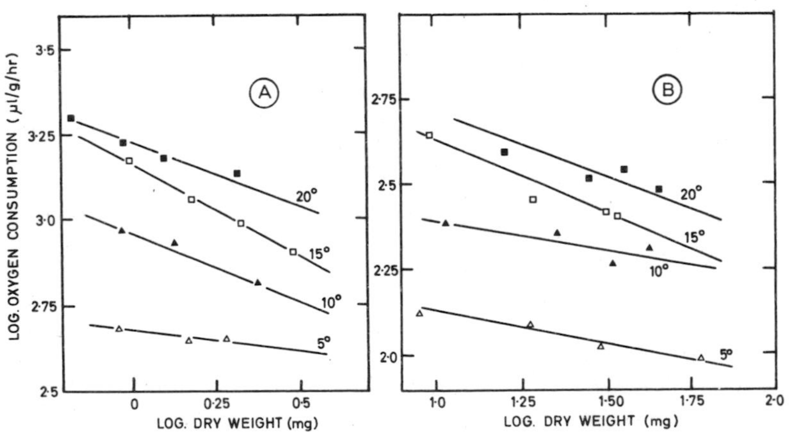

Fig. 8.32. Graphs showing the effect of temperature on the relationship between log. oxygen consumption (µl/g/hr) and log. dry weight (mg) in (A) *Balanus balanoides* collected during January and (B) *Balanus balanus* collected in January. (Redrawn after Barnes personal communication.)

specimens at low temperatures was also found to be much less dependent on body size than at higher temperatures or at other seasons of the year (Fig. 8.32B). In *Chthamalus stellatus,* although the scatter about the regression was greater than in the other two species, the slope of the lines relating respiration at 6°C with body size for animals collected in January and April was found to be much shallower ($b' =$ 0·1859) than the common regression coefficient for the other temperatures and seasons ($b' =$ 0·4073). These results therefore indicate that during January in all three species the slope of the regression line relating respiration of excised barnacles to body weight is shallower than at other times of the year, this effect being especially marked at low temperatures. On the other hand at other seasons of the year the slopes of the lines show no significant variation either with season or with temperature, and a common regression may be adopted for each species much as has been used by Davies (1966, 1967) for the respiration of *Patella vulgata.* As Barnes (personal communication) points out, the shallow slope of the regression line in all three species in January results in a decrease in the Q_{10} values with increase in size which is similar to that reported for the amphipod *Talitrus sylvaticus* (Clark, 1955). Since the habitats of the three species of barnacles are quite different (*Balanus balanoides* and *Chthamalus stellatus* are intertidal and *Balanus balanus* is sublittoral) and the state of development of the gonads is also different, it seems that the reduction in slope which occurs in January in all three species must be attributable to some seasonal factor. He suggested that the availability of food may be responsible for the low Q_{10} values recorded during the winter, which would agree with the observation of Davies (1966, also p.396) that the metabolic rate of limpets appeared to be correlated with the available food and that low Q_{10} values tended to be found in limpets of the upper shore.

There is thus abundant evidence to suggest that compensatory changes in the activity and in the metabolism of intertidal poikilotherms may occur. Such compensation may be in response to tide-dependent factors such as availability of food and temperature as well as to photoperiod (acclimatisation). Such studies have, however, been mainly concerned either with direct measurements of the activity of the organism concerned or else with its standard metabolism (cf excised barnacles, p. 384). Measurements of both active and standard rates of oxygen consumption and of the effect of seasonal and thermal acclimation on the scope for activity in intertidal organisms are scarce.

The effect of temperature on the respiration rate of certain intertidal algae is very similar to its effect on the standard rate of metabolism of a number of intertidal organisms (Newell and Pye, 1968; p. 409). Such algae include *Enteromorpha intestinalis, Ulva lactuca, Fucus*

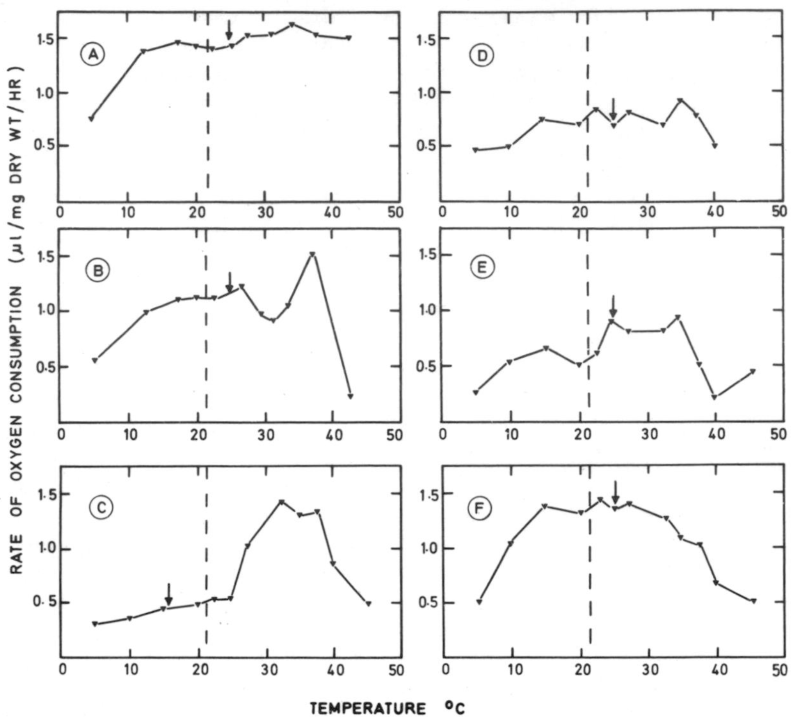

Fig. 8.33. Graphs showing the effect of temperature (°C) on the rate of oxygen consumption (µl 0_2/mg dry wt/hr) of a variety of algae collected during July 1967. The arrow indicates the temperature of the rocks (24·5°C) from which the algae were collected, except for *Ulva*, where its surface temperature (16°C) is so marked. The sea temperature (21·5°C) is indicated by a broken line. (A) *Enteromorpha intestinalis.* (B) *Fucus* (? *ceranoides*) (C) *Ulva lactuca* (D) *Chondrus crispus* (E) *Griffithsia flosculosa* (F) *Porphyra umbilicalis.* (After Newell and Pye, 1968.) Note that the slope of the RT curve is relatively shallow over the temperature range prevailing in the habitat.

(? *ceranoides*), *Porphyra umbilicalis, Griffithsia flosculosa* and *Chondrus crispus.* It has been found that the rate of respiration is relatively unaffected by temperature fluctuation within the normal environmental temperature range and in this respect resembles the standard rate of respiration of the invertebrates discussed on p. 410. However, the range of temperature over which the Q_{10} is low varies with season and with the thermal stress of the algae. The RT curves for specimens collected during July are shown in Fig. 8.33 whilst those for December are shown in Fig. 8.34. The main feature which emerges from a comparison of the two figures is that the form of the RT curves is not a fixed phenomenon characteristic of a particular species of alga,

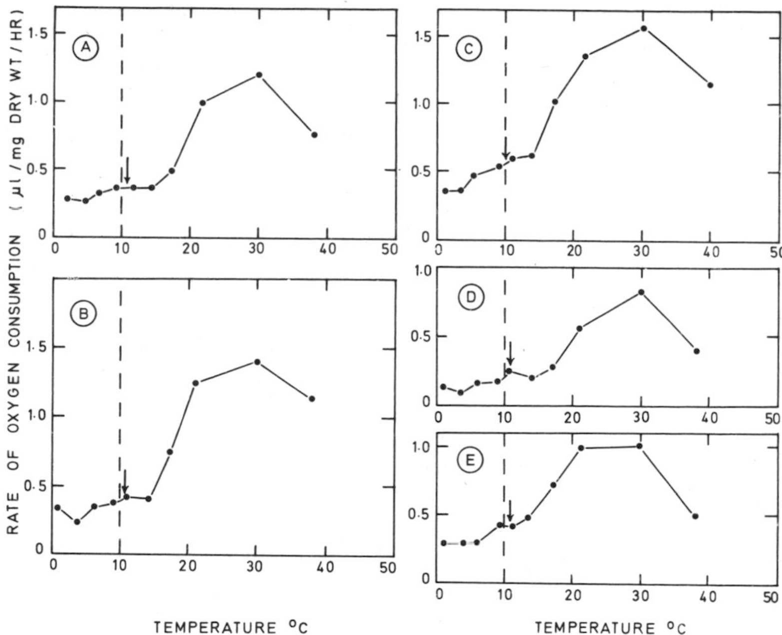

Fig. 8.34. Graphs showing the effect of temperature (°C) on the rate of oxygen consumption (µl O_2/mg dry wt/hr) of a variety of algae collected during December 1967. The arrow indicates the temperature of the rocks (10·5°C) from which the algae were collected, except for *Ulva*, where its surface temperature (10°C) is so marked. The sea temperature (10°C) is indicated by a broken line. (A) *Enteromorpha intestinalis.* (B) *Fucus*(? *ceranoides*). (C) *Ulva lactuca.* (D) *Chondrus crispus.* (E) *Griffithsia flosculosa.* (After Newell and Pye, 1968.) Note that the shallow part of the RT curve extends over the temperature range of the habitat at the time of collection (cf Fig. 8.33.)

but appears to be modifiable in such a way that the shallow region of the RT curve is appropriate to the temperatures prevailing in the habitat.

Barnes (personal communication), in the work which has in part been described above, has demonstrated a similar modification in the form of the RT curve for the respiration of excised barnacles. He has shown that in *Balanus balanoides* there is a regular change in the form of the curve relating respiration to temperature. In January there is a regular increase in the rate of respiration from 5° to 20° C, whilst in April there is an initial steep slope between 5° and 10° C which is followed by a negative Q_{10} between 10° and 15° C. In August the Q_{10} between 10° and 15° C is also low (1·20) and is followed by a steeper region, whilst in October the curve is flattened between 10° and 20° C (Fig. 8.35A). A very similar situation exists in the other intertidal

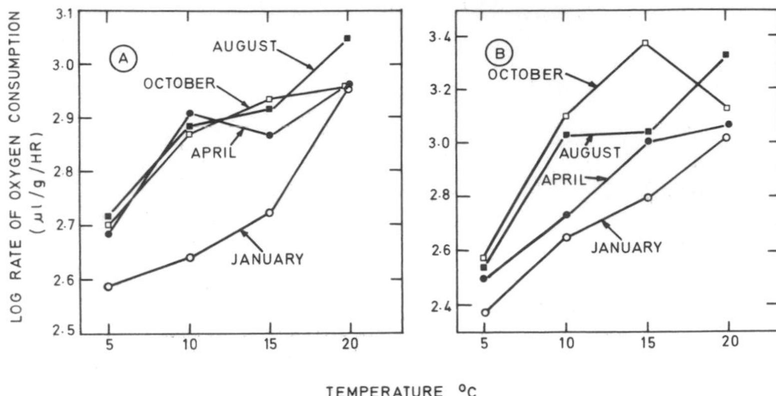

Fig. 8.35. Graphs showing the seasonal variations in the form of the curve relating log. rate of oxygen consumption (µl/g/hr) to temperature in excised specimens of (A) *Balanus balanoides* and (B) *Chthamalus stellatus* from Millport, Isle of Cumbrae. (Data from Barnes, personal communication.) Note that the form of the curve is modifiable and that in *B. balanoides* it is flattened above 10°C in April, August and October but that in January a regular increase occurs above 5°C. In *C. stellatus* the curve is flattened between 15° and 20°C in April and 10–15°C in August.

barnacle *Chthamalus stellatus,* in which there is a marked flattening of the curve between 10° and 15°C in August (Fig. 8.35B) at which time the temperature on the shore from which the animals were collected (Millport, Isle of Cumbrae, Scotland) was 14 – 20°C. In this respect the data resemble those obtained by Davies (1966) on *Patella* collected from Millport, the species examined having a low value for the Q_{10} between 10° and 15°C, and also resemble the results for the intertidal algae mentioned above. In the sublittoral *Balanus balanus,* however, there was no trace of a flattened region of the RT curve between 10° and 15°C. As Barnes pointed out, this might perhaps be expected in an organism which is not subjected to rapid fluctuations in temperature (Schlieper, 1952; Newell, 1966, 1969). Again, in the echinoid *Eucidaris tribuloides* the mean rate of oxygen uptake of well-fed animals increased fairly regularly with temperature in urchins collected during the winter, but in those collected during the summer the Q_{10} between 15·0 and 20°C was 1·0, compared with a value of 3·4 over this range in the winter animals. (McPherson, 1968).

Seasonal and temperature-induced variations in both the standard and the active rates of oxygen consumption of the winkle *Littorina littorea* and the mussel *Mytilus edulis* have recently been detected by Newell and Pye (in press). Data for the maximum and minimum rates of respiration were plotted against tissue weight as described on p. 387

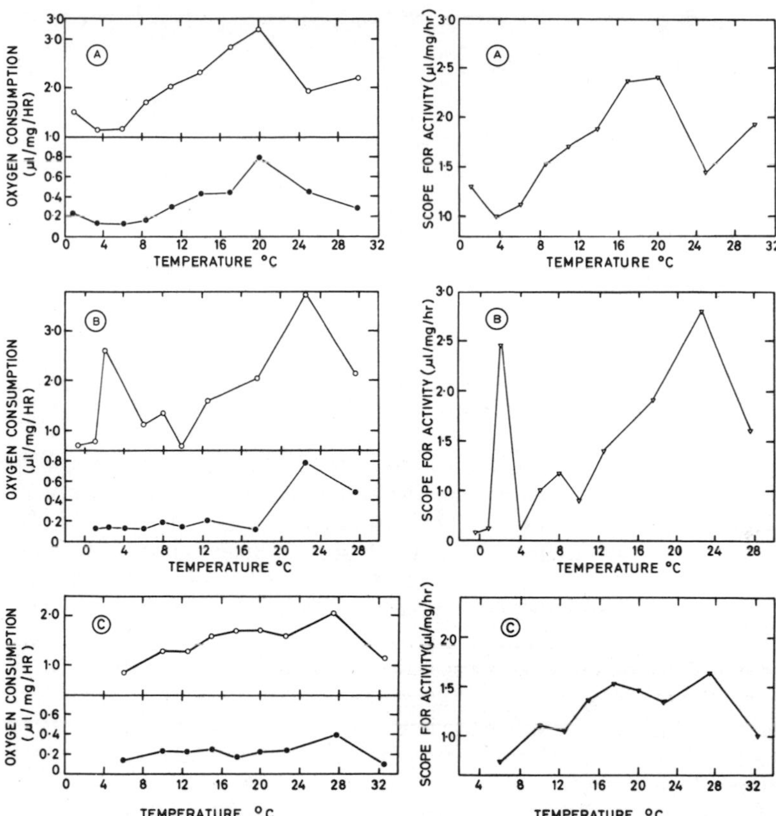

Fig. 8.36. Graphs showing the relation between temperature and the active (open circles) and standard (solid circles) rates of oxygen consumption by the winkle *Littorina littorea* during (A) January, (B) March and (C) May. The scope for activity obtained by subtraction of the standard from the active rate is shown for each season (triangles). (From Newell and Pye, in press.)

and the value for the standard rate calculated from the line relating the minima; the line relating the maxima was taken to give the active rate of respiration. The levels for the active and standard rates of respiration of *Littorina littorea* at a wide variety of temperatures for animals taken from the shore in January, March and May are shown in Fig. 8.36 together with the scope for activity (obtained by subtraction of the two rates, see p. 405). Two main features are evident from the graphs. Firstly the standard rate is much less affected by temperature than the active rate. It is particularly flattened below 10°C in the January animals, between $<1{\cdot}0^{\circ}$ and 18°C in the March animals, and between

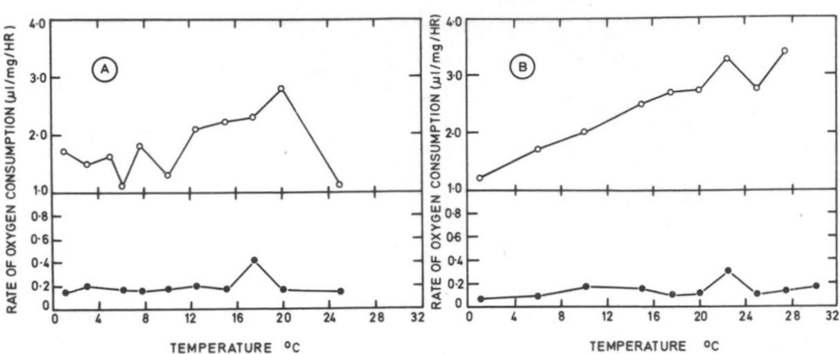

Fig. 8.37. Graphs showing the effect of temperature on the active (open circles) and standard (solid circles) rates of oxygen consumption of the mussel *Mytilus edulis* during (A) February and (B) April. (After Newell and Pye, in press.)

<6·0° and 22·5°C in the animals collected from the shore in May. On the other hand the active rate showed a general steep increase with temperature up to a point of thermal decline which coincided with the point of thermal decline of the standard rate. Superimposed upon the active rate, however, there were marked peaks coinciding with an increased level of activity at those temperatures. Thus in the animals collected from the shore in January, and to a greater extent in March, there was a pronounced peak at approximately 2°C; unfortunately recordings were not made at a low enough temperature to determine whether such a peak was present also in May. A second peak occurred in the region of 20°C in January, 22·5°C in March and 28°C in May after which a decline occurred. The general level of the active rate in May was also lower than that in January and March especially at high temperatures. Thus the standard rate appears to be relatively unaffected by temperature change, whilst the active rate shows a general increase with temperature, much as has been described for direct observations on activity (p. 406). In addition, seasonal acclimation in activity is reflected in the presence of one or more peaks in the active rate at low temperatures in January and March coupled with a general suppression of the active rate, particularly at high temperatures, in May.

Similar measurements have also been made on the mussel *Mytilus edulis*; the standard rate has been found to be nearly independent of temperature over the range 0 – 15°C in animals collected in February and from <1·0 to 20°C in animals collected in April. As with *Littorina littorea,* the active rate varied with temperature, but the animals collected in February showed a compensatory increase in the rate at low temperatures compared with animals collected during April. A further feature was that the active rate declined at temperatures above 20°C in specimens collected during February but showed no general decline below 27·5° C in the animals collected during April. In this respect the point of decline of the active rate shows a similar seasonal change to that noted in *Littorina littorea* (Fig. 8.36), and also agrees with the changes noted in the upper lethal limits of gill cilia of other bivalves (Vernberg, Schlieper and Schneider, 1963). The data for *Mytilus edulis* are illustrated in Fig. 8.37.

Throughout the seasonal measurements made on these two molluscs it was noted that the range over which the standard rate was relatively independent of temperature did not vary consistently with time of the year; it appeared to be correlated with the air and sea temperatures prevailing in the habitat during the period immediately preceding the collection of the animals from the shore. This suggested that the controlling factor in *Littorina littorea* and *Mytilus edulis* is not starvation, as is possibly the case in the barnacles and limpets discussed above, nor photoperiod as in *Hemigrapsus* sp. (p. 403), but temperature. Such a relationship with the environmental temperature was demonstrated in *Littorina littorea*, specimens of which were stored at 6·5° and 10° C for

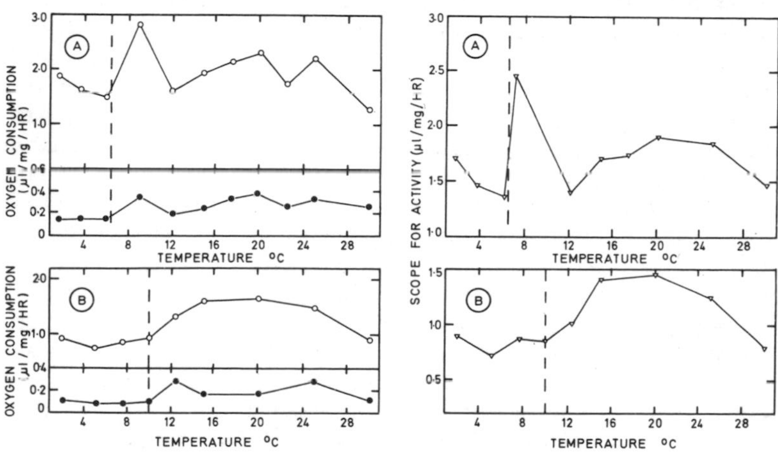

Fig. 8.38. Graphs showing the effect of temperature on the active (open circles) and standard (closed circles) rate of oxygen consumption in the winkle *Littorina littorea*. The scope for activity (triangles) is also indicated. (A) Animals stored for 3 weeks at 6·5°C. (B) Animals stored for 1 week at 10°C. (After Newell and Pye, in press.) The storage temperatures are shown by a broken line.

from 1 – 3 weeks before measurement of the rate of respiration in relation to size and temperature. The results are shown in Fig. 8.38 which indicates that the standard rate is relatively independent of temperature throughout the range at which it was measured, but the

Q_{10} is 1·0 over the range <1·5 – 6·5° C in the animals stored at 6·5° C and over the range 2·0 – 10° C in the animals stored at 10° C. The active rate in both cases showed a decline after 25° C, but also showed a clear acclimation insofar as the level of the active rate was considerably higher in the animals which had been stored at 6·5° C than in those stored at 10° C. Added to this there appeared to be a peak in the level of the active metabolism in the region of the storage temperature. The effect of these changes in the active and standard rate are reflected in the scope for activity which is also shown in Fig. 8.38 for each group of animals.

It is therefore apparent that although a number of factors may be of importance in altering the form and the level of the RT curve, nevertheless the fact of temperature-independence in the standard metabolism (and sometimes also in the active metabolism; McFarland and Pickens, 1965; Siefken and Armitage, 1968) of a great variety of organisms over the normal environmental temperature range at the time of collection has been established by a number of different experimental techniques. The significance of this has been discussed by Barcroft (1932) Schlieper (1952), Bullock (1955) and other authors (p. 407); the general conclusion was that the possession of a homeostatic mechanism of this type may offset the problems which would occur if the sequences of metabolic reactions were each to vary in their own particular way with temperature. Other factors, too, such as the well-documented temperature-independence of the internal or circadian rhythms of poikilotherms (for review Harker, 1964) argue for the presence of temperature-independent metabolic machinery within such organisms. Although the temperature-independence of the standard rate of metabolism may be influenced by a variety of factors, there is at present little evidence of the mechanism by which it is maintained. Bullock (1955), however, cites several instances in which isolated tissues of acclimated organisms may still show acclimation. Peiss and Field (1950) reported that a high level of metabolism which occurs in some arctic fishes compared with Californian species is still detectable in the isolated brain, whilst recently Gordon (1968) has demonstrated that the Q_{10} of minced preparations of the white muscle of two species of tuna lies between 1·0 and 1·2 over a wide temperature range between 5 and 35° C, very much as reported for the standard rate of metabolism of a range of intertidal invertebrates by Newell and Northcroft (1965, 1967). On the other hand Freeman (1950) found evidence of the retention of acclimation in the brain but not in muscle isolated from warm- and cold-acclimated goldfish, whilst Roberts (1952) found the reverse situation in a crab in which there was retention of acclimation in the muscle but not in the brain. Schlieper (1952) found the same temperature relations in the isolated tissues of two freshwater fishes as in the intact animals and this has been found to be true of the cell-free homogenates of a number of poikilotherms (Newell, 1966, 1967, 1969; Newell and Pye, in press). Acclimation is also known in the activity of enzyme systems (Precht, 1951a,b, Christophersen and Precht, 1952a; Kirberger, 1953; for review Bullock, 1955).

Newell (1966, 1967, 1969) showed that the rate of respiration of mitochondrial suspensions of a number of poikilotherms including the mussel *Mytilus edulis* and the anemone *Actinia equina* showed a similar independence of temperature as that described above for the standard rate of oxygen consumption of intact animals. Further, the extent of the shallow part of the RT curve appeared to be related to the environmental temperature range of the organisms concerned. In *Actinia equina* the range was from <5·0 to 12·5°C, whereas in upper shore specimens of *Mytilus edulis* the range was from 5·0 to 20°C before an increase in the slope occurred. Despite the fact that it is unlikely that in such suspensions the mitochondria were in an identical state to that in the intact organism, the similarity between the curves obtained and those of the standard metabolism of a number of invertebrates is striking (especially Collardeau, 1961; Collardeau-Roux, 1966; Roux and Roux, 1967; Newell and Northcroft, 1965, 1967; Halcrow and Boyd, 1967). Again, the slope of the curve for suspensions of mitochondria extracted from a skate suggested that in this sublittoral animal an extensive shallow region does not occur, and in this respect is similar to the data obtained for *Branchiostoma lanceolatum* and for excised specimens of the sublittoral barnacle *Balanus balanus*. Newell and Pye (in press; also Newell, 1969) have recently made an extensive survey of the effect of temperature on the respiration of cell-free homogenates of a variety of intertidal gastropods and on *Mytilus edulis* to determine whether seasonal and temperature-induced changes occur in the form of the RT curve. The results for specimens of *Littorina littorea* collected in February, March and May are shown in Fig. 8.39. It will be noticed that there is a general increase in the rate of respiration with temperature in animals collected during February, but that in cell-free homogenates of animals collected during late March the RT curve has a flattened region between 0·5°C and 12·5°C after which an increase in the slope occurs. In cell free homogenates of *Littorina littorea* collected in May, the flattened region extended to 17·5°C and was followed by a steep rise in the rate of respiration. The form of the curve is thus similar to that shown for preparations of *Actinia equina* and *Mytilus edulis,* and shows a seasonal extension in the range over which the rate of respiration is independent of temperature; in this respect the results resemble those shown for the standard rate of intact animals (Fig. 8.25). A similar seasonal modification in the form of the curve has been demonstrated in cell-free homogenates of the mussel *Mytilus edulis.* In

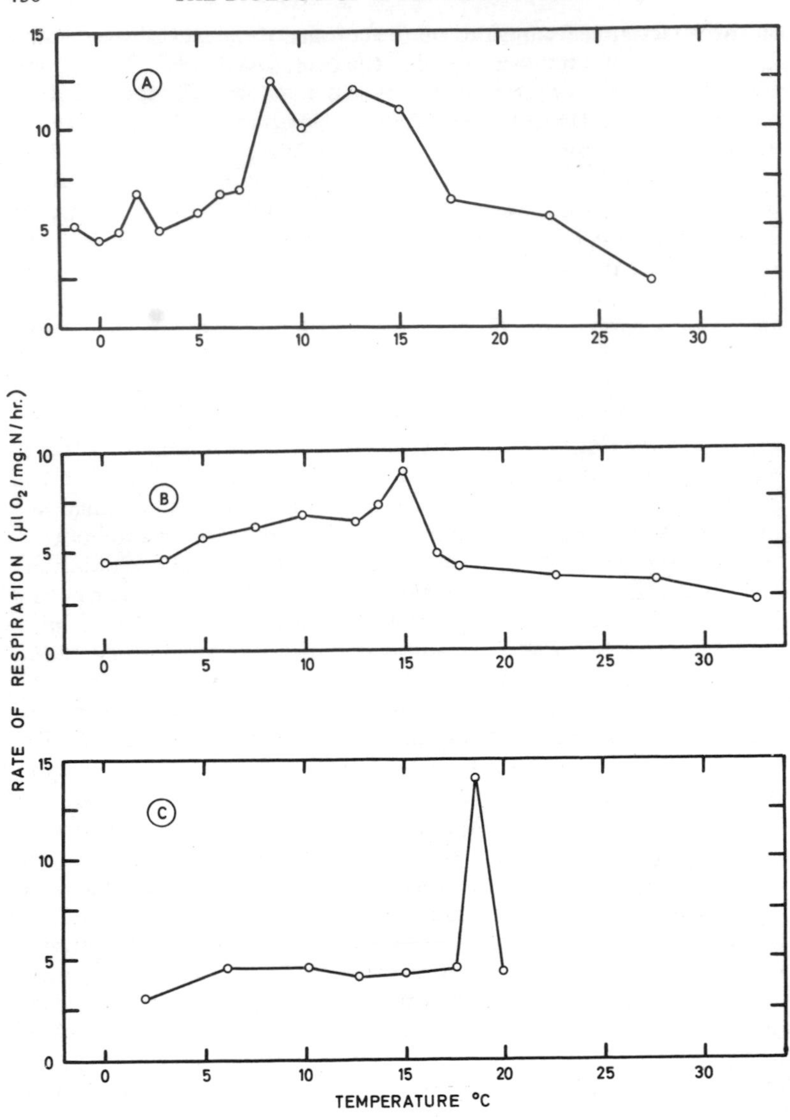

Fig. 8.39. Graphs showing the effect of temperature on the rate of respiration (μl O_2/mg N/hr) of cell-free homogenates of *Littorina littorea* collected during (A) February, (B) March and (C) May. Each point represents the mean of four respiration vessels each of which contained 1·5 ml of a mixture consisting of final concentrations of KC1 15·8mM; $MgSO_4,7H_2O$ 1·58mM; ATP 0·2mM; Cyt *c* 0·42 x 10^{-2} mM and succinate 50mM plus homogenate, the total molarity being approx. 0·44M and the *pH* 7·4. All determinations were made within 8 hours of preparation of the homogenate. (After Newell and Pye, in press.)

both *Mytilus* and *Littorina* the upper limit of the shallow slope approximates to the maximum environmental temperature at the time of collection (Fig. 8.41), and suggests that, as in the intact organisms, temperature itself may play a part in the modification of the form of the RT curves.

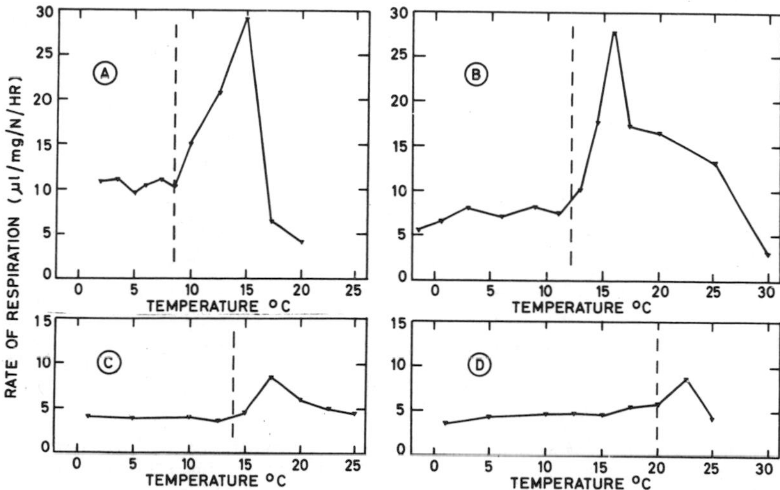

Fig. 8.40. Graphs showing the effect of temperature on the rate of respiration (μl O_2/mg N/hr) of cell-free homogenates of specimens of *Littorina littorea* which had been stored at various temperatures which are indicated by a broken line. (Experimental details as in Fig. 8.39; after Newell and Pye, in press.) (A) animals stored for 7 days at a mean temperature of 8·25°C, (B) animals stored for 14 days at a mean temperature of 13·2°C, (C) animals stored for 9 days at 14°C and D) animals stored for 9 days at 20°C. (Suspension media as in Fig. 8.39.)

Fig. 8.40 shows the RT curves obtained with cell-free homogenates of *Littorina littorea* which had been stored for from 7 to 14 days at 6·5°C, 12°C, 14°C and 20°C. It is obvious that in such preparations there is a very marked correspondence between the storage temperature and the extent of the range over which the respiration is relatively independent of temperature. In each case the region with a low Q_{10} is followed by a sharp increase in slope which is followed by a decline, the point at which decline occurs is also related in a general way to storage temperature; in addition there is an increase in the level of metabolism over the appropriate temperature range in the animals acclimated to a low temperature compared with those stored at higher temperatures. Thus acclimation in this species shows a variation in the level of metabolism as well as in the form of the RT curve. Identical results have been obtained with cell-free homogenates of *Mytilus edulis* stored at the appropriate temperatures. The relation between the upper limit of the shallow slope and the storage temperature or seasonal temperature in homogenates of both *Littorina littorea* and *Mytilus edulis* are shown in Fig. 8.41.

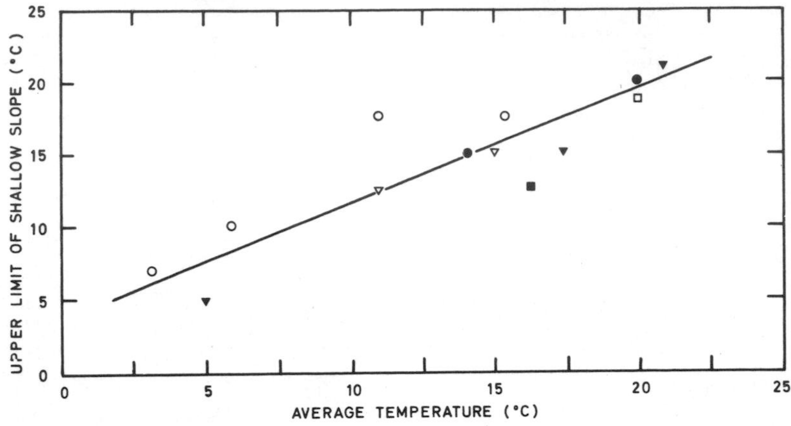

Fig. 8.41. Graph showing the relationship between the storage or environmental temperature and the upper limit of the shallow slope of the RT curve for the respiration of cell-free preparations and mitochondrial suspensions of various intertidal invertebrates. Open circles *Littorina littorea* – seasonal measurements (cell-free homogenates). Solid circles *L. littorea* – experimentally acclimated animals (cell-free homogenates). Open triangles *Mytilus edulis* – seasonal measurements (cell-free homogenates). Solid triangles *M. edulis* – experimentally acclimated animals (cell-free homogenates). Open square *M. edulis* – mitochondrial preparation of animals collected straight from shore. Solid square *Actinia equina* – mitochondrial preparation of animals collected straight from shore. (Data from Newell, 1966, 1967, 1969 and Newell and Pye, in press). Suspension media as in Fig. 8.39.

It would be premature to suggest that the observed low Q_{10} values for the standard rate of metabolism of the many poikilotherms cited on pp. 407-409 would also be demonstrable at the sub-cellular level. Nevertheless comparable results have been obtained by Gordon (1968) for minced preparations of tuna muscle, whilst the curves obtained with suspensions of sub-cellular material also correspond fairly closely with the seasonal and temperature-induced changes in the RT curve for the standard metabolism of intact organisms. Indeed, the form of the curves obtained with cell-free homogenates is identical with those described for the respiration rate of intact Trichoptera larvae (Collardeau, 1961), the upper limit of the shallow slope coinciding with the environmental temperature (Collardeau-Roux, 1966). A similar situation exists in gammarids belonging to the *pulex* group (Roux and

Roux, 1967, and in freshwater copepods such as *Diaptomus* (Siefken and Armitage, 1968). On the other hand, the preparation and storage of homogenates during the extensive series of temperatures shown in Figures 8.39 and 40 may lead to an alteration in the state of the mitochondria even though in this series of experiments all measurements were made within 8 hr of the initial preparation of the homogenate. The general occurrence of low Q_{10} values has been questioned by Tribe and Bowler (1968), who could not detect similar low Q_{10} values in the respiration of preparations of the blowfly *Calliphora* in relation to temperature. However, since their curves were based on measurements at three (and sometimes only two) temperatures it is doubtful whether this can be taken to show that the slope is steep over the whole range of tolerance of this organism; it does however, indicate that a very extensive plateau is absent in animals stored under those particular conditions. Thus in muscle preparations of flies acclimated to 10° C measurements were made at only 10° and 30° C, so that if the upper limit of the shallow slope coincided with the storage temperature, as has been shown in a number of other invertebrates (Fig. 8.40), any such region would be likely to occur below 10° C. In preparations of animals acclimated to 30° C the curve was based on measurements at 10° and 30° C, so that it is not possible to determine whether or not a flattened region exists in the region of 30° C. Tribe (personal communication) has also found no trace of a shallow slope in the respiration of excised gill of the shore crab *Carcinus maenas* at temperatures above 10° C, but since the crabs were acclimated to 10° C, a shallow slope extending above this temperature would be unlikely for similar reasons. Mangum and Sassaman (1969) have been unable to detect a shallow slope in the RT curve for the standard metabolism of the polychaete *Diopatra cuprea* even though they used an identical technique to that of Newell and Northcroft (1967). *Diopatra* is a burrowing animal and might perhaps not have been exposed to such extreme temperature fluctuations as a member of the epifauna; sand is known to buffer thermal fluctuations considerably (Chapter 1), so that perhaps this animal is not strictly comparable with those which experience the full rigours of the habitat.

It would perhaps at this stage be safe to infer that the metabolism of poikilotherms is affected by a large variety of factors, and that compensatory changes in the level, and sometimes also in the form, of the RT curve for standard metabolism occur in response to photoperiod (Dehnel, 1958), starvation related to shore level and season (Davies, 1966, 1967; Barnes, personal communication), metabolic substrate (Barnes, Barnes and Finlayson, 1963a, b) and temperature (Newell and Pye, 1967 and in press; Siefken and Armitage, 1968; Newell, 1969). In addition, such adaptive changes may occur in the isolated tissues of some species (Freeman, 1950; Roberts, 1952; Schlieper, 1952; Peiss and Field, 1950) or in the minced tissues of others (Gordon, 1968) and also in cell-free systems (Newell, 1966, 1967, 1969, Newell and Pye, in press) as well as in isolated enzyme systems (Precht, 1951a,b; Christophersen and Precht, 1951a; Kirberger, 1953). Despite the diversity of acclimation effects which have been described above, it is common to find that the level of both active and standard metabolism may be raised under "winter conditions" such as short day length (Dehnel, 1958) or low temperatures, whilst acclimation to "summer conditions" may involve a suppression of active and standard rates of metabolism over the appropriate parts of the temperature range. The phenomenon of a low Q_{10} value for the standard rate of metabolism may reflect a suppression of metabolism at high temperatures, and is possibly caused by different factors in different organisms. The result is the same, however, the metabolism of a variety of freshwater and intertidal poikilotherms, and intertidal algae being relatively independent of temperature fluctuations within the normal environmental range. In this respect the metabolism is well-suited to operate in an environment where rapid fluctuations in temperature are likely to occur. Long-term seasonal fluctuations are accommodated in a number of intertidal animals and algae by an alteration in the temperature range of the shallow slope, so that its extent is appropriate to the conditions prevailing within the habitat. Other organisms, especially sublittoral and burrowing forms, however, which live in habitats that are to some extent buffered from extreme temperature fluctuation, may not show such a flattening of the RT curve, although some of them such as *Diaptomus* (Siefken and Armitage, 1968), gammarids and Trichoptera (Collardeau, 1961; Collardeau-Roux, 1966; Roux and Roux, 1967) and tuna (Gordon, 1968) do show a plateau. It is possible that the freshwater gammarids and Trichoptera may experience thermal fluctuations in their habitat, whilst the presence of a plateau in the freshwater copepod *Diaptomus* and in the tuna might be explicable in terms of the temperature fluctuation experienced during vertical migration through the thermocline. Clearly, the significance of the presence or absence of a plateau in the RT curve can only be accounted for satisfactorily if the details of the temperature fluctuations occurring in the habitat are known. Such modifications fall into the category of "capacity adaptation" (Christophersen and Precht, 1953; also Christophersen, 1967; Prosser, 1967) which are compensatory changes permitting physiological activities within the normal environmental range. On the other hand, those adaptations which enable survival at environmental extremes, and which are discussed in the next chapter, are distinguished as ' resistance adaptations".

(iii) Some biochemical aspects of thermal acclimation

Biochemical changes accompanying thermal acclimation have received a great deal of attention in recent years, and have been extensively reviewed by Prosser (1962, 1967; also Rao, 1967). Much of such work is, unfortunately, outside the scope of this book, but some of the changes which are commonly associated with thermal acclimation may be summarised here, together with more recent data which may suggest a mechanism underlying the temperature-independent respiration of the animals cited on p. 409. Many enzymes show higher levels of activity from cold-acclimated animals, compared with those extracted from warm-acclimated ones (for review Prosser, 1962), the RT curves being displaced (or translated) to the left in cold-acclimated forms. On the other hand, different enzymes from the same tissue may be differently affected by thermal acclimation; in the bitterling, *Rhodeus,* the succinic-dehydrogenase from the muscle shows partial compensation, but lactic dehydrogenase (LDH, p. 269) shows no compensation, whilst aldolase shows inverse compensation (Kruger, 1962). Again, different tissues may show different responses to thermal acclimation (also p. 428); in the trout *Salmo gairdneri* the oxygen consumption of the liver shows over compensation, the gill tissue shows no compensation, and the brain shows complete compensation (Evans, Purdie and Hickman, 1962). We have already seen that the standard oxygen consumption of the sunfish *Lepomis gibbosus* shows perfect compensation ($Q_{10} = 1{\cdot}0$) over the range 10–17·5°C (Roberts, 1964, 1967), and a similar example of perfect compensation has been reported for the cholinesterase activity of brain tissue of the fish *Fundulus heteroclitus* at 13° and 30°C (Baslow and Nigrelli, 1964). Again, Vijayalakshmi (1964, cited by Rao, 1967) has demonstrated compensatory increases in the activity of succinic-, malic-, lactic- and pyruvic-dehydrogenase from the hepatopancreas of cold-acclimated scorpions; cytochrome *c*, cytochrome oxidase and lipase from the earthworm *Lampito mauritii* also show compensatory changes in activity in response to cold- or warm-acclimation (Saroja and Rao, 1965; also Rao, 1967). Similar compensatory changes in the activity of certain enzymes also occur in homoiotherms in response to cold-acclimation; in muscle and liver from rats acclimated to low temperatures, for example, there is a compensatory increase in the level of activity of succinic and malic dehydrogenase and in cytochrome *c* oxidase (Hannon, 1960; Hannon and Vaughan, 1960).

Most of such examples involve a translation of the RT curve to the left, resulting in a greater enzyme activity per unit protein in tissues from cold-acclimated animals compared with those from warm-acclimated ones when measurements are made at some intermediate temperature. It is clear that this implies a quantitative alteration in the enzyme concentration (Prosser, 1967). Qualitative differences in the nature of the pathways could be expected to result in an alteration of the slope (Q_{10}) of the RT curve, although this effect may also be associated with a translation to the left. Examples of simple translation without any rotation of the RT curve are numerous; Freed (1965), for instance, showed this to occur in cytochrome oxidase from goldfish muscle. Das (1965; also Rao, 1967; Prosser; 1967) has demonstrated an enhancement of protein synthesis in the tissues of cold-acclimated goldfish, and the same is true in the muscle of cold-adapted frogs (Jankowsky, 1960), in the earthworm *Lampito mauritii* (Soroja and Rao, 1965), a freshwater mussel (Rao, 1963a), and rat liver (Vaughan, Hannon and Vaughan, 1958). There is also a higher rate of incorporation of C^{14} labelled amino-acids into proteins of the muscle of cold-acclimated frogs, compared with that into warm-acclimated animals (Mews, 1957; Jankowsky, 1960). Das (1965; cited by Rao, 1967 and Prosser, 1967), using labelled amino-acids, showed that the net synthesis of proteins after 12 hr was 100 – 500% higher in 5°C acclimated fish tissues, compared with 25°C acclimated fish tissues when both were measured at 15°C. The rate of degradation of protein was higher in the warm-acclimated fish, so that part of the great difference in net protein synthesis was due to a slower rate of protein breakdown in the cold-acclimated fish, but this was also associated with an enhanced rate of synthesis. Thus there is evidence both of increased enzyme activity, and of increased protein synthesis after acclimation to low temperature, although there appears to be little direct evidence at present to show that increases in the concentrations of the enzymes themselves occur during cold-acclimation. However, the increase in the net rate of protein synthesis suggests that enzymes may show an enhanced rate of net synthesis at low temperatures; this would account for the higher levels of activity occurring in cold-acclimated animals.

Such data might be taken to indicate a non-specific increase in enzyme activity following cold-acclimation, but there is now abundant evidence which suggests that not only quantitative changes but also qualitative changes occur. That is, thermal adaptation involves not only a control of the rates of reactions, but also of their nature. Ekberg (1958) for example, using metabolic inhibitors, showed that the increased oxygen demand of goldfish during cold-compensation was due to an activation of the pentose (hexose monophosphate) shunt (p. 273). Subsequently Kanungo and Prosser (1959) showed that, although there was a 40% increase in the oxygen uptake of liver homogenates of cold-acclimated fish compared with warm-acclimated ones, the difference between the respiration of mitochondrial preparations of cold- and warm-acclimated fishes was only some 10%. The importance of the anaerobic phase in cold-acclimated fish was emphasised by the demonstration that glycolytic poisons produced more pronounced effects on homogenates from cold-acclimated specimens, whereas inhibitors of mitochondrial metabolism produced similar effects on preparations from both warm- and cold-acclimated animals. Hochachka and Hayes (1962) showed that the hexose monophosphate (pentose) shunt is of more importance in cold-acclimated fish because a C^{14} labelled glucose derivative (glycogen) breaks down to CO_2 from the C-1 position, whereas the Embden-Meyerhof (EM) pathway (glycolysis, p. 268), leading to a breakdown of glucose to CO_2 from the C-6 position, is of more importance in warm-acclimated fish. In general, cold-adaptation appears to involve the maintenance of the Krebs cycle at a constant rate and a facilitation of extramitochondrial metabolism. The changes are extremely complex in the liver, increases in glycolysis, gluconeogenesis, glycogen synthesis, lipogenesis and hexose monophosphate (HMP) shunt participation occurring. In muscle, however, the situation is more simple; Krebs cycle activity measured by acetate-1-C^{14} oxidation is somewhat decreased on cold-acclimation whilst glycolysis and HMP shunt activity are increased.

Hochachka (1967) has presented an important review of the possible role of isozymes in the control of such compensatory metabolic changes. He pointed out that the enzyme glucose-6-phosphate (G-6-P) dehydrogenase occurs in two distinct molecular forms, or isozymes (isoenzymes), and that there is some evidence that the relative participation of the HMP (pentose) shunt, and of Embden-Meyerhoff (EM) glycolysis, may be determined by such isozymes (also Rose, 1961). Although such a hypothesis has not been tested, it is possible that the production of one or other of such G-6-P dehydrogenase isozymes may be favoured by cold- or heat-acclimation. The induction of the appropriate isozyme could be brought about by the direct action of temperature itself, or by other environmental factors, by substrates, or by hormones. There is some evidence that increased participation of the HMP shunt in the liver, due to thyroid activation, may be such a case of isozyme induction by hormones (Hochachka, 1962). Again, Rao and Saroja (1963) and Saroja and Rao (1965; also Rao, 1967) have demonstrated that the addition of small quantities of body fluids from cold-acclimated earthworms to perfusion fluids containing tissues of normal worms, significantly increased oxygen consumption of the normal tissues. Conversely, the addition of body fluids from warm-acclimated worms depressed the metabolism of tissues from both normal and cold-acclimated worms. They showed that the active principle was located in nerve tissue extract, and that injection of such material from a cold-acclimated worm into intact worms was sufficient to initiate a series of biochemical changes characteristic of acclimation to low temperature. Similar enhancement of the tissue respiration of normal- and warm-acclimated scorpions following addition of extract from the sub-oesophageal ganglion of cold-acclimated animals was reported by Vijayalakshmi (1964; cited by Rao, 1967), and for a variety of fish by Rao (1963b), Jankowsky (1964) and Precht (1964). Further, Precht (1964) found that such factors could act inter-specifically, and also affected the heat resistance of the tissues (i.e. the resistance adaptation p. 434). Such results, as Rao (1967) has pointed out, suggest that a change in temperature may activate one or more hormonal systems which may release active factors that are themselves responsible for initiating the increased protein synthesis, or for activation of alternative metabolic pathways and other changes associated with thermal acclimation. There is also evidence that metabolic compensation in the muscle of eels depends upon the tonic motor discharge from the nervous system (Prosser, Precht and Jankowsky, 1965; also Precht, 1961; Schultze, 1965).

Hochachka (1965, 1966, 1967) has also shown that a series of significant changes in the lactic dehydrogenase (LDH) isozyme system occurs in association with thermal compensation in goldfish and trout, a new series of LDH isozymes being induced in tissues of cold-acclimated fishes. In mammals there are commonly five LDH isozymes which represent tetramers of two different subunits, A and B (sometimes designated as H and M; Markert, 1963; Markert and Faulhaber, 1965; Kaplan, 1964; Hochachka, 1967). These isozymes represent distinct proteins and, according to the proportion of their component subunits, may be designated A_4, A_3B_1, A_2B_2, A_1B_3 and B_4. In fishes, however, at least five subunits appear to be implicated in the formation of the fourteen LDH isozymes which are known to occur in cold-acclimated specimens of the brook trout *Salvelinus fontinalis,* and the lake trout *Salmo namaycush* (Hochachka, 1966). Three subunits, A, B and C appear to be involved in the formation of nine LDH isozymes which migrate towards the anode when separated by starch-gel electrophoresis. The five most electronegative of the bands comprise tetramers of A and B subunits to form the series A_4, A_3B_1, A_2B_2, A_1B_3 and B_4, as in mammals. Subunits B and C then form another four isozymes B_3C_1, B_2C_2, B_1C_3 and C_4, but subunits A and C do not normally form tetramers except under special circumstances (Hochachka, 1966). So far, we have nine LDH isozymes; the remaining five are made up of tetramers of the subunits D and E, forming the series D_4, D_3E_1, D_2E_2, D_1E_3 and E_4. Now the important feature of such LDH isozymes is that warm-acclimation appears to result in a loss in the concentration (activity) of the A,B,C system (LDH 1–9), which appears to be associated with the nucleus or mitochondria, whereas the activity of the D-E system (LDH 10–14), which is in the soluble fraction, appears to be unaffected by the state of thermal compensation. Cold-adaptation in fish, therefore, involves the induction of a set of LDH isozymes (ABC series; isozymes 1–9).

Hochachka (1965, 1967) has also shown that the LDH isozymes which are induced on cold-acclimation, are much more sensitive to metabolic control than the non-induced series 10–14. That is, their rate of activity can be controlled by feedback effects. The mechanism appears to be very comparable to the allosteric, or configurational, effects described on p. 322 for the positive and negative interactions between oxygen binding sites on certain respiratory pigments. The induced LDH isozymes of goldfish heart and brain and trout muscle all have sigmoid curves when the reaction velocity at $pH > 8{\cdot}0$ is plotted against substrate (pyruvate) level which suggests that positive interaction between substrate-combining sites is occurring. In the absence of such positive interaction, the curve would be a hyperbola (cf respiratory pigments, p. 316). In the presence of NAD_{red}, four pyruvates are known to bind per mole of LDH, one pyruvate probably combining with each subunit (Hochachka, 1967, Kaplan, 1964). This, as in respiratory pigments, immediately raises the possibility of positive interaction between the substrate-combining sites, the combination of substrate with one subunit facilitating further combination of substrate with other subunits, Hochachka (1967) suggested that at low temperatures when the LDH isozymes have been induced, their activity may be enhanced to compensate for the low temperature by the use of low concentrations of substrate analogues which would mimic the effect of substrate on the active sites, and produce positive interaction between them. Oxamate, which is an analogue of pyruvate, has this effect on trout muscle LDH (Hochachka, 1965) as has another analogue, oxaloacetate, on rabbit muscle LDH-5 (Fritz 1965; also Hochachka, 1967).

Another significant way in which feedback effects may control the rate of function of LDH isozymes is by activation, or inhibition, by means of effector compounds. For example Hochachka (1967) has shown in salmonid LDH B_4 and C_4, and Fritz (1965) in mammalian LDH systems, that the substrate affinity can be markedly increased by intermediates of the Krebs cycle, or by amino-acids, even though the maximum velocity of reaction at high substrate concentrations (V_{max}) is unaltered. Thus if the rate of LDH activity is plotted against substrate concentration, a positive effector such as citrate would displace the curve to the left as in Fig. 8.42, and would result in a decrease in the Michaelis constant (K_m – or substrate concentration for 50% activity), whereas a negative effector would shift the curve to the right. That is, it would reduce the affinity of the enzyme for substrate, and result in an increase in K_m even though V_{max} remained unaltered throughout.

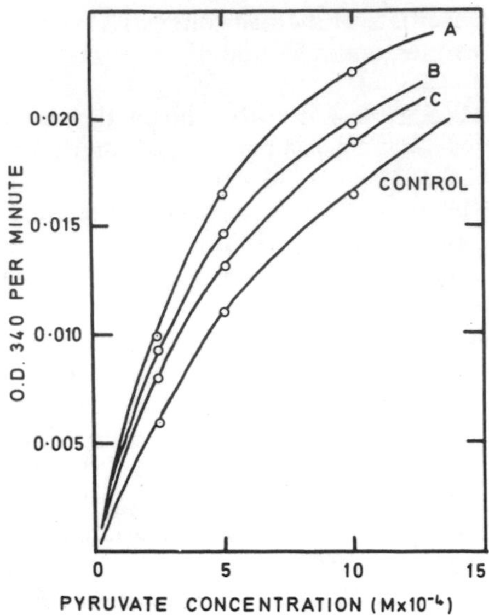

Fig. 8.42. Graphs showing the effect of citrate on the substrate affinity of trout liver B_4 LDH isozyme in 2ml of a reaction mixture consisting of 0·05 M Tris-HC1 buffer, pH 8·5; 1 x 10^{-4} M NADH; and 0·1 ml of a liver LDH preparation plus pyruvate as indicated. The reaction was started by the addition of potassium pyruvate and all measurements were made at 24°C. (After Hochachka, 1967.) A = reaction velocity in the presence of 1 x 10^{-3}M citrate. B = in the presence of 7·5 x 10^{-4} citrate. C = in the presence of 5 x 10^{-4} citrate. Control = reaction velocity in the absence of added citrate.

Such activation by citrate is probably another example of allosteric, or configurational, change resulting from interactions between substrate-combining sites on the LDH tetramer. The induction of such a set of allosteric isozymes under conditions of low temperature acclimation would undoubtedly account for the enhanced anaerobic glycolysis, compared with Krebs cycle activity, which occurs under such conditions.

Hochachka and Somero (1968) have shown that temperature itself may play an important part in acting as a positive modifier of substrate affinity of salmonid LDH isozymes. They pointed out that many studies of the thermal optima for reaction velocities, or the temperature of heat denaturation, do not correlate well with the thermal environment, and that such reaction velocities are commonly determined at high substrate levels. Since *in vivo* substrate concentrations are normally too low for enzyme activity to reach maximum velocity, it follows that under normal conditions reaction velocities will be markedly influenced by the affinity of enzyme for substrate. That is, temperature-induced changes in enzyme affinity, as indicated by the apparent K_m, are likely to be more important than thermal effects on the maximum reaction velocity (V_{max}). They showed that in LDH isozymes from brook and lake trout, temperature markedly affected the enzyme-substrate affinity, the value for K_m being at a minimum (i.e. maximum substrate affinity) at the normal environmental temperature. Fig. 8.43 shows the value of K_m for several LDH isozymes as a function of temperature. It is evident that muscle LDH has its maximum affinity for substrate between 15 and 20°C, but that both above and below this thermal range the rate of reaction would decline at low substrate concentrations because of a reduction in substrate affinity (increase in K_m). It will be remembered, however, that acclimation to low temperatures in such

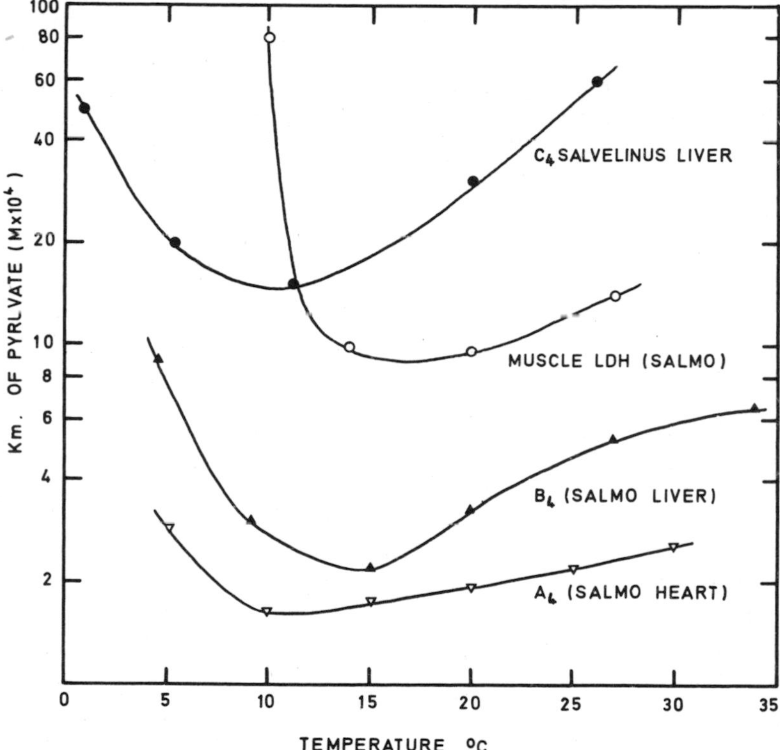

Fig. 8.43. Graphs showing the effect of temperature on the Michaelis constant (K_m) of pyruvate for several LDH isozymes from the salmonid *Salvelinus fontinalis* (brook trout) and *Salmo namaycush* (lake trout). Reaction medium was 0·1 M Tris-HC1, pH 7·5 and the reaction was measured in terms of DPNH oxidation at 340 mμ. (After Hochachka and Somero, 1968.)

salmonids results in the induction of other isozymes such as the A_4, B_4 and C_4 types, and it will be seen that they have (a) a high substrate affinity (low K_m) in the case of the A_4, B_4 isozymes, and (b) thermal optima which are displaced towards low temperatures in the case of A_4, B_4 and C_4 LDH isozymes. The induction of such isozymes with low K_m values and low thermal optima would thus serve to maintain a high level of LDH activity even at low temperatures.

Different fishes from varying thermal environments also show LDH affinity which, in contrast to their maximum velocities at high substrate concentrations (V_{max}), are clearly correlated with habitat (Hochachka and Somero, 1968). Fig. 8.44 shows the variation in substrate affinity with temperature in muscle LDH from the Antarctic fish *Trematomus borchgrevinki* (adapted to −2°C), the South American lungfish *Lepidosiren* (adapted to 27 – 30°C), and tuna, in which the muscle temperature is regulated and held above that of the heart and other organs which are exposed to ambient temperatures. In each case, the minimum value for K_m occurs at the normal environmental temperature of the organisms concerned. The minimum value for *Trematomus* muscle LDH is 0°C, that for tuna heart is 10°C, and for muscle is

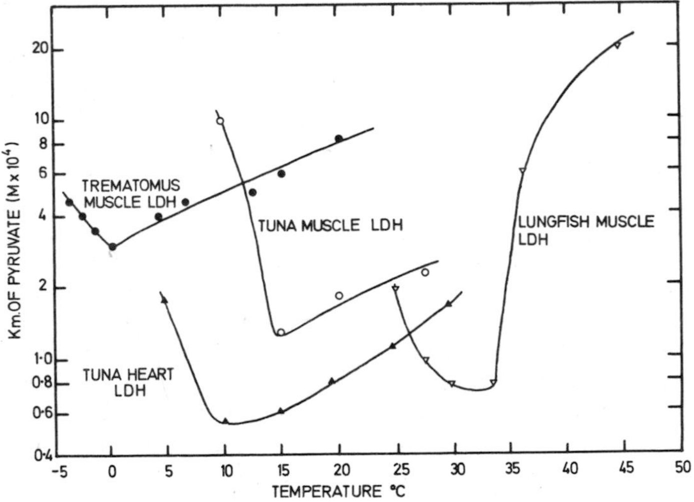

Fig. 8.44. Graphs showing the effect of temperature on the Michaelis constant (K_m) of pyruvate for LDH isozymes from the muscle of the Antarctic fish *Trematomus*, from the heart and muscle of tuna, and from lungfish muscle. Note that the maximum substrate affinity (minimum K_m) occurs at the normal environmental temperature and that a decrease in substrate affinity above this temperature would reduce the rate of reaction provided that the substrate concentration did not exceed K_m. (After Hochachka and Somero, 1968.)

15°C, whilst the minimum K_m for lungfish muscle LDH is 30°C. In contrast, Hochachka and Somero (1968) showed that V_{max} plotted against the reciprocal of the temperature (Arrhenius plot) was linear with temperatures well above the lethal range in each of the three species, and activation energies showed no correlation with environmental temperature.

One of the interesting features of such data is that it provides a possible explanation for the temperature-dependence of the active rate of metabolism in intact animals, and of the relative temperature-independence of the standard rate and of cell-free homogenates of such animals (p. 407). As Hochachka and Somero (1968) have pointed out, under conditions of low substrate, the affinity of LDH for substrate is dramatically reduced as temperatures are increased (i.e. K_m is increased, Fig. 8.44). This is especially true of lungfish muscles LDH, and would allow the reaction velocity to be held relatively constant despite an increase in temperature. On the other hand, at substrate concentrations above K_m, V_{max} increases with temperature with a Q_{10} of about 2·2, as is well-known for the effect of temperature on many enzyme systems at high substrate concentrations. When an animal is active, and substrate levels are high, enzyme activity and the active metabolism would increase with temperature in accordance with the principle of Arrhenius (p. 406). But when substrate levels are low, as during maintenance metabolism (resting or standard metabolism), enzyme-substrate affinity (indicated by K_m), rather than V_{max}, would become of importance in the control of the rate of metabolism; for the reasons given above the reaction velocity would then be relatively independent of temperature (Hochachka and Somero, 1968).

The results discussed above suggest a means by which the induction of isozymes, and the facilitation or inhibition of their substrate affinity by a variety of mechanisms, may provide a basis for thermal compensation. Clearly, long-term acclimation effects, such as the induction of a particular series of isozymes, are distinct from the effects of temperature itself on substrate affinity which provides a basis for acutely-measured thermal independence.* Both systems would be of value in an environment subjected not only to rapid fluctuations in temperature, but also to long-term seasonal change. It is thus not surprising to find that in salmonids any particular LDH isozyme shows

* Somero and Hochachka (1969) have recently shown that in leg muscle LDH of the Alaskan King Crab *Paralithodes* and in epaxial muscle LDH of the rainbow trout *Salmo gairdneri* the changes in K_m which occur as the temperature is lowered serve to activate certain isozymes appropriate to the new temperature regime. At physiological substrate concentrations of 0·01 – 0·10 mmoles, such isozymes are inactive at high temperatures. Thus both an increase in enzyme-substrate affinity and the activation of different isozymes may occur in response even to short-term temperature change.

a reduction in substrate affinity with rise in temperature (Fig. 8.44) above the environmental mean, and in this respect is compensated for transitory increases in temperature. A long-term reduction in temperature results in the development of a new series of isozymes with a raised substrate affinity and with the thermal optima displaced towards lower temperatures. Clearly, in complex multi-enzyme systems the possibilities of thermal compensation are enormously varied. However, Hochachka (1968; also Dean, 1969) has presented evidence which suggests that in the metabolism of glucose, competing pathways for glucose–6–phosphate (G–6–P) may effectively maintain carbon flow through glycolysis and the Krebs cycle constant and independent of short-term fluctuations in temperature. This would provide a basis for the thermal independence of the respiration of cell-free homogenates, and crude mitochondrial suspensions of intertidal organisms, although it does imply that purified mitochondria, from which the possibility of competing pathways for G–6–P have been removed, would not show thermal independence. As Hochachka (1968) pointed out, on the phosphorylation of glucose to G–6–P, this substance may either pass through the glycolytic sequence and Krebs cycle (p. 273), or it can pass into the hexose monophosphate (pentose) shunt (p. 273,) or else it can join the pathway to glycogen.

Hochachka (1968) has studied the fate of labelled glucose in three fishes, *Lepidosiren paradoxa, Symbranchus marmoratus* and *Electrophorus electricus,* all of which live in swamps in South America. He showed that G-6-C^{14} oxidation and acetate flux through the Krebs cycle was maintained relatively constant at all temperatures. Now when the metabolism of C^{14} glucose by liver slices from *Symbranchus* was measured at 22°C, the relative proportion of $C^{14}O_2$ derived from the C-6 position compared with the $C^{14}O_2$ from C-1 position (i.e. the C-6/C-1 ratio) is approximately 1·0, which indicates a negligible HMP shunt contribution, carbon flow occurring predominently through glycolysis and the Krebs cycle (Katz and Wood, 1960; also Hochachka, 1968). At 30°C, however, G-1-C^{14} oxidation was approximately doubled, whereas the rate of oxidation of G-6-C^{14} was unaltered, so that the C-6/C-1 ratio was greatly reduced. Such a reduction in C-6/C-1 ratio could have been brought about by a decrease in C-6 release *via* glycolysis and the Krebs cycle; since this was found to remain unaltered, it follows that the effect was due to an increase in C-1 release through the hexose monophosphate shunt. In all three fishes it was found that as higher temperatures were approached there was greater shunt activity, leading to a greater competition for G-6–P. In much the same way, at 22°, 30° and 38°C, there was an increasing incorporation of C^{14} into glycogen, and at the same time there was a marked decrease in glycogen breakdown between 30 and 38°C. At high temperatures the pathways of G-6-P into the HMP shunt, and also into glycogen, effectively compete for G-6-P and maintain glucose carbon flow through glycolysis and the Krebs cycle constant, despite an increase in temperature. Thus at low temperatures all the G-6-C^{14} passes through glycolysis and the Krebs cycle, but there is increasing shunt participation and carbon flow into glycogen with increase in temperature, resulting in a low temperature coefficient for G-6-C^{14} oxidation.

Another point at which competition for a common substrate may control the rate of carbon flow through the Krebs cycle is the regulation by citrate synthetase and acetyl-CoA-carboxylase, which catalyse the entry of acetyl-CoA into the Krebs cycle or into lipid synthesis respectively (Atkinson, 1966; Hochachka, 1968; Dean, 1969). After the formation of citrate inside the mitochondrion, some of it may either enter the Krebs cycle (p. 273) or else it may diffuse out of the mitochondrion where citrate cleavage then yields extramitochondrial acetyl-CoA, which then enters the pathway to lipid synthesis (Atkinson and Walton, 1967; Hochachka, 1968). Hochachka (1968) found that in *Symbranchus* liver slices, labelled acetate oxidation had a slightly negative temperature coefficient between 22° and 38°C, whilst oxidation by lungfish liver was independent of temperature. In homogenates of electric organ of *Electrophorus*, flux of C^{14} acetate carbon through the Krebs cycle also had a very low Q_{10}, acetate-1-C^{14} oxidation at 38°C being similar to the rate at 30°C. The ratio of C^{14} lipid/$C^{14}O_2$ gave a measure of the relative activities of the Krebs cycle and lipid biosynthesis, and it was found that in the presence of added citrate, which is known to accumulate in the presence of high endogenous acetate (Williamson, 1965), the ratio was 4·0 at 22°C, 5·0 at 30°C and > 8·0 at 38°C. As Hochachka (1968) pointed out, this implies that at high temperatures the pathway to lipid competes increasingly effectively for common substrate, compared with the Krebs cycle. In this way C^{14} flux through the Krebs cycle is maintained constant despite an increase in temperature.

Dean (1969) has studied the metabolism of red and white muscle of the trout *Salmo gairdneri,* and showed that both translation and rotation of the rate/temperature curves occurred ("type IV C" of Prosser, 1958). Palmitate, for example, was oxidised more rapidly by both red and white muscle from fishes which had been acclimated to 5°C than from fishes acclimated to 18°C. Further, the rate was dependent upon temperature in the 5°-acclimated tissues, whereas in the tissues acclimated to 18°C, the rate of oxidation of palmitate was almost independent of temperature (Fig. 8.45). This result is very similar to those described on p. 431 for the rate of respiration of cell-free homogenates of cold-acclimated, compared with warm-acclimated, intertidal invertebrates. As Dean (1969) suggested, such a

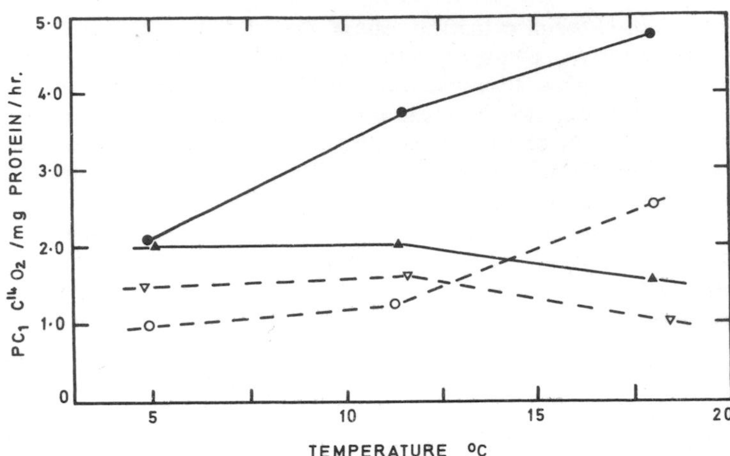

Fig. 8.45. Graphs showing the conversion of palmitate–1–C^{14} to $C^{14}O_2$ by red (solid symbols) and white (open symbols) muscle tissue of trout acclimated to 5°C (circles) and 18°C (triangles). Note that the rate of oxidation of palmitate by both red and white muscle from fish acclimated to 18°C was independent of temperature but increased with temperature in muscle from fishes acclimated to 5°C. The rate of oxidation by red muscle was also significantly higher than that by white muscle. (After Dean, 1969.)

rotation of the RT curve for palmitate oxidation might reflect a qualitative change in enzyme pathway. He found that in cold-adapted trout, acetate flow through the Krebs cycle was held constant between 5° and 11·5°C, probably due to the competitive effect of the pathway to lipid metabolism (p. 445). On the other hand, in warm-acclimated tissues the same effect was not observed, which suggests that channeling of acetate carbon between these two pathways is related also to the thermal history of the organism.

Control of carbon flow through the Krebs cycle at different temperatures is thus extremely complex but nevertheless, as Hochachka (1968) has pointed out, a small number of basic mechanisms may be operative in such multi-enzyme systems. The competitive pathway to lipid biosynthesis at high temperatures appears to be a common feature in the metabolism of fish tissues. It is also of interest to note that the recently reported thermal independence of the respiration of the copepod *Diaptomus* (Siefken and Armitage, 1968; p. 409) appears to be associated with the metabolism of lipids at low temperatures which may therefore supplement carbon flow from glycolysis, and help to maintain a constant level of respiratory activity despite a reduction in temperature. Again, changes in the Q_{10} for the metabolism of excised barnacles appear to be associated with the nature of the substrate, a

Q_{10} of 1·0 occurring in *Balanus balanoides* when lipid and protein are used as substrate, compared with a Q_{10} of 2·0 when carbohydrate is being used (Barnes, Barnes and Finlayson, 1963b; also p. 394). The presence of a pathway to lipid metabolism in the crude mitochondrial preparations, and in cell-free homogenates of invertebrates (p. 431; Newell, 1966, 1967), would also account for the thermal independence of the respiration of such systems. However, the respiration of purified mitochondria, in which the alternative pathway to lipid biosynthesis and the possibility of HMP shunt participation have been removed, would not be expected to proceed nearly independently of acute temperature fluctuation unless some additional control system is present.

2. LIMITING FACTORS

The concept of "limiting factors" was first clearly defined by Blackman (1905), who stressed that "when a process is limited as to its rapidity by a number of separate factors, the rate of the process is limited by the pace of the slowest factor" (also Burton, 1936 and Fry, 1947). Oxygen supply is one of the most important limiting factors and many animals show the phenomenon of "respiratory dependence" in which the rate of oxygen consumption declines with a fall in the external oxygen tension (p. 371). As Fry (1947) has pointed out, however, a limiting factor such as oxygen supply affects only the maximum level of oxygen uptake (the active rate), for once the limiting factor has reduced oxygen uptake to that of the standard rate the incipient lethal level has been reached and a further decline results in the accumulation of a metabolic deficit (Fig. 7.55, p. 370). In this respect, therefore, limiting factors differ from controlling factors such as temperature which act both on the active and the standard rate of metabolism.

The importance of limiting factors lies primarily in the way in which they may influence the effects of controlling factors. Fry and Hart (1948) (also Fry, 1947) illustrated this interaction between two such factors by taking for example the combined effects of oxygen tension (a limiting factor) and temperature (a controlling factor) on the metabolism of a fish. Fig. 8.46 shows the relation between the standard and active rates of metabolism in young fish and acclimation temperature, whilst the scope for activity is shown in the lower part of the figure. The "range of activity" was defined by the upper and lower incipient lethal temperatures, and extends between approximately 5° and 40° C. The oxygen was approximately at air saturation in this experiment and we may now examine the effects of a variation of a limiting factor such as oxygen tension on the metabolism. Fig. 8.47A shows the effect of oxygen tensions of 40 mm Hg, 25 mm Hg, 20 mm Hg, 15 mm Hg, and 10 mm Hg on the relation between the active rate

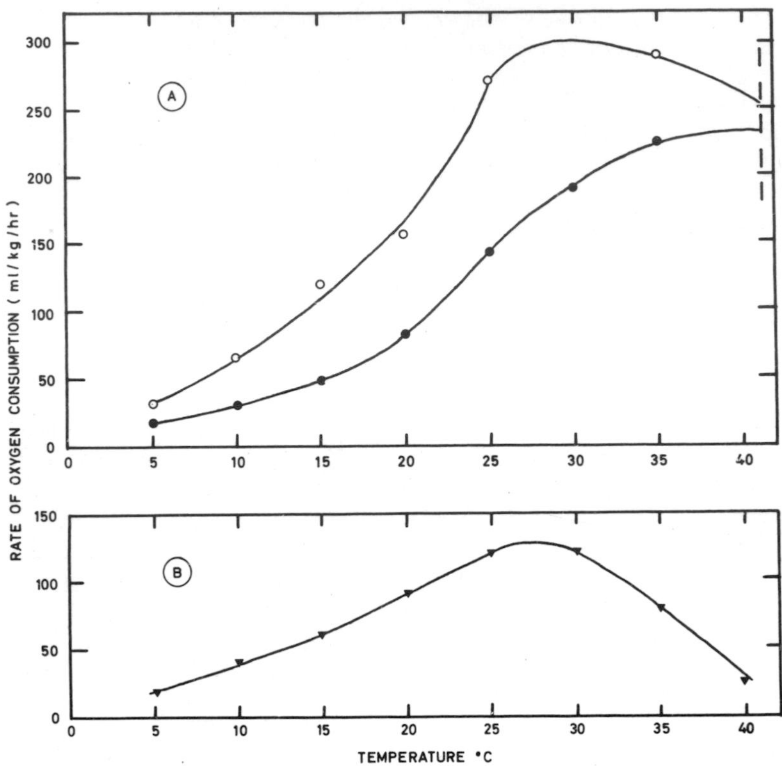

Fig. 8.46. Graphs showing the effect of temperature on (A) the active (open circles) and standard (solid circles) rate of oxygen consumption of a fish; (B) the scope for activity (obtained by subtraction of the standard from the active rate and shown by triangles). (After Fry and Hart, 1948). The upper incipient lethal temperature is indicated by a broken line.

of metabolism and temperature. The prime effect was to suppress the active rate which thus shows respiratory dependence over this range of oxygen tensions (Fig. 8.47B). Now if we replot the standard rate, which was meanwhile unaffected by the limiting factor of oxygen tension, and superimpose the active rate at oxygen tensions of 40 mm Hg, 25 mm Hg and 20 mm Hg (Fig. 8.48A, B, C.) we see that the active and standard rates meet at progressively lower temperatures as the oxygen tension becomes lower. That is, whereas activity was eliminated at the ultimate upper incipient lethal temperature in the presence of high oxygen tensions (Fig. 8.46), the scope for activity is so reduced in low oxygen tensions that activity ceases long before the upper lethal temperature is reached. It is apparent that the limiting factor of oxygen

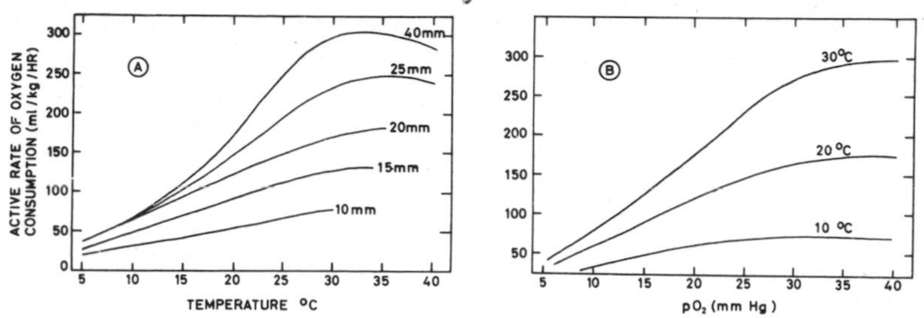

Fig. 8.47. Graphs showing the effect of oxygen tension and temperature on the active rate of oxygen consumption (ml/kg/hr) in a fish. (After Fry and Hart, 1948.) (A) The effect of temperature at various oxygen tensions. (B) The effect of oxygen tensions at various temperatures.

has two main effects when coupled with the controlling factor of temperature. Firstly the scope for activity is suppressed, and secondly, the range of activity is reduced, its extremes no longer being set by the upper and lower incipient lethal temperatures but by the points at

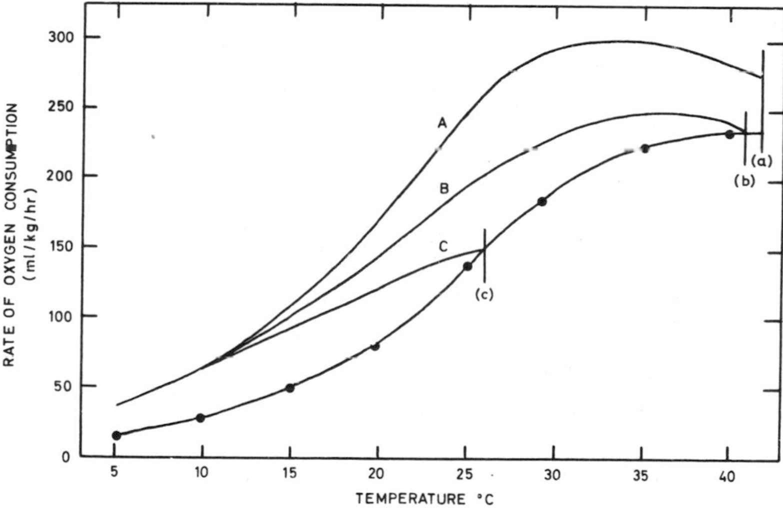

Fig. 8.48. Graphs showing the interaction between a limiting factor (oxygen tension at (A) 40 mm Hg; (B) 25 mm Hg; and (C) 20 mm Hg), and a controlling factor (temperature) on the active and standard metabolism of a fish. (Based on Fry, 1947.) Note that the limiting factor of oxygen tension affects the active rate of oxygen consumption whilst the controlling factor of temperature affects both active and standard (solid circles) rates. At high oxygen tensions activity is limited at high temperatures by the upper incipient lethal temperature (a) but at lower oxygen tensions the scope for activity is reduced to zero at temperatures below the upper incipient lethal temperature ((b) at 25 mm Hg and (c) at 20 mm Hg)

which the scope for activity is reduced to zero. Finally we may note that the cost of physiological regulation is to raise the standard metabolic rate. Thus since the scope for activity is represented by the difference between the active and standard rates, the cost of regulation would be to reduce both the scope for activity and the range, even though the level of the limiting factor (oxygen) remained constant.

As mentioned in Chapter 7, and discussed by Prosser and Brown (1961), acclimation to the limiting factor of low oxygen may occur; we have also seen that acclimation to the controlling factor of temperature is well-known. Acclimation to low oxygen primarily involves an increase in pigment synthesis as in many crustaceans, as well as the development of increased respiratory surfaces as in the polychaete *Scoloplos armiger* (p. 350), and the modification of irrigation rhythms as in many of the polychaetes (p. 293).

CHAPTER 9

Thermal stress and desiccation

A. INTRODUCTION

As we have seen in Chapter 1, a great many physical, chemical and biological factors play a part in controlling the zonational level of intertidal organisms, and it has been established that such organisms appear to be well-adapted to resist the stresses likely to be experienced at the particular shore level where they live. Two of the most obvious of such stresses are those of temperature and desiccation. Each of these in general increases towards higher shore levels in the tropics or during the summer months in temperate zones, although the temperature of the shore may also be considerably less than that of the sea in the winter in temperate zones or for much of the year in sub-arctic regions. There thus tends to be both latitudinal differences in thermal tolerance as well as differences between upper and lower shore organisms. Huntsman and Sparks (1924) and Henderson (1929), for example, demonstrated that the upper limit of thermal tolerance of a variety of marine organisms was correlated with the temperature range experienced under natural conditions, both latitudinal and vertical differences being apparent. Gowanloch and Hayes (1926) showed that even in individuals of the same species (*Littorina littorea* and *L. saxatilis*) collected from different shore levels there was a gradation of thermal tolerance, upper shore individuals having a higher limit of thermal tolerance than lower shore individuals. Similarly, Broekhuysen (1940) has demonstrated that the sequence of thermal death points in a series of six South African gastropods shows a general correspondence to their zonational sequence on the shore (Fig. 9.17). *Littorina knysnaënsis,* whose upper limit extends to approximately 12·5 ft above datum, had an upper lethal temperature of 48·6°C whilst *Oxystele variegata,* which extends up to 9·75 ft above datum, has an upper lethal temperature of 42·1°C. *Thais dubia* extends up to 9·25 ft above datum and has an upper lethal temperature of 41·7°C whilst that of *Oxystele tigrina,* which lives up to 8 ft above datum, was 39·5°C and finally that of *Oxystele sinensis,* which lives up to 5·25 ft above datum, was 39·6°C.

On the other hand Evans (1948) found no such correlation between shore level and thermal tolerance in a variety of shore animals from Cardigan Bay, Wales. He showed that in *Littorina littorea, L. saxatilis, L. littoralis, Melaraphe neritoides, Gibbula cineraria, G. umbilicalis, Patella vulgata, P. depressa, P. athletica, Thais lapillus* and *Monodonta lineata,* the thermal tolerance was certainly correlated with the normal temperature range occurring at the particular sites from which the animals were collected, but that the topography and aspect of the shore was such that no simple relationship with tidal level could be demonstrated. Edney (1961, 1962) has also found a variation in the upper limit of thermal tolerance in a variety of species of the fiddler crab *Uca* from Inhaca Island, near Lourenco Marques (Lat. 26°S). The upper lethal temperature varied between species from 43·3°C for *Uca inversa,* which lives in burrows in open sandy flats which are exposed to direct sunlight, to 40·0°C for *Uca urvillei* which lives in moist channels in mangrove swamps. However in these animals the situation is complicated by seasonal acclimation to temperature, the upper lethal temperature being raised by about 2°C in January (high summer) compared with September. Similar acclimation effects have also been noted in the upper limit of thermal tolerance of the gill cilia of the mussel *Modiolus demissus* and the oyster *Crassostrea virginica* (Vernberg, Schlieper and Schneider, 1963), and of the upper lethal temperature of *Uca* (Vernberg and Tashian, 1959). These, and other aspects of acclimation to temperature, have been discussed by Bullock (1955), Prosser (1955, 1958, 1967), Precht (1958), Vernberg (1962, 1967) and Newell (1969) (also p. 414). The effects of temperature on the thermal inactivation of proteins have also been shown to be correlated with habitat. The heat denaturation is dependent upon primary structure and often varies genetically (Langridge, 1963; Ushakov, 1967). Read (1963; 1967) has studied the thermal inactivation of aspartic-glutamic transaminase in a series of bivalves after exposure for twelve minutes to 56°C. He found that the activity of this enzyme in the salt-marsh mussel *Brachyodontes* was scarcely affected by such treatment, but that thermal inactivation was appreciable in the intertidal *Mytilus* and nearly complete in the sublittoral bivalve *Modiolus.*

In much the same way as the upper limit of thermal tolerance has been found to vary in an ecologically significant way, so too has the lower limit of thermal tolerance and resistance to freezing in intertidal compared with sublittoral animals from high latitudes. Kanwisher (1955, 1959; also Smith, 1958) has shown that the shore temperature at Wood's Hole, Mass. falls to –20° C or lower during the winter, and that the ability of intertidal invertebrates such as *Littorina littorea* and *Mytilus edulis* to survive compared with sublittoral forms was related to the resistance to ice formation. Kanwisher (1955) showed that at –15° C the percentage water frozen in *Mytilus edulis* was 62%, *Modiolus modiolus* 65%, *Littorina littorea* 59%, *Crassostrea virginica* 54% and *Littorina saxatilis* 67% whilst at –22° C the percentage of water frozen in the tissues of *Modiolus modiolus* was 71%; that in *Littorina littorea* at this temperature was 76%. Survival occurred even after several days of exposure to this temperature. Since as much as 75% of the water in the animals may be bound up as ice, the capacity to withstand such freezing must involve not only the ability to accommodate the presence of ice within the tissues but also to withstand severe dehydration of the body fluids. Many types of cells are incapable of withstanding the formation of ice crystals intracellularly (Chambers and Hale, 1932; Asahina, Aoki and Shinozaki, 1954), although in some nematodes very rapid freezing under experimental conditions may result in the vitrification of the cell contents in such a way that no ice crystals are formed and survival of the animal is possible. Under natural conditions, however, the freezing process is less rapid so that crystal formation occurs and has in fact been observed in some animals (Kanwisher, 1955, 1959). Siminovitch and Briggs (1952) have shown that the frost-hardiness of plants is related to the ability of water to migrate rapidly in and out of the cells in such a way that ice crystals are formed not intracellularly but intercellularly, where the consequent distortion can be accommodated without damage to the cells. Although direct confirmation of this hypothesis appears to be lacking in those animals which are known to be capable of withstanding the presence of ice in the tissues, it seems likely that here, too, the cell membranes may allow the rapid exit of water necessary for the formation of intercellular ice crystals.

Although there have been many investigations of the thermal tolerances of intertidal and sublittoral organisms (for example Orr, 1955; Gunter, 1957; Southward, 1958a; Fraenkel, 1960; Wilkens and Fingerman, 1965), experimental analysis of the relationship between the distribution and the upper limit of thermal tolerance in such organisms is complicated by the fact that the body temperature rarely corresponds exactly with the ambient air temperature. In some instances the tissue temperature may exceed the air temperature owing to the retention of seawater and warming by sunlight (Southward, 1958a). Similarly the tissue temperatures of the barnacle *Tetraclita squamosa*, the limpet *Fissurella barbadensis* and the gastropod *Nerita tesselata* on a shore at Barbados also exceeded the air temperature, but were nevertheless lower than those of inanimate bodies or black bodies because of the cooling effect of evaporative water loss. The importance of such cooling must vary according to temperature and to the saturation deficit of the air, but in *Tetraclita squamosa* it reduced the body temperature by 4·9° C, in *Fissurella barbadensis* by 5·5° C and in *Nerita tesselata* by 7·8° C compared with a black body (Lewis, 1960, 1963). Again, Edney (1951, 1953, 1954, 1960, 1961, 1962) has demonstrated the importance of transpiration in the control of the body temperature of supralittoral and terrestrial isopods and also of a variety of species of *Uca*. Specimens of *Ligia oceanica*, for example, were exposed to temperatures of 30° C in the interstices of the shingle in which they live. The relative humidity was >98% and 30° C was near to the lethal temperature in saturated air. Under these conditions some animals were found on the surface of the rocks and exposed to the direct effect of sunlight where the ground temperature was 34° C, but the body temperature was found to be reduced to 26° C because of the cooling effect of transpiration. In much the same way, crabs of the genus *Uca* were found to survive on the surface of sand hotter than the lethal temperature of the crabs; here again, survival was made possible by the cooling effect of transpiration in these animals. Similar results demonstrating the importance of evaporative water loss on the survival of *Uca* at high temperatures have been obtained by Wilkens and Fingerman (1965), who in addition showed that the blanching of fiddler crabs which occurs at high temperatures (Brown and Sandeen, 1948) has a thermoregulatory role, the body temperature of pale animals being approximately 2° C lower than that of dark ones.

Thus an important factor facilitating the tolerance of high environmental temperatures during the emersion period by intertidal and semi-terrestrial organisms may be the ability to withstand appreciable water loss. In this respect the stresses of temperature and desiccation are interdependent, and resistance to one factor increases the risk of damage by the other. Nevertheless interesting and important results have been obtained by laboratory analyses made independently on each of these factors. These are discussed in more detail below.

B. THERMAL STRESS

1. LETHAL TEMPERATURES

The extensive literature on the upper limits of thermal tolerance of intertidal and sublittoral organisms has been reviewed by Gunter (1957) and a general correspondence has been found between the upper and lower lethal temperature and the normal environmental temperature range to which the organism is exposed. Thus differences in thermal sensitivity may be associated with both latitudinal differences and with variations in shore level, as well as with topography and aspect of the particular shore from which the animals were collected. Most of such data, particularly those for the upper lethal temperature, have been obtained in the laboratory by slowly heating the temperature of a water bath in which the organisms were contained at a rate of approximately 1°C in 5 minutes (for example Huntsman and Sparks, 1924; Gowanloch and Hayes 1926; Gowanloch, 1926; Henderson, 1929; Broekuysen, 1940). Such a procedure was also adopted by Evans (1948) and by Southward (1958a) to facilitate comparison of their results with earlier work, much of which is summarised by Heilbrunn (1952) and by Gunter (1957). It has been pointed out by Heilbrunn (1952), Fry, Hart and Walker (1946) and Fry (1947, 1957a) as well as by Orr (1955) and by Fraenkel (1960) that although the data obtained by earlier workers form a useful basis for the comparison of the thermal tolerances of organisms from different environments, the duration of exposure to each temperature is of importance. That is, in an assessment of thermal tolerance, both temperature and time must be taken into account. It is then found that a long exposure to a low temperature may cause the same percentage mortality as a brief exposure to a high temperature. Rees (1941), for example, found that the flatworm *Monocelis fusca* died at only 37°C when heated slowly at a rate of 1°C per day, but that the lethal temperature was as high as 44°C when the rate of heating was 1°C per 30 min. Added to these complications is the fact that, rather than being a characteristic value for any particular species, the upper limit of thermal tolerance may vary according to season or storage temperature (for example, Vernberg and Tashian, 1959; Edney, 1961, 1962; Vernberg, Schlieper and Schneider, 1963).

The particular method used to compare lethal temperatures should thus take both time and acclimation temperature into account. Fry Hart and Walker (1946) and Fry (1947; also Fry, Brett and Clawson, 1942; Brett, 1944) have used a method which involves both of these parameters and which accurately defines the zone of thermal tolerance of the organism. Groups of individuals of the species in question are stored at various temperatures to acclimate. Samples from each of the acclimation vessels are then taken and placed in constant temperature baths and the percentage mortality occurring at each temperature after a given time interval is recorded. The lethal temperature may be regarded as the temperature at which more than half of the animals fail to survive when returned to the original acclimation temperature after the period of exposure to the experimental temperature. Similarly the median lethal low temperature may be determined in the same way by exposure of the acclimated animals to a series of low temperatures for a standard length of time. From such graphs, either the temperature at which 100% mortality, or better still the temperature at which 50% mortality occurs (the median lethal temperature), can be read off for each group of animals acclimated to various temperatures. The incipient lethal values so obtained, both for high and low temperatures, may then be plotted as a function of acclimation temperature. The curves obtained for the fishes, *Carassius, Ameiurus, Girella, Notemigonus* and the lobster *Homarus americanus* (after Fry, Brett and Clawson, 1942; Brett, 1944; Hart, 1952; McLeese, 1965) are superimposed in Fig. 9.1.

It will be noticed that the area enclosed by the upper and lower incipient lethal temperatures defines the "zone of tolerance" of the

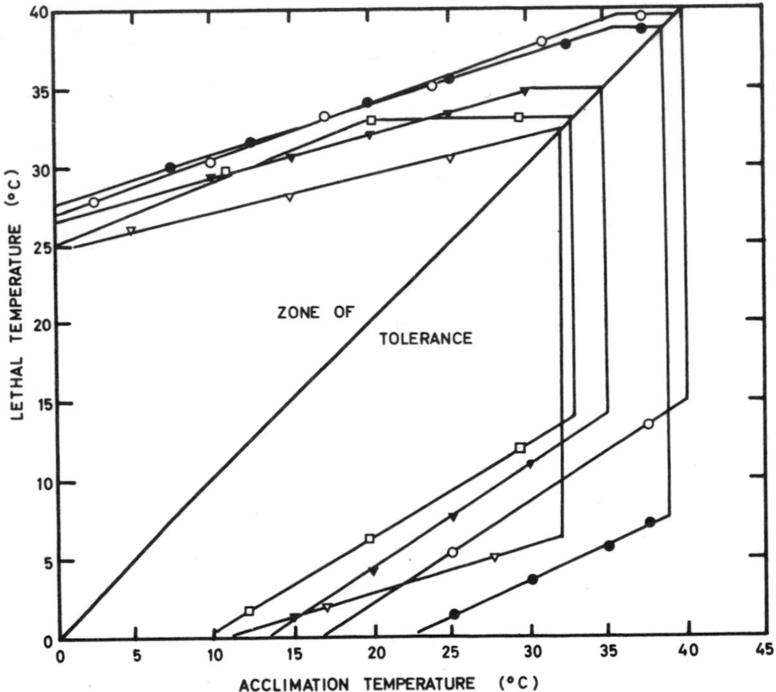

Fig. 9.1. Graphs showing the thermal tolerances of several fishes and of the lobster *Homarus americanus.* The upper and lower incipient lethal temperatures are plotted against acclimation temperature. Open circles – *Carassius auratus* (after Fry, Brett and Clawson, 1942); solid circles – *Ameiurus nebulosus*; open squares – *Girella nigricans* (after Brett, 1944); solid triangles – *Notemigonus crysoleucas* (after Hart, 1952) and open triangles – *Homarus americanus* (After McLeese, 1956.)

organism, and that the zone of tolerance for *Carassius* extends to higher temperatures than that of *Ameiurus* or *Girella*. Further, the point where the incipient upper lethal temperature joins the diagonal line on the graph indicates the temperature at which no further increase in the upper incipient lethal temperature occurs in response to thermal acclimation. That is there comes a point at high temperatures where heat-death occurs even after acclimation to near lethal temperatures. This point may be termed the "ultimate upper incipient lethal temperature" after the nomenclature of Fry, Brett and Clawson (1942), Brett (1944), Fry, Hart and Walker (1946) and Fry (1947, 1957a). There would also be an "ultimate lower incipient lethal temperature" where the lower incipient lethal temperature line joined the diagonal line; however, other factors such as the mechanical effects of ice formation, may intervene and cause death prior to this temperature being reached.

Although such a definite zone of thermal tolerance thus provides a useful measure of the adaptive capacity of the organism, it remains true that, as Fry (1947, 1957a) has pointed out, many organisms are able to survive for substantial periods at temperatures beyond the zone of tolerance. Such a zone may be defined as a "zone of resistance". The first stage in a determination of this zone is to take animals acclimated to any one temperature and to plot the percentage mortality against time for a series of lethal levels of temperature. Fry, Hart and Walker (1946) found that the use of a log time scale gave a linear relationship between mortality and time at each of the experimental temperatures.

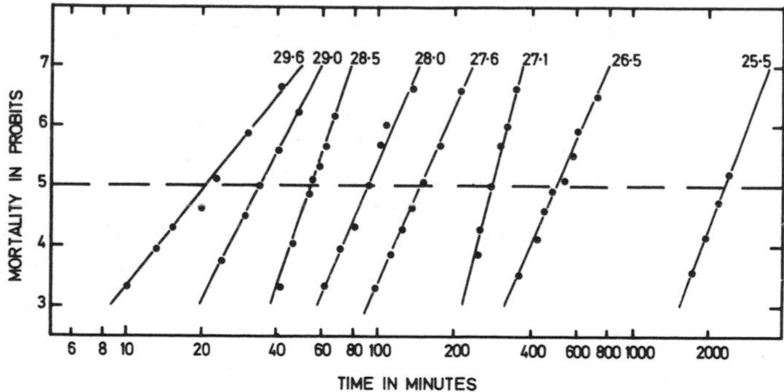

Fig. 9.2. Time-mortality curves for specimens of *Salvelinus fontinalis* acclimated to 20°C and tested at the temperatures indicated. (After Fry, Hart and Walker, 1946.)

The time-mortality curves at various temperatures for specimens of the fish *Salvelinus fontinalis* acclimated to 20°C are shown in Fig. 9.2, in which the mortality is expressed as probits (Fisher and Yates, 1943).

Similar graphs may be drawn for animals acclimated to a variety of other temperatures, and the time taken for 50% mortality (the median mortality time) read off and plotted against the lethal temperature on semi-logarithmic paper. When these resistance lines are drawn for the animals from each of the acclimation temperatures, a graph similar to that shown in Fig. 9.3 is obtained.

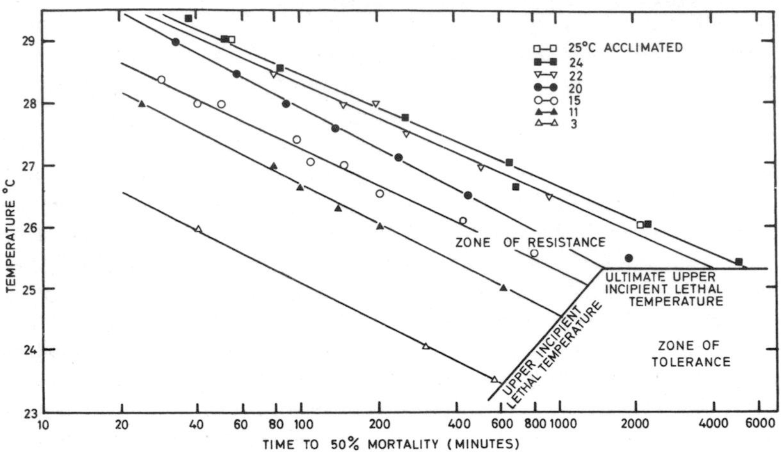

Fig. 9.3. Graphs showing the time to 50% mortality (median resistance time) at various lethal levels of temperature for *Salvelinus fontinalis* acclimated to 3°C, 11°C, 15°C, 20°C, 22°C, 24°C and 25°C. The data for animals acclimated to 24°C and 25°C fall onto the same regression line. (After Fry, Hart and Walker, 1946.)

The main feature of interest is that the lower ends of the regression lines end in two boundary lines, one of which cuts approximately at right angles to the lower four regression lines while the other runs parallel with the *x*-axis of the graph. These boundary lines mark the points beyond which an extension of the regression lines would be meaningless, since at these rather low experimental temperatures 50% mortality does not occur, and a high proportion of the animals survive irrespective of the length of time the animals are exposed to them. The zone to the right of the boundary lines thus represents the "zone of tolerance" whilst that to the left represents the "zone of resistance" because survival is for an indefinite period to the right of the boundary lines but only for a limited period of time in the zone to the left. These two zones are separated by the boundary lines which thus represent the upper incipient lethal temperatures (p. 457; Fig. 9.1) for each level of acclimation. The increase of the boundary line with temperature indicates that the incipient lethal temperature is raised by thermal acclimation, whilst a boundary line parallel to the x-axis indicates that the upper incipient lethal temperature is unchanged despite an increase in the acclimation temperature. This last boundary line is thus the ultimate upper incipient lethal temperature for this organism. As Fry, Hart and Walker (1946) point out, the slopes of the resistance lines are parallel where the upper incipient lethal temperature rises in response to an increase in acclimation temperature, but differ in slope from one another at higher acclimation temperatures. In this respect the data resemble those obtained by Doudoroff (1942, 1945) for *Girella nigricans* and *Fundulus parvipinnis* (also Fry, Hart and Walker, 1946).

Clearly, a measure of the zone of resistance for each acclimation temperature is given by the area under the appropriate time-temperature line (resistance line) in Fig. 9.3. This value may then be plotted against the appropriate acclimation temperature, the area under the curve giving an expression for the thermal resistance of the species in much the same way as the area enclosed by the upper and lower incipient lethal temperatures shown in Fig. 9.1 defines the zone of tolerance. Fry (1947) has calculated the appropriate curve for the data shown in Fig. 9.3 for *Salvelinus fontinalis;* this curve is illustrated in Fig. 9.4 and provides a useful basis for comparison with other species. A species with a high value for thermal tolerance may have a low value for thermal resistance and the reverse may equally apply, so that it is of some importance to obtain a measure of both thermal resistance and the thermal tolerance (Fry, 1947).

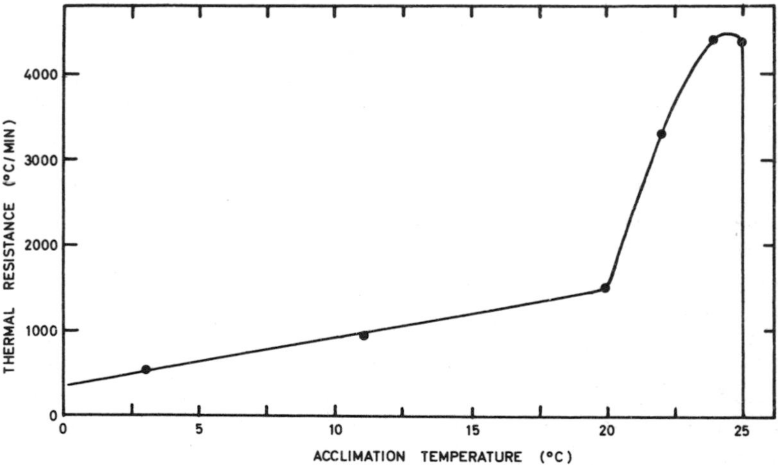

Fig. 9.4. Graph showing the thermal resistance of *Salvelinus fontinalis.* This was obtained by calculating the area under the resistance lines for each acclimation temperature shown in Fig. 9.3. (After Fry, 1947.)

Unfortunately, as in the effect of temperature on invertebrate oxygen consumption, much of the detailed work on fishes has not yet been applied to marine invertebrates. Nevertheless Orr (1955) and Fraenkel (1960) have recognised the importance of the influence of time in the determination of thermal death points and Vernberg, Schlieper and Schneider (1963) have taken acclimation temperature into account. Seasonal variations in the upper limit of thermal tolerance have also been noted in the fiddler crab *Uca* by Edney (1961, 1962). Orr (1955) studied the effect of time on the heat death of the crab *Uca pugilator,* the prawn *Palaemonetes vulgaris,* the starfish *Asterias forbesi,* the brittlestar *Ophioderma brevispinum,* the echinoid *Arbacea punctulata,* the gastropod *Nassa obsoleta* and the fishes *Fundulus heteroclitus* and *F. majalis.* He placed the animals for known lengths of time at appropriate test temperatures, and then returned them to sea-water at normal temperatures and determined the percentage showing

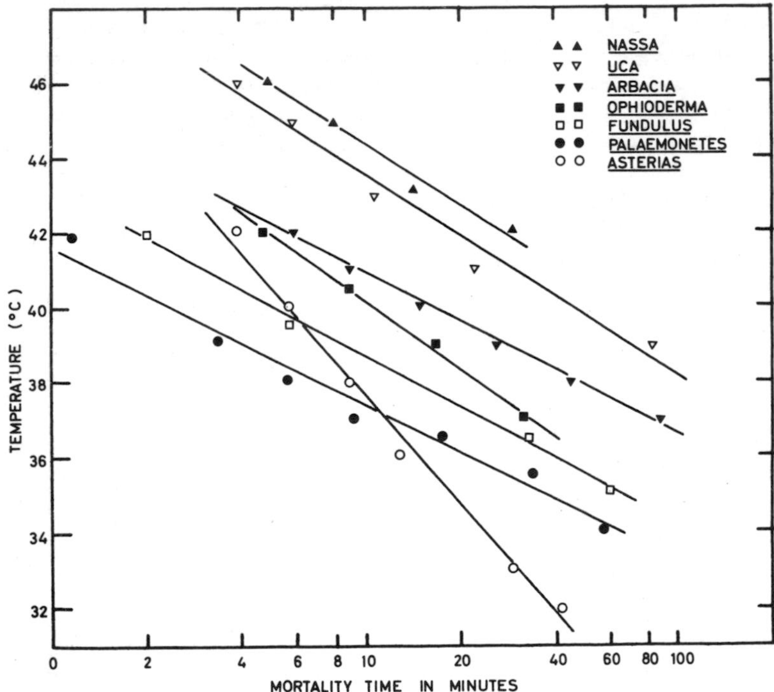

Fig. 9.5. Graphs showing the thermal resistance lines for various marine animals collected straight from the shore. The data refer to the time taken for 100% mortality except for *Palaemonetes* where the median resistance time is used. (Data replotted after Orr, 1955.)

irreversible heat damage. In most instances the temperature required to produce 100% heat death was regarded as the lethal temperature except in *Palaemonetes vulgaris* where the temperature required to induce 50% mortality was used. The curves obtained were similar, a relatively small rise in temperature resulting in a marked fall in the lethal exposure time. When the curves are replotted on a semi-logarithmic scale, as in Fig. 9.5, there is a reasonable approximation to a straight line, much as has been obtained by Fry, Hart and Walker (1946) for *Salvelinus* (Fig. 9.3).

Fraenkel (1960) has also studied survival at different temperatures as a function of time, and emphasised the importance of using proper criteria for the judgement of survival. He suggested that the resumption of normal behaviour was the best criterion for the absence of thermal damage. In *Limulus polyphemus* the resumption of digging in the sand was chosen, whilst in *Littorina littorea* the negative geotactic behaviour on vertical surfaces was used. Finally in *Pagurus longicarpus* the re-occupation of gastropod shells was used as a criterion of survival. For 1 hr exposure the lethal limit was 44°C for *Limulus*, 40–41°C for *Littorina littorea* and 36°C for *Pagurus*. The high values for *Limulus* and *Littorina littorea*, compared with those for *Pagurus* and other marine invertebrates (Gunter, 1957), may reflect the acclimation to high temperatures which would be necessary in intertidal organisms exposed to high temperatures when the tide is out.

Vernberg, Schlieper and Schneider (1963) and McLeese (1965) have taken the effect not only of time but also of acclimation temperature into account in studies on the thermal tolerances of a variety of intertidal and sublittoral organisms. These studies are therefore directly comparable with those cited on p. 456, for fishes. Vernberg *et al.* (1963) studied the temperatures lethal to the gill cilia of the intertidal mussel *Modiolus demissus* and oyster *Crassostrea virginica,* and the sublittoral bivalve *Aequipecten irradians.* The animals were acclimated to 10°C and to 22 – 25°C, after which the survival time of the cilia of isolated gills was estimated at a variety of temperatures. From the results which are shown in Fig. 9.6, two main features are apparent, the thermal tolerance after 100 min is 44°C in the intertidal species but only 37°C for the sublittoral species. These results support earlier work on the different sensitivity of intertidal and sublittoral animals to high temperatures (Henderson, 1929; Gunter 1957). Secondly, warm-acclimated *M. demissus* and *C. virginica* survived for longer at each experimental temperature than did cold-acclimated animals, whereas no such change in the upper limit of thermal tolerance occurred in the sublittoral *Aequipecten irradians.* In *Modiolus demissus* the alteration in the upper limit of thermal tolerance occurs within 17 days (Vernberg *et al.,* 1963) or over a shorter period than this; Schlieper, Flugel and

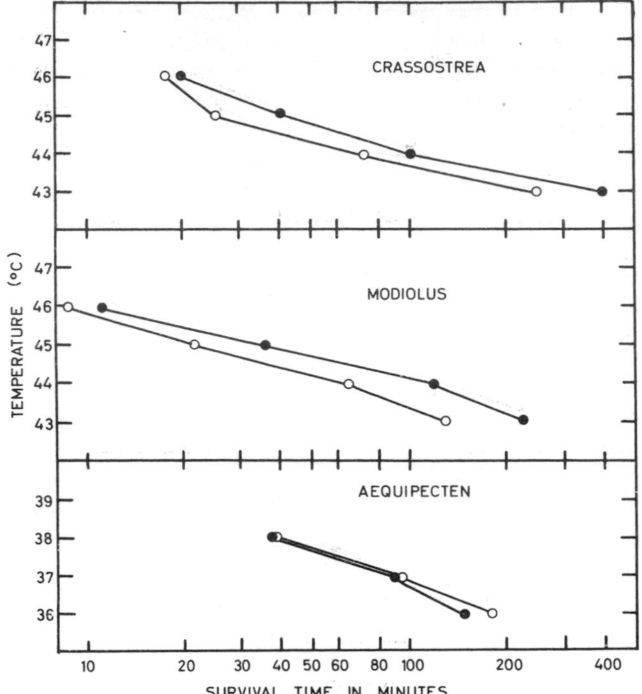

Fig. 9.6. Graphs showing the thermal resistance lines of isolated gill pieces from three species of bivalve stored at 22–25°C (solid circles) and 10°C (open circles). Note that the resistance lines of the two intertidal forms *Crassostrea* and *Modiolus* were raised in response to storage at high temperatures whereas no such change occurred in the sublittoral *Aequipecten*. (Data from Vernberg, Schlieper and Schneider, 1963.)

Rudolf (1960) found that a modification in the thermal resistance of the gill tissue of the mussel *Mytilus edulis* occurred within 4 days when the storage temperature was altered from 5 to 20°C or *vice versa.*

McLeese (1965), who studied the survival of the lobster *Homarus americanus* in moist air as a function of temperature, time and acclimation temperature, stored groups of animals at 0, 10, and 20°C and plotted the percentage mortality at a series of test temperatures as a function of time. The cumulative percentage mortality for groups of 50 specimens acclimated to 0°C and exposed in moist air at 0, 5 and 10°C is shown in Fig. 9.7A. From data such as these, the time taken for 50% mortality (median resistance time) may be estimated and plotted as a function of temperature (Fig. 9.7B); the survival time increased as the air temperature decreased until 4°C was reached, after which a further

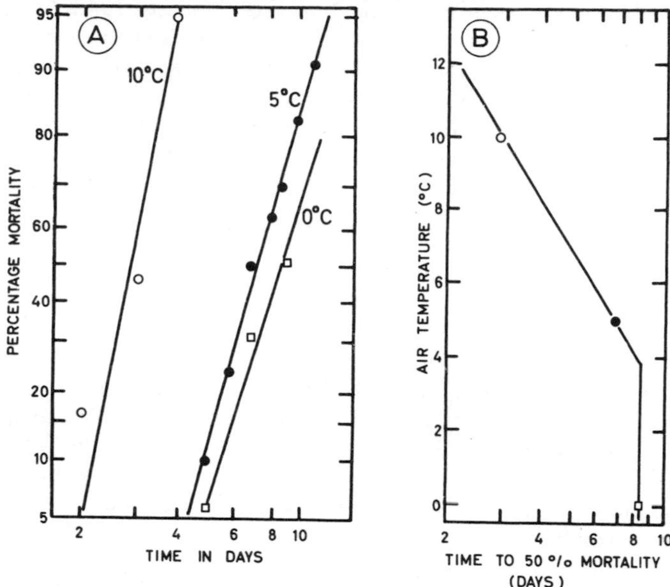

Fig. 9.7. Graphs showing the effect of temperature on the survival time of the lobster *Homarus americanus* in air. (After McLeese, 1965.) Animals were acclimated to 0°C. (A) Cumulative percentage mortality (on a probit scale) as a function of time (days – log. scale.) (B) Median thermal resistance time at various air temperatures.

lowering of air temperature resulted in no further increase of survival time. Acclimation to high temperatures was found to result in an increase of the median resistance time although the rate of metabolism of the animals also played a part in determining its length. (McLeese, 1965).

Although there is thus now a considerable body of information obtained by a variety of experimental techniques on the heat- and cold-lethal temperatures of animals, much of the data is for subtidal animals. Data which do exist for intertidal animals (for example Broekhuysen, 1940, Evans, 1948; Orr, 1955; Southward, 1958a; Fraenkel, 1960, 1961) suggest that there is a general correspondence between the heat-lethal temperature and the zonational sequence or thermal experience of the animals on the shore. However most of the data were obtained on animals immersed in water, and there is little published information on the thermal tolerances of such organisms in air where, under natural conditions, the temperatures are much more extreme. Temperatures on the surface of rocks may exceed 40°C in the summer on British shores (Evans, 1948; Nicol, 1960) whereas the maximum temperature recorded in rockpools in similar areas was 30° C (Evans, 1948; Fraenkel, 1961). Such thermal stress when the tide is out may indeed be enhanced by the warming effect of the sun on the water contained in the mantle cavity of some intertidal animals (Southward, 1958a), although evaporative water loss must also play an important part in the reduction of the temperatures of such organisms below that of a black body (Lewis, 1960, 1963). For this reason Micallef (1966) suggested that the tolerance of extremes of temperature in air may play an important part in the control of the vertical distribution of intertidal animals. He studied the effects of temperature on the mortality and zonation of a series of trochids from Plymouth, Devon, including the upper shore *Monodonta lineata,* the middle-shore *Gibbula umbilicalis,* the middle to lower shore *Gibbula cineraria* and the still lower shore trochid *Calliostoma zizyphinum* (p. 10). Micallef (1966) studied both the upper and lower limits of thermal resistance of these animals in air and in water, and demonstrated the overriding importance of the behaviour of the animals in response to temperature. This accounts for the vertical distribution of the species on the shore and for the considerable zonational overlap which occurs between the successive species.

The median resistance times of the four species in water and in air saturated with water vapour are shown in Fig. 9.8A and B. All the animals had been previously acclimated to 9–10° C for 4–5 days, and several features are apparent from the graphs. Firstly, the relation between lethal temperature and median survival time is not linear but, as has been shown for other species (Orr, 1955; Vernberg *et al.,* 1963; also p. 457), approximates to a straight line when the time axis is on a log scale. Secondly, both the heat resistance at constant time and the exposure time required to produce 50% mortality at constant temperature, parallel the zonational sequence of the animals on the shore. Thus the high-level *Monodonta lineata* has the highest thermal tolerance and longest resistance time, next comes *Gibbula umbilicalis,* then *G. cineraria,* while *Calliostoma zizyphinum* has the lowest heat tolerance and resistance time of all. The "instantaneous heat-lethal temperature" can also be estimated by extrapolation; (for convenience it may be regarded as the temperature at which 50% mortality occurs within 0·1 hr). It is clear that this value, too, under submersed conditions parallels the distribution of the animals on the shore. Finally, it should be noted that the thermal resistance of the animals in air exceeds that of the animals in water. An explanation of this effect must mainly reside in the time-lag required for the equilibration of the tissues with the air temperature, since the effect of evaporative water loss at the high humidities used in these experiments could not have been great. However, as Prosser *et al* (1950) pointed out, each gm of water evaporated at 33° C removes 580 gm calories of heat, so that the effect of evapor-

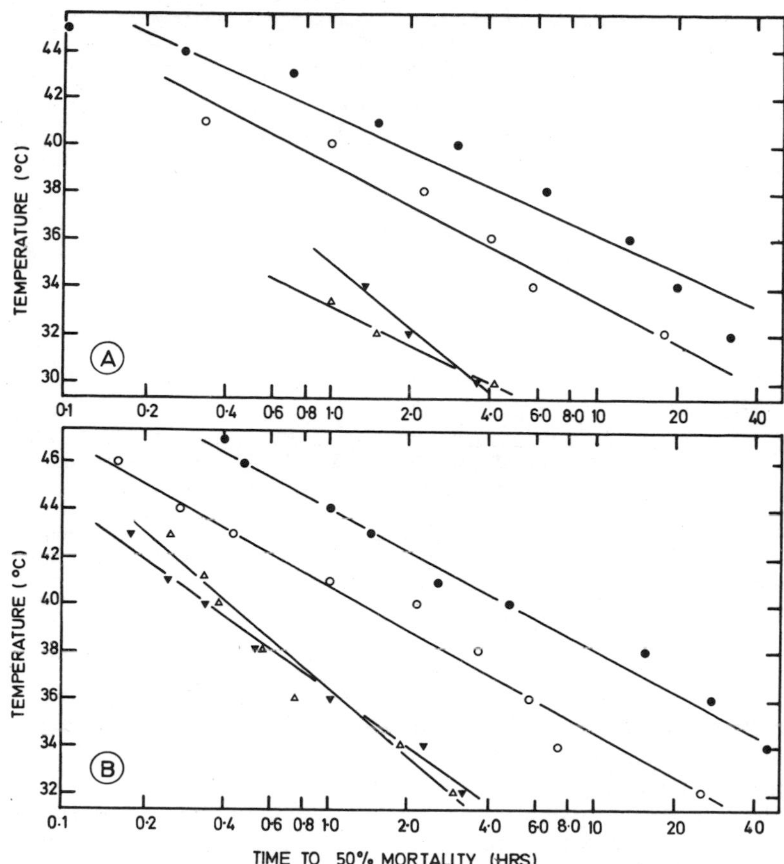

Fig. 9.8. Graphs showing the thermal resistance of a series of trochids in water (A) and in air (B). (Data from Micallef, 1966.) Solid circles – *Monodonta lineata;* open circles – *Gibbula umbilicalis;* solid triangles – *Gibbula cineraria;* open triangles – *Calliostoma zizyphinum.*

ative cooling must be of great importance under natural conditions where the humidity may often be low. Micallef (1966) also showed that a variety of other factors were important in buffering both temperature and desiccation effects; such factors include extravisceral water capacity, as has been established for *Acmaea limatula* (Segal and Dehnel, 1962), thickness and porosity of the shell and size of shell.

In much the same way, Micallef (1966) showed that both the thermal tolerance of animals and their resistance time to low temperatures was found to be related to their normal distribution on the

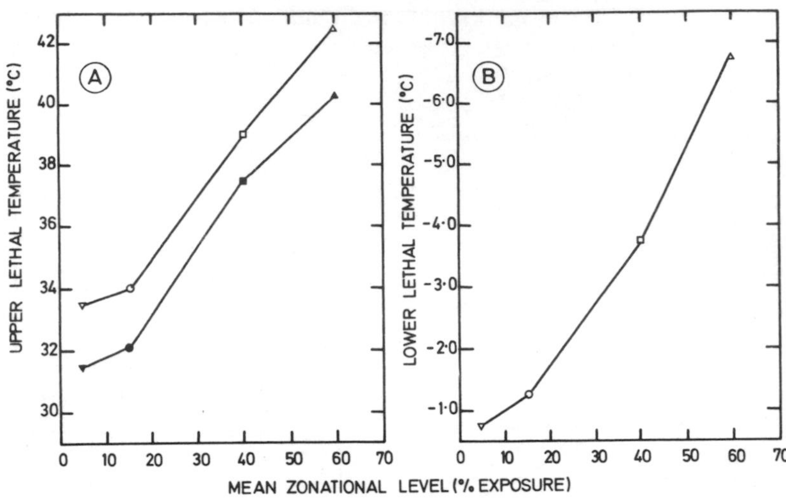

Fig. 9.9. Graphs showing (A) the relationship between the upper lethal temperature of trochids and mean zonational level in water (solid symbols) and in air (open symbols) after two hours. (B) The lower lethal temperature in air after 20 hours. Inverted triangle – *Calliostoma zizyphinum;* circle – *Gibbula cineraria;* square – *Gibbula umbilicalis;* triangle – *Monodonta lineata.*

shore. Again, the resistance data approximates to a straight line when plotted on a semi-logarithmic scale and, as in the case of the resistance to high temperatures, the lines converge towards the lower limit of thermal tolerance but diverge towards 0°C. Although the results so obtained differ in some cases substantially from those reported for upper and lower lethal temperatures by Evans (1948) and Southward (1958a), nevertheless the data of Micallef (1966) fully substantiate the general conclusion, stated clearly by Broekhuysen (1940) for South African gastropods, that the graded thermal resistance of intertidal animals is related to zonational position on the shore. It is thus possible to relate upper and lower lethal temperatures of the trochid series after a standard time to distributional level (Fig. 9.9). The "zone of tolerance" thus obtained for the trochid series acclimated to 10°C might, of course, vary with acclimation temperature, a broader zone occurring during the summer than in the winter. However, there is at present little information either on the zone of tolerance of individual intertidal species in response to acclimation temperature, or on seasonal variations in the zone of tolerance of closely related species such as those described above. It is clear, however, that the zone of tolerance of the trochids in air is greatly extended towards high temperatures compared with that in water, and that this must play an important part in facilitating survival under natural conditions on the shore. Much the same conclusion was reached by Fraenkel (1961), who found that the upper thermal tolerance of Mediterranean specimens of the winkle *Melaraphe neritoides* after 1–2 hr was as high as 46–47° C when submerged, but was further raised by approximately 2° C in air.

Although such data thus clearly support and amplify the conclusions based on other ecological sequences observed by Huntsman and Sparks (1924), Gowanloch and Hayes (1926), Henderson (1929), Broekhuysen (1940), Evans (1948) and Southward (1958a), it is obvious that in many cases the upper limit of thermal tolerance considerably exceeds that to which the animals must normally be exposed even at the height of the summer. This is particularly so when the exposure time is taken into account; *Monodonta lineata*, even in humid conditions, can survive air temperatures well in excess of 40·0°C for over 5 hr, whilst *Gibbula umbilicalis* can survive 4 hr at 38·5°C (Fig. 9.8). When the cooling effect of evaporation is taken into account, it is apparent that there is a considerable "safety factor" between the maximum temperatures likely to be experienced and the thermal tolerance of the organism. Normally the temperatures experienced would not even approach the incipient lethal temperature, and this has led to the suggestion that lethal temperature itself may not directly control the vertical zonational sequence of intertidal organisms (Broekhuysen, 1940; Evans, 1948; Southward, 1958a). Edney (1953, 1961, 1962) has shown, however, that the intertidal isopod *Ligia oceanica* may experience temperatures approaching the heat-lethal value in humid conditions, whilst fiddler crabs of the genus *Uca* are often to be found in the tropics at environmental temperatures which exceed the lethal temperature of the animal. Thus inferences on the importance of lethal temperatures in the control of the zonation of intertidal organisms may not necessarily apply to more extreme environments, such as the tropics or polar regions where upper and lower lethal tolerances may acquire an ecological significance. The situation is complicated by the fact that many motile intertidal organisms may evade thermal stress as heat- or cold-lethal temperatures are approached, so that behavioural responses to temperature may themselves exert an influence on zonational distribution. For example. Warden, Jenkins and Warner (1940) have cited evidence which suggests that temperature may be involved in the migration of the scallop *Pecten tenuicostatus,* whilst Yonge (1949) cites a number of examples of migrations of fish and crustacea into warmer waters in winter. A reversal of the geotactic response of *Littorina littoralis* in response to a change of temperature has been described by Janssen (1960), whilst Jenner (1958) and Sindermann (1961; also Moulton, 1962) have suggested that the dense aggregation of intertidal gastropods such as *Nassarius obsoletus, Thais* and littorinids, at the onset of winter in northern latitudes may preceed migration to deeper waters. Again, Naylor (1962, 1963) has described the offshore migration of the shore crab *Carcinus maenas* at the onset of winter, and Edney (1960, 1961, 1962) has described the behavioural responses of a variety of crustaceans to thermal stress.

2. THE SELECTION OF OPTIMAL TEMPERATURES

Micallef (1966; 1968) has shown that in the trochids *Monodonta lineata, Gibbula umbilicalis, G. cineraria* and *Calliostoma zizyphinum* a sequence of behavioural responses to thermal stress plays an important part in controlling the distribution of the animals. Unlike *Littorina littoralis,* which has been shown to change its response from a negative to a positive geotaxis at temperatures below 3°C and above 15°C and thus to migrate to lower shore levels (Janssen, 1960), the trochids show no active migration in response to thermal stress. Instead, a definite sequence of responses has been observed which results in the passive transport of the animals to lower tidal levels. The trochids show normal browsing and crawling responses over much of their range of thermal tolerance, but as high or low temperatures are approached the rate of crawling becomes slower and the animals finally become quiescent. Further thermal stress leads to loss of attachment to the substratum, the animals dropping off inclined surfaces and falling or being carried by turbulance to lower tidal levels where the thermal stress is less severe. The upper and lower temperatures responsible for inducing such behaviour varies only a little between the species, the upper limit in water for animals acclimated to 10°C being 30·5°C in *Monodonta lineata,* 28·0° C in *Gibbula umbilicalis,* 25·75° C in *G. cineraria,* and 26·25° C in *Calliostoma zizyphinum.* However, since the lethal temperatures of each of the species varies considerably and is higher in the upper shore species (Fig. 9.9), it follows that the "safety factor" between loss of attachment to the substratum and the lethal temperature increases in animals characteristic of upper shore levels (Table XVII).

Similarly, since heat- and cold-coma intervene between the temperature at which loss of attachment occurs and the instantaneous lethal temperatures (also Southward, 1958a; Sandison, 1966, 1967) and the coma too, varies with species, the safety factor between the behavioural response and the onset of coma is greatest in animals of the upper shore (Micallef, 1966). Sandison (cited in Lewis, 1964) has also found a gradation in the lethal temperature and that at which heat coma occurs in other intertidal gastropods; the coma temperature in air is greater than that in water and in general varies with shore level. Thus the heat coma in water was 35°C in *Littorina neritoides,* 30–31°C in

Table XVII. Table showing the instantaneous lethal temperature, coma temperature, and temperature at which loss of attachment occurs in four species of trochid submersed in water and previously acclimated to 10°C. (Data from Micallef, 1966.)

Species	*Water temp. at which loss of attachment occurs in °C*	*Coma temperature in °C*	*Instantaneous lethal temp in °C*	*Safety factor between coma and behavioural response in °C*	*Safety factor between lethal temp and behavioural response in °C*
Monodonta lineata	30·5	37·38	45·85	7–8	15·3
Gibbula umbilicalis	28·0	33–34	42·6	5–6	14·6
Gibbula cineraria	27·75	31–32	37·95	4–5	11·2
Calliostoma zizyphinum	26·25	30–31	36·7	3–4	9·45

L. saxatilis, 30–31°C in *L. littoralis*, 31°C in *L. littorea,* and 27–28°C in *Thais lapillus.* In air the corresponding temperatures for heat coma were 40°C for *L. neritoides,* 36–38°C for *L. saxatilis,* 30–32°C for *L. littoralis,* 32°C for *L. littorea* and 28°C for the low level *Thais lapillus.* Sandison noted that in air the difference between the temperature at which coma occurred and the heat lethal temperature was greatest in lower shore animals and least in the upper shore species; this trend, however, is not apparent in the corresponding values for submersed animals. Since the temperatures experienced on the shore approach the heat-coma temperatures, it seems likely that, as in the trochids, temperature stresses of less than the lethal values may play an important part in controlling the distribution of the animals on the shore.

There are thus in trochids a whole series of temperatures between the extremes set by the upper and lower instantaneous lethal temperaturs in which differing responses may occur. The first response is cessation of crawling, followed by loss of attachment which results in a passive aggregation of the animals at lower tidal levels. Further thermal stress may result in heat- or cold-coma followed by thermal death. As might be expected, the temperature at which loss of attachment occurs is not the same in air as in water. In general the temperature at which detachment from the substratum occurs is extended over a greater thermal range in air than in water. Thus the upper temperature at which loss of attachment occurs in water was 30·5° C for *Monodonta* and 26·25°C for *Calliostoma,* whereas the corresponding values in air were 34·0°C and 24·2°C. These data for the series of trochids are shown in Table XVIII. Micallef (1966) concluded, therefore, that the series of states ranging from quiescence to detachment from the substratum accounted for the semi-passive downshore migration of trochids when extreme temperatures occur (also Desai, 1959 cited by Crisp, 1964), each species having a characteristic threshold temperature for each phase, and that the temperature differences between the phases for each species were more marked at the upper thermal limits than the lower ones.

Finally, it was found that the return of the animals to their characteristic zonational position after a period of extreme temperatures, with corresponding movement to lower tidal levels, was accomplished by negative geotaxis coupled with selection of characteristic thermal optima. The maximum persistence of activity and spontaneous movements were recorded after animals had been placed in a thermal gradient and polar diagrams constructed to show the percentage of each species active over particular temperature ranges. The frequency distribution of the numbers of active animals in each zone was found to be significantly different for each of the species. The results are summarised in Table XIX.

Table XVIII Table showing the upper and lower temperatures at which loss of attachment to the substratum occurs in air and in water in a series of trochids acclimated to 10°C. (Data from Micallef 1966.)

Species	*Upper and lower water temperatures at which loss of attachment occurs in °C*		*Upper and lower air temperatures at which loss of attachment occurs in °C*	
Monodonta lineata	2·0	30·5	−1·3	34·0
Gibbula umbilicalis	2·0	28·0	−1·0	32·2
Gibbula cineraria	2·5	25·75	−0·5	28·9
Calliostoma zizyphinum	4·0	26·25	0·1	24·2

Table XIX. Table showing the percentage of active trochids in various thermal ranges. Animals were acclimated to 10°C and tested in water. (Data from Micallef, 1966.)

Species	*Percentage of animals active at the temperature range indicated*					
	20·5 – 23°C	19 – 20·5°C	17·5 – 19°C	15 – 17·5°C	12–15°C	9–12°C
Monodonta lineata	44·3	61·3	96·9	90·6	76·3	40·5
Gibbula umbilicalis	38·9	59·0	63·4	85·5	76·6	35·1
Gibbula cineraria	21·9	40·7	66·4	82·6	94·4	94·8
Calliostoma zizyphinum	4·8	17·3	38·0	50·0	94·2	94·5

It is clear that there is a good deal of overlap between the temperatures preferred by the four trochids, but that nevertheless the higher level species had higher optima than the low level species. The second feature which emerges from such data is that the selected temperature was in all cases intermediate between the maximum and minimum sea temperatures and well within the lethal limits of the organism. The overlap of the selected temperature range also suggests an explanation for the overlap of the species on the shore. Thus the results obtained by Micallef (1966) on such trochids and by Janssen (1960) and Sandison (in Lewis 1964) on littorinids underline the importance of temperature effects which operate long before the thermal limits of the organism are reached. The data of Micallef (1966) and Janssen (1960) also suggest that active aggregation in regions of optimal temperature range coupled with evasion of extreme temperatures may control the characteristic zonational sequence on the shore. Such results would also account for the lack of precise correlation between the upper limit of thermal tolerance and the normal thermal stresses to which the organisms are subjected on the shore (Broekhuysen, 1940; Evans, 1948; Southward, 1958a).

3. THE INFLUENCE OF EVAPORATION

Many other organisms have been shown to adopt particular behaviour patterns in response to thermal stress. Edney (1951, 1953; for review 1954, 1960) has shown that in the supralittoral isopod *Ligia oceanica,* the body temperature of the animal is several degrees lower than that of the environment when in dry air, but that in air saturated with water vapour, where evaporation cannot occur, the body approaches that of the environment. Fig. 9.10 shows the composite

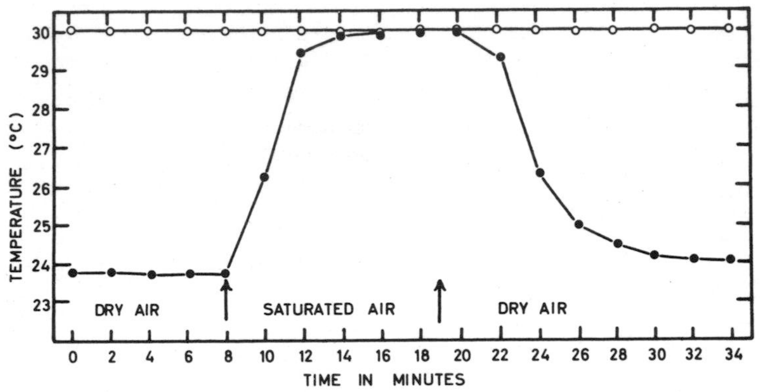

Fig. 9.10. Curve showing the body temperature of living *Ligia* (solid circles) exposed to a slowly moving stream of dry air at 30°C (open circles), alternating with a stream of air saturated with water vapour at 30°C. (After Edney, 1951.)

results of a series of experiments in which specimens of *Ligia* were exposed first to dry air at 30°C and then to water-saturated air at the same temperature. It will be noticed that during the initial period the body temperature was as much as 6·2°C below that of the air but that it reached 30° C in the saturated air. On the return of dry air to the chamber, the body temperature again fell but in this case did not become quite so low as in the initial equilibration period. These results are of interest not only in demonstrating the importance of evaporation in cooling the body of *Ligia,* but also show that the heat gain from metabolism is insignificantly small since the body temperature does not exceed that of the environment by more than 0·1°C when evaporation is prevented. Edney (1953) was then able to show that such heat loss by evaporation also occurs under natural conditions on the Pembrokeshire coast, Wales. This effect is equally apparent in moist, freshly-killed *Ligia,* but not in dead dry specimens. Evaporative cooling is thus a passive process dependent upon the presence of water, and also upon a permeable cuticle since waxed animals had body temperatures similar to those of dry specimens. The body temperatures of living, dead but moist animals and of dry dead *Ligia* are compared in Fig. 9.11

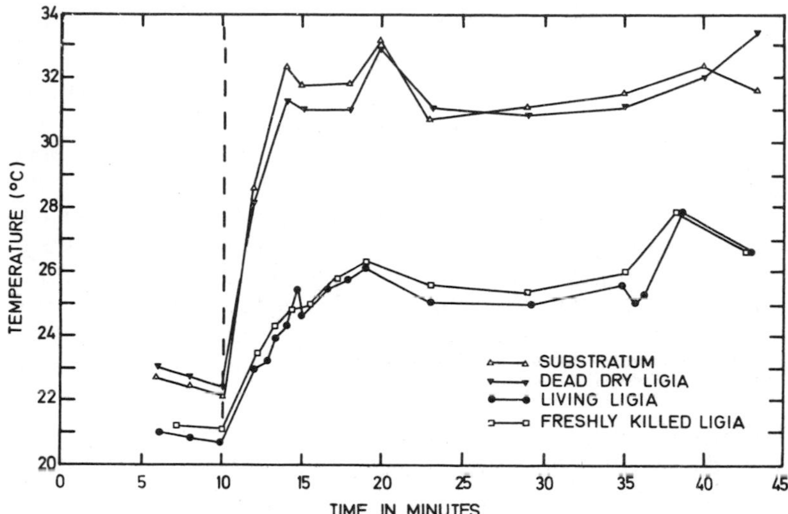

Fig. 9.11. Graphs showing the temperatures of living, freshly killed and dry dead *Ligia* as well as corresponding temperature of the substratum under natural conditions. The broken line indicates the onset of sunny conditions. (Data from Edney, 1953.)

with the corresponding values for the ground temperature. The general conclusion is, therefore, that the body temperatures of living and of freshly-killed animals are some 5–6° C below those of the dry specimens (Occasionally differences of as much as 7 – 9° C were noted for brief periods of between 10 and 20 minutes.)

The air in the interstices of the shingle in which the animals lived was 30° C and was nearly saturated with water vapour. The lethal temperature under such conditions is 32·5° C after 1 hr, so that the animals were living near their upper limit of thermal tolerance. Animals were seen to emerge onto the surface of the rocks where the temperatures varied up to a maximum of 38° C, and then the body was cooled by evaporation; for example, animals on the surface of a rock whose surface temperature was 34°C had a body temperature of 26° C, the relative humidity of the air being 60 – 70%. Thus the animals may emerge from below the shingle, where both temperature and humidity are high, and lose heat by evaporation and convection even though in so doing they are exposed to direct sunlight. Clearly, return to humid conditions is necessary at intervals to prevent death by desiccation; aggregations of *Ligia* where water was seeping from the surface of rocks were also noted by Edney (1953).

We have seen that the body temperature of specimens of *Ligia* may be reduced by evaporative water loss, and that the animals may indeed respond behaviourally by exposing themselves to conditions where transpiration is facilitated. Edney (1953) pointed out that under such circumstances the heat exchange for *Ligia,* as in resting insects (Parry, 1951), depends upon a balance between heat gained from, or lost to, the environment by radiation, conduction and convection, plus the possible heat gained from metabolism. Parry (1951) showed that in the case of insects resting in direct sunlight, neither metabolic heat gain nor evaporative heat loss contributed significantly to the total heat balance. However, in the case of crustaceans such as the wood lice, including *Ligia,* and crabs (p. 484), the cuticle is much more permeable so that evaporation acquires more importance in a calculation of the heat balance than in insects, where the cuticle is impermeable. As in insects, the heat gained by metabolism was considered insignificant by Edney (1951, 1953) since the rate of oxygen consumption in air at 25° C was 0·02 ml/cm^2/hr, which was calculated to contribute an energy gain of only 0·11 milliwatts/cm^2 (where 1 ml $O_2 \equiv 4{\cdot}775$ cals). In fact when evaporative heat loss is prevented in saturated air, the rise in temperature was never greater than 0·1°C above that of the air so that metabolic heat cannot enter significantly into the balance at high temperatures although at low temperatures the contribution made by metabolic heat may be more significant.

The heat exchange balance of a specimen of *Ligia* exposed to sunlight may therefore be simplified to:–

$$\underset{\text{(heat gain)}}{\text{radiation + conduction}} \quad = \quad \underset{\text{(heat loss)}}{\text{convection + evaporation.}}$$

It is possible now, as Edney (1953) has shown, to arrive at numerical values for radiation, convection and conduction which suggest that in *Ligia,* and probably in other crustaceans (p. 480), conduction plays rather a small part in the process of heat gain, whilst convection and evaporation are of approximately equal importance as mechanisms of heat loss. Clearly, however, the relative importance of these last two factors varies greatly with environmental conditions, including relative humidity and wind speed, so that it is convenient to take a specific example in a calculation of the heat exchange balance. Edney (1953) took as a basis for calculation the observation that at an air temperature of 21°C and a ground temperature of 35°C the temperature of living *Ligia* was 28°C, whilst that of dead dry *Ligia* was 34°C, the air speed being approximately 50 cm/sec.

An estimate of the value for radiation is based on that used by Parry (1951, also Stagg, 1950). The mean radiation load, representing the algebraic sum of input and output, upon a horizontal plate with a surface reflectivity of 50% and exposed to the sun within 2 hr of noon, was found to vary from 10 – 22 milliwatts/cm^2 according to the orientation of the plate. Edney (1953) assumed a surface reflectivity for *Ligia* of 50% and assessed the radiation load as approximately 20 mW/cm^2. Conduction represents an unknown factor but is probably small owing to the fact that contact with the ground is only at the tips of the legs and not over the whole ventral surface. The balance of the equation also suggests that conduction is small (see below). A value for convection can also be estimated from the data cited by Parry (1951) in which the convection coefficient for a disc in an air stream moving at 50 cm/sec was 1·4 mW/cm^2 for each °C by which the temperature of the disc differed from the temperature of the air moving over it. In the example we are considering, the temperature of the *Ligia* exceeded that of the air by 7°C so that the rate of heat loss by convection was 9·8 mW/cm^2. Finally, a value for heat loss by evaporation may be estimated from laboratory data coupled with a correction for the increased wind speeds occurring in the natural habitat. Edney (1953) estimated a rate of water loss in *Ligia* of 7·4 mg/cm^2/hr in dry air moving slowly at 5 cm/sec. Since the wind speed was 50 cm/sec in our example, and the rate of evaporation is approximately proportional to the square of the velocity at low wind speeds (Ramsay, 1935a,b), Edney (1953) estimated that the rate of evaporation from *Ligia* was at least 14·8 mg/cm^2/hr; this is equivalent to a heat loss of at least 9·9 mW/cm^2.

Substituting the above estimated values for heat gain by radiation (20 mW/cm^2), and heat loss by convection (9·8 mW/cm^2) and evaporation ($>$9·9 mW/cm^2) into the expression for the heat exchange balance in *Ligia,* we obtain:–

$$\underset{(20\ \mathrm{mW/cm^2})}{\text{radiation}} + \text{conduction} = \underset{(9{\cdot}8\ \mathrm{mW/cm^2})}{\text{convection}} + \underset{(9{\cdot}9\ \mathrm{mW/cm^2})}{\text{evaporation}}$$

That is, even allowing for the conservative estimate of evaporative heat loss, the contribution of conduction to the heat gain is very small, being 0·3 mW/cm^2 for the example cited above. Of course, as Edney (1953) has pointed out, if a dry *Ligia* is exposed to the same set of environmental conditions no heat loss by evaporation is possible and the temperature of the animal will rise. However, this will result in both an increase in heat loss by convection and by radiation, since the rate of heat loss from both of these sources depends upon the difference between the temperature of the body and that of the environment. The value for loss by convection will thus increase and the net radiation load will decrease since the radiation input is the same and the rate of loss has increased. Thus elimination of evaporative heat loss removes a value of 9·9 mW/cm^2 from the right hand side of the equation and the value of 9·8 mW/cm^2 for heat loss by convection increases. But it does not increase by the same amount as that originally removed by evaporation because the net radiation gain has diminished. The increase in heat loss by convection will be the value originally removed by evaporation, 9·9 mW/cm^2, minus the fall in net radiation input. If this increase in convection loss is denoted by c, and the fall in net radiation load is r, then $c = 9{\cdot}9 - r$. Thus for a dry *Ligia,* under the same set of environmental conditions as those cited above, the heat exchange balance is:–

$$\underset{(20 - r\ \mathrm{mW/cm^2})}{\text{radiation}} + \text{conduction} = \underset{(9{\cdot}8 + c\ \mathrm{mW/cm^2}).}{\text{convection}}$$

Edney (1953) has shown that it is also possible to calculate the difference in temperature which would be expected between a dry *Ligia* and one in which evaporation occurs. The relative importance of radiation and convection in accounting for the difference in temperature between the dry animal and wet one can also be calculated.

First it is necessary to derive an expression for the net radiation input, A. A body radiates energy at a rate which is proportional to the 4th power of its absolute temperature. Thus the radiation rate, R is given by:–

$$R = kT^4$$

Where k is a constant for a particular body and is dependent on both the nature and area of its surface and T is the absolute temperature of the body.

In an environment of absolute temperature T_e, the environment, too, will radiate energy so that there will be an energy exchange between it and the body. If we denote that exchange R' then:–

$$R' = k_e T_e^4 - kT^4.$$

For many situations the environment is substantially constant so that $k_e T_e^4$ may be denoted by A. Then we may write the energy exchange as:–

$$R' = A - kT^4$$

The constant k, as we have defined it, is dependent upon the area and nature of the body. In fact, more strictly, it is proportional to area, and for a perfect radiator (a black body) it is a universal constant. For a black body of unit area the constant (Stefan's constant) is $5{\cdot}735 \times 10^{-5} \frac{\text{ergs}}{\text{cm}^2.\text{sec.degree}^4}$. Since an erg is 10^7 Joules, Stefan's constant may be written as:–

$$5{\cdot}735 \times 10^{-12} \quad \text{Joules/cm}^2\ \text{sec. degree}^4$$

One joule per second is a watt, so that the constant may be written:–

$$5{\cdot}735 \times 10^{-12} \quad \text{watts/cm}^2\ \text{degree}^4$$

Very few materials even approach being black bodies; they are poor radiators but better reflectors. The extent to which they are black bodies is their Emissive Power but this value may vary widely in apparently similar substances. The Emissive Power of some common inorganic compounds is given by Hodgman (1959); for the purposes of his calculation, Edney (1953) assumed an emissivity of 0·75 in *Ligia*, a value which is approximately the mean of those cited by Hodgman (1959).

Suppose e is the emissivity of an object. Then the radiation exchange of that object with the environment for given unit area will be:–

$$R' = A - e \times 5{\cdot}735 \times 10^{-12}\ T^4 \ \text{watts/cm}^2$$

and the corresponding value for the radiation exchange of *Ligia* ($e \doteq 0{\cdot}75$) in milliwatts/cm² is:–

$$R = (A - 4{\cdot}28 \times 10^{-9}\ T^4)\ \text{mW/cm}^2$$

although it should be stressed that in using this equation A has a particular value only for a given environment and, in particular, one given environmental temperature.

Now let the energy loss due to evaporation from the wet body be E mW/cm^2, then neglecting the small effects of heat gain by conduction and metabolism, the energy balance derived from the equations on p. 476-477, for the wet body is:–

$$10^{-9}\, T^4\, (A - 4{\cdot}28) = C\,(T - T_a) + E \tag{1}$$

Where C = the convection energy loss, T = the temperature of the wet body (° Abs) and T_a = the air temperature (° Abs).

Similarly, since evaporation does not occur from the dry animal, the energy balance for the dry body is:–

$$10^{-9}\, T_1{}^4\, (A - 4{\cdot}28) = C\,(T_1 - T_a) \tag{2}$$

Where T_1 = the temperature of the dry body.

The difference in temperature between that of the wet and the dry body will thus be given by the difference between equations (1) and (2). Thus:–

$$4{\cdot}28 \times 10^{-9}\, (T_1{}^4 - T^4) = E - C\,(T_1 - T) \tag{3}$$

If $(T_1 - T) = t$, and t is much smaller than T, equation (3) becomes:–

$$4{\cdot}28 \times 10^{-9} \times 4\, T^3 t = E - Ct$$

$$\text{and } t = \frac{E}{1{\cdot}71 \times 10^{-8}\, T^3 + C}$$

Thus the temperature difference t, between the dry body and the evaporating body can be calculated, provided that the energy loss due to evaporation E and the convection energy loss C, which depends also on wind speed, can be estimated. Further, Edney (1953) has shown that since the net radiation input into a dry body is less than that into a wet body (p. 476), and this difference may be denoted by r, whilst the convection loss is increased compared with a wet body since heat removal has been eliminated (p. 476) and this increase may be denoted by c, the relative effects of radiation and convection in controlling the temperature difference between a dry and a wet animal can be calculated. From equation (3),

$$r = 4{\cdot}28 \times 10^{-9}\, (T_1{}^4 - T^4)$$

$$\text{and } c = C\,(T_1 - T).$$

Since $(T_1 - T) = t$ then:–

$$\frac{r}{c} = \frac{4{\cdot}28 \times 10^{-9} \times 4\,T^3 t}{Ct}$$

$$= \frac{1{\cdot}71 \times 10^{-8}\,T^3}{C}$$

Edney (1953) estimated a value of 1·4 for C, and if $T = 300$ (27° C),

$$\text{then} \quad \frac{r}{c} \quad 0{\cdot}33 \text{ mw/cm}^2$$

Now the significance of this is that, knowing the relative importance of radiation and convection in controlling the difference in temperature between a dry and a wet body, we can use this value to predict the increase in temperature which would be expected to occur in the dry *Ligia* compared with the wet animal in our example on p. 476. From this example, we know that:–

$$c = 9{\cdot}9 - r \text{ mw/cm}^2$$

and from the calculation just described, we know that:–

$$\frac{c}{r} = 0{\cdot}33 \text{ mw/cm}^2$$

Whence the increase in convection loss c occurring in a dry body compared with a wet one is obtained by solution of these two simultaneous equations which gives a value of $c = 7{\cdot}42$ mW/cm². Since the convection coefficient is 1·4 mW/cm²/°C, a rise in the rate of convection loss by the dry *Ligia* of 7·42 mW/cm² must be set up by an increase in temperature of 5·3° C compared with the body temperature of a wet animal. The temperature of the living *Ligia* was 28° C (p. 475) so that one would expect the body temperature of the dry one to rise to 28·0 + 5·3 = 33·3° C under the same set of environmental conditions. In fact the measured temperature was 34° C (p. 475) which is very close to the predicted value and suggests that the estimated values for radiation, conduction, convection and evaporation in the heat exchange balance for *Ligia* shown on p. 475 are not seriously in error.

Heat loss by evaporation must also play a part in enabling the survival of other organisms which possess a permeable cuticle and which are normally subjected to temperatures approaching the lethal point. Edney (1961, 1962) has made a study of the lethal temperatures of fiddler crabs belonging to the genus *Uca* and correlated these values with the temperatures where the crabs live. As mentioned on p. 452 he found that in a series of five species of *Uca,* the upper limit of thermal tolerance after 15 min exposure was related to the distribution of the species, the most terrestrial species having a higher lethal temperature than the most aquatic species. Laboratory measurements also showed that the body temperatures of living specimens of *Uca* were from 5 – 8°C below those of dead, dry animals even in rather high ambient humidities. It thus seems likely that heat loss by evaporation may, as in *Ligia,* play an important part in the heat balance of this animal.

Measurements were then made at low tide to determine whether the animals were exposed to temperatures at which the reduction of the body temperature by evaporation would aid survival. Data were also collected on the actual body temperatures under such conditions. *Uca annulipes* lives in holes in the substratum, and Edney (1961, 1962) was able to show that the temperatures in such burrows ranged from 26·0°C at approximately 25 cm depth to 35°C near the surface. The burrows also led to the water table so that the animals were not exposed to the danger of desiccation. Brief feeding excursions to the surface were made. The temperature on the ground was 44·5 – 46°C whilst that of surface pools was 35 – 38°C. The ground temperature was thus well above the lethal temperature for this crab (42°C) and heat loss by the evaporation of water would enable survival if the body temperature could be depressed below 42°C. In fact body temperatures ranged from 34·8 to 38·7°C; evaporative water losses could be restored when the animals returned into the humid burrows. Similarly, in another situation the temperature on the surface of the substratum was as high as 48·3 – 50·2°C and the body temperature reached 39·1 – 42·0°C despite the heat loss by evaporation. Under such circumstances a high lethal temperature is clearly of considerable advantage in an organism living so close to its upper lethal temperature range. *Uca chloropthalmus* has a lethal temperature of 41·25°C for 15 min exposure times and, here again, the body temperatures actually observed approached this value. The burrow temperature at 20 cm depth was 32°C, whereas that at the surface of the substratum was 46·2 – 50·0°C and the body temperature of the crab was 39·4 – 40·3°C. Finally in *Uca marionis* the body temperature ranged from 33·0 to 39·9°C.

Thus thermal stress in these organisms at Inhaca Island (lat 26°S) is not only extreme but is such that survival would not occur without heat loss by transpiration. Indeed, as Edney (1961, 1962) states, it is not surprising that the upper lethal temperature of the different species is fairly sharp and is related to the thermal stress occurring in the environment. Further, the observed lack of variation of heat lethal temperature between the individuals of a population of any one of the species would also be expected as a result of strong selection pressure leading to a homogeneous population with a high upper limit of thermal tolerance. The regular occurrence of near lethal temperatures in such organisms may thus account for the correspondence between the ecological succession and upper limit of thermal tolerance, and for the similarity of the lower lethal temperatures of all the species. Low temperatures would not be experienced by such animals and the lower limit of thermal tolerance would have little survival value. On the other hand it is tempting to suggest that in polar regions, where near-lethal low temperatures would be experienced, there would be a better correlation between ecological sequence and lower lethal temperature than with upper lethal temperatures. In temperate zones there is often a large difference between the temperatures experienced on the shore and the range of thermal tolerance (p. 467). Under such circumstances behavioural responses to the thermal conditions play an important part in determining the zonation sequence of the animals.

Wilkens and Fingerman (1965) have also demonstrated the importance of heat loss by evaporation in another species of *Uca*. They

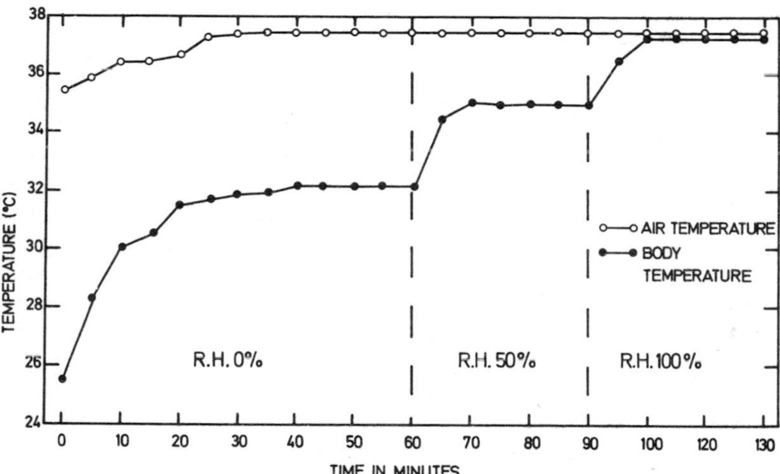

Fig. 9.12. Graphs showing the mean body temperature of specimens of *Uca pugilator* exposed to slowly moving air at the temperature indicated by the open circles and a relative humidity of 0%, 50% and 100%. (After Wilkens and Fingerman, 1965.)

showed that in *Uca pugilator* the body temperature could be related to the relative humidity of the air to which the crabs were exposed (Fig. 9.12). Further, the evaporation of water from the crab was greater at high temperatures, but this increase in evaporation was proportional to the saturation deficit of the air. They concluded that evaporative water loss was thus a physical process dependent upon saturation deficit, no control over the rate of evaporation being exerted by the crabs at any temperature. Fig. 9.13A shows the curves for the mean rate of evapor-

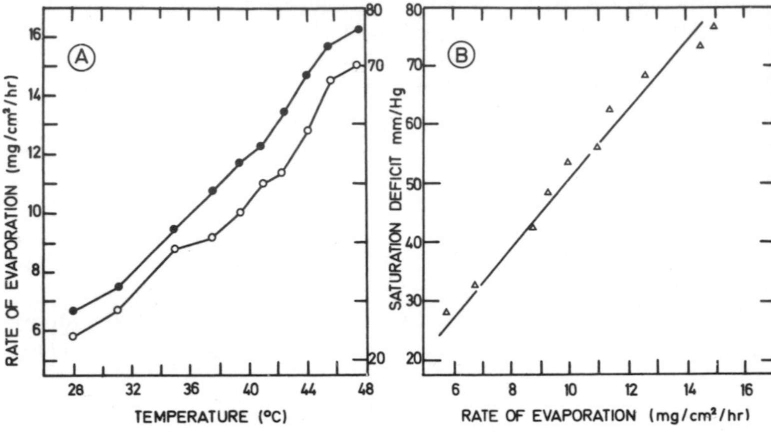

Fig. 9.13. Graphs showing (A) the relationship between temperature and rate of evaporation from *Uca pugilator* (open circles) and saturation deficit (solid circles); (B) the relationship between saturation deficit of the air and rate of evaporation from *Uca pugilator.* (Data from Wilkens and Fingerman, 1965.)

ation ($mg/cm^2/hr$) from 20 crabs in dry air from 28°C to 48°C. The relationship between the saturation deficit and temperature is also shown and the similarity between the two curves is obvious. It follows that the rate of transpiration is directly proportional to saturation deficit (Fig. 9.13B). Finally, Wilkens and Fingerman (1965) were able to demonstrate a definite increase in the heat-lethal temperature of crabs kept for 1 hr in dry air compared with those in air saturated with water vapour, as shown in Fig. 9.14. The median lethal temperature is 40·7°C in saturated air but as high as 45·1°C in dry air; the difference corresponds reasonably well with the known depression of body temperature which occurs in specimens of *Uca pugilator* in dry air compared with those in air saturated with water vapour. The values in moist air are similar to those determined by Teal (1958), who found a median lethality of 39·5°C for *Uca pugilator,* 39·9°C for *U. minax* and 40·0°C for *U. pugnax*. Although these differences between species were shown to be not significant, Vernberg and Tashian (1959) showed that the tropical *Uca rapax* was more resistant to temperatures of 42° and 44°C than the temperate *Uca pugnax;* this agrees with the sequences observed for other species by Edney (1961, 1962). Factors other than thermal tolerance and evaporation also play a part in the heat regulation of *Uca pugilator.* Wilkens and Fingerman (1965) showed, for example, that the blanching which occurs in this species at high temperatures (Brown and Sandeen, 1948) lowers the body temperature by

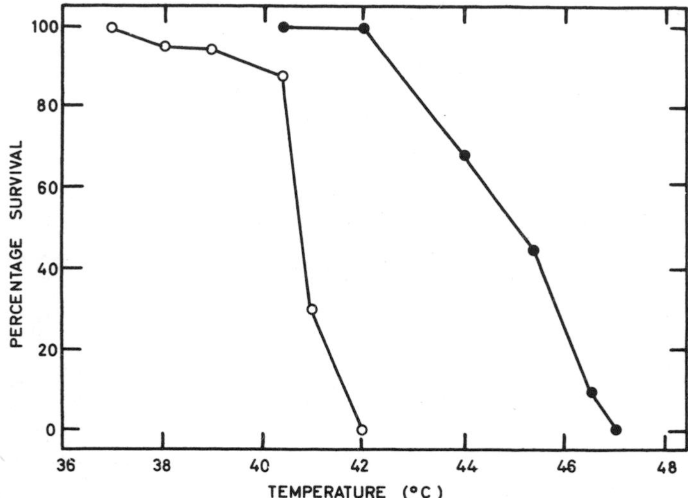

Fig. 9.14. Graphs showing the relationship between temperature and percentage survival of *Uca pugilator* after 1 hour in dry air (solid circles) and in air saturated with water vapour (open circles). (After Wilkens and Fingerman, 1965.)

2°C compared with dark crabs. Retreat into the cool burrow, too, occurred at regular intervals each 18 – 24 min and allowed the body temperature to be reduced by several degrees, and also facilitated the restoration of water lost by evaporation. Thus a combination of high thermal tolerance, heat removal by evaporation, reduction of heat absorption by blanching, coupled with evasion of thermal stress all play a part in the overall heat balance of the crab.

Other crabs also tend to lose water by evaporation and so cool themselves. Ahsanullah (1969) has shown that in both the intertidal crab *Carcinus maenas* and the sublittoral crab *Portunus marmoreus* water loss is an exponential function of the size of the crab, small specimens losing proportionally more water than large ones. Such water loss was shown to occur mainly through the gills, and for comparable sizes of crab it was greater in the sublittoral *Portunus marmoreus* than in the intertidal *Carcinus maenas* (also Gray, 1953) Thus, as would be expected, the cooling of the body at any particular humidity and temperature is greater in *Portunus* than in *Carcinus.*

These results are shown in Fig. 9.15, which also shows that the body temperature of dried animals is similar to that of air whilst that of freshly killed animals is similar to that of the living ones. In many respects, therefore, the data resemble those obtained by Edney (1951,

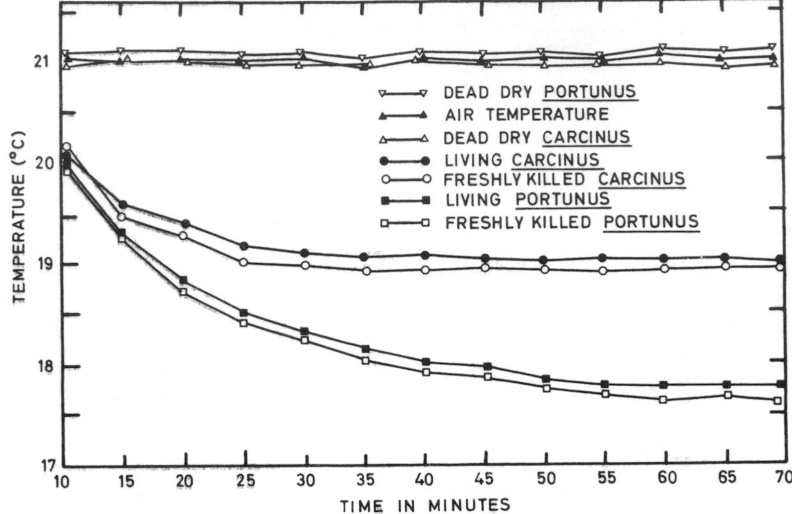

Fig. 9.15. Graphs showing the body temperature of *Carcinus maenas* and *Portunus marmoreus* after being placed in air at 21°C (solid triangles) and 68% relative humidity. (Based on Ahsanullah, 1969). Solid triangles – air temperature; open triangles – dead dry *Carcinus;* solid circles – living *Carcinus;* open circles – freshly killed *Carcinus;* inverted open triangles – dead dry *Portunus;* solid squares – living *Portunus;* open squares – freshly killed *Portunus.*

1953; also 1954, 1960, 1961, 1962) and by Wilkens and Fingerman (1965) for other crustaceans. The extent of cooling can also be shown to depend upon the saturation deficit of the air. The question arises, therefore, to what extent cooling by evaporation can be regarded as an adaptation to intertidal conditions since the sublittoral crab as well as the intertidal crab showed a depression of body temperature. Indeed, the water loss and depression of body temperature was greater in *Portunus marmoreus* than in *Carcinus maenas.* The answer to this problem must reside in the relative resistance of the two species to water loss, for although the body temperature of the sublittoral *P. marmoreus* is depressed by passive evaporation, survival occurs for only a short period of time in air owing to the susceptibility of this animal to desiccation. On the other hand *Carcinus maenas* can survive up to 25% decrease in body weight, and is thus able to survive for long periods at air temperatures which may exceed the lethal temperature in water.

The resistance lines for *Carcinus* in water and in air at a relative humidity of 60% are shown in Fig. 9.16, and it is clear that higher temperatures can be survived in air than under conditions where evaporation is prevented much as has been shown in *Uca pugilator* (Wilkens and Fingerman, 1965).

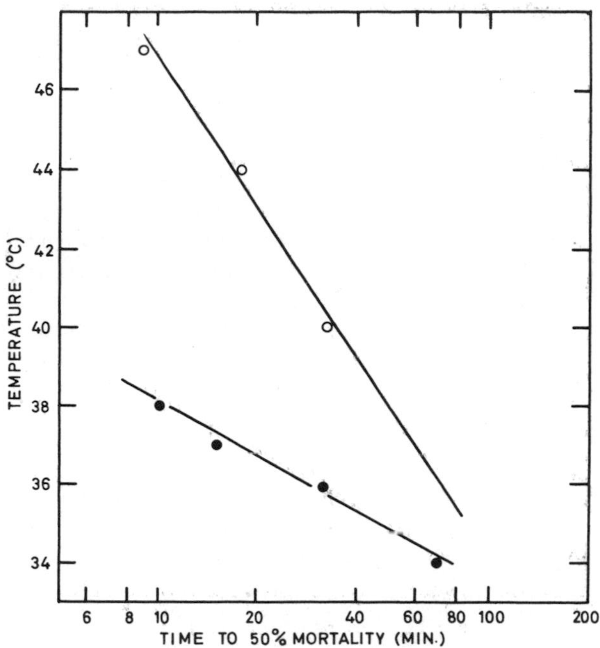

Fig. 9.16. Thermal resistance lines of *Carcinus maenas* in water (solid circles) and in air of relative humidity 60% (open circles). (Data from Ahsanullah, 1969.) Crabs acclimated to 10° C.

A second factor of importance in an assessment of the possible adaptive significance of cooling by evaporation in such organisms is whether sufficiently high temperatures or low humidities are encountered to markedly enhance survival by this means. The resistance lines shown in Fig. 9.16 for *Carcinus* acclimated to 10° C suggest that the survival time in water at 34° C is only 70 min. Now the temperature of rock pools may exceed 30° C and this value would be expected to be maintained for at least 1 hr in the summer during the exposure period. The predicted median survival time at 35° C for crabs acclimated to 10° C would be only 45 min, and although acclimation to the higher summer temperatures would be expected to raise this value, emergence to sheltered positions under rocks where a degree of evaporative cooling may occur would significantly increase the length of time during which the high temperatures could be tolerated. For example, a depression of the body temperature of as much as 1·5° C occurs even in stationary air at 88% relative humidity, which would be expected to extend survival at a temperature of 35° C from 45 min to 95 min.

Much the same argument might be applied to other intertidal organisms such as the fish *Blennius pholis.* This animal also loses water in proportion to its surface area when exposed to low humidities as one would expect in a sublittoral fish. But it differs from such fishes not only in its ability to withstand a reduction in its total fluids by evaporation of at least 30%, but also in actively seeking positions of exposure to air during periods of thermal stress when the tide is out. It is thus able, by virtue of its ability to withstand desiccation, to take advantage of the cooling which occurs on exposure to air (M.J. Daniel, personal communication). It is thus apparent that resistance to thermal stress in many intertidal animals is to some extent associated with resistance to desiccation. Further, it would be expected that a gradation of tolerance to weight loss would exist from sublittoral organisms with only a limited tolerance, through to upper shore organisms which can tolerate an extensive loss of tissue fluid.

C. DESICCATION

Although the importance of desiccation in controlling the zonational distribution of intertidal organisms is widely recognised, systematic investigations of the precise influence of this factor are surprisingly scarce. Most work has been carried out on intertidal algae. Thus Baker (1909, 1910) showed that one of the important factors controlling the distribution of intertidal algae on the upper shore was the degree of desiccation to which they were exposed, whilst Isaac (1933, 1935) and Zaneveld (1937; for review, 1969) have made comparative studies on the rate of water loss of intertidal algae in relation to shore level. Similarly Ferronière (1901) showed that polychaetes from the upper shore withstood desiccation better than those from the lower shore. Again, certain animals which characteristically occupy high levels on the shore have been shown to be remarkably resistant to conditions of desiccation. The winkles *Littorina saxatilis* and *L. neritoides* for example, have been shown to be able to withstand desiccation for 42 days (Colgan, 1910); *L. neritoides,* indeed, survived as much as 5 months (Patané, 1933). The barnacle *Chthamalus stellatus,* too, shows a remarkable ability to withstand drying, Mediterranean specimens having been shown to survive 109 days out of water (Monterosso, 1930, also p. 364). However, it was not until the pioneer study by Broekhuysen (1940) of the resistance to desiccation in a series of gastropods from different shore levels at False Bay, South Africa, that the correlation between resistance to desiccation and zonational distribution was established in detail. Later studies by Sandison (in Lewis, 1964) and by Micallef (1966) on gastropods and by Newman (1967) on barnacles, have tended to confirm and amplify his conclusions.

As has been mentioned on p. 451, Broekhuysen (1940) studied a series of gastropods ranging from the upper shore *Littorina knysnaënsis* through *Oxystele variegata, Thais dubia, Oxystele tigrina, O. cincta* to the lower shore *O. sinensis.* The intertidal distribution of these six species of prosobranchs is shown in Fig. 9.17. It is clear that the upper

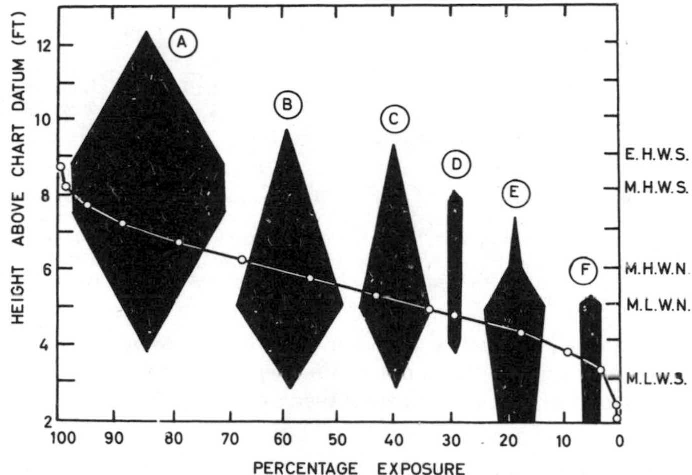

Fig. 9.17. Graph showing the relation between the distribution of gastropods on the shore at False Bay, South Africa and tidal level. The curve shows the percentage exposure at the various tidal levels. (After Broekhuysen, 1940.) A = *Littorina knysnaënsis;* B = *Oxystele variegata;* C = *Thais dubia;* D = *Oxystele tigrina;* E = *Cominella cincta;* F = *Oxystele sinensis.*

shore species, as on any other shore, would be exposed not only to greater temperature extremes (see above) but also to lower relative humidities and for a longer period of time than species characteristic of lower shore levels. In order to determine the resistance of desiccation shown by each of the six prosobranchs, Broekhuysen (1940) placed the animals in a desiccator over calcium chloride and measured both the percentage weight loss and the mortality at room temperature (14 – 27°C) and at 39 – 40°C (Fig. 9.18). It is evident that there is a general correspondence between the zonational sequence of the animals on the shore and the amount of water loss which the species can tolerate. Thus the upper shore *L. knysnaënsis* can survive 22% weight loss before the

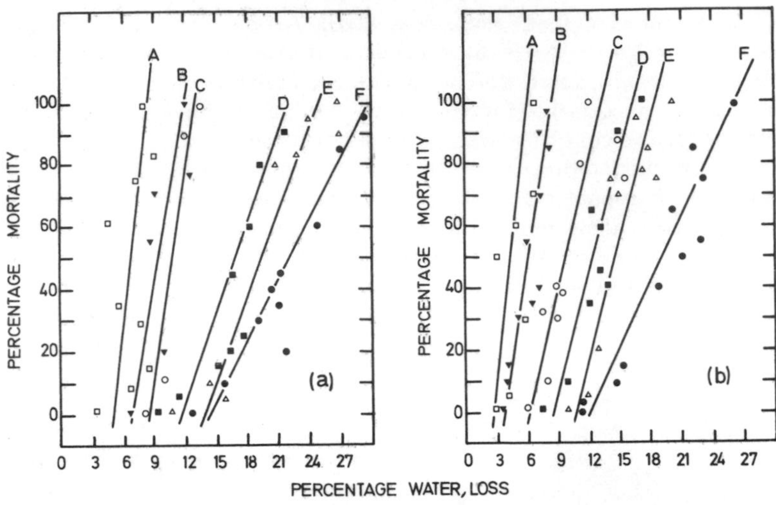

Fig. 9.18. Graphs showing the lethal water loss in six species of South African gastropods at (a) 14–27°C and (b) 39–40°C. A = *Oxystele sinensis* (open squares); B = *Oxystele tigrina* (inverted solid triangles); C = *Cominella cincta* (open circles); D = *Oxystele variegata* (solid squares); E = *Thais dubia* (open triangles); F = *Littorina knysnaensis* (solid circles.) (After Broekhuysen, 1940.)

median mortality is reached at room temperature, whereas the median mortality is reached after only 6·5% weight loss in the lower shore *Oxystele sinensis.* These values are shown on Table XX.

Table XX. Table showing the percentage weight loss required for 50% mortality in a series of gastropods from False Bay, South Africa. (Compiled from Figs. 9.17 and 9.18; data from Broekhuysen, 1940.)

Species	*Upper limit on shore (ft above datum)*	*% weight loss for 50% mortality at 14 – 17°C*	*% weight loss for 50% mortality at 39 – 40°C*
Littorina knysnaënsis	12·25	22·0	19·0
Oxystele variegata	9·75	17·5	13·0
Thais dubia	9·25	19·0	14·5
Oxystele tigrina	8·0	9·0	6·0
Cominella cincta	7·25	11·0	9·5
Oxystele sinensis	5·25	6·5	4·0

The main exception to this generalisation is *Oxystele tigrina* which, although it lives at fairly high levels on the shore, is less tolerant of weight loss than *Cominella cincta* which lives at a lower level on the shore. The explanation of this anomaly is that *O. tigrina* is normally restricted to damp situations or to pools so that its high zonational position is little reflection of the degree of desiccation to which it is normally subjected. Broekhuysen (1940) showed that under natural conditions on the shore, the relative humidity at the upper limit of the *Littorina* zone was 74–76%, 80% at the upper limit of the balanoid zone and 83% at the lower limit of the balanoid zone. There was thus relatively little difference between the degree of desiccation to which each zone was subjected, but the duration of exposure would be of considerable importance in restricting the distribution of the animals to their characteristic zones on the shore.

In much the same way, Micallef (1966) has shown that a graded resistance to weight loss exists in the trochids *Monodonta lineata, Gibbula umbilicalis* and *G. cineraria,* although the lower shore *Calliostoma zizyphinum* appears to have a higher tolerance of desiccation than its zonational level would suggest. This implies that in *Calliostoma* some other factor, such as lowered temperature tolerance, is of importance in controlling the upper limit of distribution. The time taken for 50% mortality in each of these species when placed in a desiccator over calcium chloride at 3·5°, 5·0°, 10·0° and 20°C is shown in Table XXI.

Table XXI Table showing the relationship between zonational level and the median resistance time of a series of trochids in a desiccator containing $CaCl_2$ at 3·5, 5·0, 10·0, and 20°C. (Data from Micallef, 1966.)

Species	*Zonational level % exposure to air*	*Time in days to 50% mortality*			
		3·5°C	5·0°C	10·0°C	20·0°C
Monodonta lineata	35 – 85	12	12	9·7	3·25
Gibbula umbilicalis	5 – 75	3	3	3·1	1
Gibbula cineraria	0 – 30	1·5	1·6	1·5	1·5
Calliostoma zizyphinum	0 – 10	3	3	3	1

Apart from the unexpectedly high resistance of the lower shore *Calliostoma zizyphinum* to weight loss, it is apparent that *Gibbula umbilicalis* is much less tolerant than *Monodonta lineata* even though these species overlap on the shore. A possible explanation of this lies in the fact that *Gibbula umbilicalis* is characteristically found in rockpools or damp situations in the upper parts of its range and, like *Oxystele tigrina,* is therefore not normally exposed to such severe desiccation as its zonational level would imply. This result is rather similar to that obtained by Colgan (1910) who found that although *Littorina littorea* and *L. littoralis* occupy similar zonational levels on the shore, *L. littorea* could withstand emersion for 23 days, as compared with only 6 days for *L. littoralis.* His result agrees fairly well with that of Gowanloch and Hayes (1926), who found a value of 28 days for *Littorina littorea.* The difference between *L. littorea* and *L. littoralis* is due to the difference in habitat of the two species; *L. littorea* occupies damp situations and also open rock, while *L. littoralis* is confined to fucoids (p. 21). Similar variations in percentage water loss in littorinids have also been noted by Sandison (cited in Lewis, 1964).

Broekhuysen (1940) was unable to demonstrate any clear correlation between rate of water loss and zonational sequence in the six species of gastropods which he studied. However after a period of desiccation the high level *Littorina knysnaënsis* was able to decrease its rate of water loss to a very low level by closure of the operculum, whilst Gowanloch (1926) had earlier reported the absence of such behaviour in the sublittoral whelk *Buccinum undatum* when it was occasionally exposed by exceptionally low tides. Similarly, the upper shore *Littorina neritoides* closes the operculum in dry air whereas *L. littoralis* remains extended and even makes browsing movements in dry air (Sandison, cited in Lewis, 1964). Micallef (1966) also emphasised the importance of the behavioural responses of trochids to desiccation and showed that, except for the upper shore *Monodonta lineata* which soon closed the shell by means of the tightly-fitting operculum, the behaviour of *Gibbula umbilicalis, G. cineraria* and *Calliostoma zizyphinum* was such that death by desiccation was hastened. However, there was a gradation in the withdrawal response with the lower shore *Calliostoma* remaining open for several hours. Other factors such as the permeability of the shell and operculum as well as surface area/volume ratio, which is of course greater in the small species than in the larger ones, also contributed to the graded resistance of the trochids to water loss.

The resistance to desiccation of barnacles too, appears to be related partly to structural features and partly to behavioural factors. Thus Barnes and Barnes (1957) showed that the behaviour of intertidal and of sublittoral barnacles was very different from each other when the animals were removed from water. The intertidal species, as in the high-level gastropods described above, tend to adjust their behaviour in such a way that water loss is minimised. The intertidal *Chthamalus stellatus, C. fragilis, C. dalli, Balanus balanoides* and *B. glandula,* for example, withdraw the cirri and close the valves. On the other hand the sublittoral species *Balanus crenatus* and *B. balanus* do not withdraw the cirri and soon become desiccated. This lack of withdrawal is obviously rather similar to the behaviour of the lower shore trochids and of *Littorina littoralis* which also remain extruded and so lose water rapidly. Newman (1967) has recently studied the osmoregulatory ability and resistance to desiccation of *Balanus amphitrite, Balanus glandula* and *Balanus improvisus.* He showed that the relative rates of desiccation in these three species could be correlated with their osmoregulatory ability. *Balanus amphitrite* and *B. glandula* tend to keep the fluid of the mantle cavity isotonic with the blood and hypertonic to seawater, when the latter is diluted more than half. *Balanus improvisus,* however, conforms and the blood is maintained only slightly hypertonic to the external medium right down to only 3% seawater. At these dilutions survival is almost indefinite. Fig. 9.19.

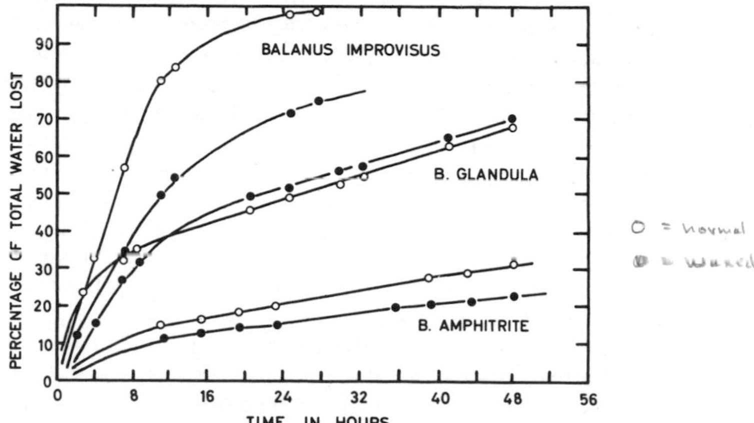

Fig. 9.19. Graphs showing the percentage of the total water lost by *Balanus improvisus, B. glandula* and *B. amphitrite* in air of vapour pressure deficit of 14–17 mm Hg as a function of time. Solid circles indicate the results for individuals in which the operculum had been sealed with wax; open circles indicate normal individuals. (After Newman, 1967.)

shows the percentage of the total water content of each of the species at a vapour pressure deficit of between 14 and 17 mm Hg over a period of 48 hr. Comparison of the slopes of the graphs between 3·75 and 11·25 hr for *Balanus improvisus,* and from 19·5 hr onwards in the closely related *Balanus amphitrite,* gives a rate of water loss of 5·47 µl/hr for *B. improvisus* and 1·69 µl/hr for *B. amphitrite.* These rates were calculated by Newman (1967) to be equivalent to a water loss of 5%/hr in *B. improvisus* and 0·34%/hr in *B. amphitrite.* One of the reasons for the difference in the rate of water loss is undoubtedly the fact that *B. improvisus* is smaller than *B. amphitrite.* But this does not account for all of the great difference in water loss, for when the relative sizes are taken into account it would be expected that for a rate of water loss of 0·005 ml/hr, the small *B. improvisus* would lose 5% of its initial volume of water per hr whereas the larger *B. amphitrite* would lose only 3%. In fact the discrepancy between the water loss of *B. improvisus* and *B. amphitrite* is 5% and 0·34% (i.e. approximately 10 – 15 times rather than about double) so that increased permeability of the integument, as well as smaller size, accounts for the greater susceptibility to desiccation. This greater permeability is correlated with the lack of osmotic regulation in dilute media in *Balanus improvisus* and perhaps accounts for the absence of this species high in the intertidal zone where more thick-shelled barnacles such as *Balanus balanoides, Chthamalus stellatus* and *Elminius modestus* can flourish.

Davies (1969) has recently studied the effects of desiccation on upper and lower shore *Patella vulgata* and on the lower shore *P. aspera.* He measured the percentage water loss over a 12 hr period in a large number of individuals of each of the three populations of limpets; the values obtained were then plotted on a logarithmic scale against the logarithm of the weight of the tissues. In all three groups there was an exponential inverse relationship between percentage water loss and body weight, much as one would expect if water loss was from the general body surface. In fact the slope of the lines was found to have a value of –0·44 in *P. aspera* and the lower shore *P. vulgata.* (Thus percentage water loss varies with body weight$^{-0·44}$). If water loss was proportional to surface area, a slope of –0·33 would have been expected and the fact that the slope is steeper than this suggests a relative decrease in evaporative surface in larger animals. This appears to be due to an increase in shell angle which occurs during growth (Ebling *et al.,* 1962) and leads to a decrease in shell circumference. In the upper shore *P. vulgata* the slope of the line was steeper than in the other two groups (–0·55) so that large individuals of this species had a much lower rate of water loss than large individuals of *P. aspera* or low-level *P. vulgata.* This reduction in the rate of water loss of large individuals suggests an explanation for the fact that in *P. vulgata* the small individuals tend to be confined to damp and shaded positions on the lower parts of the shore whilst the more exposed rock surfaces and upper parts of the shore are inhabited by larger animals (Davies, 1969).

Comparison of the levels of the regression lines indicated that the percentage water loss after 12 hr was greater in *P. aspera* than in low-level *P. vulgata,* which in turn lost water at a faster rate than upper-shore *P. vulgata.* The lethal water loss also showed a clear correlation with the intertidal distribution of the three groups of limpets, although all were able to survive more than 20% weight loss. Thus high level *P. vulgata* was found to tolerate greater weight loss than low-level *P. vulgata,* which in turn had a greater tolerance than *P. aspera.* (Table XXII). Recovery from such weight loss was initially rapid and depended upon the difference between the osmotic pressure of the internal medium and sea water. However, it is obvious that the rate of diffusion would become progressively slower as the concentration of the tissue fluids approached that of seawater, so that recovery may take as long as 4–5 hrs when more than 28% weight-loss has occurred.

Table XXII Table showing the tolerance of water loss in two populations of *Patella vulgata* and in *P. aspera.* Note the interspecific and intraspecific differences which occur. (After Davies, 1969.)

	High-level *P. vulgata*	Low-level *P. vulgata*	*P. aspera*
Percentage weight-loss which all survive	60	44	24
Percentage weight-loss for 50% mortality	60–65	50–55	30–35
Percentage weight-loss for 100% mortality	65	55–60	40

Davies (1969) found that specimens of *P. vulgata* on the shore rarely showed evidence of extensive desiccation, values for the depression of freezing point of the blood indicating a value of 3–5% weight loss being common. However, under extreme conditions a mean weight loss of 10% occurred and in one instance a small specimen showed a 19% weight loss. Extreme weight loss may occur in animals which fail to return to their "home" and in which the shell fits the rocks badly. These interesting measurements suggest that the amount of water loss actually experienced is usually well within the limits of tolerance of the limpets although the effect of size is of great importance (Davies, 1969). Thus under conditions where a 21 mm long high-level *P. vulgata* (5 gm wet tissue weight) would lose 15% of its body weight, a small limpet of 15 mm shell length (0·15 gm wet tissue weight) would lose as much as 65% of its body weight. The effects of desiccation itself are therefore likely to be limiting mainly in small animals. Also of great importance ecologically is the suggestion that it is the relation between weight loss and recovery time which may limit the distribution of limpets. Davies (1969) showed that an individual which had lost 23% of its body weight would require 3·5 hr submersion to regain its initial blood concentration. Thus at high shore levels submersion time, as well as the extent of desiccation, is of importance. Those individuals which occur at high levels would be expected to be confined to the damper situations, so that weight loss is minimised and can be restored during the brief period of submersion.

It is apparent that, as in the resistance of intertidal organisms to thermal stress, the susceptibility to desiccation is related in a general way to zonational distribution, and is affected not only by structural features such as the impermeability of the shell and operculum, but also by behavioural responses which vary in a clearly adaptive way with shore level. Indeed such behavioural adaptations may be far more than mere closure, for in some gastropods such as *Cerithium,* active aggregation occurs during the intertidal period and this results both in a reduction in the temperature and in the desiccation stress to which the group is subjected (Moulton, 1962). Added to these considerations is the fact that since small organisms have a larger surface/volume ratio than large ones, the effects of desiccation are likely to be more severe in small than in large animals. Micallef (1966) noted the preferential survival of large trochids in low humidities whilst the same appears to be true amongst the barnacles studied by Newman (1967) although in these, as in trochids, the permeability of the shell is also of importance. In *Patella vulgata* the smaller individuals tend to be found in damp and shaded situations and are more common on the lower shore (Davies, 1969). There is also, as we have seen above, an increase in the shell angle during growth which leads to a relative decrease in circumference, so reducing the rate of water loss in the large individuals of the upper shore (Davies, 1969; also Ebling, Sloane, Kitching and Davies, 1962). Large thick-shelled animals with an appropriate behavioural response to desiccation would therefore be expected on the upper shore. It is perhaps not surprising that in some motile organisms such as the winkle *Littorina littorea,* the initial settlement of juveniles is on the lower shore, the upper tidal levels being colonised later by the larger individuals (Smith and Newell, 1955).

REFERENCES

AHSANULLAH, M. (1969) *A comparative study of the effects of desiccation and thermal stress on the intertidal crab* Carcinus maenas (L.) *and the sublittoral crab* Portunus marmoreus *Leach.* Ph.D. Thesis, University of London.

ALEEM, A. A. (1949) The diatom community inhabiting the mud-flats at Whitstable. *New Phytol.* **49**, 174–188.

ALLEN, J. A. (1958) On the basic form and adaptations to habitat in the Lucinacea (Eulamellibranchia). *Phil Trans. roy. Soc. (Lond.) B,* **241**, 421–484.

ALLEN, J. A. (1962) Preliminary experiments on the feeding and excretion of bivalves using *Phaeodactylum* labelled with ^{32}P. *J. mar. biol. Ass. U.K.,* **42**, 609–623.

AMBLER, M. P. and CHAPMAN, V. J. (1950) A quantitative study of some factors affecting tide pools. *Trans. roy. Soc. N.Z.,* **78**, 394–409.

ANDERSON, J. G. and MEADOWS, P. S. (1965) Micro-organisms and organic matter attached to the surfaces of marine sand grains. *J. gen. Microbiol.,* **41**, 21.

ANKEL, W. E. (1936a) Prosobranchia. Eds. G. Grimpe and E. Wagler, *Die Tierwelt der Nord und Ostsee. IX,* **1**, Akademische Verlagsgesellschaft, Leipzig.

ANKEL, W. E. (1936b) Die Fresspuren von *Helcion* und *Littorina* und die Funktion der Radula. *Verh. dtsch. Zool. Ges. 38, Zool. Anz. suppl.,* **9**, 174–182.

ANKEL, W. E. (1938) Erwerb und Aufnahme der Nahrung bei Gastropoden. *Verh. dtsch. Zool. Ges. 40, Zool. Anz. Suppl.,* **11**, 223–295.

ANSELL, A. D. (1961) The functional morphology of the British species of Veneracea (Eulamellibranchia). *J. mar. biol. Ass. U.K.,* **41**, 489–515.

ANSELL, A. D. and TRUEMAN, E. R. (1967) Observations on burrowing in *Glycymeris glycymeris* (L.) (Bivalvia, Arcacea). *J. exp. mar. Biol. Ecol.,* **1**, (1). 65–75.

AREY, L. B. and CROZIER, W. J. (1918) The homing habits of the pulmonate mollusk *Onchidium. Proc. Nat. Acad. Sci. Wash.,* **4**, 319–21.

ASAHINA, E., AOKI, K., and SHINOZAKI, J. (1954) The freezing process of frost-hardy caterpillars. *Bull. entomol. Res.,* **45**, 329–339.

ATKINS, D. (1936) On the ciliary mechanisms and interrelationships of lamellibranchs I. Some new observations on sorting mechanisms in certain lamellibranchs. *Quart. J. micr. Sci.,* **79**, 181–308.

ATKINS, D. (1937a) On the ciliary mechanisms and interrelationships of lamellibranchs II. Sorting devices on the gills. *Quart. J. micr. Sci.,* **79**, 339–73.

ATKINS, D. (1937b) On the ciliary mechanisms and interrelationships of lamellibranchs III. Types of lamellibranch gills and their food currents. *Quart. J. micr. Sci.,* **79**, 375–421.

ATKINS, D. (1938) On the ciliary mechanisms and interrelationships of lamellibranchs VII. Latero-frontal cilia of the gill filaments and their phylogenetic value. *Quart. J. micr. Sci.,* **80**, 345–436.

ATKINS, D. (1943) On the ciliary mechanisms and interrelationships of lamellibranchs VIII. Notes on gill musculature in Microciliobranchia. *Quart. J. micr. Sci.,* **84**, 187–256.

ATKINSON, D. E. (1966) Regulation of enzyme activity. *Ann. Rev. Biochem.,* **35**, (1) 85–124.

ATKINSON, D. E. and WALTON, G. M. (1967) Adenosine triphosphate conservation in metabolic regulation. Rat liver citrate cleavage enzyme. *J. biol. Chem.,* **242**, 3239–3241.

AYERS, J. C. (1938) Relationship of habitat to oxygen consumption by certain estuarine crabs. *Ecology* **19**, 523–527.

BADER, R. G., HOOD, D. W., and SMITH, J. B. (1960) Recovery of dissolved organic matter in seawater and organic sorption by particulate material. *Geochim, cosmochim. Acta.,* **19**, 236–243.

BADOUIN, R. (1939) Sur l'habitat d'*Aëpophilus bonnairei* Signoret en doux stations des côtes de France. *Bull. Soc. zool. Fr.,* **64**, 18.

BADOUIN, R. (1947) Contributions to the ecology of *Aëpophilus bonnairei* Signoret and of certain other air-breathing Arthropoda in the intertidal zone. *Bull. Soc. zool. Fr.* **71**, 109.

BAERENDS, G. P. (1956) Aufbau des tierischen Verhaltens. *Handbuch der Zoologie.* Ed. W. Kukenthal, de Gruyter, Berlin, **8**, 1–32.

BAIER, C. R. (1935) Studien zur Hydrobakteriologie stehender Binnengewässer. *Arch. Hydrobiol.* **29**, 183–264.

BAINBRIDGE, R., and WATERMAN, T. H. (1957) Polarised light and the orientation of two marine Crustacea. *J. exp. Biol.*, **34**, 342–364.

BAINBRIDGE, R., and WATERMAN, T. H. (1958) Turbidity and the polarised light orientation of the crustacean *Mysidium. J. exp. Biol.*, **35**, 487–493.

BAKER, S. M. (1909) On the causes of the zoning of brown seaweeds on the seashore. *New. Phytol.*, **8**, 196–202.

BAKER, S. M. (1910) On the causes of zoning of brown seaweeds on the seashore II. The effect of periodic exposure on the expulsion of gametes and on the germination of the oospore. *New Phytol.*, **9**, 54–67.

BALDWIN, E. (1963) *Dynamic aspects of Biochemistry.* 4th ed. Cambridge University Press, London, 554 pp.

BALLANTINE, W. J. (1961) A biologically-defined exposure scale for the comparative description of rocky shores. *Field Studies,* **1**, 19 pp.

BALLANTINE, D. and MORTON, J. E. (1956) Filtering, feeding and digestion in the lamellibranch *Lasaea rubra. J. mar. biol. Ass. U.K.*, **35**, 241–74.

BARBER, S. B. (1961) Chemoreception and thermoreception. *The Physiology of Crustacea* **2**, Ed. T. H. Waterman, Academic Press, New York and London, 109–131.

BARCROFT, J. (1932) "La fixité du milieu intérieur est la condition de la vie libre" (Claude Bernard). *Biol. Rev.*, **7**, 24–87.

BARCROFT, J. (1934) *Features in the architecture of physiological function.* Cambridge Unversity Press, London.

BARCROFT, J. and BARCROFT, H. (1924) The blood pigment of *Arenicola. Proc. roy. Soc. B.*, **96**, 28–42.

BARD, P. (Ed) (1961) *Medical Physiology.* 11th ed. Mosby St. Louis, Missouri, 1339 pp.

BARKMAN, J. J. (1955) On the distribution and ecology of *Littorina obtusata* (L) and its subspecific units. *Arch. Néerl. Zool.* **11**, (1), 22–86.

BARNES, H. (1959) Stomach content and micro-feeding of some common cirripedes. *Canad. J. Zool.*, **37**, 231–236.

BARNES, H. and BARNES, M. (1957) Resistance to desiccation in intertidal barnacles. *Science,* **126**, 358.

BARNES, H. and BARNES, M. (1958) Note on the opening response of *Balanus balanoides* (L) in relation to salinity and certain inorganic ions. *Veröff. Inst. f. Meeresforsch., Bremerhaven,* **5**, 160–164.

BARNES, H. and BARNES, M. (1959) Studies on the metabolism of cirripedes. The relation between body weight, oxygen uptake and species habitat. *Veröff. Inst. f. Meeresforsch., Bremerhaven* **6**, 515–523.

BARNES, H. and BARNES, M. (1964) Some relations between the habitat, behaviour and metabolism on exposure to air of the high-level intertidal cirripede *Chthamalus stellatus* (Poli). *Helgol. wiss. Meeresunters,* **10**, 19–28.

BARNES, H., BARNES, M., and FINLAYSON, D. M. (1963a) The seasonal changes in body weight, biochemical composition and oxygen uptake of two common boreo-arctic cirripedes, *Balanus balanoides* (L) and *Balanus balanus* (L). *J. mar. biol. Ass. U.K.*, **43**, 185–211.

BARNES, H., BARNES, M., and FINLAYSON, D. M. (1963b) The metabolism during starvation of *Balanus balanoides, J. mar. biol. Ass. U.K.*, **43**, 213–233.

BARNES, H., CRISP, D. J., and POWELL, H. T. (1951) Observations on the orientation of some species of barnacles. *J. anim. Ecol.*, **20**, 227–241.

BARNES, H., FINLAYSON, D. M. and PIATIGORSKY, J. (1963) The effect of desiccation and anaerobic conditions on the behaviour, survival and general metabolism of three common cirripedes. *J. anim. Ecol.*, **32**, 233–252.

BARNES, H. and POWELL, H. T. (1950a) Some observations on the effect of fibrous glass surfaces upon the settlement of certain sedentary marine organisms. *J. mar. biol. Ass. U.K.*, **29**, 299–302.

BARNES, H. and POWELL, H. T. (1950b) The development, general morphology and subsequent elimination of barnacle populations, *Balanus crenatus* and *Balanus balanoides*, after a heavy initial settlement. *J. anim. Ecol.*, **19**, 175–179.

BARNES, H. and POWELL, H. T. (1953) The growth of *Balanus balanoides* (L) and *B. crenatus* Brug. under varying conditions of submersion. *J. mar. biol. Ass. U.K.*, **32**, 107–127.

BARNES, H. and POWELL, H. T. (1954) *Onchidoris fusca* (Muller) a predator of barnacles. *J. anim. Ecol.*, **23**, 361–363.

BARNES, H. and REESE, E. S. (1959) Feeding in the pedunculate cirripede *Pollicipes polymerus* J. B. Sowerby. *Proc. zool. Soc. Lond.* **132**, 569–585.

BARNES, J. H. (1966) Studies on three venomous cubomedusae. *The Cnidaria and their evolution* Ed. W. J. Rees *Symp. zool. Soc. Lond.* (*16*) Academic Press London and New York, 307–331.

BARNES, T. C. (1937) *Textbook of General Physiology*. Blakiston, Philadelphia.

BARRINGTON, E. J. W. (1967) *Invertebrate structure and function.* Nelson, London, 549 pp.

BASCOM, W. N. (1951) The relation between sand size and beach face slope. *Trans. Amer. Geophys. Union.*, **32**, (6) 866–874.

BASLOW, M. H. and NIGRELLI, R. F. (1964) Effects of thermal acclimation on brain cholinesterase activity of the killifish, *Fundulus heteroclitus. Zoologica*, **49**, 41–51.

BATHAM, E. J. (1945) *Pollicipes spinosus* Quoy and Gaimard. I. Notes on the biology and anatomy of the adult barnacle. *Proc. roy. Soc. N.Z.*, **74**, 359–374.

BAYLOR, E. R. and SMITH, F. E. (1953) The orientation of Cladocera to polarised light. *Am. Nat.*, **87**, (833) 97–101.

BAYNE, B. L. (1969) The gregarious behaviour of the larvae of *Ostrea edulis* L. at settlement. *J. mar. biol. Ass. U.K.*, **49**, 327–356.

BEADLE, L. (1961) Adaptations of some aquatic animals to low oxygen levels and anaerobic conditions. Mechanisms of Biological Competition. *Symp. Soc. exp. Biol.*, **15**, 120–131.

BEAMISH, F. W. H. (1964) Respiration of fishes with special emphasis on standard oxygen consumption II. Influence of weight and temperature on respiration of several species. *Canad. J. Zool.*, **42**, 177–188.

BEAMISH, F. W. H. and MOOKHERJII, P. S. (1964) Respiration of fishes with special emphasis on standard metabolism I. Influence of weight and temperature on respiration of goldfish *Carassius auratus* L. *Canad. J. Zool.*, **42**, 161–175.

BEANLAND, F. L. (1940) Sand and mud communities in the Dovey estuary. *J. mar. biol. Ass. U.K.*, **24**, 589–611.

BĚLEHRÁDEK, J. (1930) Temperature coefficients in biology. *Biol. Rev.*, **5**, 30–58,

BĚLEHRÁDEK, J. (1935) Temperature and living matter. *Protoplasma Monographien*, **8**, Berlin.

BERG, T. and STEEN, J. B. (1965) Physiological mechanisms for aerial respiration in the eel. *Comp. Biochem. Physiol.*, **15**, 469–484.

BERG, T. and STEEN, J. B. (1966) Regulation of ventilation in eels exposed to air. *Comp. Biochem. Physiol.*, **18**, 511–516.

BERRILL, N. J. (1950) *The tunicata, with an account of the British species.* Ray Society of London.

BERTALANFFY, L. von. (1957) Quantitative laws in metabolism and growth. *Quart. Rev. Biol.*, **32**, 217–231.

BISHOP, M. W. H. (1947) Establishment of an immigrant barnacle in British coastal waters. *Nature, Lond.*, **159**, 501.

BLACKMAN, F. F. (1905) Optima and limiting factors. *Ann. Bot.*, **19**, 281–295.

BLEGVAD, H. (1914) Undersøgelser over Naering og Ernaeringsforhold hos Havbundens invertebrate Dyresamfund i danske Farvande. *Ber. danske biol. Sta.*, **22**, 37–132.

BLOK, J. W. de and GEELEN, H. J. F. M. (1958) The substratum required for the settling of mussels (*Mytilus edulis* L.). *Arch. néerl. zool.* **13**, 446–460.

BOHLIN, K. (1897) Zur Morphologie and Biologie einzelliger Algen. *Öfvers, Kongl. Vedensk. Akad.*, **6**, 507–539.

BONE, Q. (1958) Nervous control of cilia in *Amphioxus* (*Branchiostoma*). *Nature, Lond.*, **181**, 193.

BONE, Q. (1961) The organisation of the atrial nervous system of *Amphioxus* (*Branchiostoma*) *lanceolatum* (Pallas). *Phil. Trans. roy. Soc. B.*, **243**, 241–269.

BONEY, A. D. (1966) *A biology of marine algae.* Hutchinson, London, 216 pp.

BORDEN, M. A. (1931) A study of the respiration and of the function of haemoglobin in *Planorbis corneus* and *Arenicola marina. J. mar. biol. Ass. U.K.*, **17**, 709–738.

BOSSANYI, J. (1957) A preliminary survey of the small natant fauna in the vicinity of the sea floor off Blyth, Northumberland. *J. anim. Ecol.*, **26**, 353–368.

BOUCHET, C. (1961) Le contrôle de la discharge nématocystique chez Phydre. *C. r. hebd. Seanc. Acad. Sci., Paris,* **252**, 327–328.

BOVBJERG, R. V. (1953) Dominance order in the crayfish *Orconectes virilis* (Hagen). *Physiol. Zoöl.*, **26**, 173–178.

BRADY, F. (1943) The distribution of the fauna of some intertidal sands and muds on the Northumberland coast. *J. anim. Ecol.*, **12**, 27–41.

BRAFIELD, A. E. (1963) The effects of oxygen deficiency on the behaviour of *Macoma balthica* (L). *Anim. Behav.* **11**, 345–346.

BRAFIELD, A. E. (1964) The oxygen content of interstitial water in sandy shores. *J. anim. Ecol.*, **33**, 97–116.

BRAFIELD, A. E. and NEWELL, G. E. (1961) The behaviour of *Macoma balthica* (L). *J. mar. biol. Ass. U.K.*, **41**, 81–87.

BRAMBELL, F. W. R. and COLE, H. A. (1939) *Saccoglossus cambriensis* sp. n., a new enteropneust occurring in Wales. *Proc. zool. Soc. Lond.*, **109**, 211–236.

BREGENZER, A. (1916) Anatomie und Histologie von *Bythinella dunkeri. Zool Jb.* (*Anat. Ont.*), **39**, 237–292.

BRENNER, W. (1916) Strandzoner i Nylands skärgård. *Bot. Notiser for ar* 1916.

BRETT, J. R. (1944) Some lethal temperatures of Algonquin Park fishes. *Publ. Ont. Fish. Res. Lab.*, **63**, 1–49.

BRETT, J. R. (1962) Some considerations in the study of respiratory metabolism in fish, particularly salmon. *J. Fish, Res. Bd. Canada*, **19**, 1025–1038.

BRETT, J. R. (1963) The energy required for swimming by young sockeye salmon with a comparison of the drag force on dead fish. *Trans. roy. Soc. Canada IV,* **1**, 441–457.

BRODY, S. (1945) *Bioenergetics and Growth.* Hafner, New York, 1023 pp.

BRODY, S. and PROCTER, R. C. (1932) Relation between basal metabolism and mature body weight in different species of mammals and birds. *Missouri Agr. Exp. Sta. Res. Bull.*, **166**, 89–101.

BROEKHUYSEN, C. J. (1940) A preliminary investigation of the importance of desiccation, temperature and salinity as factors controlling the vertical distribution of certain marine gastropods in False Bay, South Africa. *Trans. roy. Soc. S. Africa,* **28**, 255–292.

BROWN, F. A. (1960) Response to pervasive geophysical factors and the biological clock problem. *Cold Spring. Harb. Symp. Quant. Biol.*, **25**, 57.

BROWN, F. A., BENNETT, M. F. and WEBB, H. M. (1954) Persistent daily and tidal rhythms of oxygen consumption in fiddler crabs. *J. cell. comp. Physiol.*, **44**, 477–505.

BROWN, F. A., BENNETT, M. F. and WEBB, H. M. (1960) An organismic magnetic compass response. *Biol. Bull. Wood's Hole* , **119**, 65–74.

BROWN, F. A., BRETT, W. J., BENNETT, M. F. and BARNWELL, F. H. (1960) Magentic response of an organism and its solar relationships. *Biol. Bull. Wood's Hole,* **118**, 367--381.

BROWN, F. A., FREELAND, M. and RALPH, C. L. (1955) Persistent rhythms of oxygen consumption in potatoes, carrots and a seaweed *Fucus. Plant Physiol.*, **30**, 280–292.

BROWN, F. A. and SANDEEN, M. I. (1948) Responses of the chromatophores of the fiddler crab, *Uca*, to light and temperature *Physiol. Zoöl.*, **21**, 361–371.

BROWN, F. A., WEBB, H. M. and BENNETT, M. F. (1958) Comparison of some fluctuations in cosmic radiation and organismic activity during 1954, 1955 and 1956. *Amer. J. Physiol.*, **195**, 237.

BROWN, F. A., WEBB, H. M., BENNETT, M. F. and SANDEEN, M. I. (1955) Evidence for an exogenous contribution to persistent diurnal and lunar rhythmicity under so-called constant conditions. *Biol. Bull Wood's Hole*, **109**, 238–254.

BROWN, F. A., WEBB, H. M. and BRETT, W. J. (1960) Magnetic response of an organism and its lunar relationships. *Biol. Bull. Wood's Hole*, **118**, 382–392.

BROWN, H. T. and ESCOMBE, F. (1900) Static diffusion of gases and liquids in relation to the assimilation of carbon and translocation in plants. *Phil. Trans. roy. Soc. B*, **193**, 233–291.

BRUCE, J. R. (1928a) Physical factors on the sandy beach I. Tidal, climatic and edaphic. *J. mar. biol. Ass. U.K.*, **15**, (2) 535–552.

BRUCE, J. R. (1928b) Physical factors on the sandy beach II. Chemical changes. *J. mar. biol. Ass. U.K.*, **15**, 553–565.

BUCHANAN, J. B. (1966) The biology of *Echinocardium cordatum* (Echinodermata; Spatangoidea) from different habitats. *J. mar. biol. Ass. U.K.* **46**, 97–114.

BUDDENBROCK, W. von. (1938) Einige Beobachtungen über die Tatigkeit der Wasserlungen der Holothurien. *z. vergl. Physiol.*, **26**, 303.

BUDDENBROCK, W. von. (1956) *Vergleichende Physiologie III. Ernährung Wasserhaushalt und Mineralhaushalt der Tiere*, Birkhäuser, Basel.

BUDINGTON, R. A. (1937) The normal spontaneity of movement of the respiratory muscles of *Thyone briareus. Physiol. Zoöl.*, **10**, 141.

BULEY, H. M. (1936) Consumption of diatoms and dinoflagellates by the mussel. *Bull. Scripps Instn. Oceanogr. tech.*, **4**, 19–27.

BULLOCK, T. H. (1951) To what degree are aquatic and cold-bloods independent of temperature? Cited by Rao and Bullock (1954) from ms on *Factors in distribution of intertidal organisms.* Western Society of Naturalists, Claremont, California.

BULLOCK, T. H. (1955) Compensation for temperature in the metabolism and activity of poikilotherms. *Biol. Rev.*, **30**, 311–342.

BURDEN-JONES, C. (1952) Development and biology of the larva of *Saccoglossus horsti* (Enteropneusta). *Phil Trans. roy. Soc. B*, **236**, 553–590.

BURDEN-JONES, C. and CHARLES, G. H. (1958a) Light reactions of littoral gastropods. *Nature, Lond.*, **181**, 129–131.

BURDEN-JONES, C. and CHARLES, G. H. (1958b) Light responses of littoral gastropods. *Proc. XV Int. Congr. Zool. Lond.*, 889–891.

BURNETT, A. L., LENTZ, T. and WARREN, M. (1960) The nematocyst of *Hydra* Part I. The question of control of the nematocyst discharge reaction by fully-fed *Hydra. Annls. Soc. r. zool. malacol. Belg.*, **90**, 247–267.

BURROWS, E. M., CONWAY, E., LODGE, S. M. and POWELL, H. T. (1954) The raising of intertidal algal zones on Fair Isle. *J. Ecol.*, **42**, (2) 283–288.

BURTON, A. C. (1936) The basis of the principle of the master reaction in biology. *J. Cell. Comp. Physiol.*, **9**, 1–14.

CALLAME, B. (1961) Contribution a l'étude du milieu meuble intercotidal (Côtes Charentaises). *Trav. Cent. Rech. Etud. Oceanogr.*, **4**, 116 pp.

CARLGREN, O. (1940) A contribution to the knowledge of the structure and distribution of the cnidae in the Anthozoa. Acta Univ. Lund., **36**, 1–62.

CARRIKER, M. R. (1955) Critical review of the biology and control of the oyster drills *Urosalpinx* and *Eupleura. Spec. Sci. Rep. U.S. Dept. Inst. Fish.*, **148**, 1–150.

CARRIKER, M. R. (1958) Additional information on the mechanical-chemical nature of drilling by the gastropods *Urosalpinx* and *Eupleura. Assoc. South-Eastern Biologists Bull.*, **5**, 5.

CARRIKER, M. R. (1959) Comparative functional morphology of the drilling mechanism in *Urosalpinx* and *Eupleura* (Muricid Gastropods) *Proc. XVth Int.*

Congr. Zool. London, 1958, 373–376.

CARRIKER, M. R. (1961) Comparative functional morphology of boring mechanisms in gastropods. *Am. Zoologist*, **1**, 263–266.

CARTHY, J. D. (1957a) *An introduction to the behaviour of invertebrates*, Allen and Unwin, London, 380 pp.

CARTHY, J. D. (1957b) Polarised light and animals. *Discovery, March 1957*, 105–109.

CASPERS, H. (1940) Über Nahrungswerb und Darmverlauf bei *Nucula. Zool. Anz.*, **129**, 48–55.

CHAMBERS, R. and HALE, H. P. (1932) The formation of ice in protoplasm. *Proc. roy. Soc. Lond. B*, **110**, 336–352.

CHAPMAN, G. (1949) The thixotropy and dilatancy of a marine soil. *J. mar. biol. Ass. U.K.*, **28**, (1) 123–140.

CHAPMAN, G. and NEWELL, G. E. (1947) The role of the body fluid in relation to movement in soft-bodied invertebrates. I. The burrowing of *Arenicola. Proc roy. Soc. B*, **134**, 431–455.

CHAPMAN, G. B. (1961) The fine structure of the stenoteles of *Hydra. The Biology of Hydra* pp. 131–151. Eds. H. M. Lenhoff and W. F. Loomis, University of Miami Press, Florida.

CHARLES, G. H. (1961) The orientation of *Littorina* species to polarised light. *J. exp. Biol.*, **38**, 189–202.

CHARLES, G. H. (1966) Sense organs (less Cephalopods). *The Physiology of Mollusca* **2**, Eds. K. M. Wilbur and C. M. Yonge, Academic Press. New York and London, 455–519.

CHIBA, K. and OSHIMA, Y. (1957) Effects of suspended particles on the pumping of marine bivalves, especially of the Japanese neck-clam. *Bull. Jap. Soc. Sci. Fish.*, **23**, 348–353.

CHIPMAN, W. A. and HOPKINS, J. G. (1954) Water filtration by the bay scallop, *Pecten irradians*, as observed with the use of radioactive plankton. *Biol. Bull. Wood's Hole*, **107**, 80–91.

CHRISTOPHERSEN, J. (1967) Adaptive temperature responses of micro-organisms. Molecular mechanisms of temperature adaptation. Ed. C. L. Prosser, *Amer. Assoc. adv. Sci. Symp. (84)*, Washington D.C., 327–348.

CHRISTOPHERSEN, J. and PRECHT, H. (1952a) Untersuchungen uber die Bedeutung des Wassergehaltes von Hefezellen für Temperaturanpassungun. *Arch. Mikrobiol.*, **18**, 32–48.

CHRISTOPHERSEN, J. and PRECHT, H. (1952b) Untersuchungen zum Problem der Hitzresistenz I. Versuche an Karauschen (*Carassius vulgaris* Nils.) *Biol. Zbl.*, **71**, 313–326.

CHRISTOPHERSEN, J. and PRECHT, H. (1953) Die Bedeutung des Wassergehaltes der Zelle für Temperaturanpassungen. *Biol. Zentralblatt.*, **72**, 104–119.

CLARK, A. J. (1927) *Comparative physiology of the heart.* Cambridge University Press.

CLARK, D. P. (1955) The influence of body weight, temperature and season upon the rate of oxygen consumption of the terrestrial amphipod *Talitrus sylvaticus. Biol. Bull. Wood's Hole*, **108**, 253–257.

CLARK, M. E. (1968) *The ecology of supralittoral rockpools with special reference to the copepod fauna.* Ph.D. thesis. University of Aberdeen, Scotland.

CLARK, R. B. (1956) The eyes and photonegative behaviour of *Nephtys* (Annelida Polychaeta). *J. exp. Biol.*, **33** (3) 461–477.

CLARK, R. B. (1960a) Habituation of the polychaete *Nereis* to sudden stimuli I. General properties of the habituation process. *Anim. Behav.* **8**, 82–91.

CLARK, R. B. (1960b) Habituation of the polychate *Nereis* to sudden stimuli II. Biological significance of habituation. *Anim. Behav.*, **8**, 92–103.

CLARK, R. B. (1962) Observations on the food of *Nephtys. Limnol. and Oceanogr.*, **7**, (3) 380–385.

CLARK, R. B. (1964) The learning abilities of nereid polychaetes and the role of the supra-oesophageal ganglion. Learning and associated phenomena in invertebrates. *Anim. Behav. Suppl. I,* 89–99.

COE, W. R. (1948) Nutrition, environmental conditions, and growth of marine bivalve molluscs. *J. mar. Res.,* **7,** 586–601.

COHEN, M. J. and DIJKGRAAF, S. (1961) Mechanoreception. *The Physiology of Crustacea.* Ed. T. H. Waterman, Academic Press, New York and London, **2,** 65–108.

COHEN, P. P. and BROWN, G. W. Jr. (1960) Ammonia metabolism and urea biosynthesis. Comparative Biochemistry, **2,** Eds. M. Florkin and H. S. Mason, Academic Press, London and New York, 161–244.

COLE, H. A. and KNIGHT-JONES, E. W. (1939) Some observations and experiments on the setting behaviour of larvae of *Ostrea edulis. J. Cons. int. Explor. Mer.,* **14,** 86–105.

COLE, H. A. and KNIGHT-JONES, E. W. (1949) The setting behaviour of larvae of the European flat oyster *Ostrea edulis* L., and its influence on methods of cultivation and spat collection. *Fish. Invest. Ser., 2,* **17** (3) 1–39.

COLGAN, N. (1910) Notes on the adaptability of certain littoral molluscs. *Irish Nat.,* **19,** 127–133.

COLLARDEAU, C. (1961) Influence de la température sur la consommation d'oxygène de quelques larves de Trichoptères. *Hydrobiologia,* **18,** (3) 252–264.

COLLARDEAU-ROUX, C. (1966) Influence de la température sur la consommation d'oxygène de *Micropterna testacea* (Gmel) (Trichoptera, Limnophilidae). *Hydrobiologia,* **27,** 385–394.

COLMAN, J. (1933) The nature of the intertidal zonation of plants and animals. *J. mar. biol. Ass. U.K.* **18,** (2) 435–476.

COLMAN, J. (1940) On the faunas inhabiting intertidal seaweeds. *J. mar. biol. Ass. U.K.,* **24,** 129–183.

COLMAN, J. S. and SEGROVE, F. (1955) The fauna living in Stoupe Beck sands, Robin Hood's Bay (Yorkshire, North Riding). *J. anim. Ecol.,* **24,** (2) 426–444.

COLMAN, J. S. and STEPHENSON, A. (1966) Aspects of the ecology of a "tideless" shore. *Some Contemporary studies in Marine Science.* Ed. H. Barnes, George, Allen, Unwin and Lond. 163–170.

CONWAY, E. (1946) Browsing of *Patella. Nature, Lond.,* **158,** 752.

COONFIELD, B. R. (1931) The cilia of *Nephthys bucera. Proc. Nat. Acad. Sci. Wash.* **17,** 416.

CORNELIUS, P. F. S. (1968) *Activity of intertidal animals in relation to position on the shore.* Ph.D Thesis, University of London.

COURTNEY, W. A. M. and NEWELL, R. C. (1965) Ciliary activity and oxygen uptake in *Branchiostoma lanceolatum* (Pallas). *J. exp. Biol.,* **43,** 1–12.

COURTNEY, W. A. M. and WEBB, J. E. (1964) The effect of the cold winter 1962/3 on the Helgoland population of *Branchiostoma lanceolatum* Pallas. *4th Mar. Biol. Symp. Helgol. Helgol. wiss. Meeresunters* **10,** 301–312.

CRISP, D. J. (1955) The behaviour of barnacle cyprids in relation to water movement over a surface. *J. exp. Biol.,* **32,** 569–590.

CRISP, D. J. (1958) The spread of *Elminius modestus* Darwin in northwest Europe. *J. mar. biol. Ass. U.K.,* **37,** 483–520.

CRISP, D. J. (1961) Territorial behaviour in barnacle settlement. *J. exp. Biol.,* **38,** 429–446.

CRISP, D. J. (1964) The effects of the severe winter of 1962–63 on marine life in Britain. *J. anim. Ecol.,* **33,** 165–210.

CRISP, D. J. and AUSTIN, A. P. (1960) The action of copper in antifouling paints. *Ann. appl. Biol.,* **48,** 787–799.

CRISP, D. J. and BARNES, H. (1954) The orientation and distribution of barnacles at settlement, with particular reference to surface contour. *J. anim. Ecol.,* **23,** 142–162.

CRISP, D. J. and MEADOWS, P. S. (1962) The chemical basis of gregariousness in cirripedes. *Proc. roy. Soc. B.,* **156,** 500–520.

CRISP, D. J. and MEADOWS, P. S. (1963) Adsorbed layers; the stimulus to settlement in barnacles. *Proc. roy. Soc. B.,* **158,** 364–387.

CRISP, D. J. and RITZ, D. A. (1967a) Changes in temperature tolerance of

Balanus balanoides during its life-cycle. *Helgol. wiss. Meeresunters,* **15,** 98–115.

CRISP, D. J. and RITZ, D. A. (1967b) Temperature acclimation in barnacles. *J. exp. mar. Biol. Ecol.,* **1,** 236–256.

CRISP, D. J. and RYLAND J. S. (1960) Influence of filming and of surface texture on the settlement of marine organisms. *Nature, Lond.,* **185,** 119.

CRISP, D. J. and SOUTHWARD, A. J. (1961) Different types of cirral activity of barnacles. *Phil. Trans. roy. Soc. B.,* **243,** (705) 271–308.

CRISP, D. J. and WILLIAMS, G. B. (1960) Effects of extracts from fucoids in promoting settlement of epiphytic polyzoa. *Nature Lond.,* **185,** 1206–1207.

CROZIER, W. J. (1924) On biological oxidations as a function of temperature. *J. gen. Physiol.,* 7, 189–216.

CUTRESS, C. (1955) An interpretation of the structure and distribution of cnidae in the Anthozoa. *Syst. Zool.,* **4,** 129–137.

DAHL, E. (1948) On the smaller Arthropoda of marine algae, especially in the polyhaline waters off the Swedish west coast. *Undersökninger över Öresund,* **35,** Lund.

DAKIN, W. J. (1928) The eyes of *Pecten, Spondylus, Amussium* and allied lamellibranchs, with a short discussion on their evolution. *Proc. roy. Soc. B.,* **103,** 355–365.

DALES, R. P. (1955) Feeding and digestion in terebellid polychaetes. *J. mar. biol. Ass. U.K.,* **34,** 55–79.

DALES, R. P. (1957) The feeding mechanism and morphology of the gut of *Owenia fusiformis* delle Chiaje. *J. mar. biol. Ass. U.K.,* **36,** 81–89.

DALES, R. P. (1958) Survival of anaerobic periods by two intertidal polychaetes, *Arenicola marina* (L) and *Owenia fusiformis* Delle Chiaje. *J. mar. biol. Ass. U.K.,* **37,** (2) 521–529.

DALES, R. P. (1961) Oxygen uptake and irrigation of the burrow by three terebellid polychaetes: *Eupolymnia, Thelepus* and *Neoamphitrite. Physiol. Zool.,* **34,** (4) 306–311.

DALES, R. P. (1963) *Annelids.* Hutchinson, London, 200 pp.

DALYELL, J. (1853) *Powers of the Creator.* London, **1,** 191–192.

DAM, L. van (1935) On the utilisation of oxygen by *Mya arenaria. J. exp. Biol.,* **12,** 86–94.

DAM, L. van (1937) Über die Atembewegungen und das Atemvolumen von *Phyganea*-Larven, *Arenicola marina,* und *Nereis virens,* sowie über der Sauerstoffausnutzung bei *Anodonta cygnea, Arenicola marina* und *Nereis virens. Zool. Anz.,* **118,** 122–128.

DAM, L. van (1938) *On the utilisation of oxygen and the regulation of breathing in some aquatic animals. (Dissertation).* Gröningen, Drukkerij "Volharding", 143 pp.

DAM, L. van (1940) On the mechanism of ventilation in *Aphrodite aculeata. J. exp. Biol.* **17,** 1–7.

DAM, L. van (1954) On the respiration in scallops (Lamellibranchia). *Biol. Bull. Wood's Hole* **107,** 192–202.

DANIEL, A. (1957) Illumination and its effect on the settlement of barnacle cyprids *Proc. zool. Soc. Lond.* **129,** 305–313.

DANIELS, J. M. and ARMITAGE, K. B. (1969) Temperature acclimation and oxygen consumption in *Physa, Lawnii* Lea (Gastropoda, Pulmonata). *Hydrobiologia,* **33,** (1) 1–13.

DARNELL, R. M. (1964) Organic detritus in relation to secondary production in aquatic communities. *Verh. Internat. Verein. Limnol.,* **15,** 462–470.

DARNELL, R. M. (1967a) The organic detritus problem. *Estuaries. Amer. Assoc. adv. Sci.,* 374–375.

DARNELL, R. M. (1967b) Organic detritus in relation to the estuarine ecosystem. *Estuaries, Amer. Assoc. adv. Sci. Symp. (83) Washington, D.C.,* 376-382.

DARWIN, C. (1854) A monograph on the sub-class Cirripedia with figures of all the species. London.

DAS, A. B. (1965) *Protein and RNA synthesis in thermal acclimation of goldfish tissue.* Ph.D Thesis, University of Illinois, Urbana.

DAVENPORT, D., ROSS, D. M. and SUTTON, L. (1961) The remote control of nematocyst discharge in the attachment of *Calliactis parasitica* to shells of hermit crabs. *Vie Milieu,* **12,** 197–209.

DAVID, H. M. (1943) Studies on the autecology of *Ascophyllum nodosum* Le Jol. *J. Ecol.* **31,** 178–199.

DAVIES, P. S. (1965) Environmental acclimation in the limpet *Patella vulgata* L. *Nature, Lond.* **205** (4974) 924.

DAVIES, P. S. (1966) Physiological ecology of *Patella* I. The effect of body size and temperature on metabolic rate. *J. mar. biol. Ass. U.K.,* **46,** 647–658.

DAVIES, P. S. (1967) Physiological ecology of *Patella II.* Effect of environmental acclimation on the metabolic rate. *J. mar. biol. Ass. U.K.,* **47,** 61–74.

DAVIES, P. S. (1969) Physiological ecology of *Patella* III. Desiccation effects. *J. mar. biol. Ass. U.K.,* **49,** 291–304.

DAVIS, H. C. (1953) On food and feeding of larvae of the American oyster, *C. virginica.* Biol. Bull. Woods Hole **104,** 334–350.

DAY, J. H. and WILSON, D. P. (1934) On the relation of the substratum to the metamorphosis of *Scolecolepis fuliginosa* (Claparède). *J. mar. biol. Ass. U.K.* **19,** 655–661.

DEAN, J. M. (1969) The metabolism of tissues of thermally acclimated trout (*Salmo gairdneri*). *Comp. Biochem. Physiol.* **29,** 185–196.

DEHNEL, P. A. (1955) Rate of growth of gastropods as a function of latitude. *Physiol. Zoöl.,* **28,** 115–144.

DEHNEL, P. A. (1958) Effect of photoperiod of the oxygen consumption of two species of intertidal crabs. *Nature,Lond.,* **181,** 1415–1417.

DENNELL, R. (1933) The habitats and feeding mechanism of the amphipod *Haustorius arenarius* Slabber. *J. Lin. Soc. (zool.)* **38,** 363–388.

DENNIS, M. J. (1965) Lateral inhibition in a simple eye. *Am. Zool.,* **5,** 651.

DENNIS, M. J. (1967) Interactions between the fine receptor cells of a simple eye. *Invertebrate nervous systems.* Ed. C.A.G. Wiersma, University of Chicago Press.

DESAI, B. N. (1959) Ph.D Thesis, University of Wales. (cited by D.J. Crisp, 1964).

DESHPANDE, R. D. (1957) Ph.D. Thesis, University of Reading. (cited by Fretter, V and Graham, A, 1962).

DIEHL, M. (1956) Die Raubschnecke *Velutina velutina* das Feind und Bruteinmieter der Ascidie *Styela coriacea. Kieler Meeresforsch,* **12,** 180–185.

DIGBY, P. S. B. (1967) Pressure sensitivity and its mechanism in the shallow marine environment. *Symp. zool. Soc. Lond.,* (19) 159–188.

DINAMANI, P. (1964) Feeding in *Dentalium conspicuum. Proc. malac. Soc. Lond.,* **36,** 1–5.

DODGSON, R. W. (1928) Report on mussel purification. *Gt. Brit. Fishery Invest. Ser.,* 11, **10,** (1) 498 pp.

DONGEN, A. van. (1956) The preference of *Littorina obtusata* for Fucacae. *Arch. Néerl. Zool.,* **11,** (3) 373–386.

DOOCHIN, H. D. (1951) The morphology of *Balanus improvisus* Darwin and *Balanus amphitrite niveus* Darwin during initial attachment and metamorphosis. *Bull. Mar. Sci. Gulf Caribbean,* **1,** 15–39.

DOTY, M. S. (1946) Critical tide factors that are correlated with the vertical distribution of marine algae and other organisms along the Pacific coast. *Ecology,* **27,** 315–328.

DOUDOROFF, P. (1942) The resistance and acclimatisation of marine fishes to temperature changes I. Experiments with *Girella nigricans* (Ayres). *Biol. Bull. Wood's Hole,* **83,** 219–244.

DOUDOROFF, P. (1945) The resistance and acclimatisation of marine fishes to temperature changes. 2. Experiments with *Fundulus* and *Atherinops. Biol. Bull. Wood's Hole,* **88,** (2) 194–206.

DROOP, M. R. (1953) On the ecology of flagellates from some brackish and freshwater rockpools of Finland. *Acta. bot. Fenn.,* **51,** 1–52.

DUCHÂTEAU, G. and FLORKIN, M. (1954) Sur la composition de l'arthropodine et de la scléroprotéine cuticulaires de deux crustacés décapodes

(*Homarus vulgaris*, Edwards, *Callinectes sapidus* Rathbun). *Physiol. comp.*, **3**, 365–369.

DUERDEN, J. E. (1902) Aggregated colonies in madreporarian corals. *Am. Nat.*, **36**, 461–471.

DUGAL, L. P. (1939) The use of calcareous shell to buffer the product of anaerobic glycolysis in *Venus mercenaria*. *J. cell. comp. Physiol.*, **13**, 235–251.

—DUNCAN, A. (1966) The oxygen consumption of *Potamopyrgus jenkinsi* (Smith) (Prosobranchiata) in different temperatures and salinities. *Verh. int. Ver. Limnol.*, **16**, 1739–1751.

DUVAL, M. and PORTIER, P. (1927) Sur la teneur en gaz carbonique total du sang des invertébrés d'eau douce et des invertébrés marins. *C.R.Acad. Sci. Paris*, **184**, 1594–1596.

EAKIN, R. M. (1963) Lines of evolution of photoreceptors. *General physiology of cell specialisation*. Eds. D. Mazia and A. Tyler, McGraw-Hill, N.Y.

EAKIN, R. M. (1965) Evolution of photoreceptors. Cold Spring *Harb. Symp. quant. Biol.*, **30**, 363–370.

EBLING, F. J., SLOANE, J. F., KITCHING, J. A. and DAVIES, H. M. (1962) The ecology of Lough Ine XII. The distribution and characteristics of *Patella* species *J. Anim ecol.*, **37**, 457–470

EDMONDSON, C. H. (1929) Growth of Hawaiian corals. *Bull. Bernice P. Bishop Mus.* (58), 1–38.

EDNEY, E. B. (1951) The body temperature of woodlice. *J. exp. Biol.*, **28**, 271–280.

EDNEY, E. B. (1953) The temperature of woodlice in the sun. *J. exp. Biol.*, **30**, 331–349.

EDNEY, E. B. (1954) Woodlice and the land habitat. *Biol. Rev.*, **29**, 185–219.

EDNEY, E. B. (1957) *The water relations of terrestrial arthropods.* Cambridge University Press, London, 109 pp.

EDNEY, E. B. (1960) Terrestrial adaptations. *The physiology of Crustacea,* **1**, Ed. T. H. Waterman, Academic Press, New York and London, 367–363. 367–363.

EDNEY, E. B. (1961) The water and heat relationships of fiddler crabs (*Uca* spp). *Trans. roy. Soc. S. Afr.*, **36**, (2) 71–91.

EDNEY, E. B. (1962) Some aspects of the temperature relations of fiddler crabs (*Uca* spp). *Biometeorology*, 79–85.

EDNEY, E. B. and SPENCER, J. O. (1955) Cutaneous respiration in woodlice. *J. exp. Biol.*, **32**, 256–269.

– EDWARDS, G. A. (1946) The influence of temperature upon the oxygen consumption of several arthropods. *J. cell. comp. Physiol.* **27**, 53–64.

– EDWARDS, G. A. and IRVING, L. (1943a) The influence of temperature and season upon the oxygen consumption of the sand crab, *Emerita talpoida* Say. *J. cell. comp. Physiol.*, **21**, 169–182.

– EDWARDS, G. A. and IRVING, L. (1943b) The influence of season and temperature upon the oxygen consumption of the beach flea, *Talorchestia megalopthalma*. *J. cell. comp. Physiol.*, **21**, 183-189.

EIGENBRODT, H. (1941) Untersuchungen über die Funktion der Radula einiger Schnecken. *Z. Morph. Ökol. Tiere*, **37**, 735–791.

EISMA, D., DAS, H. A., HOEDE, D., van RAAPHORST, J. G. and ZONDERHUIS, J. (1966) Iron and trace elements in Dutch coastal sands. *Neth. J. Sea Res.*, **3**, 68-94.

– EKBERG, D. R. (1958) Respiration in tissues of goldfish adapted to high and low temperatures. *Biol. Bull. Wood's Hole*, **114**, 308–316.

– ELLENBY, C. (1951) Body size in relation to oxygen consumption and pleopod beat in *Ligia oceanica* L. *J. exp. Biol.*, **28**, 492–507.

ELLENBY, C. (1953) Oxygen consumption and cell size. A comparison of the rate of oxygen consumption of diploid and triploid prepupae of *Drosophila melanogaster* Meigen. *J. exp. Biol.*, **30**, 475–491.

– ELLENBY, C. and EVANS, D. A. (1956) On the relative importance of body weight and surface area measurements for the prediction of the level of oxygen consumption of *Ligia oceanica* L, and prepupae of *Drosophila melanogaster* Meig, *J. exp. Biol.*, **33**, 134–141.

ENDERS, H. E. (1909) A study of the life history and habits of *Chaetopterus variopedatus* Renier and Claparède. *J. Morphology,* **20** 479–532.

EVANS, F. (1961) Responses to disturbance of the periwinkle *Littorina punctata* (Gmelin) on a shore in Ghana. *Proc. zool. Soc. Lond.*, **137**, 393–402.

EVANS, F. (1965) The effect of light on the zonation of the four periwinkles *Littorina littorea* (L), *L. obtusata* (L) and *Melarapha neritoides* (L) in an experimental tidal tank. *Netherlands J. Sea Res.*, **2**, (4) 556–565.

EVANS, F. G. C. (1951) An analysis of the behaviour of *Lepidochitona cinereus* in response to certain physical features of the environment. *J. anim. Ecol.*, **20**, (1) 1–10.

EVANS, R. G. (1947a) The intertidal ecology of Cardigan Bay. *J. Ecol.*, **34**, 273–309.

EVANS, R. G. (1947b) The intertidal ecology of selected localities in the Plymouth neighbourhood. *J. mar. biol. Ass. U.K.*, **27**, 173–218.

EVANS, R. G. (1948) The lethal temperatures of some common British littoral molluscs. *J. anim. Ecol.*, **17**, (2) 165–173.

EVANS, R. M., PURDIE, F. C. and HICKMAN, C. P. (1962) Effect of temperature and photoperiod on respiratory metabolism of rainbow trout (*Salmo gairdneri*). *Can. J. Zool.*, **40**, 107–108.

EVANS, S. M. (1963a) The effect of brain extirpation on learning and retention in nereid polychaetes. *Anim. Behav.*, **11**, 172–178.

EVANS, S. M. (1963b) The behaviour of the polychaete *Nereis* in T-mazes. *Anim. Behav.*, **11**, 379–392.

EVANS, S. M. (1966a) Non-associative avoidance learning in nereid polychaetes. *Anim. Behav.*, **14**, 102–106.

EVANS, S. M. (1966b) Non-associative behavioural modifications in the polychaete *Nereis diversicolor. Anim. Behav.*, **14**, 107–119.

EWER, R. F. (1947) On the function and mode of action of the nematocysts of Hydra. *Proc. zool. Soc. Lond.*, **117**, 365–376.

EWER, R. F. and FOX, H. M. (1940) On the function of chlorocruorin. *Proc. roy. Soc. B,* **129**, 137–153.

FERRONIÈRE, G. (1901) Études biologique sur les zones supralittorales de la Loire-Inférieure. *Bull. Soc. Sci. Nat.*, **2**, (1) 1–451.

FISCHER-PIETTE, E. (1935) Histroire d'une moulière. *Bull. biol. France et Belg.*, **69**, (2) 152–177.

FISHER, R. A. and YATES, F. (1943) *Statistical tables for biological, agricultural and medical research.* Oliver and Boyd, London.

FLINT, P. (1965) The effect of sensory deprivation on the behaviour of the polychaete *Nereis* in T-mazes. *Anim. Behav.* **13**, 187–193.

FLORKIN, M. (1933) Recherches sur les hémérythrins. *Arch. int. Physiol.* **36**, 247–328.

FLORKIN, M. (1934) La fonction respiratoire du "milieu intérieur" dans la série animale. *Ann. Physiol. Physicochim. biol.* **10**, 599–684.

FLORKIN. M. (1949) *Biochemical evolution. Trans. S. Morgulis.* Academic Press, New York and London, 157 pp.

FORREST, J. E. (1953) On the feeding habits and the morphology and mode of functioning of the alimentary canal in some littoral dorid nudibranchiate Mollusca. *Proc. Linn. Soc. Lond.*. **164**, 225–236.

FORSMAN, B. (1951) Studies on *Gammarus duebeni* Lillj. with notes on some rockpool organisms in Sweden. *Zool. Bidr. Uppsala,* **29**, 215–236.

FOX, C. J. J. (1907) On the coefficients of absorption of the atmospheric gases in distilled water and in sea water, 1. Nitrogen and oxygen. *Pub. Circ. Cons. Explor. Mer* (41), 23 pp.

FOX, D. L. (1950) Comparative metabolism of organic detritus by inshore animals. *Ecology*, **31**, (1) 100–108.

FOX, D. L. and COE, W. R. (1943) Biology of the California sea mussel (*Mytilus californianus*) II. Nutrition, metabolism, growth and calcium deposition. *J. exp. Zool.*, **93**, 205–249.

FOX, D. L. ISAACS, J. D. and CORCORAN, E. F. (1952) Marine leptopel, its recovery, measurement and distribution. *J. mar. Res.*, **11**, 29–46.

FOX, H. M. (1932) The oxygen affinity of chlorocruorin. *Proc. roy. Soc. B.* **111**, 356–363.

– FOX, H. M. (1936) The activity and metabolism of poikilothermal animals in different latitudes I. *Proc. zool. Soc. Lond.*, (2), 945–955

FOX, H. M. (1938a) On the blood circulation and metabolism of sabellids. *Proc. roy. Soc. B.*, **125**, 554–569.

FOX, H.M. (1938b) The activity and metabolism of poikilothermal animals in different latitudes III. *Proc. zool. Soc. Lond.* **108**, 501–505.

FOX, H. M. (1939) The activity and metabolism of poikilothermal animals in different latitudes V. *Proc. zool. Soc. Lond.*, **109**, 141–156.

FOX, H. M. (1955) The effect of oxygen on the concentration of haem in invertebrates. *Proc. roy. Soc. B* **143**, 203–214.

– FOX, H. M. and TAYLOR, A. E. R. (1954) Injurious effect of air-saturated water on certain invertebrates. *Nature, Lond.*, **174**, 312.

– FOX, H. M. and TAYLOR, A. E. R. (1955) The tolerance of oxygen by aquatic invertebrates. *Proc. roy. Soc. B* **143**, 214–225

FOX, H. M. and VEVERS, G. (1960) *The nature of animal colours.* Sidgwick and Jackson, London, 246 pp.

– FOX, H. M. and WINGFIELD, C. A. (1937) The activity and metabolism of Poikilothermal animals in different latitudes H. *Proc. zool. Soc. Lond.*, **107**, 275–282.

FRAENKEL, G. (1927) Beiträge zur Geotaxis und Phototaxis von *Littorina. Z. vergl. Physiol.*, **5**, 585–597.

FRAENKEL, G. S. (1929) Uber die Geotaxis von *Convoluta roscoffensis. Z. wiss. Biol. Abt. C.Z. vergl. Physiol.*, **10**, 237–247.

– FRAENKEL, G. (1960) Lethal high temperatures for three marine invertebrates: *Limulus polyphemus, Littorina littorea* and *Pagurus longicarpus. Oikos,* **11**, (2) 171–182.

– FRAENKEL, G. S. (1961) Resistance to high temperatures in a Mediterranean snail, *Littorina neritoides. Ecology,* **42**, 604–606.

FRAENKEL, G. S. and GUNN, D. L. (1940) *The orientation of animals. Kineses, Taxes, and Light-compass reactions.* Oxford University Press, London and New York, 352 pp.

FRASER, J. H. (1932) Observations on the fauna and constituents of an estuarine mud in a polluted area. *J. mar. Biol. Ass. U.K.*, **18**, 589–611.

FRASER, J. H. (1936) The occurrence ecology and life history of *Tigriopus fulrus* (Fischer). *J. mar. biol. Ass. U.K.*, **20**, 523–536.

FREED, J. 1965. Changes in activity of cytochrome oxidase during adaptations of goldfish to different temperatures. *Comp. Biochem. Physiol.*, **14**, 651–659.

FREEMAN, J. A. (1950) Oxygen consumption, brain metabolism and respiratory movements of goldfish during temperature acclimatisation with special reference to lowered temperatures. *Biol. Bull. Wood's Hole,* **99**, 416–424.

FREEMAN, R. F. H. (1963) Haemoglobin in the digenetic trematode *Proctoeces subtenuis* (Linton). *Comp. Biochem. Physiol.*, **10**, 253–356.

FRETTER, V. (1951) Some observations on British cypraeids. *Proc. malac. Soc. Lond.*, **29**, 14–20.

FRETTER, V. and GRAHAM, A. (1962) *British Prosobranch Molluscs.* Ray Society of London, 755 pp.

FREUNDLICH, H. (1935) *Thixotropy.* Hermann, Paris.

FREUNDLICH, H. and JONES, A. D. (1936) Sedimentation volume, dilatancy, thixotropy and plastic properties of concentrated suspensions. *J. Phys. Chem.*, **40**, 1217–1236.

FREUNDLICH, H. and RÖDER, H. L. (1938) Dilatancy and its relation to thixotropy. *Trans. Faraday Soc.* **34**, 308–316.

FRITZ, P. J. (1965) Rabbit muscle lactate dehydrogenase–5, a regulatory enzyme. *Science,* **150**, 364–366.

FRY, F. E. J. (1947) Effects of the environment on animal activity. Univ. Toronto Stud. Biol. 55 (Publ. Ontario Fish. Res. Lab.), **68**, 1–62.

FRY, F. E. J. (1957a) The lethal temperature as a tool in taxonomy. *Ann. Biol.*,

33, (5–6) 205–219.

FRY, F. E. J. (1957b) The aquatic respiration of fish. *The physiology of fishes,* **1.** Ed. M. E. Brown, Academic Press. N. Y. 1–63.

–FRY, F. E. J. (1958) Temperature compensation. *Ann. Rev. Physiol.,* **20,** 207–224.

FRY, F. E. J., BRETT, J. R. and CLAWSON, G. H. (1942) Lethal limits of temperature for young goldfish. *Rev. Can. Biol.,* **1,** 50–56.

FRY, F. E. J. and HART, J. S. (1948) The relation of temperature to oxygen consumption in the goldfish. *Biol. Bull. Wood's Hole,* **94,** 66–77.

FRY, F. E. J., HART, J. S. and WALKER, K. F. (1946) Lethal temperature relations for a sample of young speckled trout, *Salvelinus fontinalis. Univ. Toronto Stud. Biol. 54, Publ. Ont. Fish. Res. Lab.,* **66,** 9–35.

GAIL, F. W. (1919) Hydrogen ion concentration and other factors affecting the distribution of *Fucus. Publ. Puget Sd. Mar. Biol. St.,* **2,** 287.

GALTSOFF, P. S. (1928) The effect of temperature on the mechanical activity of the gills of the oyster (*Ostrea virginica* Gm). *J. gen. Physiol.,* **11,** 415–431.

GAMBLE, F. W. and KEEBLE, F. (1904) The bionomics of *Convoluta roscoffensis. Quart. J. micr. Sci.,* **47.** 363–431.

–GANAPATI, P. N. and PRASADA RAO, D. G. V. (1960) Studies on the respiration of barnacles. Oxygen uptake and metabolic rate in relation to body size in *Balanus amphitrite communis* (Darwin). *J. anim. Morph. Physiol.,* **7,** 27–31.

GANNING, B. (1966) Short time fluctuations of the microfauna in a rockpool in the northern Baltic proper. *Veröff. Inst. Meeresforsch. Bremerh.* (*sonderbd*), **2,** 149–154.

GANNING, B. (1967) Laboratory experiments in the ecological work on rock-pool animals with special notes on the ostracod *Heterocypris salinus. Helgol. wiss. Meeresunters.,* **15,**27–40.

GANNING, B. and WULFF, F. (1966) A chamber for offering alternative conditions to small motile aquatic animals. *Ophelia,* **3,** 151–160.

GANNING, B. and WULFF, F. (1969) The effects of bird droppings on chemical and biological dynamics in brackish water rockpools. *Oikos,* **20,** 274–286.

GANSEN, P. van (1960) Adaptations structurelles des animaux filtrants. *Ann. Soc. Roy. Zool. Belg.,* **90,** 161–231.

GARBARINI, P. (1936) Le choix du support pour les larvae de *Spirorbis borealis* Daudin. *C. R. Soc. Biol. Paris,* **122,** 158–160.

GEE, J. M. (1965) Chemical stimulation of settlement in larvae of *Spirorbis rupestris* (Serpulidae). *Anim. Behav.,* **13,** 181–186.

GEE, J. M. and KNIGHT-JONES, E. W. (1962) The morphology and larval behaviour of a new species of *Spirorbis* (Serpulidae). *J. mar. biol. Ass. U.K.,* **42,** 641–654.

–GHIRETTI, F. (1966) Respiration. *The physiology of Mollusca* Eds. K. M. Wilbur and C. M. Yonge. Academic Press, New York and London, **2,** 175–208.

GIESE, A. C. (1952) Myoglobin in muscles of chitons. *Anat. Rec.,* **113,** 103.

GLYNNE-WILLIAMS, J. and HOBART, J. (1952) Studies on the crevice fauna of a selected shore in Anglesey. *Proc. zool. Soc. Lond.,* **122,** (III) 794–824.

GORDON, M. S. (1968) Oxygen consumption of red and white muscles from tuna fish. *Science,* **159,** 87–90.

GOWENLOCH, J. N. (1926) Contributions to the study of marine gastropods II. The intertidal life of *Buccinum undatum,* a study in non-adaptation. *Contr. Canad. Biol. N.S.,* **3,** 169–177.

GOWENLOCH, J. N. and HAYES, F. R. (1926) Contributions to the study of marine gastropods I. The physical factors, behaviour and intertidal life of *Littorina. Contr. Canad. Biol. N.S.,* **3,** 133–166.

GRAHAM, A. (1931) The morphology, feeding mechanism and digestion of *Ensis siliqua* (Schumacher). *Trans. roy. Soc. Edinb.,* **56,** 725–751.

GRAHAM, A. (1938a) On a ciliary process of food-collecting in the gastropod *Turritella communis* Risso. *J. mar. biol. Ass. U.K.,* **26,** 377–380.

GRAHAM, A. (1938b) The structure and function of the alimentary canal of aeolid molluscs with a discussion on their nematocysts. *Trans. roy. Soc.*

Edinb., **59**, 267–307.

GRAHAM, A. and FRETTER, V. (1947) The life-history of *Patina pellucida* (L). *J. mar. biol. Ass. U.K.*, **26**, 590–601.

GRAINGER, F. and NEWELL, G. E. (1965) Aerial respiration in *Balanus balanoides*. *J. mar. biol. Ass. U.K.*, **45**, 469–479.

GRAY, I. E. (1953) Comparative study of gill area in crabs. *Anat. Rec.*, **117**, 567–568.

GRAY, J. (1923) The mechanism of ciliary movement III. The effect of temperature. *Proc. roy. Soc. Lond.* (*B*), **95**, 6–15.

GRAY, J. S. (1966) The attractive factor of intertidal sands to *Protodrilus symbioticus*. *J. mar. biol. Ass. U.K.*, **46**, 627–645.

GREGG, J. H. (1945) Background illumination as a factor in the attachment of barnacle cyprids. *Biol. Bull. Wood's Hole*, **88**, 44–49.

GROSS, J. and KNIGHT-JONES, E. W. (1957) The settlement of *Spirorbis borealis* on algae. *Rep. Challenger Soc.* **3**, (9) 18.

GUILLARD, R. R. L. (1959) Further evidence of the destruction of bivalve larvae by bacteria. *Biol. Bull. Wood's Hole.*, **117**, 258–266.

GUNTER, G. (1957) Temperature. *Treatise on Marine Ecology and Paleoecology.* (*Ecology*) Ed. J. W. Hedgpeth, Mem, (67) Geol. Soc. Amer., **1**, 159–184.

GWILLIAM, G. F. (1963) The mechanism of the shadow reflex in Cirripedia. *Biol. Bull. Wood's Hole*, **125**, 470–485.

HALCROW, K. and BOYD, C. M. (1967) The oxygen consumption and swimming activity of the amphipod *Gammarus oceanicus* at different temperatures. *Comp. Biochem. Physiol.*, **23**, 233–242.

HALL, V. E. (1931) Muscular activity and oxygen consumption of *Urechis caupo*. *Biol. Bull. Wood's Hole*, **61**, 400–416.

HANCOCK, D. A. (1955) The feeding behaviour of starfish on Essex oyster beds. *J. mar. biol. Ass. U.K.*, **34**, 313–331.

HANCOCK, D. A. (1957) The feeding behaviour of the sea urchin *Psammechinus miliaris* (Gmelin) in the laboratory. *Proc. zool. Soc. Lond.*, **129**, (2) 255–262.

HAND, C. (1961) Present state of nematocyst research: types, structure and function. *The Biology of Hydra* Eds. H. M. Lenhoff and W. F. Loomis, University of Miami Press, Florida, 187–202.

HANNON, J. P. (1960) Effect of prolonged cold exposure on components of the electron transport system. *Am. J. Physiol.*, **198**, 740–744.

HANNON, J. P. and VAUGHAN, D. A. (1960) Effect of prolonged cold exposure on the glycolytic enzymes of liver and muscle. *Am. J. Physiol.*, **198**, 375–380.

HARKER, J. E. (1964) *The physiology of diurnal rhythms.* Camb. Monogr. in Exp. Biol., **13**, Cambridge University Press, London, 114 pp.

HARLEY, M. B. (1950) The occurrence of a filter-feeding mechanism in the polychaete *Nereis diversicolor. Nature, Lond.*, **165**, 734.

HART, J. S. (1952) Geographic variations of some physiological and morphological characters in freshwater fish. *Univ. Toronto Stud. Biol. 60, Publ. Ontario Fish. Res. Lab.*, **72**, 1–79.

HART, T. J. (1930) Preliminary notes on the bionomics of the amphipod *Corophium volutator* (Pallas). *J. mar. Biol. Ass. U.K.*, **16**, 761–789.

HARVEY, E. (1928) The oxygen consumption of luminous bacteria. *J. gen. Physiol.*, **11**, 469–475.

HASEMAN, J. D. (1911) The rhythmical movements of *Littorina littorea* synchronous with ocean tides. *Biol. Bull. Wood's Hole*, **21**, 113–121.

HATTON, H. and FISCHER-PIETTE, E. (1932) Observations et expériences sur le peuplement des côtes rocheuses par les Cirripèdes. *Bull. Inst. oceanogr. Monaco*, **592**, 1–15.

HAZELHOFF, E. H. (1938) Über die Ausnutzung des Sauerstoffs bei Vershiedenen Wassertieren. *Z. vergl. Physiol.*, **26**, 306–372.

HEILBRUNN, L. V. (1952) *An outline of General Physiology.* 3rd ed. Saunders, Philadelphia, 818 pp.

HELM, M. M. and TRUEMAN, E. R. (1967) The effect of exposure on the heart rate of the mussel *Mytilus edulis* L. *Comp. Biochem. Physiol.*, **21**, 171–177.

HEMMINGSEN, A. M. (1950) The relation of standard (basal) energy metabolism to total fresh weight of living organisms. *Rep. Steno. Hosp. Copenh.*, **4**, 7–58.

HEMMINGSEN, A. M. (1960) Energy metabolism as related to body size and respiratory surfaces and its evolution. *Rep. Steno Hosp. Copenh.*, **9**, 7–110.

HEMMINGSEN, E. (1962) Accelerated exchange of oxygen–18 through a membrane containing oxygen saturated haemoglobin. *Science*, **135**, 733–734.

HENDERSON, J. T. (1929) Lethal temperatures of Lamellibranchiata *Contr. Can. Biol. and Fish.*, **4**, (25) 399–411.

HENDERSON, L. J. (1928) *Blood, a study in General Physiology*. Yale University Press, New Haven, Conn.

HIRASAKA, K. (1927) Notes on *Nucula. J. mar. biol. Ass. U.K.*, **14**, 629–645.

HOAGLAND, H. (1935) *Pacemakers and the control of behaviour.* Macmillan, New York.

HOAR, W. S. (1966) *General and Comparative Physiology*. Prentice-Hall, New York, 815 pp.

HOCHACHKA, P. W. (1962) Thyroidal effects on pathways for carbohydrate metabolism in a teleost. *Gen. comp. Endocrinol.*, **2**, 499–505.

HOCHACHKA, P. W. (1965) Isoenzymes in metabolic adaptation of a poikilotherm: subunit relationships in lactic dehydrogenases of goldfish. *Arch. Biochem. Biophys.*, **111**, 96–103.

HOCHACHKA, P. W. (1966) Lactic dehydrogenases in poikilotherms: definition of a complex isozyme system. *Comp. Biochem. Physiol.*, **18**, 261–269.

HOCHACHKA, P. W. (1967) Organisation of metabolism during temperature compensation. *Molecular mechanisms of temperature adaptation.* Ed. C. L. Prosser. Amer. Assoc. adv. Sci. (84) Washington D.C. 177–203.

HOCHACHKA, P. W. (1968) Action of temperature on branch points in glucose and acetate metabolism. *Comp. Biochem. Physiol.*, **25**, 107–118.

HOCHACHKA, P. W. and HAYES, F. R. (1962) The effect of temperature acclimation on pathways of glucose metabolism in the trout. *Can. J. Zool.*, **40**, 261–270.

HOCHACHKA, P. W. and SOMERO, G. N. (1968) The adaptation of enzymes to temperature. *Comp. Biochem. Physiol.*, **22**, 659–668.

HODGMAN, C. D. (Ed) (1959) *Handbook of Chemistry and Physics. Chemical Rubber Publishing Company*, **40**, 3456 pp.

HOLME, N. A. (1949) The fauna of sand and mud banks near the mouth of the Exe estuary. *J. mar. biol. Ass. U.K.*, **28**, 189–237.

HOPKINS, A. E. (1933) Experiments on the feeding behaviour of the oyster *Ostrea gigas. J. exp. Zool.*, **64**, 469–494.

HOPKINS, A. E. (1935a) Temperature optima in the feeding mechanism of the oyster *Ostrea gigas. J. exp. Zool.*, **71**, 195–208.

HOPKINS, A. E. (1935b) Attachment of the larvae of the Olympia oyster (*Ostrea lurida*) to plane surfaces. *Ecology, N. Y.* **16**, 82–87.

HORRIDGE, G. A. (1959) Analysis of the rapid response of *Nereis* and *Harmothoë* (Annelida). *Proc. roy. Soc. B*, **150**, 245–262.

HOWELLS, H. H. (1936) The structure and function of the alimentary canal of *Aplysia punctata. Quart. J. micr. Sci.*, **83**, 357–397.

HUGHES, G. M. (1963) *The Comparative Physiology of Vertebrate Respiration.* Heinmann, London, 145 pp.

HUGHES, G. M. and SHELTON, G. (1962) Respiratory mechanisms and their nervous control in fish. *Adv. comp. Physiol. Biochem.*, **1**, 275–364.

HULBERT, G. C. E. B. and YONGE, C. M. (1937) A possible function of the osphradium in the Gastropoda. *Nature, Lond.*, **139**, 840.

HUMPHREY, R. R. and MACY, R. W. (1930) Observations on some of the probable factors controlling the size of some tide-pool snails. *Publ. Puget Sd. Mar. Biol. Sta.*, 7, 205.

HUNT, O. D. (1925) The food of the bottom fauna of the Plymouth fishing grounds. *J. mar. biol. Ass. U.K.* **13**, 560–599.

HUNTSMAN, A. G. and SPARKS, M. I. (1924) Limiting factors for marine animals III. Relative resistance to high temperatures. *Contr. Canad. Biol. N.S.*, **2**, (6) 97–114.

ISAAC, W. E. (1933) Some observations and experiments on the drought resistance of *Pelvetia canaliculata. Ann. Bot.*, **47**, (186) 343–348.

ISAAC, W. E. (1935) A preliminary study of the water loss of *Laminaria digitata* during the intertidal exposure. *Ann. Bot.*, **49**, (193) 109–117.

JACOBSEN, V. H. (1967) The feeding of the lugworm *Arenicola marina* (L). Quantitative studies. *Ophelia*, **4**, 91–109.

JÄGERSTEN, G. (1940) Die Abhängigkeit der Metamorphose vom Substrate des Biotops bei *Protodrilus. Ark. Zool.*, **32**, 1–12.

JANISCH, E. (1925) Uber die Temperaturabhängigkeit biologischer Vorgange und ihre kurvenmässige Analyse. *Pflügers Archiv.*, **209**, 414–436.

JANKOWSKY, H. D. (1960) Uber die Hormonale Beinflussung der Temperaturadaptation beim Grasfrosch (*Rana temporaria*). *Z. vergl. Physiol.*, **43**, 392–410.

JANKOWSKY, H. D. (1964) Der Einfluss des Blutes auf den Sauerstoffverbrauch des isolierten Muskelgewebes von Schleien (*Tinca tinca* L). *Zool. Anz.*, **172**, 233–239.

JANSSEN, B. O. (1966) On the ecology of *Derocheilocaris remanei* Delamere and Chappuis (Crustacea, Mystocarida). *Vie Milieu*, **17**, 143–186.

JANSSEN, C. R. (1960) The influence of temperature on geotaxis and phototaxis in *Littorina obtusata* (L). *Arch. néerl. Zool.*, **13**, 500–510.

JÄRNEFELT, H. (1940) Beobachtungen über die Hydrologie einiger Schärentümpel. *Verh. int. Ver. Limnol.*, **9**, 79–101.

JENNER, C. E. (1958) An attempted analysis of schooling behaviour in the marine snail *Nassarius obsoletus. Biol. Bull. Wood's Hole*, **115**, 337–338.

JENSEN, P. B. (1915) Studies concerning the organic matter of the sea bottom. *Rep. Danish Biol. Sta.*, **22**, 3–39.

JENSEN, P. B. and PETERSEN, C. G. J. (1911) Valuation of the sea, 1. Animal life on the sea bottom, its food and quantity. *Rep. Danish Biol. Sta.*, **20**, 1–81.

JOHANSSON, J. (1939) Anatomische studien über die Gastropodenfamilien Rissoidae und Littorinidae. *Zool. Bidr. Uppsala*, **18**, 289–296.

JOHNSON, D. S. and SKUTCH, A. F. (1928) Littoral vegetation on the headland of Mt. Desert Island, Maine II. Tide pools and the environment and classification of submersible plant communities. *Ecology*, **9**, 307–338.

JOHNSON, D. W. (1919) *Shore processes and shoreline development.* Wiley, New York (Chapman Hall, London) 584 pp.

JOHNSON, F. B. and LENHOFF, H. A. (1958) Histochemical study of purified *Hydra* nematocysts. *J. Histochem. Cytochem.*, **6**, 34.

JONES, C. (1947) The control and discharge of nematocysts in hydra. *J. exp. Zool.*, **105**, 25–61.

JONES, J. D. (1954) Observations on the respiratory physiology and on the haemoglobin of the polychaete genus *Nephthys* with special reference to *N. hombergii* (Aud. et M-Edw). *J. exp. Biol.*, **32**, (1) 110–125.

JONES, J. D. (1963) The functions of the respiratory pigments in invertebrates. *Problems in Biol.*, **1**, 11–89.

JONES, N. E. and DEMETROPOULOS, A. (1968) Exposure to wave action: measurements of an important ecological parameter on rocky shores on Anglesey. *J. exp. mar. Biol. Ecol.*, **2**, 46–63.

JONES, N. S. (1948) Observations and experiments on the biology of *Patella vulgata* at Port St. Mary, Isle of Man. *Proc. L'pool biol. Soc.* **56**, 60–77.

JORDAN, H. J. and HIRSCH, G. C. (1927) Einige Vergleichendphysiologische Probleme der Verdauung bei Metazoen. *Handb. norm. path. Physiol.* 3B, (2)

JØRGENSEN, C. B. (1949a) Feeding rates of sponges, lamellibranchs and ascidians. *Nature, Lond.*, **163**, 912.

JØRGENSEN, C. B. (1949b) The rate of feeding by *Mytilus* in different kinds of suspension. *J. mar. Biol. Ass. U.K.*, **28**, 333–344.

JØRGENSEN, C. B. (1952) On the relation between water transport and food-requirements in some marine filter-feeding invertebrates. *Biol. Bull. Wood's Hole*, **103**, 356–363.

JØRGENSEN, C. B. (1955) Quantitative aspects of filter feeding in invertebrates. Biol. Rev., **30**, 391–454.

JØRGENSEN, C. B. (1960) Efficiency of particle retention and rate of water transport in undisturbed lamellibranchs. *J. Cons. Int. Explor. Mer.*, **26**, 94–116.

JØRGENSEN, C. B. (1966) *The Biology of Suspension Feeding.* Pergamon Press, Oxford, 357 pp.

JØRGENSEN, C. B. and GOLDBERG, E. D. (1953) Particle filtration in some ascidians and lamellibranchs. *Biol. Bull. Wood's Hole*, **105**, 477–489.

KALMUS, H. (1959) The orientation of animals to polarised light. *Nature, Lond.*, **184**, 228–230.

KANUNGO, M. S. and PROSSER, C. L. (1959) Physiological and biochemical adaptation of goldfish to cold and warm temperatures. II Oxygen consumption of liver homogenate and oxidative phosphorylation of liver mitochondria. *J. cell. comp. Physiol.*, **54**, 265–274.

KANWISHER, J. W. (1955) Freezing in intertidal animals. *Biol. Bull. Wood's Hole*, **109**, 56–63.

KANWISHER, J. W. (1959) Histology and metabolism of frozen intertidal animals. *Biol. Bull. Wood's Hole*, **116**, 258–264.

KAPLAN, N. O. (1964) Lactate dehydrogenase – structure and function. *Brookhaven Symp. Biol.*, **17**, 131–153.

KATZ, J. and WOOD, H. G. (1960) The use of glucose-C^{14} for the evaluation of the pathways of glucose metabolism. *J. Biol. Chem.* **235**, 2165–2177.

KELLOGG, J. L. (1915) Ciliary mechanisms of lamellibranchs with descriptions of anatomy. *J. morph.*, **26**, 625–701.

KENNEDY, D. (1960) Neural photoreception in a lamellibranch mollusc. *J. gen. Physiol.*, **44**, 277–299.

KILIAN, E. F. (1952) Wasserströmung und Nahrungsaufnahme beim Süsswasserschwamm *Ephydatia fluviatilis. Z. vergl. Physiol.*, **34**, 407–447.

KINNE, O. (1958) Adaptation to salinity variation – some facts and problems. *Physiological adaptation.* Ed. C. L. Prosser, Ronald Press, New York, 92–105.

KINNE, O. (1963) The effect of temperature and salinity on marine and brackish water animals I. Temperature. *Oceanogr. mar. Biol. ann. Rev.*, **1**, 301–340.

KINNE, O. (1964) Non-genetic adaptation to temperature and salinity. *Helgol. wiss. Meeresunters*, **9**, 433–458.

KIRBERGER, C. (1953) Untersuchungen über die Temperaturabhängigkeit von Lebensprozessen bei verschiedenen wirbellosen. *Z. vergl Physiol.*, **35**, 175–198.

KITCHING, J. A., MACAN, T. T. and GILSON, H. C. (1934) Studies in sublittoral ecology I. A submarine gulley in Wembury Bay, South Devon. *J. mar. biol. Ass. U.K.*, **19**, 677–705.

KLEIBER, M. (1932) Body size and metabolism. *Hilgardia*, **6**, 315–353.

KLEIBER, M. (1947) Body size and metabolic rate. *Physiol. Rev.*, **27**, 511–541.

KLEKOWSKI, R. Z. and DUNCAN, A. (1966) The oxygen consumption in saline water of young *Potamopyrgus jenkinsi* (Smith) (Prosobranchiata). *Verh. int. Ver. Limnol.*, **16**, 1753–1760.

KLINE, E. S. (1961) Chemistry of nematocyst capsule and toxin of *Hydra littoralis. The Biology of Hydra.* Eds. H. M. Lenhoff, and W. F. Loomis. University of Miami Press, Florida, 153–166.

KLUGH, A. B. (1924) Factors controlling the biota of rock pools. *Ecology*, **5**, 192–196.

KNIGHT-JONES, E. W. (1951a) Aspects of the settling behaviour of larvae of *Ostrea edulis* on Essex oyster beds. *Rapp. & Proc. verb. du Conseil*, **128**, 30–34.

KNIGHT-JONES, E. W. (1951b) Gregariousness and some other aspects of the setting behaviour of *Spirorbis. J. mar. biol. Ass. U.K.*, **30**, 201–222.

KNIGHT-JONES, E. W. (1952) On the nervous system of *Saccoglossus cambrensis* (Enteropneusta) *Phil. Trans. roy. Soc. B*, **236**, 315–354.

KNIGHT-JONES, E. W. (1953a) Feeding in *Saccoglossus* (Enteropneusta). *Proc. zool. Soc. Lond.*, **123**, 637–654.

KNIGHT-JONES, E. W. (1953b) Laboratory experiments on gregariousness during setting in *Balanus balanoides* and other barnacles. *J. exp. Biol.*, **30**, 584–598.

KNIGHT-JONES, E. W. (1953c) Some further observations on gregariousness in marine larvae. *Brit. J. Anim. Behav.*, **1**, 81–82.

KNIGHT-JONES, E. W. (1953d) Decreased discrimination during setting after prolonged planktonic life in larvae of *Spirorbis borealis* (Serpulidae). *J. mar. biol. Ass. U.K.*, **32**, 337–345.

KNIGHT-JONES, E. W. (1955) The gregarious setting reaction of barnacles as a measure of systematic affinity. *Nature, Lond.*, **174**, 266.

KNIGHT-JONES, E. W. and CRISP, D. J. (1953) Gregariousness in barnacles in relation to the fouling of ships and to antifouling research. *Nature, Lond.*, **171**, 1109.

KNIGHT-JONES, E. W. and MOYSE, J. (1961) Intraspecific competition in sedentary marine animals. (Mechanisms in Biological Competition) *Symp. Soc. exp. Biol.* (15), 72–95.

KNIGHT-JONES, E. W. and STEVENSON, J. P. (1950) Gregariousness during settlement in the barnacle *Elminius modestus* Darwin. *J. mar. biol. Ass. U.K.*, **29**, 281–297.

KORRINGA, P. (1941) Experiments and observations on swarming, pelagic life and setting in the European flat oyster, *O. edulis. Arch. néerl Zool.*, **5**, 1–249.

KRASNE, F. B. and LAWRENCE, P. A. (1966) Structure of the photoreceptors in the compound cyespots of *Branchiostoma vesiculosum. J. Cell. Sci.*, **1**, 239–248.

KREPS, E. (1929) Untersuchungen über den respiratorischen Gaswechsel bei *Balanus crenatus* bei verschiedenem Salzgehalt des Aussenmilieus. I. Uber den Sauerstoffverbrauch im wassermilieu bei verschiedenem Salzgehalt. *Pflüg. Arch. ges. Physiol.*, **222**, 215–233.

KROGH, A. (1914) The quantitative relation between temperature and standard metabolism in animals. *Int. Zeits. physik-chem. Biol.*, **1**, 491–508.

KROGH, A. (1919) The rate of diffusion of gases through animal tissues, with some remarks of the coefficient of invasion. *J. Physiol.*, **52**, 391–408.

KROGH, A. (1941) *The comparative physiology of respiratory mechanisms* (William J. Cooper Foundation Lectures, 1939) University of Pennsylvania Press, Philadelphia. 172 pp.

KRUGER, F. (1959) Zur Ernährungsphysiologie von *Arenicola marina* L. *Zool. Anz. Suppl.*, **22**, 115–120.

KRUGER, G. (1962) Uber die Temperaturadaptation des Bitterlings (*Rhodeus amarus* Bloch). *Z. Wiss. Zool. Abt. A*, **167**, 87–102.

KRULL, H. (1935) Anatomische Untersuchungen an einheimischen Prosobranchiern Beiträge zur Phylogenie der Gastropoden. *Zool. Jb.* (*Anat. Ont.*), **60**, 399–464.

KRUMBEIN, W. C. (1939) Graphic presentation and statistical analysis of sedimentary data. (Recent Marine Sediments) Ed. P. D. Trask,. Tulsa *Amer. Assoc. Petrol Geol.*, 558–591.

KRUMBEIN, W. C. (1944/47) Shore processes and beach characteristics. *U. S. Army Erosion Board. Tech. Mem.* (3), 35 pp.

KUENEN, D. J. (1939) Zur Kenntnis der Stehenden Kleingewässer auf den Schäreninseln *Mem. Soc. F. Fl. Fenn.*, **15**, 16–22.

KUENZLER, E. J. (1961) Structure and energy flow of a mussel population in a Georgia salt marsh. *Limnol. Oceanogr.*, **6**, 191–204.

LANCE, J. (1963) The salinity tolerance of some estuarine planktonic copepods. *Limnol. Oceanogr.*, **8**, 440–449.

LAND, M. F. (1965) Image formation by a concave reflector in the eye of the scallop *Pecten maximus. J. Physiol Lond.*, **179**, 138–153.

LAND, M. F. (1966a) Activity in the optic nerves of *Pecten maximus* in response to changes in light-intensity, and to pattern of movement in the optical environment. *J. exp. Biol.*, **45**, 83–99.

LAND, M. F. (1966b) A multilayer interference reflector in the eye of the scallop, *Pecten maximus. J. exp. Biol.*, **45**, 433–447.

LAND, M. F. (1968) Functional aspects of the optical and retinal organisation of the mollusc eye. *Symp. zool. Soc. Lond.* (23), 75–96.

LANE, C. E. (1961) *Physalia* nematocysts and their toxin. *The Biology of Hydra.* Eds. H. M. Lenhoff and W. F. Loomis, University of Miami Press, Florida,

169–178.

LANGRIDGE, J. (1963) Biochemical aspects of temperature response. *Ann. Rev. Plant. Physiol.*, **14**, 441–462.

LEE, D. L. and SMITH, M. H. (1965) Haemoglobins of parasitic animals. *Exp. Parasitol.*, **16**, 392–424.

LEHMANN, H. and HUNTSMAN, R. G. (1961) Why are red cells the shape they are? The evolution of the human red cell. *Functions of the Blood.* Blackwell, Oxford.

LENHOFF, H. M., KLINE, E. and HURLEY, R. (1957) A hydroxyproline-rich intracellular, collagen-like protein of hydra nematocysts. *Biochem. Biophys. Acta.*, **26**, 204–205.

LENT, C. (1968) Air-gaping by the ribbed mussel *Modiolus demissus* (Dillwyn): effects and adaptive significance. *Biol. Bull. Wood's Hole,* **134**, (1) 60–73.

LENTZ, T. L. (1966) *The Cell Biology of Hydra.* North Holland Publ. Co., Amsterdam, 199 pp.

LENTZ, T. L. and BARRNETT, R. J. (1961a) Enzyme histochemistry of hydra. *J. exp. Zool.*, **147**, 125–150.

LENTZ, T. L. and BARRNETT, R. J. (1961b) Enzyme histochemistry of hydra nematocysts. *Anat. Rec.*, **139**, 312.

LENTZ, T. L. and BARRNETT, R. J. (1962) The effects of enzyme substrates and Pharmacological agents on nematocyst discharge. *J. exp. Zool.*, **149**, (1) 33–38.

LENTZ, T. L. and WOOD, J. G. (1964) Amines in the nervous system of coelenterates. *J. Histochem. Cytochem,* **12**, 37.

LEVANDER, K. M. (1900) Zur Kenntnis des Lebens in stehanden Kleingewassern den Skäreninseln. *Acta Soc. F. Fl. Fenn.*, **18**, (6) 1–107.

LEWIS, J. B. (1960) The fauna of rocky shores of Barbados, West Indies. *Canad. J. Zool.*, **38**, 391–435.

LEWIS, J. B. (1963) Environmental and tissue temperatures of some tropical intertidal marine animals. *Biol. Bull. Wood's Hole,* **124**, 277–284.

LEWIS, J. R. (1954a) The ecology of exposed rocky shores of Caithness. *Trans. roy. Soc. Edinb.*, **62**, 695–723.

LEWIS, J. R. (1954b) Observations on a high-level population of limpets. *J. anim. Ecol.*, **23**, 85–100.

LEWIS, J. R. (1955) The mode of occurrence of the universal intertidal zones in Great Britain; with a comment by T. A. and Anne Stephenson. *J. Ecol.*, **43**, 270–290.

LEWIS, J. R. (1961) The littoral zone on a rocky shore – a biological or a physical entity? *Oikos,* **12**, 280–301.

LEWIS, J. R. (1964) *The ecology of rocky shores.* English University Press, London, 323 pp.

LILLY, M. M. (1953) The mode of life and structure and functioning of the reproductive ducts of *Bithynia tentaculata* (L). *Proc. malac. Soc. Lond.*, **30**, 87–110.

LINDBERG, H. (1944) Okoligish-geographische Untersuchungen zu Insectenfauna der Felsentumpel an den Küsten Finnlands. *Acta. zool. Fenn.*, **41**, 1–178.

LINDROTH, A. (1938) Studien über die respiratorischen Mechanismen von *Nereis virens* Sars. *Zool. Bidr. Uppsala,* **17**, 367–497.

LINKE, O. (1939) Die Biota des Jadebusenwattes. *Helgol. wiss. Meeresunters,* **1**, 201–348.

LONGBOTTOM, M. R. (1968) *Nutritional factors affecting the distribution of Arenicola marina L.* Ph.D. Thesis, University of London.

LOOSANOFF, V. L. (1949) On the food-selectivity of oysters. *Science,* **110**, 122.

LOOSANOFF, V. L. and NOMEJKO, C. A. (1946) Feeding of oysters in relation to tidal stages and to periods of light and darkness. *Biol. Bull. Wood's Hole,* **90**, 244–264.

LOOSANOFF, V. L. and TOMMERS, F. D. (1948) Effect of suspended silt and other substances on the rate of feeding of oysters. *Science,* **107**, 69–70.

LORENZ, K. (1953) Die Entwicklung der vergleichenden Verhaltensforschung in den letzen 12 Jahren. *Zool. Anz. Suppl.*, **17**, 37–58.

LOWE, M. E. (1956) Dominance-subordinance relationships in *Cambarellus shufeldtii. Tulane Stud. Zool.*, **4**, 139–170.

LUND, E. J. (1957) A quantitative study of clearance of a turbid medium, and feeding by the oyster. *Inst. mar. Sci.*, **4**, 296–312.

LUTHER, W. (1930) Versuche Über die Chemorezeptoren der Brachyuren. *Z. vergleich. Physiol.*, **12**, 177–205.

MACGINITIE, G. E. (1932) The role of bacteria as food for bottom animals. *Science*, **76**, 490.

MACGINITIE, G. E. (1939a) The method of feeding of *Chaetopterus. Biol. Bull. Wood's Hole*, **77**, 115–118.

MACGINITIE, G. E. (1939b) The method of feeding of tunicates. *Biol. Bull. Wood's Hole*, **77**, 443–447.

MACGINITIE, G. E. (1941) On the method of feeding of four pelecypods. *Biol. Bull. Wood's Hole*, **80**, 18–25.

MACGINITIE, G. E. and MACGINITIE, N. (1949) *Natural History of Marine Animals.* McGraw-Hill, New York and London.

MACKIE, G. O. (1960) Studies on *Physalia physalis* (L) 2. Behaviour and physiology. *Discovery Rep.*, **30**, 369–407.

MANGUM, C. P. (1964) Activity patterns in the metabolism and ecology of polychaetes. *Comp. Biochem. Physiol.*, **11**, 239–256.

MANGUM, C. P. and SASSAMAN, C. (1969) Temperature sensitivity of active and resting metabolism in a polychaetous annelid *Comp. Biochem. Physiol.* **30**, 111–116.

MANWELL, C. (1958) The oxygen-respiratory pigment equilibrium of the haemocyanin and myoglobin of the amphineuran mollusc *Cryptochiton stelleri, J. cell. comp. Physiol.*, **52**, 341–352.

MANWELL, C. (1959) Alkaline denaturation and oxygen equilibrium of annelid haemoglobins. *J. cell. comp. Physiol.*, **53**, 61–74.

MANWELL, C. (1960a) Comparative physiology, blood pigments. *Ann. Rev. Physiol.*, **22**, 191–244.

MANWELL, C. (1960b) Histological specificity of respiratory pigments 1. Comparisons of the coelom and muscle haemoglobins of the polychaete worms *Travisia pupa* and the echiuroid worm *Arychite pugettensis. Comp. Biochem. Physiol.* **1**, 267–276.

MANWELL, C. (1964) Chemistry, genetics and function of invertebrate respiratory pigments. Configurational changes and allosteric effects. *Oxygen in the animal organism.* MacMillan (Pergamon), New York, 49–119.

MARE, M. F. (1942) A study of a marine benthic community with special reference to the micro-organisms. *J. mar. biol. Ass. U.K.* **25**, 517–553.

MARKERT, C. L. (1963) Epigenetic control of specific protein synthesis in differentiating cells. *Symp. Soc. Study Develop. Growth*, **21**, (1962) 64–84.

MARKERT, C. L. and FAULHABER, I. (1965) Lactic dehydrogenase isozyme patterns of fish. *J. exp. Zool.*, **159**, 319–332.

MATTERN, C. F., PARK, H. D. and DANIEL, W. A. (1965) Electron microscope observations on the structure and discharge of the stenotele of hydra. *J. cell. Biol.*, **27**, 621–638.

MATTHEWS, A. (1916) The development of *Alcyonium digitatum* with some notes on the early colony formation. *Quart. J. micr. Sci.*, **62**, 43–94.

MATUTANI, K. (1961) Studies on the heat resistance of *Tigriopus japonicus. Publ. Seto. mar. biol. Lab.*, **9**, 379–411.

MAYER, A. G. (1914) The effects of temperature on tropical marine animals. *Papers from the Tortugas Lab., Carnegie Inst.*, **6**, 3–24.

McFARLAND, W. N. and PICKENS, P. E. (1965) The effects of season, temperature and salinity on standard and active oxygen consumption of the grass shrimp *Palaemonetes vulgaris* (Say). *Canad. J. Zool.*, **43**, 571–585.

McLEESE, D. W. (1965) Survival of Lobsters, *Homarus americanus,* out of water. *J. Fish. Res. Bd. Canada,* **22**, (2) 385–394.

McLEESE, D. W. (1956) Effects of temperature, salinity and oxygen on the survival of the American lobster. *J. Fish. Res. Bd. Canada,* **13**, (2) 247–272.

McPHERSON, B. F. (1968) Feeding and oxygen uptake of the tropical sea urchin

Eucidaris tribuloides (Lamarck). *Biol. Bull. Wood's Hole,* **135,** 308–321.

MEADOWS, P. S. (1964) Experiments on substrate selection by *Corophium* species: films and bacteria on sand particles. *J. exp. Biol.,* **41,** 499–511.

MEADOWS, P. S. (1965) Attachment of marine and freshwater bacteria to solid surfaces. *Nature, Lond.,* **207,** 1108.

MEADOWS, P. S. and ANDERSON, J. G. (1966) Micro-organisms attached to marine and freshwater sand grains. *Nature, Lond.,* **212,** 1059–1060.

MEADOWS, P. S. and ANDERSON, J. G. (1968) Micro-organisms attached to marine sand grains. *J. mar. biol. Ass. U.K.,* **48,** 161–175.

MEADOWS, P. S. and REID, A. (1966) The behaviour of *Corophium volutator* (Crustacea, Amphipoda). *J. zool. Lond.,* **150,** 387–399.

MEADOWS, P. S. and WILLIAMS, G. B. (1963) Settlement of *Spirorbis borealis* Daudin larvae on surfaces bearing films of micro-organisms. *Nature, Lond.,* **198,** 610–611.

MENZEL, R. W. (1955) Some phases of the biology of *Ostrea equistris* Say and a comparison with *Crassostrea virginica* (Gmelin). *Publ. Inst. Marine Sci. Univ. Texas,* **3,** 69–153.

MEWS, H. H. (1957) Uber die temperaturadaptation der eiweisspaltenden und synthetisierenden Zellfermente von Froschen. *Z. vergl. Physiol.,* **40,** 356–362.

MICALLEF, H. (1966) *The ecology and behaviour of selected intertidal gastropods.* Ph.D Thesis, University of London.

MICALLEF, H. (1968) The activity of *Monodonta lineata* in relation to temperature as studied by means of an aktograph. *J. Zool. Lond.,* **154,** (2) 155–159.

MICALLEF, H. and BANNISTER, W. H. (1967) Aerial and aquatic oxygen consumption of *Monodonta turbinata* (Mollusca, Gastropoda). *J. Zool. Lond.,* **151,** 479–482.

MIKHAILOFF, S. (1923) Experiences reflexologiques. *Bull. inst. océanogr.* (422), 1–16.

MILLER, W. H. (1960) Visual photoreceptor structures. *The Cell.* **4,** 325–364. Eds. J. Brachet, and A. E. Mirsky, Academic Press, London and New York.

MILLOTT, N. (1954) Sensitivity to light and the reactions to changes in light intensity of the echinoid *Diadema antillarum* Philippi. *Phil. Trans. roy. Soc. B,* **238,** 187–220.

MILLOTT, N. and YOSHIDA, M. (1959) The photosensitivity of the sea urchin *Diadema antillarum* Philippi. Responses to increase in light intensity. *Proc. zool. Soc. Lond.,* **133,** 67–71.

MILLOTT, N. and YOSHIDA, M. (1960a) The shadow reaction of *Diadema philippi.* I. The spine response and its relation to the stimulus *J. exp. Biol.,* **37,** 363–375.

MILLOTT, N. and YOSHIDA, M. (1960b) The shadow reaction of *Diadema philippi* II. Inhibition by light. *J. exp. Biol.,* **37,** 376–389.

MONTEROSSO, B. (1928a) Studi cirrepedologici II. Anabiosi nei Ctamalini. *R. C. Accad. Lincei Ser.* (6) **7,** 939–944.

MONTEROSSO, B. (1928b) Studi cirrepedologici III. Persistenza dei fenomeni respiratori nei ctamalini mantenuti in ambiente subaereo. *Boll. Soc. Biol. sper.,* **3,** 1067–1070.

MONTEROSSO, B. (1928c) Studi cirrepedologici IV. Fenomeni che precedono l'anabiosi nei Ctamalini. *R. C. Accad. Lincei Ser.* (6), **8,** 91–96.

MONTEROSSO, B. (1930) Studi cirrepedologici VI. Sul comportamento di *"Chthamalus stellatus"* in diverse condizione sperimentale. *R. C. Accad. Lincei Ser.* 6, **11,** 501–505.

MOORE, H. B. (1931) The muds of the Clyde Sea Area III. Chemical and physical conditions; rate and nature of sedimentation; and fauna. *J. mar. biol. Ass. U.K.,* **17,** 325–358.

MOORE, H. B. (1934a) The biology of *Balanus balanoides* I. Growth rate and its relation to size, season and tidal level. *J. mar. biol. Ass. U.K.,* **19,** (2) 851–868.

MOORE, H. B. (1934b) The relation of shell growth to environment in *Patella vulgata. Proc. malac. Soc. Lond.,* **21,** (3) 217–222.

MOORE, H. B. (1935a) The biology of *Balanus balanoides* IV. Relation to environmental factors. *J. mar. biol. Ass. U.K.*, **20**, (2) 297–307.

MOORE, H. B. (1935b) A comparison of the biology of *Echinus esculentus* in different habitats II. *J. mar. biol. Ass. U.K.*, **20**, (1) 109–128.

MOORE, H. B. (1936) The biology of *Balanus balanoides* V. Distribution in the Plymouth area. *J. mar. biol. Ass. U.K.*, **20**, 701–716.

MOORE, H. B. (1958) *Marine Ecology*, John Wiley and Son Inc, New York and London, 493 pp.

MOORE, H. B. and KITCHING, J. A. (1939) The biology of *Chthamalus stellatus* (Poli). *J. mar. biol. Ass. U.K.*, **23**, (2) 521–541.

MOORE, L. B. (1944) Some intertidal sessile barnacles of New Zealand. *Trans. roy. Soc. N.Z.*, **73**, 315–334.

MORGAN, E. (1965) The activity rhythm of the amphipod *Corophium volutator* (Pallas) and its possible relationship to changes in hydrostatic pressure associated with the tides. *J. anim. Ecol.*, **34**, 731–746.

MORGANS, J. F. C. (1956) Notes on the analysis of shallow water soft substrata. *J. anim. Ecol.*, **25**, 367–387.

MORTENSEN, T. H. (1921) *Studies on the development and larval forms of Echinoderms.* Gad. København, Copenhagen, 266 pp.

MORTENSEN, T. H. (1938) Contributions to the study of the development and larval forms of Echinoderms IV. *Kgl. Danske Vidensk. Selsk. Skrifter Naturv. og Math. Afd. Rackke 9*, **7**, 1–59.

MORTON, J. (1954) The crevice fauna of the upper intertidal zone at Wembury. *J. mar. biol. Ass. U.K.*, **33**, 187–224.

MORTON, J. E. (1959) The habits and feeding organs of *Dentalium entalis*. *J. mar. biol. Ass. U.K.*, **38**, 225–238.

MORTON, J. E. (1960) The responses and orientation of the bivalve. *Lasaea rubra* Montagu. *J. mar. biol. Ass. U.K.*, **39**, 5–26.

MORTON, J. E., BONEY, A. D. and CORNER, E. D. S. (1957) The adaptations of *Lasaea rubra* (Montagu), a small intertidal lamellibranch. *J. mar. biol. Ass. U.K.*, **36**, 383–405.

MOULTON, J. M. (1962) Intertidal clustering of an Australian gastropod. *Biol. Bull. Wood's Hole*, **123**, 170–178.

MUIR, B. S., NELSON, G. J. and BRIDGES, K. W. (1965) A method for measuring swimming speed in oxygen consumption studies on the aholehole, *Kuhlia sandvicensis. Trans. Am. Fish. Soc.*, **94**, 378–382.

NAYLOR, E. (1955) The diet and feeding-mechanism of *Idotea. J. mar. biol. Ass. U.K.*, **34**, 347–355.

NAYLOR, E. (1962) Seasonal changes in population of *Carcinus maenas* (L) in the littoral zone. *J. anim. Ecol.* **31**, 601–609.

NAYLOR, E. (1963) Temperature relationships of the locomotor rhythm of *Carcinus. J. exp. Biol.*, **40**, 669–679.

NAYLOR, E. and SLINN, D. J. (1958) Observations on the ecology of some brackish water organisms in pools at Scarlett Point, Isle of Man. *J. anim. Ecol.*, **27**, 15–25.

NELSON, T. C. (1960) The feeding mechanism of the oyster II. On the gills and palps of *Ostrea edulis, Crassostrea virginica* and *C. angulata. J. Morph.*, **107**, 163–191.

NEU, W. (1933) Der einfluss des Farbtons der Unterlage auf die Beseidlung mit *Balanus* und *Spirorbis. Int. Rev. Hydrobiol.*, **28**, 228–246.

NEWELL, G. E. (1948) A contribution to our knowledge of the life history of *Arenicola marina* L. *J. mar. biol. Ass. U.K.*, **27**, 554–579.

NEWELL, G. E. (1949) The later larval life of *Arenicola marina* L. *J. mar. biol. Ass. U.K.*, **28**, 635–640.

NEWELL, G. E. (1958a) The behaviour of *Littorina littorea* (L) under natural conditions and its relation to position on the shore. *J. mar. biol. Ass. U.K.*, **37**, 229–239.

NEWELL, G. E. (1958b) An experimental analysis of the behaviour of *Littorina littorea* (L) under natural conditions and in the laboratory. *J. mar. biol. Ass. U.K.*, **37**, 241–266.

NEWELL, G. E. (1965) The eye of *Littorina littorea*. *Proc. zool. Soc. Lond.*, **144**, 75–86.

NEWELL, G. E. (1966) Physiological aspects of the ecology of intertidal molluscs. *The Physiology of Mollusca*. Eds. K. M. Wilbur and C. M. Yonge, Academic Press, New York and London, 59–81.

NEWELL, R. (1960) The behaviour of *Hydrobia ulvae*. *Ann. Rept. Challenger Soc.*, **3**, (2).

NEWELL, R. (1962) Behavioural aspects of the ecology of *Peringia* (=*Hydrobia*) *ulvae* (Pennant) (Gasteropoda, Prosobranchia). *Proc. zool. Soc. Lond.*, **138**, 49–75.

NEWELL, R. (1964) Some factors controlling the upstream distribution of *Hydrobia ulvae* (Pennant) (Gastropoda, Prosobranchia). *Proc. zool. Soc. Lond.*, **142**, (1) 85–106.

NEWELL, R. (1965) The role of detritus in the nutrition of two marine deposit-feeders, the prosobranch *Hydrobia ulvae* and the bivalve *Macoma balthica*. *Proc. zool. Soc. Lond.*, **144**, 25–45.

NEWELL, R. C. (1966) The effect of temperature on the metabolism of poikilotherms. *Nature, Lond.*, **212**, 426–428.

NEWELL, R. C. (1967) Oxidative activity of poikilotherm mitochondria as a function of temperature. *J. Zool. Lond.*, **151**, 299–311.

NEWELL, R. C. (1969) Adaptations to temperature fluctuation in intertidal animals. *Amer. Zool.*, **9**, 293–307.

NEWELL, R. C. and COURTNEY, W. A. M. (1965) Respiratory movements in *Holothuria forskali* Delle Chiaje. *J. exp. Biol.* **42**, 45–57.

NEWELL, R. C. and NORTHCROFT, H. R. (1965) The relationship between cirral activity and oxygen uptake in *Balanus balanoides*. *J. mar. biol. Ass. U.K.*, **45**, 387–403.

NEWELL, R. C. and NORTHCROFT, H. R. (1967) A re-interpretation of the effect of temperature on the metabolism of certain marine invertebrates. *J. zool. Lond.*, **151**, 277–298.

NEWELL, R. C. and PYE, V. I. (1968) Seasonal variations in the effect of temperature on the respiration of certain intertidal algae. *J. mar. biol. Ass. U.K.*, **48**, 341–348.

NEWELL, R. C. and PYE, V. I. (1970) Seasonal changes in the effect of temperature on the oxygen consumption of the winkle *Littorina littorea* (L.) and the mussel *Mytilus edulis* L. *Comp. Biochem. Physiol. (in press).*

NEWELL, R. C. and PYE, V. I. (1970) The influence of thermal acclimation on the relation between oxygen consumption and temperature in *Littorina littorea* (L.) and *Mytilus edulis* L. *Comp. Biochem. Physiol.* (in press).

NEWMAN, W. A. (1967) On physiology and behaviour of estuarine barnacles. Proc. Symp. Crustacea III. *Mar. Biol. Ass. India*, 1038–1066.

NICHOLLS, A. G. (1931a) Studies on *Ligia oceanica* I. A. Habitat and change of environment on respiration. B. Observations on moulting and breeding. *J. mar. biol. Ass. U.K.*, **17**, 655–673.

NICHOLLS, A. G. (1931b) Studies on *Ligia oceanica* II. The process of feeding, digestion and absorption, with a description of the structure of the foregut. *J. mar. biol. Ass. U.K.*, **17**, 675–707.

NICOL, E. A. T. (1930) The feeding mechanism, formation of the tube, and physiology of digestion in *Sabella pavonina*. *Trans. roy. Soc. Edin.*, **56**, (23) 537–596.

NICOL, E. A. T. (1932) The feeding habits of the Galatheidae. *J. mar. biol. Ass. U.K.*, **18**, 87–106.

NICOL, E. A. T. (1935) The ecology of a salt marsh. *J. mar. biol. Ass. U.K.* **20**, 203–261.

NICOL, J. A. C. (1950) The responses of *Branchiomma vesiculosum* (Montagu) to photic stimulation. *J. mar. biol. Ass. U.K.*, **29**, 303–320.

NICOL, J. A. C. (1960) *The Biology of Marine Animals*. Pitman, London, 707 pp.

NICOLLE, M. and ALILARE, E. (1909) Note sur la production en grand des corps bactériens et sur leur composition chimique *Annls. Inst. Pasteur, Paris*, **23**, 547–556.

NISBET, R. H. (1953) *The structure and function of the buccal mass of some gastropod molluscs I. Monodonta lineata* (da Costa). Ph. D. Thesis, University of London.

NOMOURA, S. (1926) The influence of oxygen tension on the rate of oxygen consumption in *Caudina. Sci. Rep. Tohoku Univ.*, (Ser. IV) **2**, 133–138.

NYHOLM, K. (1950) Contributions to the life-history of the Ampharetid, *Melinna cristata. Zool. Bidr. Uppsala.*, **29**, 79–91.

ORR, A. P. (1933) Brit. Mus. (Nat. Hist) Gt. Barrier Reef Exped. 1928–29. *Sci. Repts.*, **2**, 87.

ORR, P. R. (1955) Heat death I. Time-temperature relationships in marine animals. *Physiol. Zool.*, **28**, 290–294.

ORTON, J. H. (1912) The mode of feeding of *Crepidula*, with an account of the current-producing mechanisms in the mantle cavity, and some remarks on the mode of feeding in gastropods and lamellibranchs. *J. mar. biol. Ass. U.K.*, **9**, 444–478.

ORTON, J. H. (1913a) On ciliary mechanisms in brachiopods and some polychaetes, with a comparison of the ciliary mechanisms on the gills of molluscs. Protochordata, brachiopods and some cryptocephalous polychaetes, and an account of the endostyle of *Crepidula* and its allies. *J. mar. biol. Ass. U.K.*, **10**, 283–311.

ORTON, J. H. (1913b) The ciliary mechanisms on the gill and the mode of feeding in *Amphioxus*, ascidians and *Solenomya togata. J. mar. biol. Ass. U.K.*, **10**, 19–49.

ORTON, J. H. (1926) The mode of feeding of the hermit crab *Eupagurus bernhardus* and some other Decapoda. *J. mar. biol. Ass. U.K.*, **14**, 909–921

ORTON, J. H. (1933) Studies on the relation between organism and environment. *Proc. L'pool. Biol. Soc.*, **46**, 1–16.

ORTON, J. H., SOUTHWARD, A. J. and DODD, J. M. (1956) Studies on the biology of limpets H. The breeding of *Patella vulgata* L. in Britain. *J. mar. biol. Ass. U.K.*, **35**, 149–176.

OWEN, G. (1961) A note on the habits and nutrition of *Solemya parkinsoni* (Protobranchia; Bivalvia). *Quart. J. micr. Sci.*, **102**, 15–21.

OWEN, G. (1966) Feeding. *Physiology of Mollusca*, **2**, Eds. K. M. Wilbur and C. M. Yonge, Academic Press, New York and London, 1–51.

PANTIN, C. F. A. (1924) On the physiology of ameboid movement II. The effect of temperature. *J. exp. Biol.*, **1**, 519–538.

PANTIN, C. F. A. (1932) Physiological adaptation. *J. Linn. Soc.* (Zool.), **37**, 705–711.

PANTIN, C. F. A. (1942) The excitation of nematocysts. *J. exp. Biol.*, **19**, 294–310.

PANTIN, C. F. A. and SAWAYA, P. (1953) Muscular action in *Holothuria grisea. Bol. Fac. Filos. Ciênc. S. Paulo (Zool.)*, **18**, 51–59.

PAPI, F. (1955) Experiments on the sense of time in *Talitrus saltator. Experimentia*, **11**, 201.

PAPI, F. and PARDI, L. (1953) Ricerch sull'orientamento di *Talitrus saltator* (Montagu) (Crustacea, Amphipoda) II. Sui fattori chez regolano le variazione dell'angolo di orientamento nel corso del giorne. L'orientamento diurno di altre popolazione. *Z. vergl. Physiol.*, **35**, 490–518.

PARDI, L. (1960) Innate components on the solar orientation of littoral amphipods. *Cold Spring. Harb. Symp. quant. Biol.*, **25**, 395–401.

PARDI, L. and GRASSI, M. (1955) Experimental modification of direction finding in *Talitrus saltator* (Montagu) and *Talorchestia deshayesei* (Aud) (Crustacea, Amphipoda). *Experimentia*, **11**, 202–203.

PARDI, L. and PAPI, F. (1952) Die Sonne als Kompass bei *Talitrus saltator* (Montagu) (Amphipoda, Talitridae). *Naturwiss*, **39**, 262–263.

PARDI, L. and PAPI, F. (1953) Ricerche sull'orientamento di *Talitrus saltator* (Montagu) (Crustacea-Amphipoda). I. L'orientamento durante il giorno in una popolazione del litorale tirrenico. *Z. vergl. Physiol.*, **35**, 459–489.

PARDI, L. and PAPI, F. (1961) Kinetic and Tactic responses. *The Physiology of Crustacea*, **2**, Ed. T. H. Waterman, Academic Press, New York and London. 365–399.

PARRY, D. A. (1951) Factors determining the temperature of terrestrial arthropods in sunlight. *J. exp. Biol.*, **28**, 445–462.

PASSANO, L. M. (1960) Molting and its control. *The Physiology of Crustacea.* Ed. T. H. Waterman, Academic Press, New York and London.

PATANÈ, L. (1933) Sul comportamento di *Littorina neritoides* (L) mantenuta in ambiente subaero ed in altre condizione sperimentale. *R. C. Accad. Lincei (Ser. 6)*, **17**, 961–967.

PATANÉ, L. (1946a) Anaerobiosi in *Littorina neritoides* (L). *Boll. Soc. ital. biol. sper.*, **21**, (7) 928–929.

PATANÉ, L. (1946b) Anaerobiosi in *Littorina neritoides* (L). *Boll. Soc. ital. biol. sper.*, **22**, (7) 929–930.

PATANÉ, L. (1955) Cinesi e tropismi, anidro-e anaerobiosi in *Littorina neritoides* (L). *Boll. accad. sci. nat. Gioenia sper. IV,* **3**, 65–73.

PEARSE, A. S., HUMM, H. J. and WHARTON, G. W. (1942) The ecology of *Elisha Mitchell Sci. Soc.*, **44**, 230–237.

PEARSE, A. S. (1929b) Observations on certain littoral and terrestrial animals at Tortugas, Florida, with special reference to migrations from marine to terrestrial habitats. *Papers Tortugas Lab.*, **26**, 205–223.

PEARSE, A. S. HUMM, H. J. and WHARTON, G. W. (1942) The ecology of sandy beaches at Beaufort. *Ecol Monogr.*, **12**, 135–190.

PEISS, C. N. and FIELD, J. (1950) The respiratory metabolism of excised tissues of warm- and cold-acclimated fishes. *Biol. Bull. Wood's Hole,* **99**, 213–224.

PICKEN, L. E. R. (1953) A note on the nematocysts of *Corynactis viridis. Quart. J. micr. Sci.*, **94**, 203–227.

PICKEN, L. E. R. and SKAER, R. J. (1966) A review of researches on nematocysts. (*The Cnidaria and their evolution*) Ed. W. J. Rees. *Symp. zool. Soc. Lond. (16)*., Academic Press, London and New York, 19–49.

PICKENS, P. E. (1965) Heart rates of mussels as a function of latitude, intertidal height and acclimation temperature. *Physiol. Zool.*, **38**, 390–405.

PIÉRON, H. (1909a) Contribution à la biologie de la Patelle et de la Calyptrée I. L'ethologie, les phénomènes sensoriels. *Bull. Sci. Fr. Belg.*, **43**, 183–202.

PIÉRON, H. (1909b) II. Le sens du retour et la mémoire topographique. *Arch. Zool. exp. gen. (5) Notes et Revues,* 18–29.

PIÉRON, H. (1933) Notes éthologiques sur les gastéropodes perceurs et leur comportement avec utilisation de méthodes statistiques. *Arch. Zool. exp. gén.*, **75**, 1–20.

PILKINGTON, R. (1957) The ways of the sea. Routledge, London.

POMERAT, C. M. and GREGG, J. H. (1942) Attachment of marine and sedentary organisms to black and white glass plates in the horizontal position. *J. Alabama Acad. Sci.*, **14**, 57.

POMERAT, C. M. and REINER, E. R. (1942) The influence of surface angle and of light on the attachment of barnacles and other sedentary organisms. *Biol. Bull. Wood's Hole,* **82**, 14–25.

POMERAT, C. M. and WEISS, C. M. (1946) The influence of texture and composition of surface on the attachment of sedentary marine organisms. *Biol. Bull. Wood's Hole,* **91**, 57–65.

POPHAM, E. J. (1966) The littoral fauna of the Ribble estuary, Lancashire, England. *Oikos,* **17**, 19–32.

POTTS, F. A. (1915) *Hapalocarcinus,* the gall-forming crab. *Publ. Carnegie Inst. Washington. Dept. mar. Biol.*, (212) 33–69.

POTTS, W. T. W. and PARRY, G. (1964) *Osmotic and ionic regulation in animals.* Pergamon Press, New York.

POWERS, E. B. (1920) The variation of the condition of seawater, especially the hydrogen ion concentration, and its relation to marine organisms. *Publ. Puget Sd. Mar. Biol. Sta.*, **2**, 369.

PRASADA RAO, D. G. V. and GANAPATI, P. N. (1969) Oxygen consumption in relation to body size in the barnacle *Balanus tintinnabulum tintinnabulum* (L.). *Comp. Biochem. Physiol.*, **28**, 193–198.

PRECHT, H. (1951a) Der Einfluss der Temperatur auf das Fermentsystem. *Verh.* dtsch. zool. Ges., **376**, 179.

PRECHT, H. (1951b) Der Einfluss der Temperatur auf die Atmung und auf einige

Fermente beim Aal (*Anguilla vulgaris* L). *Biol. Zbl.*, **70**, 71–85.

PRECHT, H. (1958) Concepts of temperature adaptation of unchanging reaction systems of cold-blooded animals. *Physiological Adaptation.* Ed. C. L. Prosser, Am. Physiol. Soc. Washington D. C., 50–78.

PRECHT, H. (1961) Beiträge zur Temperaturadaptation des Aales (*Anguilla vulgaris* L.). *Z. vergl. Physiol.*, **44**, 451–462.

PRECHT, H. (1964) Über die Bedeutung des Blutes für die Temperaturadaptation von Fischen. *Zool. Jahb. Abt. Allgem. Zool. Physiol. Tiere*, **71**, 313–327.

PRECHT, H., CHRISTOPHERSEN, J. and HENSEL, H. (1955) *Temperatur und Leben.* Springer-Verlag, Berlin, 514 pp.

PROSSER, C. L. (1955) Physiological variation in animals. *Biol. Rev.*, **30**, 229–262.

PROSSER, C. L. (1958) The nature of physiological adaptation. *Physiological adaptation.* Ed. C. L. Prosser, Amer. Physiol. Soc. Washington D. C., 167–180.

PROSSER, C. L. (1962) Acclimation of poikilothermic vertebrates to low temperatures. *Comparative physiology of temperature regulation III.* Eds. J. P. Hannon and E. Viereck, Arctic Aeromedical Lab. Fort Wainwright, Alaska, 1–44.

PROSSER, C. L. (1967) Molecular mechanisms of temperature adaptation in relation to speciation. *Molecular mechanisms of temperature adaptation.* Ed. C. L. Prosser, Amer. Assoc. adv. Sci. Symp. (84) Washington D. C. 351–376.

PROSSER, C. L. and BROWN, F. A. (1961) *Comparative Animal Physiology.* 2nd Ed. Saunders, Philadelphia, 688 pp.

PROSSER, C. L., BROWN, F. A., BISHOP, D. W., JAHN, T. L. and WULFF, V. J. (1950) *Comparative Animal Physiology.* Saunders, Philadelphia.

PROSSER, C. L., PRECHT, H. and JANKOWSKY, H. D. (1965) Nervous control of metabolism during temperature acclimation of fish. *Naturwiss.*, **52**, 168–169.

PUMPHREY, R. J. (1961) Concerning vision. *The cell and the organism.* Eds. J. A. Ramsay and V. B. Wigglesworth, Cambridge University Press, 193–208.

PURCHON, R. D. (1955) The structure and function of the British Pholadidae (rock-boring Lamellibranchia). *Proc. zool. Soc. Lond.*, **124**, 859–911.

PYEFINCH, K. A. (1943) The intertidal ecology of Bardsey Island, North Wales, with special reference to the recolonisation of rock surfaces and the rockpool environment. *J. anim. Ecol.*, **12**, 82–108.

PYEFINCH, K. A. (1948) Notes on the biology of cirripedes. *J. mar. biol. Ass. U.K.*, **27**, 464–503.

PYEFINCH, K. A. and DOWNING, F. S. (1949) Notes on the general biology of *Tubularia larynx* Ellis and Solander. *J. mar. biol. Ass. U.K.*, **28**, 21–43.

RAMSAY, J. A. (1935a) Methods of measuring the evaporation of water from animals. *J. exp. Biol.*, **12**, 355–372.

RAMSAY, J. A. (1935b) The evaporation of water from the cockroach. *J. exp. Biol.*, **12**, 373–383.

RAO, K. P. (1953a) Rate of water propulsion in *Mytilus californianus* as a function of latitude. *Biol. Bull. Wood's Hole*, **104**, 171–181.

RAO, K. P. (1953b) Shell weight as a function of intertidal height in a littoral population of pelecypods. *Experimentia*, **9**, 465.

RAO, K. P. (1954) Tidal rhythmicity of the rate of water propulsion in *Mytilus* and its modifiability by transplantation. *Biol. Bull. Wood's Hole*, **106**, 353–359.

RAO, K. P. (1958) Oxygen consumption as a function of size and salinity in *Metapenaeus monoceros* Fab. from marine and brackish-water environments. *J. exp. Biol.*, **35**, 307–313.

RAO, K. P. (1963a) Physiology of low-temperature adaptation in tropical poikilotherms IV. Quantitative changes in the nucleic acid content of the tissues of the freshwater mussel *L. marginalis. Proc. Ind. Acad. Sci. B*, **58**, 11–13.

RAO, K. P. (1963b) Some biochemical mechanisms of low temperature acclimation in tropical poikilotherms. *The cell and environmental temperature.*

Ed. A. S. Troshin U.S.S.R. Academy of Sciences, Moscow and Pergamon Press, New York, 73–81.

RAO, K. P. (1967) Biochemical correlates of temperature acclimation. *Molecular mechanisms of temperature adaptation.* Ed. C. L. Prosser, Amer, Assoc. adv. Sci. Symp. (84) Washington D. C. 227–244.

RAO, K. P. and BULLOCK, T. H. (1954) Q_{10} as a function of size and habitat temperature in poikilotherms. *Amer. Nat.*, **87**, (838) 33–43.

RAO, K. P. and SAROJA, K. (1963) Physiology of low temperature acclimation in tropical poikilotherms V. Changes in the activity of neurosecretory cells in the earthworm *Lampito mauritii* and evidence for a humoral agent influencing metabolism. *Proc. Ind. Acad. Sci. B*, **58**, 14–18.

RASMONT, R. (1961) Une technique de culture des éponges d'eau douce en milieu controlé. *Ann. Soc. Roy. Zool. Belg.*, **91**, 147–156.

READ, K. R. H. (1962) The haemoglobin of the bivalved mollusc *Phacoides pectinatus* (Gmelin). *Comp. Biochem. Physiol.*, **15**, 137–158.

READ, K. R. H. (1963) Thermal inactivation of preparations of aspartic/glutamic transaminase from species of bivalved molluscs from the sublittoral and intertidal zones. *Comp. Biochem. Physiol.*, **9**, 161–180.

READ, K. R. H. (1966) Molluscan haemoglobin and Myoglobin. *The Physiology of Mollusca* Eds. K. M. Wilbur and C. M. Yonge. Academic Press, New York and London, **2**, 209-232.

READ, K. R. H. (1967) Thermostability of proteins in poikilotherms. *Molecular mechanisms of temperature adaptation.* Ed. C. L. Prosser. Amer. Assoc. adv. Sci. Symp. (84) Washington D. C. 93–106.

REDFIELD, A. C., COOLIDGE, T. and HURD, A. L. (1926) The transport of oxygen and carbon dioxide by some bloods containing haemocyanin. *J. biol. Chem.*, **69**, 475–509.

REDFIELD, A. C. and FLORKIN, M. (1931) The respiratory function of the blood of *Urechis caupo. Biol. Bull. Wood's Hole*, **61**, 185–210.

REDFIELD, A. C. and INGALLS, E. N. (1933) The oxygen dissociation curves of some bloods containing haemocyanin. *J. cell. comp. Phsyiol.*, **3**, 169–202.

REDMOND, J. R. (1955) The respiratory function of the haemocyanin in Crustacea. *J. cell. comp. Physiol.*, **46**, 209–247.

REES, G. (1941) The resistance of the flatworm *Monocelis fusca* to changes in temperature and salinity under natural and experimental conditons. *J. anim. Ecol.*, **10**, 121–145.

REID, D. M. (1930) Salinity interchange between seawater in sand and overflowing freshwater at low tide. *J. mar. biol. Ass. U.K.*, **16**, (2) 609–614.

REID, D. M. (1932) Salinity interchange between salt water in sand and overflowing freshwater at low tide. 11. *J. mar. biol. Ass. U.K.*, **18**, 299–306.

REMANE, A. (1933) Verteilung und Organisation der benthonischen Microfauna der Kieler Bucht. Wiss. Meeresunters. *Pr. Komm. Abt. Kiel.*, **21**, 163–221.

REUSZER, H. W. (1933) marine bacteria and their role in the cycle of life in the sea III. The distribution of bacteria in the ocean waters and muds about Cape Cod. *Biol. Bull. Wood's Hole*, **65**, 480–497.

REYNOLDS, O. (1885) On the dilatancy of media composed of rigid particles in contact, with experimental illustrations. *Phil. Mag.*, **20**, 469–481.

RICE, T. R. and SMITH, R. J. (1958) Filtering rates of the hard clam, *Venus mercenaria,* determined with radioactive phytoplankton. *Fish. Bull. U.S.*, **58**, 73–82.

RICKETTS, E. F. and CALVIN, J. (1948) *Between Pacific Tides.* Stanford University Press, California, 365 pp.

RIETZ, G. E. du (1940) *Wellengrenzen als ökologische Äquivalente wasserstrandlinien.* Zool. Bidr. Uppsala.

RIGGS, A. (1951) The oxygen equilibrium of the haemoglobin of the eel, *Anguilla rostrata. J. gen. Physiol.*, **35**, 41–44.

RITZ, D. A. and FOSTER, B. A. (1968) Comparison of the temperature responses of barnacles from Britain, South Africa and New Zealand, with special reference to temperature acclimation in *Elminius modestus. J. mar. biol. Ass. U.K.*, **48**, 545–559.

ROBERTS, J. L. (1952) *Studies on acclimation of respiration to temperature in the lined shore crab* Pachygrapsus crassipes Randall. Ph.D. dissertation. University of California. Los Angeles.

ROBERTS, J. L. (1957) Thermal acclimation of metabolism in the crab *Pachygrapsus crassipes* Randall. I The influence of body size, starvation and molting. *Physiol. Zool.*, **30**, 232–242.

ROBERTS, J. L. (1964) Metabolic responses of freshwater sunfish to seasonal photoperiods and temperatures. *Helgol. wiss. Meeresunters.*, **9**, 459–473.

ROBERTS, J. L. (1967) Metabolic compensations for temperature in sunfish. (*Molecular mechanisms of temperature adaptation*) Ed. C. L. Prosser, Amer. Assoc. adv. Sci. Symp. (84), Washington D.C. 245–262.

ROBSON, E. A. (1953) Nematocysts of *Corynactis.* The activity of the filament during discharge. *Quart. J. micr. Sci.*, **94**, 229–234.

ROSE, I. A. (1961) The use of kinetic isotope effects in the study of metabolic control. I Degradation of glucose-1-D by the hexose monophosphate pathway. *J. biol. Chem.*, **236**, 603–609.

ROSS, D. M. (1960) The association between the hermit crab *Eupagurus bernhardus* (L) and the sea anemone *Calliactis parasitica* (Couch). *Proc. zool. Soc. Lond.*, **134**, 43–57.

ROUX, C. and ROUX, A. L. (1967) Température et métabolisme respiratoire d'espèces sympatriques de gammares du groupe *pulex* (Crustacés, Amphipodes). *Ann. de Limnol.*, **3**, 3–16.

RUCK, P. (1961) Electrophysiology of the insect dorsal ocellus. *J. gen. Physiol.*, **44**, 605–657.

RULLIER, F. (1959) Étude bionomique de l'Aber de Roscoff. *Trav. Stn. biol. Roscoff.*, **10**, 1–350.

RUNHAM, N. W. and THORNTON, P. R. (1967) Mechanical wear of the gastropod radula; a scanning electron microscope study. *J. Zool. Lond.*, **153**, 445–452.

RUNNSTROM, S. (1925) Zur Biologie und Entwicklung von *Balanus balanoides. Univ. Bergen. Årb. naturv. R.*, **5**, 1–46.

RUSSELL, R. C. H. and MACMILLAN, D. H. (1952) *Waves and Tides,* Hutchinson, London, 348 pp.

RYLAND, J. S. (1959) Experiments on the selection of algal substrates by Polyzoa larvae. *J. exp. Biol.*, **36**, 613–631.

SANDERS, H. L. (1956) Oceanography of Long Island Sound 1952–1954. X. Biology of marine bottom communities. *Bull. Bingham Oceanogr. Coll.*, **15**, 345–414.

SANDISON, E. E. (1966) The oxygen consumption of some intertidal gastropods in relation to zonation. *J. Zool. Lond.*, **149**, 163–173.

SANDISON, E. E. (1967) Respiratory response to temperature and temperature tolerance of some intertidal gastropods. *J. exp. mar. Biol. Ecol.*, **1**, 271–281.

SAROJA, K. and RAO, K. P. (1965) Some aspects of the mechanism of thermal acclimation in the earthworm *Lampito mauritii. Z. vergl. Physiol.*, **50**, 35–54.

SAUNDERS, R. L. (1962) The irrigation of the gills in fishes II. Efficiency of oxygen uptake in relation to respiratory flow activity and concentrations of oxygen and carbon dioxide. *Canad. J. Zool.*, **40**, 817–862.

SAUNDERS, R. L. (1963) Respiration of the Atlantic Cod. *J. Fish. Res. Bd. Canada,* **20**, 373–386.

SAXENA, D. B. (1963) A review on ecological studies and their importance in the physiology of air-breathing fishes. *Ichthyologica,* **11**, 116–128.

SCHÄFER, H. (1952) Ein Beitrag zur Ernährungsbiologie von *Bithynia tentaculata* L. (Gastropoda, Prosobranchia). *Zool. Anz.*, **148**, 299–303.

SCHÄFER, H. (1953a) Beobachtungen zur Ökologie von *Bithynia tentaculata. Arch. Molluskenk,* **82**, 67–70.

SCHÄFER, H. (1953b) Beiträge zur Ernährungsbiologie einheimischer Süsswasserprosobranchier. *Z. Morph Ökol. Tiere,* **41**, 247–264

SCHELTEMA, R. S. (1961) Metamorphosis of the veliger larvae of *Nassarius obsoletus* (Gastropoda) in response to bottom sediment. *Biol. Bull. Wood's Hole,* **120**, 92–109.

SCHIJFSMA, K. (1939) Preliminary notes on early stages in the growth of colonies of *Hydractinia echinata* (Flem). *Arch. néerl. Zool.*, **4**, 93–102.

SCHLIEPER, C. R. (1952) Über die Temperatur-stoffwechselrelation einiger eurythermer Wassertiere. *Verh. Deutsh. Zool. Suppl.* **16**, **5**, 267–272.

SCHLIEPER, C. (1955) Die Regulation der Herzschlages der Miesmuschel *Mytilus edulis* L. bei geöffneten und bei geschlossenen Schalen. *Kieler Meeresforsch*, **11**, 139–148.

SCHLIEPER, C. (1957) Comparative study of *Asterias rubens* and *Mytilus edulis* from the North Sea (30 per 1000 S) and the western Baltic Sea (15 per 1000 S). *Anée. biol.* (*fasc 3–4*) **33**, 117–127.

SCHLIEPER, C. R., FLUGEL, H. and RUDOLF, J. (1960) Temperature and salinity relationships in marine bottom invertebrates. *Experimentia*, **16**, 470–477.

SCHLIEPER, C. R., KOWALSKI, R. and ERMAN, P. (1958) Beitrag zur ökologisch-zellphysiologischen Characterisierung des borealen lamel – libranchiers *Modiolus modiolus* L. *Kieler Meeresforsch*, **14**, 3–10.

SCHOLANDER, P. F., FLAGG, W., WALTERS, V. and IRVING, L. (1953) Climatic adaptation in arctic and tropical poikilotherms. *Physiol. Zool.*, **26**, 67–92.

SCHÖNE, H. (1961a) Complex Behaviour. *The Physiology of Crustacea II. Sense organs, integration and behaviour.* Academic Press, New York and London.

SCHÖNE, H. (1961b) Learning in the spiny lobster *Panulirus argus. Biol. Bull. Wood's Hole* **121**, 354–365.

SCHÖNE, H. (1964) Release and orientation of behaviour and the role of learning as demonstrated in Crustacea. (Learning and associated phenomena in invertebrates) *Anim. Behav. Suppl. 1*, 135–143.

SCHULTZE, D. (1965) Beiträge zur Temperaturadaptation des Aales (*Anguilla vulgaris* L.). *Z. wiss. Zool. Abt. A*, **172**, 104–133.

SCHULZE, P. (1917) Neue Beiträge zur einer Monographie der Gattung *Hydra Arch. Biontol. Berl.*, **4**, 39.

SEGAL, E. (1956) Microgeographic variation as thermal acclimation in an intertidal mollusc. *Bull. Biol. Wood's Hole.*, **111**, 129–152.

SEGAL, E. (1961) Acclimation in molluscs. *Am. Zool.*, **1**, 235–244.

SEGAL, E. (1962) Initial response of heart rate of a gastropod *Acmaea limatula* to abrupt changes in temperature. *Nature, Lond.*, **195**, 674–675.

SEGAL, E. and DEHNEL, P. A. (1962) Osmotic behaviour in an intertidal limpet *Acmaea limatula. Bull. Biol. Wood's Hole*, **122**, 417–430.

SEGAL, E., RAO, K. P. and JAMES, T. W. (1953) Rate of activity as a function of intertidal height within populations of some littoral molluscs. *Nature, Lond.*, **172**, 1108–1109.

SEKI, H. and TAGA, N. (1963) Microbiological studies on the decomposition of chitin in the marine environment. I–V. J. *Oceanogr. Soc. Japan*, **19**, (2) 27–81.

SHEPARD, M. P. (1955) Resistance and tolerance of young speckled trout (*Salvelinus fontinalis*) to oxygen lack, with special reference to low oxygen acclimation. *J. Fish. Res. Bd. Canada*, **12**, 387–446.

SHEPARD, P. S. (1948) *Submarine Geology*, Harper and Roe, New York, 348 pp.

SIEFKEN, M. and ARMITAGE, K. B. (1968) Seasonal variations in metabolism and organic nutrients in three *Diaptomus* (Crustacea: Copepoda). *Comp. Biochem. Physiol.*, **24**, 591–609.

SILÈN, L. (1954) Developmental biology of Phoronidae of the Gullmer Fiord area (west coast of Sweden). *Acta Zoologica*, **35**, 215–257.

SILVA, P. M. D. M. de (1962) Experiments on choice of substrata by *Spirorbis* larvae (Serpulidae). *J. exp. Biol.* **39**, 483–490.

SILVA, P. M. D. M. de and KNIGHT-JONES, E. W. (1962) *Spirorbis corallinae* n.sp. and some other Spirorbinae (Serpulidae) common on British shores. *J. mar. biol. Ass. U.K.* **42**, **601–608**.

SIMINOVITCH, D. and BRIGGS, D. R. (1952) Studies on the chemistry of the living bark of the black locust in relation to its frost hardiness. *Plant Physiol.*, **28**, 15–34.

SINDERMANN, C. J. (1961) The effect of larval trematode parasites on snail migrations. *Amer. Zoologist,* **1,** 389.

SIZER, I. W. (1943) The effects of temperature on enzyme kinetics. *Adv. Enzymol.,* **3,** 35–62.

SKAER, R. J. and PICKEN, L. E. R. (1965) The structure of the nematocyst thread and geometry of discharge in *Corynactis viridis* Allmann, *Phil. Trans. roy. Soc. B,* **250,** 131–163.

SLAUTTERBACK, D. B. (1961) Nematocyst development. *The Biology of Hydra.* Eds. H. M. Lenhoff and W. F. Loomis, University of Miami Press, Florida, 77–129.

SMIDT, E. L. B. (1951) Animal production in the Danish Waddensea. *Medd. Komm. Danmarks Fisk. Havunders. Ser. Fiskeri,* **11,** (6) 1–151.

SMIT, H. (1965) Some experiments on the oxygen consumption of goldfish (*Carassius auratus* L) in relation to swimming speed. *Canad. J. Zool.,* **43,** 623–633.

SMITH, A. U. (1958) The resistance of animals to cooling and freezing. *Biol. Rev.,* **33,** 197–253.

SMITH, J. E. and NEWELL, G. E. (1955) The dynamics of the zonation of the common periwinkle (*Littorina littorea* (L)), on a stony beach. *J. anim. Ecol.,* **24,** (1) 35–56.

SOMERO, G. N. and HOCHACHKA, P. W. (1969) Isoenzymes and short-term temperature compensation in poikilotherms: Activation of lactate dehydrogenase isoenzymes by temperature decreases. *Nature Lond.* **223,** 194–195.

SOUTHWARD, A. J. (1952) Organic matter in littoral deposits. *Nature, Lond.,* **169,** 888.

SOUTHWARD, A. J. (1953a) The fauna of some sandy and muddy shores in the south of the Isle of Man. *Proc. L'pool Biol. Soc.,* **59,** 51–71.

SOUTHWARD, A. J. (1953b) The ecology of some rocky shores in the south of the Isle of Man. *Proc. L'pool Biol. Soc.,* **59,** 1–50.

SOUTHWARD, A. J. (1955a) Feeding of barnacles. *Nature Lond.,* **175,** 1124.

SOUTHWARD, A. J, (1955b) On the behaviour of barnacles I. The relation of cirral and other activities to temperature. *J. mar. biol. Ass. U.K.,* **34,** 403–422.

SOUTHWARD, A. J. (1955c) On the behaviour of barnacles II. The influence of habitat and tidal level on cirral activity. *J. mar. biol. Ass. U.K.,* **34,** 423–433.

SOUTHWARD, A. J. (1956) The population balance between limpets and seaweeds on wave beaten rocky shores. *Rep. Mar. biol. Sta. Pt. Erin.,* **68,** 20–29.

SOUTHWARD, A. J. (1957) On the behaviour of barnacles III. Further observations on the influence of temperature and age on cirral activity. *J. mar. biol. Ass. U.K.,* **36,** 323–334.

SOUTHWARD, A. J. (1958a) Note on the temperature tolerance of some intertidal animals in relation to environmental temperatures and geographical distribution. *J. mar. biol. Ass. U.K.,* **37,** 49–66.

SOUTHWARD, A. J. (1958b) The zonation of plants and animals on rocky shores. *Biol. Rev.,* **33,** 137–177.

SOUTHWARD, A. J. (1962) On the behaviour of barnacles IV. The influence of temperature on cirral activity and survival of some warm-water species. *J. mar. biol. Ass. U.K.,* **42,** 163–177.

SOUTHWARD, A. J. (1964) The relationship between temperature and rhythmic cirral activity in some Cirripedia considered in connection with their geographical distribution. *Helgol. wiss. Meeresunters,* **10,** 391–403.

SOUTHWARD, A. J. (1965) On the metabolism and survival of cirripedes at high temperatures. *Trav. Cent. Rech. Étud. Oceanogr.,* **6,** 441–446.

SOUTHWARD, A. J. (1965) *Life on the sea shore.* Heinemann, 153 pp.

SOUTHWARD, A. J. and CRISP, D. J. (1956) Fluctuations in the distribution and abundance of intertidal barnacles. *J. mar. biol. Ass. U.K.,* **35,** 211–229.

SOUTHWARD, A. J. and CRISP, D. J. (1965) Activity rhythms of barnacles in relation to respiration and feeding. *J. mar. biol. Ass., U.K.* **45,** 161–185.

SPANGENBERG, D. B. and HAM, R. G. (1960) The epidermal nerve net of *Hydra. J. exp. Zool.,* **143,** 195–201.

SPOONER, G. M. and MOORE, H. B. (1940) The ecology of the Tamar estuary

VI. An account of the intertidal muds. *J. mar. biol. Ass. U.K.*, **24**, 283–330.

SPOOR, W. A. (1946) A quantitative study of the relationship between the activity and oxygen consumption of goldfish, and its application to the measurement of respiratory metabolism in fishes. *Biol. Bull. Wood's Hole*, **91**, 312–325.

STAGG, J. M. (1950) Solar radiation at Kew Observatory. *Geophys. Mem.*, **11**, (86) 1–37.

STASEK, C. R. (1961) The ciliation and function of the labial palps of *Acila castrensis* (Protobranchia, Nuculidae) and an evaluation of the role of the protobranch organs of feeding in the evolution of the Bivalvia. *Proc. zool. Soc. Lond.*, **137**, 511–538.

STASEK, C. R. (1965) Feeding and particle sorting in *Yoldia ensifera* (Bivalvia; Protobranchia) with notes on other nuculanids. *Malacologia*, **2**, 349–366.

STELFOX, A. W. (1916) *Otina otis* on the Co. Down coast. *Proc. malac. Soc. Lond.*, **12**, 318.

STEPHEN, A. C. (1929) Studies on the Scottish marine fauna I. The fauna of the sandy and muddy areas of the tidal zone. *Trans. roy. Soc. Edin.*, **56**, 291–306.

STEPHEN, A. C. (1930) Studies on the Scottish marine fauna. Additional observations on the fauna of sandy and muddy areas of the tidal zone. *Trans. roy. Soc. Edin.*, **56**, 521–535.

STEPHEN, A. C. (1952) Life on some sandy shores. *Essays in Marine Biology*. Oliver and Boyd, 144 pp.

STEPHENS, G. C., FINGERMAN, M. and BROWN, F. A. (1953) The orientation of *Drosophila* to plane polarised light. *Ann Entomol. Soc. Amer.*, **46**, 75–83.

STEPHENSON, T. A. (1927) *The British Sea Anemones.* **1**, Ray Society of London.

STEPHENSON, T. A. (1942) The causes of the vertical and horizontal distribution of organisms between tidemarks in South Africa. *Proc. Linn. Soc. Lond.*, **154**, 219.

STEPHENSON, T. A. and STEPHENSON, A. (1949) The universal features of zonation between tidemarks on rocky coasts. *J. Ecol.* **38**, 289–305.

STEPHENSON, T. A. and STEPHENSON, A. (1961) Life between tide marks in North America. IVa Vancouver Island. *J. Ecol.*, **49**, 1–29.

✓STEPHENSON, T. A., ZOOND, A. and EYRE, J. (1934) The liberation and utilisation of oxygen by the population of rockpools. *J. exp. Biol.*, **11**, 162–172.

STUBBINGS, H. G. (1950) Earlier records of *Elminius modestus* Darwin in British waters. *Nature, Lond.*, **166**, 272–278.

TEAL, J. M. (1958) Distribution of fiddler crabs in Georgia salt marshes. *Ecology*, **39**, 185–193.

THAMDRUP, H. M. (1935) Beiträge zur Okologie der Wattenfauna. *Meddr. Komn. Danm. Fisk. og Havunders, Fiskeri*, **10**, (2) 1–125.

THOMAS, H. J. (1954) The oxygen uptake of the lobster (*Homarus vulgaris* Edw). *J. exp. Biol.*, **31**, 228–251.

THOMPSON, T. E. (1958) The natural history, embryology larval biology and post-larval development of *Adalaria proxima* (Alder and Hancock). Gastropoda, Opisthobranchia. *Phil. Trans. roy. Soc. Lond. B*, **242**, 1–58.

THOMPSON, T. E. and SLINN, D. J. (1959) On the biology of the opisthobranch *Pleurobranchus membranaceus. J. mar. biol. Ass. U.K.*, **38**, 507–524.

THORPE, W. H. (1956) *Learning and instinct in animals.* Methuen, London, 493 pp.

THORSON, G. (1946) Reproduction and larval development of Danish marine bottom invertebrates with special reference to the planktonic larvae in the sound (Oresund). *Medd. Komm. Danm. Fiskeri-og Havunders. Kbh. Ser. Plankt.*, **4**, (1) 523 pp.

THORSON, G. (1950) Reproductive and larval ecology of marine bottom invertebrates. *Biol. Rev.*, **25**, 1–45.

THORSON, G. (1957) Bottom communities (sublittoral or shallow shelf) *Treatise on Marine Ecology and Paleoecology.* **1**, (*Ecology*) Ed. J. W. Hedgpeth, Geol. Soc. Amer. Mem. 67, 461–534.

TINBERGEN, N. (1951) *The study of instinct.* Oxford University Press, London and New York, 228 pp.

TODD, E. S. and EBELING, A. W. (1966) Aerial respiration in the longjaw mudsucker *Gillichthys mirabilis* (Teleosti: Gobiidae). *Biol. Bull. Wood's Hole,* **130,** (2) 265–288.

TOPPE, O. (1909) Über die Wirkungsweise der Nesselkapseln von *Hydra. Zool. Anz.,* **33,** 798.

TRIBE, M. A. and BOWLER, K. (1968) Temperature dependence of "standard metabolic rate" in a poikilotherm. *Comp. Biochem. Physiol.,* **25,** 427–436.

TRIM, A. R. (1941) Studies in the chemistry of the insect cuticle. I. Some general observations on certain arthropod cuticles with special reference to the characterisation of the proteins. *Biochem. J.,* **35,** 1088–1098.

TRUEMAN, E. R. (1967) The activity and heart-rate of bivalve molluscs in their natural habitat. (cited by Helm and Trueman, 1967.)

TRUEMAN, E. R. (1968) The burrowing process of *Dentalium* (Scaphopoda). *J. Zool. Lond.,* **154,** (1) 19–27.

ULLMAN, A. and BOOKHOUT, C. G. (1949) The histology of the digestive tract of *Clymenella torquata* (Leidy). *J. morph.,* **84,** (1) 31–48.

ULLYOTT, P. (1936a) The behaviour of *Dendrocoelum lacteum* I. Responses at light and dark boundaries. *J. exp. Biol.,* **13,** 253–264.

ULLYOTT, P. (1936b) The behaviour of *Dendrocoelum lacteum* II. Responses in non-directional gradients. *J. exp. Biol.,* **13,** 265–278.

USHAKOV, B. P. (1967) Coupled evolutionary changes in protein thermostability. *Molecular mechanisms of temperature adaptation.* Ed. C. L. Prosser, Amer. Assoc. adv. Sci. Symp. (84) Washington D.C., 107–129.

VAUGHAN, D. A., HANNON, J. P. and VAUGHAN, L. N. (1958) Associated effects of diet, environmental temperature and duration of exposure on the major constituents of the liver of rats. *Am. J. Physiol.,* **194,** 441–445.

VERNBERG, F. J. (1962) Comparative physiology: latitudinal effects on physiological properties of animal populations. *Ann. Rev. Physiol.* **24,** 517–546.

VERNBERG, F. J. (1967) Future problems in the physiological ecology of estuarine animals. In: Estuaries. pp. 554–557. Ed. G. H. Lauff, Amer. Ass. Adv. Sci. (83) Washington D.C.

VERNBERG, F. J., SCHLIEPER, C. and SCHNEIDER, D. E. (1963) The influence of temperature and salinity on ciliary activity of excised gill tissue of molluscs from North Carolina *Comp. Biochem. Physiol.,* **8,** 271–285.

VERNBERG, F. J. and TASHIAN, R. E. (1959) Studies on the physiological variation between tropical and temperate zone fiddler crabs of the genus *Uca.* I Thermal death limits. *Ecol.* **12,** 589–593.

VERNON, H. M. (1897) The relation of the respiratory exchange of cold-blooded animals to temperature. *J. Physiol. Lond.,* **21,** 443–496.

VERWEY, J. (1952) The ecology of the distribution of cockle and mussel in the Dutch Waddensea. Their role in sedimentation and the source of their food supply. With a short review of the feeding behaviour in bivalve molluscs. *Arch. néerl. Zool.,* **10,** 171–239.

VERWEY, J. (1954) De mossel in zijn eisen. Averdruk uit. Faraday 24e Jaargang. (2) 13 pp. Groningen.

VERWEY, J. (1957) Discussion. Coll. intern. biol. mar. St. Roscoff. *Année Biol.* **(3), 33,** 238.

VIJAYALAKSHMI, C. (1964) *Studies on the metabolism of scorpions with special reference to thermal acclimation.* Ph.D Thesis Sri Venkateswara University, Tirupati, India. (cited by Rao, 1967).

VISSCHER, J. P. (1928) Reactions of the cyprid larvae of barnacles at the time of attachment. *Biol. Bull. Wood's Hole,* **54,** 327–335.

VON BRAND, T. (1946) *Anaerobiosis in invertebrates.* Biodynamica Monographs, **4,** 328 pp.

WALLENGREN, H. (1905) Zur Biologie der Muscheln I. Die Wasserströmungen. II. Die Nahrungsaufnahme. *Lunds. Univ. Årsskr.* (*N.F.*), **2,** (1) 1–64 and (2) 1–59.

WALNE, P. R. (1958) The importance of bacteria in laboratory experiments on rearing the larvae of *Ostrea edulis* (L). *J. mar. biol. Ass. U.K.*, **37**, 415–425.

WALTON SMITH, F. G. (1948) Surface illumination and barnacle attachment. *Biol. Bull. Wood's Hole*, **94**, 33–39,

WARDEN, C. J., JENKINS, T. N. and WARNER, L. H. (1940) *Comparative Psychology*. Ronald Press, New York.

WATERMAN, T. H. (1954) Polarised light and the angle of stimulus incidence in the compound eye of *Limulus. Proc. nat. Acad. Sci. U.S.*, **40**, 258–262.

WATERMAN, T. H. (1959) Animal navigation in the sea. *Gunma J. Medic. Sci.*, **8**, (3) 243–262.

WATERMAN, T. H. (1961) Light sensitivity and vision. *The Physiology of Crustacea*, **2**, Ed. T. H. Waterman, Academic Press, New York and London, 1–64.

WATKIN, E. E. (1939) The pelagic phase in the life-history of the amphipod genus *Bathyporeia. J. mar. biol. Ass. U.K.*, **23**, 467–481.

WATKIN, E. E. (1941a) Observations on the night tidal migrant crustacea of Kames Bay. *J. mar. biol. Ass. U.K.*, **25**, 81–96.

WATKIN, E. E. (1941b) The male of the amphipod *Haustorius arenarius* Slabber. *J. mar. biol. Ass. U.K.*, **25**, 303–305.

WATKIN, E. E. (1942) The macrofauna of the intertidal fauna of Kames Bay, Millport, Buteshire. *Trans. roy. Soc. Edinb.*, **60**, 543–561.

✓ WEBB, H. M. and BROWN, F. A. (1958) The repitition of pattern in the respiration of *Uca pugnax. Biol. Bull. Wood's Hole*, **115**, 303–318.

WEBB, J. E. (1969) Biologically significant properties of submerged marine sands. Proc. roy. Soc. (Lond) B, **174**, 355–402.

WEBB, J. E. and HILL, M. B. (1958) The ecology of Lagos Lagoon. *Phil. Trans. roy. Soc. Lond. B*, **241**, 307–419.

WEEL, P. B. van. (1940) Beiträge zur Ernährungsbiologie der Ascidien. *Pubbl. Staz. Zool. Napoli.*, **18**, 50–79.

WEILL, R. (1934) Contribution à l'etude des Cnidaires et de leurs nematocystes. *Trav. Stat. zool. Wimereux*, **10**, 1–347.

WELLS, G. P. (1944) Mechanism of burrowing in *Arenicola marina* (L). *Nature, Lond.*, **154**, 396.

WELLS, G. P. (1945) The mode of life of *Arenicola marina* L. *J. mar. biol. Ass. U.K.*, **26**, 170–207.

WELLS, G. P. (1949a) Respiratory movements of *Arenicola marina* L. Intermittent irrigation of the tube, and intermittent aerial respiration. *J. mar. biol. Ass. U.K.*, **28**, 447–464.

WELLS, G. P. (1949b) The behaviour of *Arenicola marina* L. in sand, and the role of the spontaneous activity cycles. *J. mar. biol. Ass. U.K.*, **28**, 465–478.

WELLS, G. P. (1950) Spontaneous activity cycles in polychaete worms. *Symp. Soc. Exp. Biol.* **4**, (Physiological mechanisms in animal behaviour.) Cambridge University Press, London, 127–142.

WELLS, G. P. (1951) On the behaviour of *Sabella. Proc. roy. Soc. B*, **138**, 278–299.

WELLS, G. P. (1952) The respiratory significance of the crown in the polychaete worms *Sabella* and *Myxicola. Proc. roy. Soc. Lond. B.* **140**, 70–82.

WELLS, G. P. (1953) Defaecation in relation to the spontaneous activity cycles of *Arenicola marina* L. *J. mar. biol. Ass. U.K.*, **32**, 51–63.

WELLS, G. P. (1959) Worm autobiographies *Sci. Amer.*, **200**, (6) 132–141.

WELLS, G. P. and ALBRECHT, E. B. (1951a) The integration of activity cycles in the behaviour of *Arenicola marina* L. *J. exp. Biol.*, **28**, 41–50.

WELLS, G. P. and ALBRECHT, E. B. (1951b) The role of the oesophageal rhythms in the behaviour of *Arenicola ecaudata* Johnson, *J. exp. Biol.*, **28**, 51–56.

WELLS, G. P. and DALES, R. P. (1951) Spontaneous activity patterns in animal behaviour; the irrigation of the burrow in the polychaetes *Chaetopterus variopedatus* Renier and *Nereis diversicolor* O. F. Muller. *J. mar. biol. Ass. U.K.*, **29**, 661–680.

WELLS, M. J. (1966) Cephalopod sense organs. *The Physiology of Mollusca* Eds.

K. M. Wilbur and C. M. Yonge, Academic Press, New York and London, 523–545.

WELSH, J. H. (1960) 5-hydroxytryptamine in coelenterates. *Nature, Lond.*, **186**, 811–812.

WELSH, J. H. (1961) Compounds of pharmacological interest in coelenterates. *The Biology of Hydra.* Eds. H. M. Lenhoff and W. F. Loomis, University of Miami Press, Florida, 179–186.

WERNER, B. (1952) Ausbildungsstufen der Filtrationmechanismen bei filtreieenden Prosobranchiern. *Verhandl. deut. zool. Ges.* (*zool. Anz. Suppl. 17),* 529–546

WERNER, B. (1953) Uber den Nahrungswerb der Calyptraeidae (Gastropoda, Prosobranchia) Morphologie, Histologie und Function der Nahrungswerb beteiligten Organe. *Helgoland wiss Meeresunters,* **4**, 260–315.

WERNER, B. (1959) Das Prinzip des endlosen Schleimfilters beim Nahrungswerwerb wirbelloser Meerestiere. *Int. Rev. ges. Hydrobiol. Hydrogr.*, **44**, 181–216.

WERNER, E. and WERNER, B. (1954) Uber den mechanismus des Nahrungserwerbs der Tunicaten speciell der Ascidien. *Helgol. wiss Meeresunters.*, **5**, 57–92.

WESTFALL, J. A. (1964) Fine structure and development of nematocysts in the tentacle of *Metridium. Am. Zool.*, **4**, 435.

WESTFALL, J. A. and HAND, C. (1962) Fine structure of nematocysts in a sea anemone. *5th Int. Congr. for Electron Microscopy,* **2**, (M 13) Ed. S. S. Breese, Academic Press, New York and London.

✓ WIENS, A. W. and ARMITAGE, K. B. (1961) The oxygen consumption of the crayfish *Orconectes immunis* and *Orconectes nais* in response to temperature and oxygen saturation. *Physiol. Zool.*, **34**, 39–54.

WIESER, W. (1952) Investigations on the microfauna inhabiting seaweeds on rocky coasts. IV Studies on the vertical distribution of the fauna inhabiting seaweeds below the Plymouth laboratory. *J. mar. biol. Ass. U.K.*, **31**, 145–173.

WIESER, W. (1954) Untersuchungen über die algenbewohnende Mikrofauna mariner Hartböden. *Hydrobiologica,* **6**, 144–224.

WIESER, W. (1956) Factors influencing the choice of substratum in *Cumella vulgaris* Hart (Crustacea, Cumacea). *Limnol. Oceanogr.*, **1**, 274–285.

WIESER, W. (1959) The effect of grain size on the distribution of small invertebrates inhabiting the beaches of Puget Sound. *Limnol. Oceanogr.* **4**, 181–194.

✓WIESER, W. and KANWISHER, J. (1959) Respiration and anaerobic survival in some seaweed-inhabiting invertebrates. *Biol. Bull. Wood's Hole*, **117**, 594–600.

WILKENS, J. L. and FINGERMAN, M. (1965) Heat tolerance and temperature relationships of the fiddler crab *Uca pugilator,* with reference to body coloration. *Biol. Bull. Wood's Hole,* **128**, (1) 133–141.

WILLIAMS, G. B. (1964) The effect of extracts of *Fucus serratus* in promoting the settlement of larvae of *Spirorbis borealis* (Polychaeta). *J. mar. biol. Ass. U.K.*, **44**, 397–414.

WILLIAMS, G. B. (1965a) *Substrate specificity in marine invertebrates.* Ph.D. Thesis, University of Wales.

WILLIAMS, G. B. (1965b) Settlement inducing factors in fucoids. *Rep. Challenger Soc.*, **3** (17).

WILLIAMS, G. B. (1965c) Observations on the behaviour of the planulae larvae of *Clava squamata. J. mar. biol. Ass. U.K.*, **45**, 257–273.

WILLIAMSON, D. I. (1951) Studies on the biology of the Talitridae (Crustacea, Amphipoda). Visual orientation in *Talitrus saltator. J. mar. biol. Ass. U.K.*, **30**, 91–99.

WILLIAMSON, D. I. (1953) Landward and seaward movement of the sandhopper *Talitrus saltator. How animals find their way about.* Repts. Brit. Assoc. Section D, 71–73.

WILLIAMSON, J. R. (1965) Glycolytic control mechanisms I. Inhibition of glycolysis by acetate in the isolated perfused rat heart. *J. biol. Chem.*, **240**, 2308–2321.

WILSKA, A. and HARTLINE, H. K. (1941) The origin of the "off-responses" in the optic pathway. *Am. J. Physiol.*, **133**, 491–492.

WILSON, D. P. (1928) The larvae of *Polydora ciliata* Johnston and *Polydora hoplura Claparède*. *J. mar. biol. Ass. U.K.* **15**, 567–603.

WILSON, D. P. (1932) On the mitraria larva of *Owenia fusiformis* Delle Chiaje. *Phil. Trans. roy. Soc. Lond. B.*, **221**, 231–334.

WILSON, D. P. (1937) The influence of the substratum on the metamorphosis of *Notomastus* larvae. *J. mar. biol. Ass. U.K.*, **22**, 227–243.

WILSON, D. P. (1948) The relation of the substratum to the metamorphosis of *Ophelia* larvae. *J. mar. biol. Ass. U.K.*, **27**, 723–760.

WILSON, D. P. (1951) Larval metamorphosis and the substratum. *Ann. Biol. (Paris)* **27**, 259–269.

WILSON, D. P. (1952) The influence of the nature of the substratum on the metamorphosis of the larvae of marine animals, especially the larvae of *Ophelia bicornis*, Savigny. *Ann. Inst. oceanogr. Monaco.*, **27**, 49–156.

WILSON, D. P. (1953a) The settlement of *Ophelia bicornis* Savigny larvae. The 1951 experiments. *J. mar. biol. Ass. U.K.*, **31**, 413–438.

WILSON, D. P. (1953b) The settlement of *Ophelia bicornis* Savigny larvae. The 1952 experiments. *J. mar. biol. Ass. U.K.*, **32**, 209–233.

WILSON, D. P. (1954) The attractive factor in the settlement of *Ophelia bicornis* Savigny. *J. mar. biol. Ass. U.K.*, **33**, 361–380.

WILSON, D. P. (1955) The role of micro-organisms in the settlement of *Ophelia bicornis* Savigny. *J. mar. biol. Ass. U.K.*, **34**, 513–543.

WILSON, D. P. (1968) The settlement behaviour of the larvae of *Sabellaria alveolata* (L). *J. mar. biol. Ass. U.K.*, **48**, 387–435.

WINTERSTEIN, H. (1909) Uber die Atmung der Holothurien. *Arch. Fisiol.*, **7**, 87–93.

WISELY, B. (1960) Observations on the settling behaviour of larvae of the tubeworm *Spirorbis borealis* Daudin (Polychaeta). *Aust. J. mar. freshw. Res.*, **2**. 55–72.

WOLVEKAMP, H. P. (1961) The evolution of oxygen transport. *Functions of the Blood.* Blackwell, Oxford.

WOLVEKAMP, H. P. and VREEDE, M. C. (1941) On the gas binding properties of the blood of the lugworm *Arenicola marina* L. *Arch. néerl. Physiol.*, **25**, 265–276.

✓WOLVEKAMP, H. P. and WATERMAN, T. H. (1960) Respiration. *The Physiology of Crustacea 1*, Ed. T. H. Waterman, Academic Press, New York and London.

WOOD, E. J. F. (1964) Studies in the microbial ecology of the Australasian region V. Microbiology of some Australian estuaries. *Nova Hedwiga.*, **8**, 461–527.

WOOD, E. J. F. (1965) *Marine Microbial Ecology.* Chapman and Hall, London, 243 pp.

YANAGITA, T. M. (1943) Discharge of nematocysts. *J. Fac. Sci. Imp. Univ., Tokyo*, **6**, 97–108.

YANAGITA, T. M. and WADA, T. (1954) Effects of trypsin and thioglycollate upon the nematocysts of the sea anemone. *Nature, Lond.*, **173**, 171.

YONGE, C. M. (1926) The structure and physiology of the organs of feeding and digestion in *Ostrea edulis*. *J. mar. biol. Ass. U.K.*, **14**, 295–386.

YONGE, C. M. (1928) Feeding mechanisms in invertebrates. *Biol. Rev.*, **3**, 21–76.

YONGE, C. M. (1931) Digestive processes in marine invertebrates and fishes. *J. Cons. int. Explor. Mer.*, **6**, 175–212.

YONGE, C. M. (1937a) Evolution and adaptation in the digestive system of the Metazoa. *Biol. Rev.* **12**, 87–115.

YONGE, C. M. (1937b) The biology of *Aporrhais pes-pelicani* (L) and *A. serresiana* (Mich.). *J. mar. biol. Ass. U.K.*, **21**, 687–704.

YONGE, C. M. (1938) Evolution of ciliary feeding in the Prosobranchia with an account of feeding in *Capulus ungaricus*. *J. mar. biol. Ass. U.K.*, **22**, 453–468.

YONGE, C. M. (1939) The Protobranchiate Mollusca; a functional interpretaion of their structure and evolution. *Phil. Trans. roy. Soc. Lond. B*, **230**, 79–147.

YONGE, C. M. (1946) On the habits of *Turritella communis* Risso. *J. mar. biol. Ass. U.K.*, **26**, 377–380.

YONGE, C. M. (1948) Cleansing mechanisms and the function of the fourth pallial aperture in *Spisula subtruncata* (Da Costa) and *Lutraria lutraria* (L). *J. mar. biol. Ass. U.K.*, **27**, 585–596.

YONGE, C. M. (1949) *The sea shore.* Collins, London.

YONGE, C. M. (1950a) On the structure and adaptations of the Tellinacea, deposit-feeding Eulamellibranchia. *Phil. Trans. roy. Soc. Lond. B,* **234**, 29–76.

YONGE, C. M. (1950b) Life on Sandy Shores. *Sci. Progr.,* **151**, 430–443.

YONGE, C. M. (1952) Aspects of life on muddy shores. *Essays in Marine Biology.* pp 29–41. Oliver and Boyd. 144 pp.

YONGE, C. M. (1956) Marine bottom substrata and their fauna. Proc. XIV int. Congr. Zool. Copenh. (1953).

YOSHIDA, M. (1962) The effect of light on the shadow reaction of the sea urchin, *Diadema setosum* (Leske). *J. exp. Biol.,* **39**, 589–602.

YOSHIDA, M. and MILLOTT, N. (1960) The shadow reaction of *Diadema antillarum* Philippi III. Re-examination of the spectral sensitivity. *J. exp. Biol.,* **37**, 390–397.

ZANEVELD, J. S. (1937) The littoral zonation of some Fucacea in relation to desiccation. *J. Ecol.,* **25**, (2) 431–468.

ZANEVELD, J. S. (1969) Factors controlling the delimitation of littoral benthic marine algal zonation. *Amer. Zool.* **9**, 367–391.

ZEUTHEN, E. (1947) Body size and metabolic rate in the animal kingdom. *C. R. Lab. Carlsb. Sér. chim.,* **26**, 15–161.

ZEUTHEN, E. (1953) Oxygen uptake as related to body size in organisms. *Quart. Rev. Biol.,* **28**, 1–12.

ZIEGELMEIER, E. (1954) Beobachtungen über den Nahrungserwerb bei der Naticide *Lunatia nitida* Donovan (Gastropoda, Prosobranchia). *Helgol. wiss. Meeresunters.,* **5**, 1–33.

ZOBELL, C. E. (1936) Periphytic habits of some marine bacteria. *Proc. Soc. exp. Biol. Med.,* **35**, 270–273.

ZOBELL, C. E. (1938) Studies on the bacterial flora of marine bottom sediments. *J. sedim. Petrol.,* **8**, 10–18.

ZOBELL, C. E. (1943) The effect of solid surfaces on bacterial activity. *J. Bact.,* **46**, 39–56.

ZOBELL, C. E. (1946) Marine Microbiology. *A monograph on Hydrobacteriology.* Chromica Botanica Co. Waltham, Mass. 240 pp.

ZOBELL, C. E. and ANDERSON, D. Q. (1936) Observations on the multiplication of bacteria in different volumes of stored seawater and the influence of oxygen tension and solid surfaces. *Biol. Bull. Wood's Hole,* **71**, 324–342.

ZOBELL, C. E. and FELTHAM, C. B. (1942) The bacterial flora of a marine mud flat as an ecological factor. *Ecology,* **23**, 69–78.

ZOBELL, C. E. and LANDON, W. A. (1937) The bacterial nutrition of the California mussel. *Proc. Soc. exp. Biol. N.Y.,* **36**, 607–609.

AUTHOR INDEX

SUBJECT INDEX